Microbiota Unveiled: Science, Health, and Future Therapies

Ranbir Chander Sobti • Ram Krishan Negi
Anita Rana

Microbiota Unveiled: Science, Health, and Future Therapies

Springer

Ranbir Chander Sobti
Department of Biotechnology
Panjab University
Chandigarh, Chandigarh, India

Ram Krishan Negi
Department of Zoology
University of Delhi
Delhi, India

Anita Rana
Department of Biosciences (UIBT)
Chandigarh University
Mohali, Punjab, India

ISBN 978-981-95-9030-8 ISBN 978-981-95-9031-5 (eBook)
https://doi.org/10.1007/978-981-95-9031-5

This Springer imprint is published by the registered company Springer Nature Singapore Pte Ltd.
The registered company address is: 152 Beach Road, #21-01/04 Gateway East, Singapore 189721, Singapore

Preface

The past two decades have witnessed an unprecedented expansion in our understanding of the microbial world and its intimate relationship with human health, disease, and the environment. Once perceived primarily as agents of infection, microorganisms are now recognized as indispensable partners in biological systems, shaping immunity, metabolism, development, and resilience across ecosystems. This paradigm shift from a pathogen-centric view of microbiology to a holistic understanding of the microbiota and microbiome has fundamentally altered how we approach medicine, agriculture, food systems, and environmental sustainability. The present volume was conceived in response to this scientific transformation and the growing need for an integrated, interdisciplinary resource that captures both the depth and breadth of contemporary microbiota research.

This book was written to address a critical gap in the existing literature. While numerous texts and reviews focus on isolated aspects of microbiome science such as gut health, probiotics, cancer, or synthetic biology, there is a relative scarcity of comprehensive works that connect these domains into a coherent framework. The editors envisioned this volume as a unifying platform that brings together foundational concepts, mechanistic insights, translational applications, and policy considerations related to microbiota-based science. By spanning topics from the historical origins of microbiology to next-generation probiotics, microbiota-based vaccines, cancer therapeutics, aging, sustainable agriculture, food production, and regulatory frameworks, this book aims to reflect the true multidimensional nature of the microbiome field.

The primary audience for this book includes graduate and postgraduate students, doctoral researchers, clinicians, biomedical scientists, microbiologists, biotechnologists, environmental scientists, and professionals working at the interface of health, agriculture, and policy. In addition, this volume is intended to serve as a valuable reference for educators, industry researchers, and regulatory stakeholders seeking a scientifically rigorous yet accessible overview of microbiota-based approaches. Recognizing the interdisciplinary readership, the chapters have been structured to progress from fundamental principles to advanced applications, allowing readers from diverse backgrounds to engage meaningfully with the material.

The organization of the book reflects a deliberate and logical progression. The opening chapters lay the conceptual foundation by tracing the evolution of microbiology and microbiome science, introducing omics technologies, and outlining the composition and functional roles of the human microbiota. These chapters establish a common scientific language and framework essential for understanding later discussions. Subsequent sections focus on the role of microbiota in health and disease, with particular emphasis on cancer biology, immune modulation, therapeutic responsiveness, and systemic disorders. Dedicated chapters explore microbiota-driven interventions in cancer therapy, aging, and infectious disease prevention, highlighting both mechanistic pathways and clinical implications.

Beyond human health, this volume expands its scope to encompass the role of microbiomes in sustainable agriculture, food production, and environmental resilience. These chapters underscore the interconnectedness of human, animal, plant, and environmental microbiomes, reinforcing the concept of a shared microbial ecosystem. The inclusion of synthetic biology and microbiota engineering reflects the rapid technological advances shaping the future of the field, while later chapters addressing safety, standardization, ethics, and regulatory policy emphasize the responsibility that accompanies innovation. Collectively, the structure of the book is designed to guide the reader from discovery to application and from promise to practical implementation.

The selection of contributors was guided by a commitment to scientific excellence, diversity of expertise, and global perspective. As editors, we believe that microbiota-based science represents one of the most transformative frontiers in modern biology and medicine. The ability to modulate microbial ecosystems for disease prevention, therapeutic enhancement, environmental sustainability, and personalized health holds immense promise but also demands careful evaluation, ethical responsibility, and robust regulatory oversight. It is our hope that this book will stimulate informed discussion, inspire interdisciplinary collaboration, and support the responsible translation of microbiome research into real-world solutions.

We extend our sincere gratitude to all contributing authors for their dedication and scholarly contributions and to the reviewers and publishing team for their invaluable support. Most importantly, we hope that readers will find this volume not only informative but also motivating encouraging new ideas, innovative research, and thoughtful engagement with the microbial world that so profoundly shapes life on Earth.

RCS is thankful to Indian National Science Academy, New Delhi, and Panjab University, Chandigarh, for providing an opportunity as Honorary Scientist and Emeritus Professorship to continue academic pursuits.

He is also thankful to Dr. Archna and Dr. Halima of Ramaia Institute, Bangalore, Dr. Deepak Saahi Department of Microbiology for giving academic input to the manuscript of the book.

Gratitude to Dr. Vipin Sobti, Er. Aditi, and Dr. Aastha for providing help as and whenever needed. RKN is thankful to Padma Dolkar, Sneha Siwach, Pankaj Yadav, Apoorva Atri, Aarzoo Yadav, Meenu Chaurasia, L. Themchuirin, Nongmaithem Sonia, Shrutti Kapoor, Vikrant Rana, Vyakhya Singh for giving academic input to the manuscript of this book.

Chandigarh, Chandigarh, India — Ranbir Chander Sobti
Delhi, India — Ram Krishan Negi
Mohali, Punjab, India — Anita Rana

Contents

Abbreviations

AhR	Aryl hydrocarbon receptor
AI	Artificial intelligence
AIEC	Adherent invasive *Escherichia coli*
AMF	Arbuscular mycorrhizal fungi
AMPs	Antimicrobial peptides
APAP	N-acetyl-p-aminophenol (Paracetamol)
APCs	Antigen-presenting cells
ART	Antiretroviral therapy
ATMPs	Advanced therapy medicinal products
BAC	Bacterial artificial chromosome
BFT	*Bacteroides fragilis* toxin/Biofloc technology
BV	Bacterial vaginosis
CAR-T	Chimeric antigen receptor T-cell
CCF	Commensal colonization factors
CDI	*Clostridioides difficile* infection
CHO	Chinese hamster ovary (cells)
CNS	Central nervous system
COVID-19	Coronavirus disease 2019
CRC	Colorectal cancer
CRISPR	Clustered regularly interspaced short palindromic repeats
CTA	Clinical trial application
CTLA-4	Cytotoxic T-lymphocyte–associated protein 4
DCs	Dendritic cells
DNA	Deoxyribonucleic acid
DSHEA	Dietary supplement health and education act
EFSA	European Food Safety Authority
EMA	European Medicines Agency
ENS	Enteric nervous system
EPA	Eicosapentaenoic acid
EPS	Exopolysaccharides
EU	European Union
EVs	Extracellular vesicles
FAO	Food and Agriculture Organization

FDA	Food and Drug Administration
FeF	Fucoidan extract fraction
FeHMP	Fucoidan enzymatically hydrolyzed modified polysaccharide
FMT	Fecal microbiota transplantation
FoxP3	Forkhead box P3
GABA	Gamma-aminobutyric acid
GALT	Gut-associated lymphoid tissue
GCs	Goblet cells
GM-CSF	Granulocyte-macrophage colony-stimulating factor
GMO	Genetically modified organism
GPCRs	G-protein-coupled receptors
GRAS	Generally recognized as safe
HCT/P	Human cell, tissue, or cellular- or tissue-based product
HDAC	Histone deacetylase
HIV	Human immunodeficiency virus
HNC	Head and neck cancer
HSP27	Heat shock protein 27
IBD	Inflammatory bowel disease
IBS	Irritable bowel syndrome
IEC(s)	Intestinal epithelial cell(s)
IFN-γ	Interferon-gamma
IgA	Immunoglobulin A
IgG	Immunoglobulin G
IgY	Immunoglobulin Y
IL	Interleukin
ILCs	Innate lymphoid cells
IND	Investigational new drug
LAB	Lactic acid bacteria
LPS	Lipopolysaccharide
MAMPs	Microbe-associated molecular patterns
MAPK	Mitogen-activated protein kinase
MHC	Major histocompatibility complex
mRNA	Messenger ribonucleic acid
MSR	Macrophage scavenger receptor
NF-κB	Nuclear factor kappa-B
NGPs	Next-generation probiotics
NLRs	NOD-like receptors
NO	Nitric oxide
OMVs	Outer membrane vesicles
PAMPs	Pathogen-associated molecular patterns
PCs	Plasma cells
PGE2	Prostaglandin E2
PRRs	Pattern recognition receptors
QS-21	Quillaja saponaria-21
RA	Retinoic acid

rCDI	Recurrent *Clostridioides difficile* infection
ROS	Reactive oxygen species
RS	Resistant starch
SCFAs	Short-chain fatty acids
sIgA	Secretory immunoglobulin A
SLPs	Surface-layer proteins
T6SS	Type VI secretion system
TGA	Therapeutic Goods Administration
Th cells	T helper cells
TLRs	Toll-like receptors
TNF-α	Tumor necrosis factor-alpha
Tregs	Regulatory T cells
WGS	Whole-genome sequencing
WHO	World Health Organization

About the Authors

Ranbir Chander Sobti is Senior Scientist at Indian National Science Academy (INSA) and Emeritus Professor at Panjab University. He is former Vice-Chancellor of Panjab University, Chandigarh, and Babasaheb Bhimrao Ambedkar University (Central), Lucknow. He started his career as a cytogeneticist and then moved on to molecular biology, including genomics, to understand the susceptibility and disease process of cancer, COPD, AIDS, metabolic syndrome, and kidney diseases. He has also used stem cells and nanoparticles to understand the process of tissue organ development through a designed de-cellularization protocol. He has published more than 350 research articles in journals of international repute and has published about five dozen books by many international publishers. He is a Fellow of the Third World Academy of Sciences, National Academy of Sciences India, Indian National Science Academy, National Academy of Medical Sciences, the National Academy of Agricultural Sciences, the Canadian Academy of Cardiovascular Diseases, and a few others. He was the General President of the Indian Science Congress for the 102nd session held at the University of Jammu in 2014. He is the recipient of many prestigious awards like the INSA Young Scientist Medal, UGC Career Award, Punjab Rattan Award, JC Bose Oration and Sriram Oration Awards, and Lifetime Achievement Awards of the Punjab Academy of Sciences, Zoological Society of India, the Environment Academy of India, and Padmashree by the Government of India in 2009.

Ram Krishan Negi is Professor of Zoology, University of Delhi, Delhi. He is the Fellow of Royal Soceity of Biology, London, Fellow of Indian Academy of Environmental sciences (IAES), Honorary Fellow of Fellow of Himalayan Forum for Science and Technology Communication (FHFSTC). He started his career as an Assistant Professor in Gurukul Kangri University, Haridwar, and served for nearly 12 years and became Professor at the Department of Zoology, University of Delhi. He has over 25 years longstanding experience in teaching, administration, and research. He specialized himself in the area of fish biodiversity, molecular taxonomy, aquatic biodiversity of river, lakes, reservoir and hill streams of Western Himalayas, genomic and metagenomics, fish gut microbiome, extremophile environmental microbial diversity, genomics, and metagenomics. He has extensively explored microbial diversity and published three novel bacterial species. He is serving as a reviewer in various leading national and international scientific journals. He has over 100 publications in peer-reviewed journalsand 13 edited book chapters. He is a life member of several national and international societies and has been honored with the Prof. Harswarup memorial medal for outstanding research and acedemic contribution as well as the Gold medal from the Indian Academy of Environmental Science for his significant contribution to environment and aqautic research.

Anita Rana is an assistant professor in biosciences at Chandigarh University, Mohali. She is an educator and researcher dedicated to advancing education, research, and community service. Dr. Anita Rana has demonstrated exceptional academic proficiency by qualifying multiple competitive exams, including ICMR-JRF (AIR 2), CSIR-NET JRF (AIR 62), GATE (AIR 375), HPSET, HTET and 1st rank holder in PU & PGI Ph.D. Entrance Examination. Her research focuses on apicultural products, microbiota, and biomedical applications, with 40+ publications, multiple patents, and editorial contributions to international books. She has received several prestigious awards, including the Young Scientist Award (2016, 2018, 2021), Research Luminary Award (2023, 2024, 2025), Women Innovator Award (2025), Women in Science Award (2026), and has successfully supervised numerous postgraduate dissertations.

Introduction to Microbiota

1

Abstract

Microbiomes—the collective assemblages of microorganisms and their genomes—form highly dynamic ecosystems that have broad impacts on host health and environmental balance. The concept of the microbiome has evolved from early microscopic observations to the modern system-level understanding made possible through culture-independent and high-throughput sequencing methods. Techniques such as 16S rRNA amplicon sequencing, shotgun metagenomics, metatranscriptomics, metabolomics, and metaproteomics have transformed our ability to characterize microbial communities, revealing their diversity, dynamics, and functional roles. The composition of microbiomes is affected by various factors that includes age, diet, host genetics, environmental conditions, and anthropogenic pressures such as antibiotics and pollutants. Dysbiosis, a disturbance in this balance, is linked to wide-ranging consequences across systems—from plant and animal disease, through compromised aquaculture production, to human disorders like colorectal cancer, obesity, allergies, inflammatory bowel disease, and neurodegeneration. This chapter provides an overview of microbiome discovery and evolution, details the major techniques used to study microbial communities, examines the factors that shape their structure, and highlights the role of dysbiosis in human health and disease.

Keywords

Microbiome · Microbiota · Metagenomics and multiomics · Host–microbe interactions · Dysbiosis and disease associations · Immune modulation · Microbiome-based therapeutics

R. C. Sobti et al., *Microbiota Unveiled: Science, Health, and Future Therapies*,
https://doi.org/10.1007/978-981-95-9031-5_1

1.1 Tracing the Origins of the Microbiology and Microbiome Concepts

Microbes, the foundation of microbiology, represent a vast array of microscopic life forms that include bacteria, archaea, protozoa, fungi, and viruses, each playing critical roles in ecosystems, industry, and health. These organisms inhabit nearly all of Earth's ecological habitats. In order to understand the contemporary realm of microbiome research, one should recognize the extensive and intricate history of microbiology. The term *microbiota* (mī′′-krō-bī′-ō-′tə) was introduced in the late nineteenth century to describe the collective community of microorganisms in a given place. It originates from Greek words *micro* that means small and *-bios* that means life, later adapted through Latin into modern scientific usage to describe the collective assemblage of microorganisms that includes bacteria, archaea, protozoa, fungi, and viruses inhabiting a specific environment (Cabrera and Monte 2024). The term microbiome (mī′′-krō-bī′-ōm) was first introduced by Whipps et al. in 1988, who used it to describe a characteristic microbial community in a defined habitat along with its "theatre of activity." It combines microbe (small organism) and biome (bios = life, -ome = complete system or mass), reflecting the concept of a microbial ecosystem rather than individual organisms alone. The definition captured that microbes do not necessarily need to be examined as distinct entities but as interactive networks operating within their environment. Subsequently, in the early 2000s, Joshua Lederberg, a Nobel laureate molecular biologist, popularized and redefined the term. He emphasized that the microbiome is not just the microbial community but also its collective genomes, metabolic potential, and interactions with the host. Lederberg's concept of the microbiome brought to prominence its significance to health and disease, which was instrumental in the start of the Human Microbiome Project (HMP) in 2007, a systems biology and genomics milestone (Cabrera and Monte 2024). The concepts of microbiology and the microbiome, now cornerstones of modern biological and biomedical sciences, have evolved through a complex and fascinating history marked by speculation, discovery, and technological innovation. Although the microbial world remained invisible and largely unknown for most of human history, its effects—both beneficial and devastating—were keenly felt. Patrick Collard (1976) in his book "The Development of Microbiology" categorized the development of microbiology into four major historical phases: the *Era of Speculation* (5000 BC to 1675), where microbial processes were observed but not understood; the *Era of Observation* (1675 to mid-nineteenth century), characterized by the advent of microscopy and the first glimpses of microbial life; the *Era of Cultivation* (mid-nineteenth to early twentieth century), where techniques for isolating and growing microbes were developed; and the *Era of Physiological Study* (early twentieth century onward), which explored microbial metabolism, structure, and function. To these foundational stages, modern scientists have added a fifth and sixth phase—*the Era of Molecular Study* and *the Era of Rapid Methods*—reflecting the powerful influence of genomics, high-throughput sequencing, and computational biology in the past few decades (Collard 1976; Guardino 2005). This section revisits the genesis of microbiology as a formal science, starting with early microscopic observations and culminating in the conceptualization of the "microbiome." The journey begins in the seventeenth century with pioneering visualizations of microbial life, unfolds through the golden age of microbiology in the nineteenth century, and ultimately transitions into the molecular era, where the study of microbial communities within host organisms emerged as a central focus. The following Sects. 1.1.1 and 1.1.2 explain these critical developments that laid the foundation for contemporary microbiome science.

1.1.1 From Microscopes to Microbes (Seventeenth to Nineteenth Century)

The scientific journey into the microbial world began with the invention of the microscope in seventeenth century. In the mid-seventeenth century, pioneering work with simple microscopes revealed a hidden world of life. Robert Hooke's *Micrographia* (1665) contained the first published drawings of microscopic organisms (e.g., the fungal mold *Mucor*). Shortly thereafter, in the 1670s, the Dutch draper Antonie van Leeuwenhoek crafted extremely high-quality single-lens microscopes with up to ~500× magnification. Antonie van Leeuwenhoek is known as the Father of Microbiology. He was the first to observe and describe bacteria, protozoa, and fungi microscopic organisms he termed as "animalcules," in pond water. Over the following decades, he examined materials like blood, saliva, water, dust, and documented diverse microbes, such as cocci, bacilli, and spirilla, revealing for the first time the vast presence (Fig. 1.1) of microscopic life in everyday substances. Although Leeuwenhoek did not link microbes to disease, his detailed reports to the Royal Society of London marked the first recorded observations of

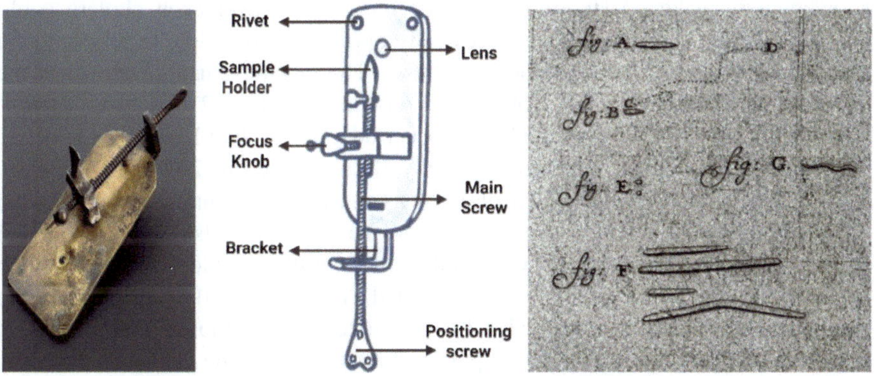

Fig. 1.1 Leeuwenhoek's microscope replica and early observations of oral microbes. A Leiden-built replica of Antony van Leeuwenhoek's basic microscope is shown. The instrument consists of a single convex lens mounted between brass plates, a focusing screw for adjusting clarity, and a specimen pin to secure samples close to the lens. Alongside are illustrations of Leeuwenhoek's original drawings from dental plaque, where he initially identified several types of bacteria, such as cocci (e), fusiform bacteria (f), and spirochetes (g)

bacteria and protozoa, confirming the existence of microscopic life and opening a new era in biology. In the mid-eighteenth century, John Turberville Needham argued that microbes could still emerge in boiled broths even when the containers were sealed, which appeared to support the theory of abiogenesis. The Italian naturalist Lazzaro Spallanzani (1729–1799) challenged Needham's findings by boiling broths for different durations and then hermetically sealing some flasks while using corked ones as controls. He observed that only the sealed flasks remain free of microbial growth, directly contradicting Needham and undermining the idea of abiogenesis (Mancini et al. 2007). In the seventeenth–eighteenth centuries, microbiology remained largely observational. However, advances in optics during the nineteenth century transformed it into an experimental science. Giovanni Amici's achromatic objectives enabled clearer, higher-magnification imaging (~600×), while Ernst Abbe later corrected optical aberrations and, in 1878, introduced the oil-immersion lens, significantly enhancing resolution.

A turning point came with the work of Louis Pasteur (1822–1895). Louis Pasteur, originally trained as a chemist, revolutionized microbiology by demonstrating that fermentation is driven by living organisms. He showed that bacteria produced lactic acid and yeast produced alcohol, overturning the idea of purely chemical fermentation, and introduced pasteurization (gentle heating at ~60–68 °C) to prevent spoilage without altering flavor (Cavaillon and Legout 2022; Berche 2012). Through his swan-neck flask experiments, Pasteur disproved spontaneous generation, proving that microbes arise from environmental contamination rather than de novo formation. This research served as a cornerstone for the development of the germ theory of disease (Cavaillon and Legout 2022). He also demonstrated that microbes could cause disease in animals, in the 1860s, Pasteur identified the parasitic cause of silkworm disease and devised control methods by excluding contaminated eggs. Pasteur's most famous achievement was developing vaccines by attenuation. He showed that weakened strains of *Bacillus anthracis* and chicken cholera could protect animals, and in 1885 he successfully applied the principle to rabies, using attenuated virus to immunize exposed patients. These achievements established the principle of preventive vaccination (Cavaillon and Legout 2022).

A parallel revolution in methodology occurred under Robert Koch (1843–1910), shifting microbiology toward pathogenicity (Sravanthi et al. 2024). In the early 1870s, Koch identified *Bacillus anthracis* as the cause of anthrax. He observed rod-shaped cells in infected sheep blood, grew the organism outside the animal, and demonstrated that pure cultures could infect healthy mice, which succumbed with anthrax and the same bacilli were reisolated (Lakhtakia 2014). This sequence fulfilled what became known as Koch's postulates: "the pathogen must be found in all cases of the disease, isolated in pure culture, reproduce the disease in a healthy host, and be re-isolated." Koch applied these postulates to discover major pathogens. In 1882, he announced the discovery of the tubercle bacillus (*Mycobacterium tuberculosis*), growing it on coagulated blood serum. After long incubation, only *Mycobacterium tuberculosis* (TB) colonies appeared, and he reproduced disease in animals (Lakhtakia 2014). He also identified the comma-shaped *Vibrio cholerae* from cholera patients in India, linking it to polluted water. These achievements established that specific microbes cause specific diseases. He noted that in liquid broths different bacterial species mixed and could not be separated. Early on, he grew cultures on boiled potato slices and observed that individual colonies of one color belonged to single species. To refine this, Koch's assistants invented the "streak plate" method on solid medium. Koch also experimented with gelatin as a solidifying agent, but gelatin liquefied at incubator temperatures. In 1882, Angelina Fanny Hesse suggested agar-agar, a seaweed-derived jelly which remained solid and resisted degradation, making it an ideal culture medium. Koch adopted agar and reported cultivation of *M. tuberculosis* on agar (Koch 1882). Soon after, Julius Petri (1887) designed the covered glass dish—the Petri dish—reducing contamination and condensation (Mahajan 2021). These innovations stabilized bacteriology as a discipline.

While Leeuwenhoek had revealed microbes, it was not until the mid-nineteenth century that they were linked to human health. One of the earliest reports came in 1842 when John Goodsir identified *Sarcina ventriculi* in stomach fluids, proposing it caused illness, though skeptics like Friedrich von Frerichs disagreed (Farré-Maduell and Casals-Pascual 2019). By the late nineteenth century, Beale, Hallier, Woodward, and Uffelmann confirmed bacteria in feces, laying the groundwork for the idea of the human gut microbiota. Pioneering contributions soon followed. Theodor Escherich (1885, 1886) described *Bacterium coli commune* (later *Escherichia coli*) in infants, linking gut microbes to digestion (Escherich 1885; Escherich 1886). Henry Tissier (1899) introduced anaerobic culture techniques and identified *Bifidobacterium* in healthy infants, observing its decline during diarrheal disease and proposing therapeutic restoration of gut flora. Charles-Jacques Bouchard (1887) advanced the "autointoxication theory," arguing that gut microbes produced systemic toxins, anticipating later concepts such as microbial translocation. The Nobel laureate Élie Metchnikoff extended these ideas. Noting longevity in Bulgarian populations consuming fermented milk, he proposed that lactic acid bacteria could suppress harmful gut microbes and prolong life (Farré-Maduell and Casals-Pascual 2019). His influential book *The Prolongation of Life* popularized fermented milk as a health-promoting food and laid theoretical foundations for probiotics. Alfred Nissle in 1917 isolated *E. coli* Nissle 1917 from a soldier resistant to dysentery during World War I in Balkan outbreak. He developed this strain into Mutaflor©, a therapy for intestinal infections that remains in clinical use today (Linton 2010). By the end of the nineteenth and early twentieth centuries, these discoveries had transformed microbiology from descriptive microscopy into an experimental science linking microbes to fermentation, disease, and gut physiology. Together, the contributions of Pasteur, Koch, Escherich, Tissier, Metchnikoff, and Nissle sensitized science to the idea of the microbiota origin and established the conceptual foundations for today's microbiome research. All these pioneer scientists aided the foundation for today's microbiome research, which continues to uncover the vital roles microbes play in everything from immunity to mental health (Fig. 1.2).

Fig. 1.2 Pioneers of microbiology. Top row (left to right): Antonie van Leeuwenhoek, Robert Koch, Louis Pasteur, and Theodor Escherich. Bottom row (left to right): Élie Metchnikoff, Charles-Jacques Bouchard, Stamen Grigorov, and Alfred-Nissle-Gesellschaft

1.1.2 The Emergence of the Microbiome Concept

The twentieth century witnessed a major shift in microbiology research. Instead of isolating specific bacteria, researchers began to view the microbes as complex communities present in different environments. The terms *microbiota* and *microbiome* have evolved through complex scientific and conceptual histories. The usage of the term *microbiome* for the first time is often wrongly credited to Nobel Laureate Joshua Lederberg, who in 2001 highlighted how microbial communities affect human health and disease (Prescott 2017). But the term was actually used earlier, in 1988, when Whipps and colleagues defined the microbiome in their book as stated, *"A convenient ecological framework in which to examine biocontrol systems is that of the microbiome. This may be defined as a characteristic microbial community occupying a reasonably well-defined habitat which has distinct physio-chemical properties. This term thus not only refers to the microorganisms involved but also encompasses their 'theatre of activity'."* (Whipps et al. 1988). This definition focused on the microbiome as a living system shaped by interactions between microbes and their surroundings, not just a list of species or genes. In the early 2000s, Joshua Lederberg broadened and redefined the concept of the microbiome. He stressed that it encompasses not only the resident microbial communities but also their combined genomes, functional capacities, and dynamic interactions with the host. By framing the microbiome in this way, Lederberg highlighted its central role

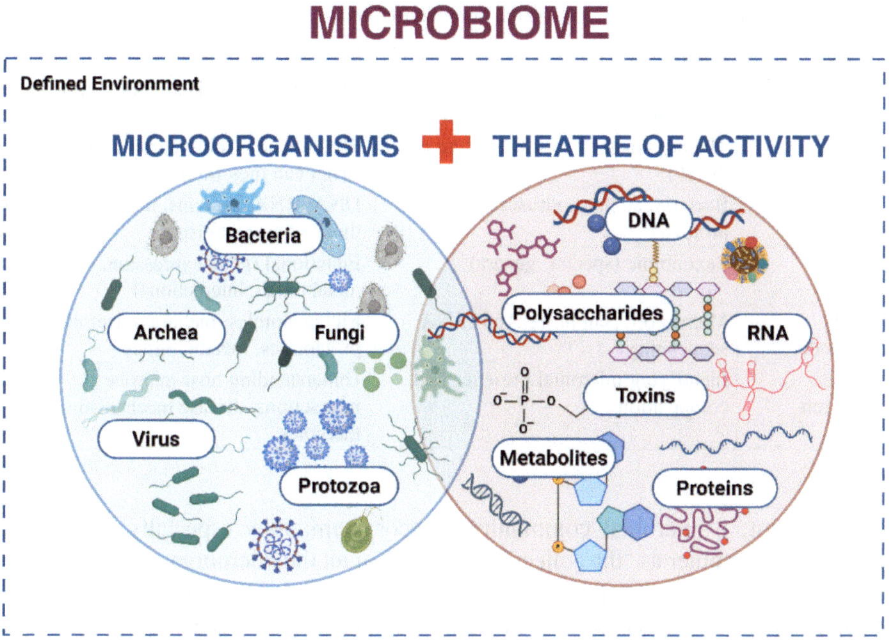

Fig. 1.3 Conceptual illustration of the microbiota and microbiome. This diagram differentiates the microbiota, which refers to the assemblage of living microorganisms—including bacteria, archaea, fungi, and viruses inhabiting a specific environment—from the microbiome, which includes the organisms themselves along with their collective genomes, biochemical products, and environmental interactions (the "theatre of activity"). The microbiome encompasses microbial structural elements like proteins, polysaccharides, nucleic acids, lipids, and mobile genetic elements such as bacteriophages, as well as microbial metabolites (signaling molecules, toxins, organic compounds). Internal and external factors—both intrinsic (e.g., host genetics, immunity) and extrinsic (e.g., diet, antibiotics)—regulate the structure and function of these microbial communities

in both health and disease. Later, as microbiome research grew, some definitions began to center on microbial DNA, often making a distinction between "microbiome" and "microbiota" (Fig. 1.3). Still, the original concept by Whipps and the ecological approach of early scientists like Sergei Winogradsky continue to be central to how we understand these communities today.

While "microbiota" refers to microbial populations inhabiting a specific environment, "microbiome" also encompasses their collective genetic material, including their functional genes and metabolic products (Höll and Bossert 2022). While microbiota answers the question "who is there?" in a given niche, the microbiome seeks to understand "what are their functions?" which is a crucial distinction in both clinical and ecological microbiology (Table 1.1). According to the Oxford English Dictionary (2020), the word *microbiome* combines *micro*, meaning "very small," and *biome*, which means "the plant and animal community of a major climatic region or habitat." Interestingly, the dictionary offers two fundamentally distinct definitions of *microbiome*: one as "a population of microorganisms inhabiting a specific

Table 1.1 Difference between microbiota and microbiome

Feature	Microbiota	Microbiome
Definition	The community of microorganisms living in a specific environment	The complete collection of genes/genomes of the microbiota
Focus	The organisms themselves (who is there?)	The functions and genetic potential (what can they do?)
Components	Bacteria, archaea, viruses, fungi, protozoa	DNA, RNA, proteins, metabolites from those microorganisms
Scope	Taxonomic (species, genera)	Functional (gene expression, metabolism, interactions)
Study methods	Microscopy, culturing, 16S rRNA sequencing	Metagenomics, metatranscriptomics, proteomics, metabolomics
Use in research	Identifying microbial presence and composition	Understanding host-microbe interactions, disease mechanisms, therapies

environment; a microbial community or ecosystem, now especially that of the body," and the other as "the collective genomes of all the microorganisms inhabiting a specific environment, especially that of the body" (Berg et al. 2020). Even though there is a difference between describing microbial communities as organisms versus as their genetic content, both interpretations are shaped by a predominantly ecological perspective. This framing reflects a broader transformation in biomedical understanding, i.e., shifting from a combative view of microbes as invaders to embracing them as indispensable collaborators within a finely tuned, coevolved ecosystem of mutual coexistence. While initial efforts were focused on identifying and classifying microorganisms associated with the human body, contemporary research now emphasizes function, dynamics, and therapeutic potential of microbial communities within their host environments. Contemporary research recognizes that only a small fraction of microbes is pathogenic. Most are symbiotic, commensal, or beneficial, influencing host development, immunity, metabolism, and even mental health. This realization underpins the One Health framework, which sees the microbiome as central to the health of humans, animals, plants, and ecosystems. The microbiome is no longer viewed in isolation but as a dynamic component of a broader ecological and evolutionary network.

1.2 The Omics Era of Microbiome

1.2.1 From Culture-Dependent to Culture-Independent Studies

Early human microbiome research relied on culture-dependent methods, meaning scientists needed to cultivate microbes in the laboratory before they can study them. The first bacterium cultured was *Bacillus anthracis*, cultured by Robert Koch in 1876, beginning pure culture microbiology. The first commensal human gut bacterium isolated was *Escherichia coli* in 1885 by Theodor Escherich. Then, in the

1970s, human intestinal microbiota was analyzed using gram-staining and microscopy, which showed Gram-negative bacteria to be the most common, whereas culture counts led one to believe Gram-positive anaerobes were most common. This disparity was called as the "great plate count anomaly," explaining the disparity between microbes seen on the microscope and microbes that could be grown in vitro, with possibly only 1% being easily cultured. Hungate's roll tube method in 1969 pushed the boundaries further by allowing very oxygen-sensitive bacteria to be cultured, and this opened up the way for the isolation of very ubiquitous genera like *Bacteroides, Clostridium, Ruminococcus,* and *Bifidobacterium*. Prior to the development of molecular methods, the gut microbiota was estimated to comprise 400–500 different species, though many were partially characterized because of the limitations of culture-based investigations (Hugon et al. 2017). This approach revealed some important human-associated bacteria but was severely limited—by standard techniques, only ~1% of environmental bacteria could be cultivated in-vitro (Hofer 2018). Many microorganisms require specific nutrients, growth factors, or strict anaerobic conditions that are difficult to replicate in vitro. As a result, culture-based approaches provide a biased view of microbial diversity, often overrepresenting fast-growing and robust species while missing slow-growing or oxygen-sensitive taxa. In the human microbiome, the vast majority of microbes remained the so-called "unculturable" organisms that eluded isolation. The advent of culture-independent techniques transformed this field by bypassing the need for lab cultivation. In the 1970s, Carl Woese and George Fox revolutionized phylogenetic analysis by introducing 16S rRNA as a molecular marker, a breakthrough that ultimately led to the identification of Archaea as a distinct domain of life. In the 1980s, Norman R. Pace and colleagues revolutionized microbial ecology by developing culture-independent approaches using 16S rRNA sequencing directly from environmental samples such as Yellowstone hot springs. This method revealed vast uncultured microbial diversity and laid the foundation for modern microbiome research (DeLong 2021). Methods like DNA sequencing of microbial communities using 16S rRNA gene sequencing and shotgun metagenomics allowed direct detection of microbes from samples, thereby overcoming great plate count anomaly and revealing an enormous hidden diversity. With these sequencing-based approaches, researchers discovered that most bacterial and archaeal species in human-associated habitats had never been cultured—for example, only ~28–31% of sequences in gut metagenomes could be matched to known cultured species, and just ~4–5% of active community transcripts correspond to cultured microbes (Hofer 2018). This shift from culture-dependent to culture-independent studies opened a new vista on the human microbiome, uncovering a "microbial dark matter" of previously unknown organisms. It also highlighted functional potential: even uncultured microbes contribute to key metabolic and immunological processes in the host. The transition to molecular profiling (amplicon sequencing, whole-community shotgun sequencing, etc.) thus marked a turning point, enabling comprehensive surveys of microbiome composition and gene functions without cultivation (Yang et al. 2025). Early sequencing surveys profiled the microbiota across body sites and life stages—for instance, 16S rRNA gene studies showed that two bacterial phyla, Bacteroidetes and

Firmicutes, dominate the healthy adult gut (>90% of the community) (Qin et al. 2010), and they identified alterations in microbiota composition that correlate with aging and various disease conditions (Yang et al. 2025). In early life, the gastrointestinal flora mainly consists of facultative anaerobes, e.g., *Escherichia coli* and *Streptococcus* becoming predominantly composed of obligate anaerobes such as *Bacteroides* and *Firmicutes* in adulthood. As people age and develop diseases like diabetes, inflammatory bowel disease (IBD), and colorectal cancer (CRC), obesity, their microbiome frequently exhibits dysbiosis and decreased diversity. Comparative studies across children, adults, healthy elderly, and geriatric patients with *Clostridium difficile* diarrhea revealed clear age- and disease-related shifts: children harbored higher numbers of enterobacteria and a less mature microbial community, while patients with *C. difficile* infection exhibited markedly reduced species diversity, overgrowth of facultative anaerobes, and depletion of protective *Bifidobacterium* and *Bacteroides* (Yang et al. 2025). Such findings validated the power of culture-independent methods to characterize the human microbiome's taxonomic landscape and spurred efforts to explore its functional roles in health and disease. Incorporation of various "omics" data layers provides a holistic view of microbiome function and host interactions. The following section illustrates how metagenomic (DNA) data, metatranscriptomics (RNA) data, metaproteomic (protein) data, and metabolomic (metabolite) data can be combined to map the flow of biological information from microbial genes to enzymes to metabolic products, and how these intersect with host processes. Given the complementary insights offered by each omics technique, the current era of microbiome research is increasingly multiomics. Rather than examining the genome, transcriptome, proteome, or metabolome in isolation, scientists are now integrating these data to build a comprehensive systems-level model of the human microbiome.

1.2.2 Culturomics: High-Throughput Culturing Renaissance

Interestingly, the genomic era of microbiome research has also led to a renaissance of advanced culture methods termed culturomics (Fig. 1.4). Culturomics involves massively parallel culturing under diverse conditions, combined with modern identification techniques (like Matrix Assisted Laser Deaborption/Ionization-Time of Flight (MALDI-TOF) is a powerful mass spectrometry techniques used to identify microorganisms, analyse biomolecules and 16S sequencing), to recover far more microbial species from a sample than traditional culture allowed (Dickson 2017; Lagier et al. 2018). This approach, pioneered in gut microbiome studies, has dramatically expanded the catalogue of cultivable human microbes. For example, Lagier and colleagues in 2012 applied 212 different culture conditions to human stool and isolated 340 distinct bacterial species, comprising 174 species that have not been previously documented within the human gut. This includes 31 previously unidentified species or genera whose genomes have been sequenced (Lagier et al. 2012). These initial culturomics experiments uncovered novel taxa (even from rarely encountered phyla) and thousands of new genes, proving that many "unculturable"

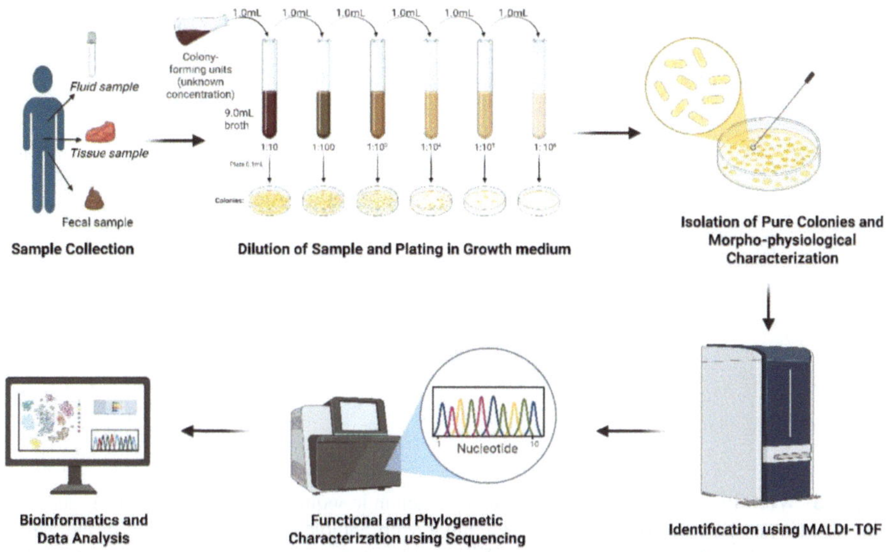

Fig. 1.4 Workflow of culturomics in human microbiome research. This diagram illustrates the culturomics approach, beginning with sample collection from various human body sites, followed by cultivation under diverse environmental conditions. High-throughput identification methods like 16S rRNA gene sequencing, MALDI-TOF MS, and whole genome sequencing are then applied. The resulting data undergo bioinformatics analysis for taxonomic and functional insights. Culturomics enables the recovery of live microbial strains and complements metagenomic approaches, facilitating novel species discovery, antimicrobial resistance profiling, and microbiota-based therapeutic development

microbes could indeed be cultured with the right techniques (Wan et al. 2023). Subsequent refinement of culturomics protocols—such as using fresh samples, extended incubation to detect slow-growing colonies, and specialized media for anaerobes or extremophiles—yielded even greater success (Sood et al. 2021; Diakite et al. 2020). Culturomics complements sequencing-based analysis by providing live isolates for experimentation and by assigning names and functions to sequences that were previously unassigned. In other words, culturomics has helped bridge the gap between metagenomic "dark matter" and formal microbial taxonomy, enabling functional studies (e.g., phenotypic assays, pathogen virulence tests, probiotic potential) on newly isolated strains. This high-throughput culturing strategy continues to discover new human-associated microbes and enrich our understanding of the microbiome's composition, offering a valuable counterpart to strictly culture-independent approaches.

1.2.3 Metagenomics—Amplicon and Shotgun Sequencing

Jo Handelsman and colleagues were the first to introduce the term metagenomics. Through her research on biosynthetic gene clusters and their associated natural

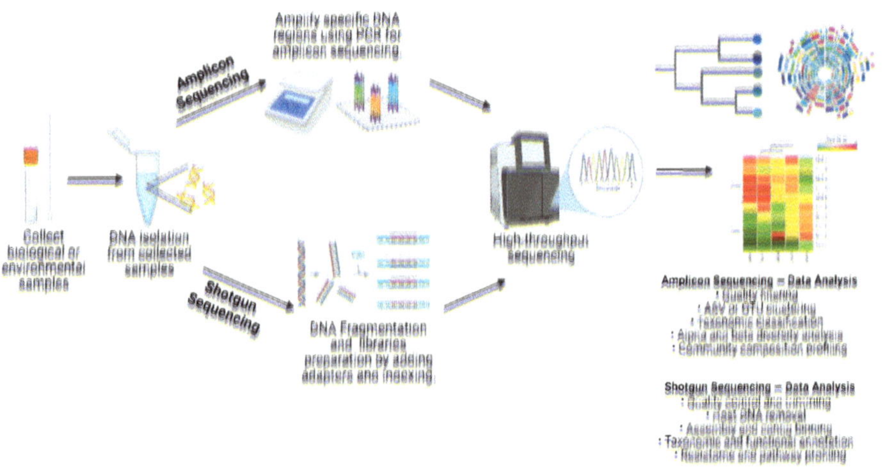

Fig. 1.5 Workflow of amplicon and shotgun metagenomic sequencing approaches. The schematic illustrates key steps in both methods, beginning with sample collection and DNA extraction. Amplicon sequencing involves targeted PCR amplification of marker genes like 16S rRNA, followed by sequencing and taxonomic profiling. In contrast, shotgun sequencing bypasses targeted amplification, involving random fragmentation of total DNA, enabling comprehensive sequencing of all genomic content. Downstream analysis differs, with amplicon data yielding taxonomic insights based on reference databases, while shotgun data allows for both taxonomic and functional characterization, including genome assembly and gene annotation

products, they proposed that analyzing DNA directly from entire environmental samples could help predict whether microbial communities harbor novel bioactive small molecules (Handelsman et al. 1998). Metagenomics includes both amplicon metagenomics and shotgun metagenomics. The introduction of 16S rRNA amplicon sequencing and whole-genome shotgun sequencing has revolutionized culture-independent profiling of microbiomes (Fig. 1.5).

The 16S rRNA gene is a highly conserved genetic marker found in all bacteria and archaea, with species-specific variable regions that serve as barcodes for identification. By PCR amplifying and sequencing 16S genes from a community DNA sample, researchers can obtain a census of the microbial taxa present and their relative abundances without needing to culture any organisms. This amplicon sequencing approach became a workhorse in human microbiome research due to its cost-effectiveness and robust reference databases of 16S sequences for taxonomic assignment. Using 16S surveys, scientists defined the core taxonomic composition of sites like the gut, skin, oral cavity, and vagina. For instance, as previously highlighted, 16S rRNA-based profiling has demonstrated that the gut microbiota of healthy adults is largely dominated by the Bacteroidetes and Firmicutes phyla, which together constitute more than 90% of the microbial sequences detected in stool samples. (Qin et al. 2010). Such studies also highlighted interpersonal and age-related differences: infants harbor very different gut communities than adults, and even among healthy adults, there is substantial person-to-person variability in microbiome composition. Clinically, 16S surveys provided initial links between microbiome dysbiosis and disease. Reductions in normal beneficial groups (like *Bifidobacterium* or certain *Bacteroides*) and overgrowth of opportunists were observed in conditions like antibiotic-associated *Clostridioides difficile* infection, metabolic disorders and inflammatory bowel disease (IBD) (Yang et al. 2025). The 16S method thus allowed large-scale profiling of microbial communities that was previously impossible, enabling direct comparison of microbiota across hundreds of individuals or samples. By the late 2000s, international projects like the Human Microbiome Project (HMP) and MetaHIT had employed 16S sequencing on thousands of samples, constructing a broad picture of the "who is there" in the human microbiome (Human Microbiome Project Consortium 2012). The limitation of 16S amplicon sequencing, however, is its focus on a single gene it identifies organisms (sometimes only to the genus level) but provides limited insight into their functional capacities or strain-level differences. Despite this, 16S profiling remains a valuable, rapid first-pass tool in clinical microbiome studies (for example, to detect shifts in diversity of gut microbiota or the detection of specific pathogens) and continues to be widely used for microbial community surveillance.

To move beyond taxonomy and explore the functional gene content of microbiome, researchers turned to shotgun sequencing. In shotgun metagenomics, all DNA extracted from a sample is sequenced typically with next-generation sequencing, producing millions of reads that represent random fragments of all genomes in the community. This approach captures not just the 16S rRNA gene, but all genes allowing identification of microbial species and, importantly, the virulence factors, metabolic pathways, antibiotic resistance genes, and other functional elements present in the microbiome. Shotgun metagenomics was a pivotal step in microbiome research because it revealed the genomic and metabolic potential of the microbiota, shedding light on what these microbes could be doing in the host environment (Joseph and Pe'er 2021). One hallmark early finding was that the collective gene repertoire of the gut microbiome is immense: the MetaHIT project's 2010 gene catalogue identified about 3.3 million unique microbial genes in the gut, that is ~150 times greater than the human genome's gene count (Qin et al. 2010). This illustrates that our microbiome encodes a vast array of functions far beyond our own genetic capabilities. In fact, based on gene content, researchers estimated on the order of 1000 bacterial species reside in the gut and each individual harbors around 150–160 species at any given time. Shotgun metagenomic studies also allowed assembly of genomes directly from microbiome data and the discovery of new microbial species (the so-called metagenome-assembled genomes or MAGs). Over the years, sequencing of more samples and deeper coverage has continually expanded these gene catalogues and revealed previously unknown taxa. Importantly, shotgun metagenomics provided functional insights: it became possible to link microbiome composition with metabolic pathways and clinical phenotypes (Lindner et al. 2024). For example, in a systematic review and meta-analysis by Pinart et al. (2021), investigated differences in gut microbiome composition between obese and non-obese adults using high-throughput sequencing. A total of 32 cross-sectional studies were included. While 9 out of 22 studies reported that alpha diversity

(Shannon index) reduced significantly in obese individuals, the pooled meta-analysis showed a nonsignificant mean difference. At the phylum level, several studies observed higher Firmicutes and lower Bacteroidetes in obese individuals, with meta-analysis confirming significantly higher Firmicutes but nonsignificantly lower Bacteroidetes. At the genus level, obesity was linked to a reduced presence of *Eggerthella* and *Bifidobacterium*, alongside elevated levels of genera such as *Prevotella, Fusobacterium, Streptococcus, Escherichia-Shigella, Dorea, Roseburia, Dialister, Acidaminococcus*, and others.". The findings suggest some consistent differences in microbial composition, but high heterogeneity across studies prevents firm conclusions (Pinart et al. 2021). Other studies identified disease-associated gene signatures in the microbiome e.g., IBD patients often show increases in genes for oxidative stress and inflammation, whereas individuals with metabolic syndrome may have fewer genes for beneficial short-chain fatty acid (SCFA) production (Yang et al. 2025). Landmark projects like the HMP (USA) and MetaHIT (Europe) applied shotgun metagenomics at large scale, characterizing thousands of reference microbial genomes and millions of genes from the human microbiome. The HMP, for instance, sequenced ~800 cultured reference strains in parallel with community metagenomes, generating a comprehensive resource of microbial genomic data (Human Microbiome Project Consortium 2012). These efforts demonstrated that while each person's microbiome has a unique species makeup, there is a conserved core of microbial genes (e.g., for nutrient metabolism, vitamin synthesis, etc.) shared across healthy individuals. The use of shotgun sequencing in human research has produced massive datasets due to the quick improvement in our capacity to resolve whole genomes within consortia. Pasolli and colleagues in 2019 published a study in which they analyzed 46 datasets and reported over 150,000 genomes from roughly 10,000 human metagenomes (Pasolli et al. 2019). Shotgun metagenomics thus established the foundation of microbiome genomics, allowing researchers to catalogue not just who is present, but what biological capabilities the microbiome possesses. This capability underpins many microbiome discoveries related to drug metabolism, diet–microbe interactions, and pathways by which gut bacteria influence host physiology.

1.2.4 Metatranscriptomics (Community-Wide Gene Expression)

DNA-based methods like 16S and shotgun sequencing show the genetic potential of microbiome, but they do not indicate which genes are actually active at any given time. Metatranscriptomics addresses this by sequencing the collective mRNA transcripts from a microbial community. The term "metatranscriptome" describes the entire collection of gene transcripts (RNA copies of genes) that are present in a microbial community at a given time, when regarded as a single entity. The metatranscriptome varies according to temporal and environmental circumstances because gene expression is dynamic. Metatranscriptomics is the study of these transcripts utilizing high-throughput techniques to track the dynamics of gene expression and capture the community's complete expression profile. When combined with other "omics" techniques, this method offers valuable insights into the composition, roles, and adaptation strategies of intricate microbial communities. It can also be applied broadly in biotechnology, medicine, and microbial ecology (Parro and Paz 2023). Transcriptomics emerged in the 1980s with the advent of techniques like northern blotting and reverse transcription–PCR (RT-PCR), which allowed for the detection of individual RNA transcripts. The field advanced significantly in the 1990s with the introduction of microarray technology (Schena et al. 1995) that was a breakthrough, capable of examining hundreds or thousands of genes at the same time. This formed the basis of single-organism transcriptome analysis (Liu et al. 2003) and, ultimately, the complex environmental transcriptomes (Parro et al. 2007). The actual breakthrough for metatranscriptomics was Leininger et al. (2006), with the report of the first metatranscriptome of an ammonia-oxidizing prokaryotic community in soils, thanks to the invention of high-throughput sequencing platforms. Metatranscriptomics has since snowballed, not only enabling large-scale functional profiling of microbial communities but also the identification of small noncoding RNAs and thus transforming our understanding of microbial activity in natural and host-associated ecosystems (Parro and Paz 2023). In a metatranscriptomic analysis, total RNA is extracted from a sample and the microbial messenger RNA (after rRNA removal) is converted to cDNA and sequenced (Fig. 1.6). The result is a profile of genes that are being actively expressed by the community under the sampled conditions. This approach can identify the functions that microbes are performing in real-time and the metabolic state of the microbiome. For instance, metatranscriptomic studies have shown that only a subset of the microbiome's genes is highly expressed at any moment, and expression can shift with environmental or host factors. One notable application was the discovery of how gut microbes respond to and even modify medications: in an integrated metagenomic-transcriptomic study, scientists found specific bacterial genes that were upregulated when metabolizing a cardiac drug, revealing a mechanism of gut microbial drug inactivation. This aligns with earlier findings that certain gut bacteria like *Eggerthella* can inactivate drugs such as digoxin—metatranscriptomics pinpointed the responsible genes. Bashiardes et al. (2016) have used metatranscriptomics to uncover community responses to host cues for example, Toll-like receptor 5 knockout mice (lacking a key immune sensor of flagella) showed increased expression of bacterial motility and chemotaxis genes in their gut microbiome, indicating the microbes were adjusting their behavior in the altered host environment (Cullender et al. 2013). Metatranscriptomics has also highlighted unique microbial processes in specialized niches: one study found that deep-sea vent microbiomes express genes for phosphonate metabolism that are largely silent in shallow environments, illustrating how transcriptomics can reveal niche-specific activities. On a broader scale, projects like the Men's Lifestyle Validation Study (part of iHMP) applied metatranscriptomics in healthy individuals and identified a core set of transcripts (e.g., glycolysis and nucleotide biosynthesis genes) that were consistently expressed in all individuals, alongside more personalized transcripts reflecting each person's unique microbiome composition (Abu-Ali et al. 2018). There are technical challenges to metatranscriptomic profiling bacterial mRNA is

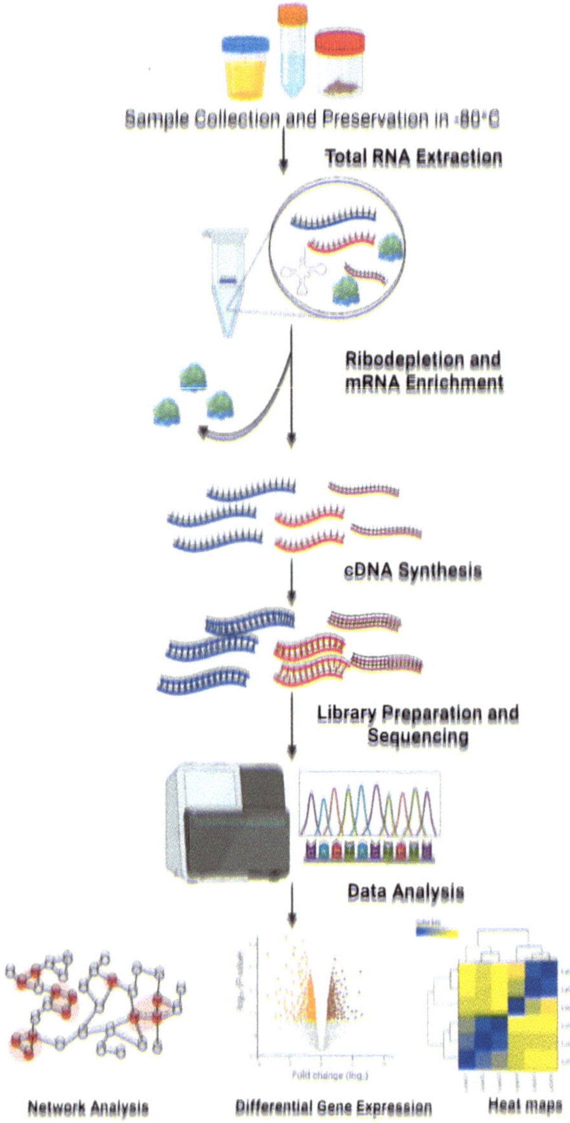

Fig. 1.6 Overview of the metatranscriptomics workflow for studying the microbiome. The process involves eight key steps: (1) collection of biological samples containing microbial communities, (2) extraction of total RNA, (3) removal of abundant ribosomal RNA to enrich for messenger RNA, (4) synthesis of complementary DNA (cDNA) from mRNA, (5) preparation of sequencing libraries, (6) high-throughput sequencing, (7) bioinformatic analysis to identify and quantify microbial gene expression, and (8) functional interpretation to understand the active roles of microbes in the host environment

only ~1–5% of total RNA and lacks the poly-A tail used for eukaryotic mRNA enrichment, so special rRNA-depletion and enrichment protocols are needed to capture microbial transcripts. Additionally, mRNA is short-lived, and not all transcripts lead to protein production. Despite these hurdles, metatranscriptomics provides a crucial layer of functional insight by identifying which microbial genes are actively being transcribed. It often reveals a different perspective than DNA alone; for example, genes for certain carbohydrate-degrading enzymes might be present in a microbiome (metagenomics) but only expressed when specific dietary substrates are available (metatranscriptomics confirms the diet-dependent expression). By combining metatranscriptomic data with metagenomic data, researchers can distinguish mere genetic potential from actual gene expression, thereby gaining a more dynamic view of host-microbe interactions.

1.2.5 Metaproteomics (Community Proteins and Enzymatic Functions)

Metaproteomics evolved from proteomics, which emerged in the 1990s with advances like two-dimensional gel electrophoresis and mass spectrometry for large-scale protein analysis. The term metaproteomics was first introduced by Wilmes and Bond (2004) in a study of acid mine drainage biofilms, demonstrating that proteins from entire microbial communities could be profiled rather than from single species. With the advent of shotgun proteomics (LC-MS/MS) between 2006 and 2010 and integration with metagenomics, metaproteomics rapidly expanded to environmental and host-associated microbiomes. By the 2010s, it was recognized as a distinct discipline, applied in fields ranging from ecology to medicine and biotechnology. Today, with high-resolution MS and multiomics integration, metaproteomics provides direct characterization of the functional dynamics within microbial populations, making it a central tool in microbiome research (Wilmes and Bond 2004; Wilmes et al. 2015). While transcripts indicate gene expression, the metaproteome reveals the actual proteins present and functioning in the microbiome. Metaproteomics involves extracting and identifying the proteins from a microbial community, typically using high-resolution mass spectrometry to detect peptide fragments and match them to databases of predicted proteins from metagenomes (Fig. 1.7). This approach shows which microbial proteins (enzymes, structural proteins, toxins, etc.) are produced and potentially active, thus directly linking genes to biochemical function (Van Den Bossche et al. 2021). Metaproteomic studies of the human microbiome have uncovered which microbial pathways are translated into proteins and how stable or variable the protein expression profiles are over time. One pioneering study demonstrated that individuals have a subject-specific yet stable gut proteome: over a one-year period, each person's gut microbiota expressed a characteristic protein profile that remained relatively consistent, including a fundamental collection of approximately 1000 proteins primarily engaged in carbohydrate metabolism and interactions between hosts and microbes. This finding of temporally stable protein expression is striking given the taxonomic turnover of the microbiome, and it suggests a functional redundancy where different microbes may fulfil similar roles to maintain key functions (Kolmeder et al. 2012). Metaproteomics has also been useful in identifying proteins that correlate with disease states or environmental changes. For example, targeted metaproteomic analyses have identified bacterial enzymes and virulence factors linked to inflammation: in patients suffering from inflammatory bowel disease, there was overabundance of certain proteases and enzymes that degrade the mucus barrier, as well as

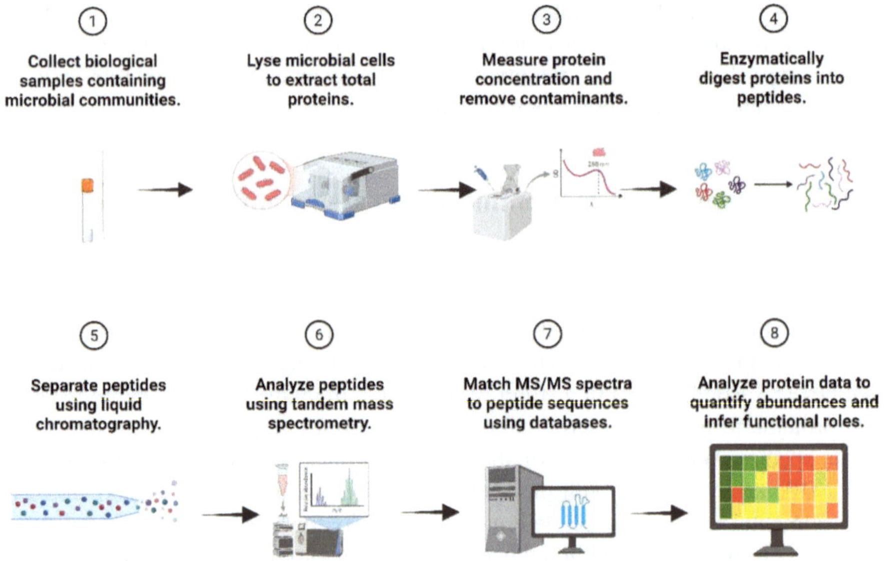

Fig. 1.7 Workflow of metaproteomics for analysis of microbial communities in the human microbiome. The process starts with sample collection, followed by protein extraction and cleanup. Proteins are enzymatically digested into peptides, which are separated using liquid chromatography and subsequently analyzed by tandem mass spectrometry (LC-MS/MS); resulting spectra are matched against protein databases for identification and quantification. Finally, bioinformatic analyses provide insights into microbial composition and functional activity based on expressed proteins

stress-response proteins, reflecting an active microbial contribution to the inflamed gut environment.

Novel techniques like Activity-Based Protein Profiling (ABPP) have further enhanced metaproteomics by focusing on functionally active enzymes. ABPP uses chemical probes that bind to the active sites of enzymes (such as serine hydrolases) in a sample, allowing those active enzymes to be isolated and identified (Fig. 1.8). Applied to gut microbiome samples, ABPP has revealed specific microbial hydrolases and esterases that are active under host-like conditions (Ortega et al. 2016). This is important because it distinguishes the mere presence of a protein from actual catalytic activity. Using such approaches, researchers have discovered microbial enzymes that modulate host signaling (for instance, bacterial proteases that cleave host cytokines) and have noted increased activity of certain microbial enzymes during disease flares in IBD (Mayers et al. 2017). Challenges remain in metaproteomics, notably the incomplete coverage of reference protein databases (many detected peptides might belong to uncharacterized proteins) and the dynamic range of proteins (abundant proteins can mask detection of low-abundance ones). Moreover, protein levels do not always correlate with mRNA levels due to post-transcriptional regulation. Nonetheless, metaproteomics adds a critical dimension by

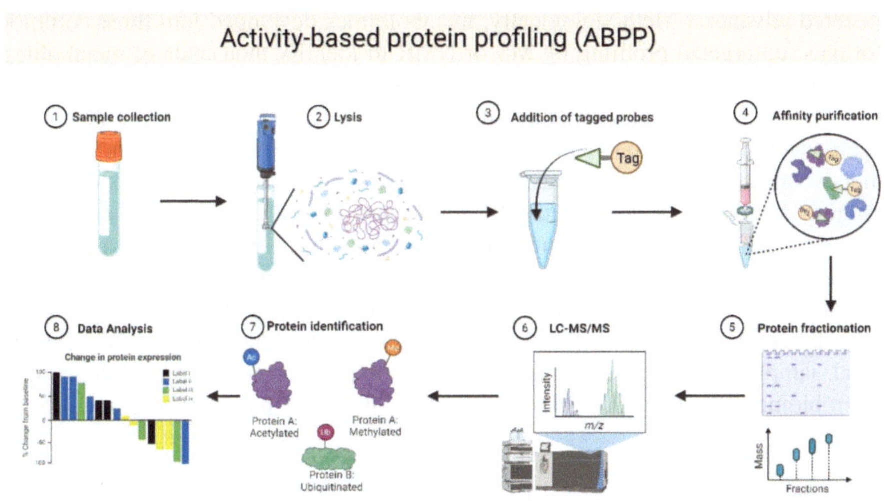

Fig. 1.8 Workflow of Activity-Based Protein Profiling (ABPP). Proteins from biological samples are extracted by cell lysis and incubated with activity-based probes that covalently bind to the active sites of specific enzymes. Tagged proteins are then enriched by affinity purification, fractionated, and analyzed using LC-MS/MS. Identified proteins are subjected to bioinformatic data analysis, allowing functional annotation of enzyme activity within complex proteomes

verifying which gene products are present and functional. It complements metatranscriptomics by confirming the translation of highly expressed genes, and it can highlight post-translational modifications or enzyme activities that are invisible to nucleic-acid-based methods.

1.2.6 Metabolomics (Microbiome-Related Metabolites)

At the end of the omics cascade is metabolomics, which profiles the small-molecule metabolites produced or modified by the microbiome. Metabolomics arose in starting of 2000s as one aspect of the larger omics technique, providing a data-driven complement to hypothesis-driven biology. Metabolomics reduced the gap of the understanding the complexity of biological systems by producing biomarker sets of the metabolome in basal and perturbed states that could both guide hypothesis generation and reveal unexpected findings. The term *metabolome* was first defined by Oliver et al. in 1998, followed shortly by Nicholson et al.' introduced Nuclear Magnetic Resonance (NMR)-based metabonomics. Nicholson was the one who coined the term metabonomics to define "the dynamic metabolic response of living systems to biological stimuli" (Aharoni et al. 2023). Fiehn and others laid the foundation for the discipline metabolomics. In 2001, the first GC-MS–based metabolomic analysis in plants was demonstrated by Fiehn et al., marking a key milestone in the field, while Raamsdonk et al. applied NMR-based metabolomics in yeast. In the United States, the 2003 NIH Metabolomics Roadmap Initiative further fostered advances. Methodologically, metabolomics developed into three common formats: untargeted profiling by MS or NMR to identify thousands of metabolites, targeted analysis of a known group of metabolites, and approaches that involve isotope labelling to monitor metabolite concentrations and fluxes. The metabolome itself encompasses not only endogenous molecules but also exogenous molecules of dietary origin, microbiota, drugs, and environment exposures, and thus the wide-ranging biological and biomedical significance of the discipline is established (Zhang et al. 2011). The gut microbiota, in particular, functions like a chemical factory in the human body—fermenting dietary components, producing short-chain fatty acids, synthesizing vitamins, transforming bile acids, and generating numerous other bioactive compounds. Metabolomic analysis of biofluids (like serum, urine) or fecal samples allows researchers to measure these metabolites and link them to microbiome activity (Fig. 1.9). A key advantage is that metabolomic profiles

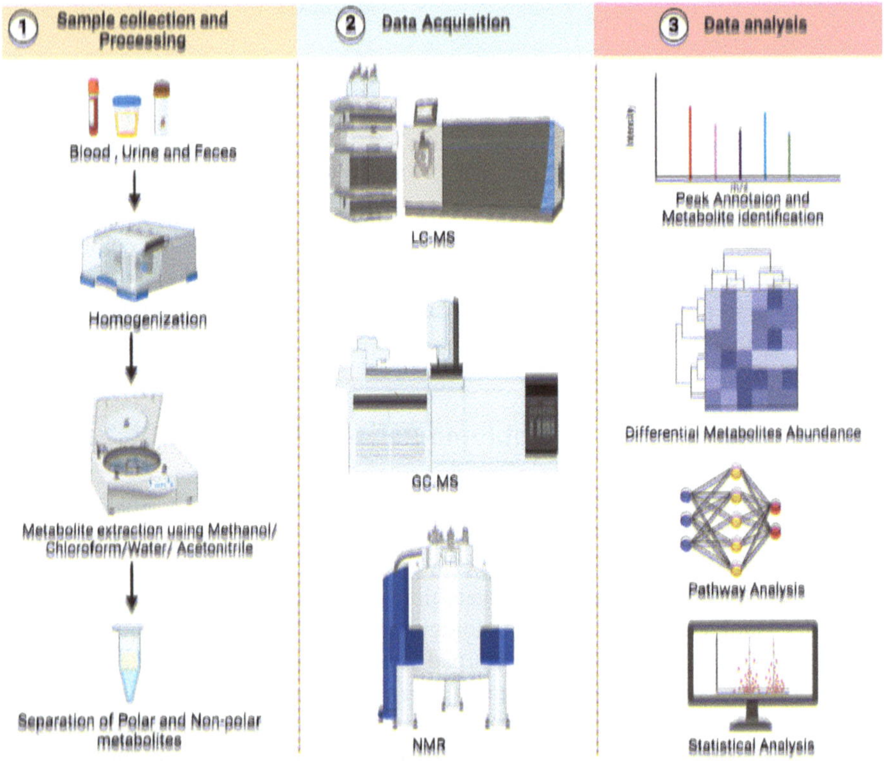

Fig. 1.9 Metabolomics workflow illustrating essential steps for analyzing metabolic profiles in biological samples. The process begins with sample collection and preparation, followed by metabolite extraction using solvent-based methods. Instrumental analysis is performed using techniques such as mass spectrometry or NMR. The resulting data are acquired, processed, and metabolites are identified using reference databases. Finally, statistical analysis and biological interpretation are conducted to understand the metabolic activity and its implications in the biological system

represent the actual functional output of the microbiome in real time (Yang et al. 2025). Unlike DNA or RNA, which indicate potential or ongoing gene expression, metabolites are the end products that directly influence host physiology. One of the landmark discoveries via microbiome metabolomics was the link between gut microbial metabolism of dietary nutrients and cardiovascular disease. In a now-classic study, researchers found that gut bacteria convert dietary phosphatidylcholine (from foods like red meat and eggs) in trimethylamine-N-oxide (TMAO) which is a metabolite that promotes atherosclerosis (Kahraman Ilıkkan 2025). This finding highlighted how a microbial metabolite can contribute to a common human disease, leading to intense interest in the gut microbial metabolome as a source of potential biomarkers and therapeutic targets.

Metabolomics has also illuminated beneficial compounds from commensal microbes for instance, fermentation of fiber by intestinal microbes leads to the formation of key SCFAs, including acetate, propionate, and butyrate. These SCFAs play crucial roles in maintaining colonic health and modulating immune responses. Clinical metabolomic studies have observed that patients with IBD or metabolic syndrome often have depleted levels of butyrate-producing bacteria and lower fecal butyrate; consistent with this, multiomics analyses in IBD cohorts found that butyrate-producing species (e.g., *Faecalibacterium prausnitzii, Roseburia*) and their associated metabolites are significantly reduced in IBD patients (Ning et al. 2023). Such correlations underscore the importance of microbial metabolites in disease. Early microbiome metabolomics work, often using NMR spectroscopy, demonstrated complex interactions among diet, the microbiome, and host metabolism for example, differences in urinary metabolite profiles between germ-free and conventional mice revealed which metabolites were of microbial origin. In recent years, the field has advanced with high-resolution mass spectrometry and large metabolite databases, enabling the detection of thousands of compounds in stool or blood. By applying machine learning to these data, scientists can identify metabolite patterns associated with diseases or drug responses (Wishart et al. 2018). A challenge with metabolomics is that it provides a snapshot of numerous metabolites without immediately indicating which microbial genes or pathways produced them. Therefore, integrating metabolomic data with other omics (especially metagenomics and metatranscriptomics) is essential to map metabolites back to the responsible microbes or enzymes. Despite this, metabolomics is indispensable for capturing the chemical impact of the microbiome on the host. It directly informs clinical research—for example, measuring microbial metabolites in patients can potentially predict disease risk (as with TMAO for heart disease) or stratify patients (as certain metabolite profiles indicate dysbiosis). Overall, metabolomics links microbial community activity to host phenotype in tangible chemical terms, rounding out the functional portrait provided by DNA, RNA, and protein analyses.

1.3 The Microbial World in Human Health

The human body is a host to a highly diverse and intricately structured microbial ecosystem that colonizes nearly every surface and cavity (Fig. 1.10). These organisms are not merely passive inhabitants; rather, they are essential partners in health and disease. The human microbiota, which is often referred to as a "hidden organ or second genome," consisting of over 100 trillion symbiotic microorganisms inhabiting both the surface and internal environments of the human body (Ley et al. 2006). These microbial communities contribute significantly to human physiology, particularly through their extensive genetic repertoire, which is estimated to be over 150 times greater than that of human genome (Zhu et al. 2010). Although early estimates provided a 10:1 ratio of microbial cells to human cells (Savage 1977), recent estimates put this into perspective and the abundance of human and microbial cells in the human body is roughly equivalent (Sender et al. 2016). From birth, the human body is colonized with microbes, and this is determined by mode of delivery, early diet, and environmental exposures. With age, this microbial colonization is individualized and relatively fixed by three years of age, although it is still

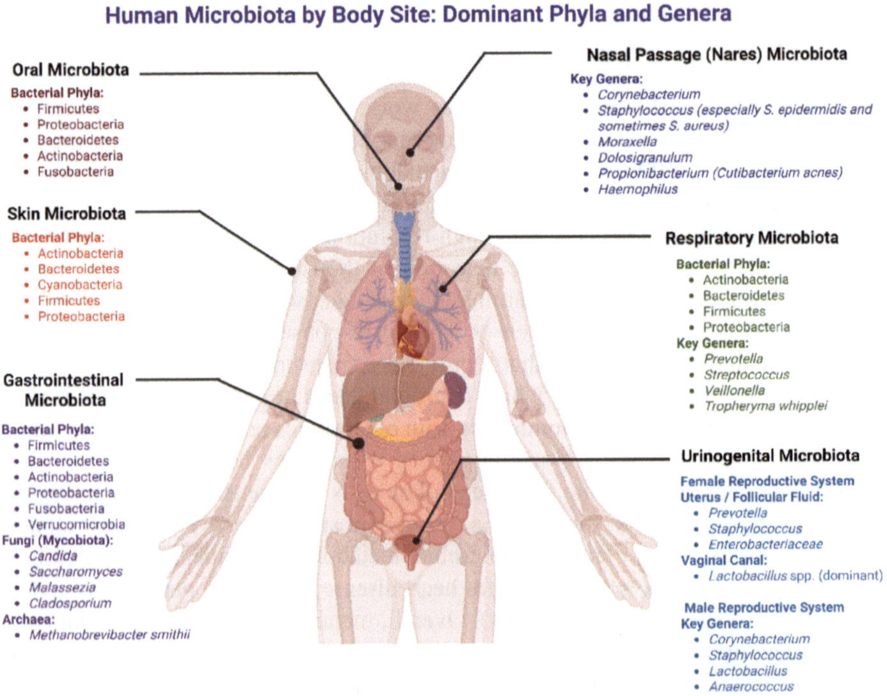

Fig. 1.10 Anatomical distribution of dominant human microbiota across body sites. This diagram illustrates the site-specific colonization of key microbial taxa throughout the human body, highlighting major bacterial phyla and representative genera involved in gut, oral, lung, skin, nasal, and reproductive tract health. Each anatomical region harbors distinct microbial communities shaped by environmental conditions like pH, oxygen levels, and host factors

modifiable throughout life due to genetics, antibiotics, diet, and lifestyle (Gilbert et al. 2018). These communities of microbes are not distributed randomly throughout the body. The anatomical variations are large, dependent on oxygen concentration, pH, hydration status, and nutrient availability. The gut microbiota is extremely well-characterized and of utmost significance, playing a crucial role in preserving human health by processes including food fermentation, protection of the host from virulent pathogens, modulating immune responses, and vitamin production. It is primarily composed of six bacterial phyla. Among the major bacterial phyla inhabiting the human body, Firmicutes and Bacteroidetes are most abundant, accompanied by Actinobacteria, Proteobacteria, Fusobacteria, and Verrucomicrobia (Rinninella et al. 2019). The gut also harbors a fungal community called the mycobiota, which is broadly represented by genera like *Candida*, *Saccharomyces*, *Malassezia*, and *Cladosporium* (Auchtung et al. 2018). Beyond bacterial taxa, the gut microbiome also harbors viruses, bacteriophages, and archaea, like *Methanobrevibacter smithii* (Lozupone et al. 2012). Less densely populated than the gut, the oral cavity harbors the second-most dense microbial community of the human body. This region harbors several distinct microhabitats, such as saliva, the tongue, teeth, gums, inner cheeks, the palate, as well as subgingival and supragingival plaque. The microbial community in these niches differs based on microbial interaction, genetic determinants, and pH fluctuation. The most common bacterial phyla in the oral microbiota are Firmicutes, Proteobacteria, Bacteroidetes, Actinobacteria, and Fusobacteria (Santacroce et al. 2023). Previously thought to be devoid of microorganisms, the lung has its own resident microbiota, including Bacteroidetes, Firmicutes, Actinobacteria, and Proteobacteria. Its composition is characterized by microbial invasion from the upper airways, host defense mechanisms, and the ability of microorganisms to survive and reproduce in the pulmonary niche. The lungs contain a unique low density microbiota comprising genera including *Veillonella*, *Streptococcus*, and *Prevotella*. The microbiota is vital for the modulation of airway immune tolerance. Additionally, the lungs are a unique ecological niche for *Tropheryma whipplei* (Segal et al. 2014). The skin microbiota is remarkably diversified and deeply dependent on the anatomical site. Alteration in glandular distribution, moisture, and exposure to the external environment leads to characteristic microbial communities. The skin is characterized by the dominance of Actinobacteria, Bacteroidetes, Cyanobacteria, Firmicutes, and Proteobacteria. Species such as *Staphylococcus epidermidis* act as benign residents that also maintain the skin barrier and modulate immunity (Severn and Horswill 2023). Different bacteria are found in different parts of the reproductive system. *Prevotella, Staphylococcus,* and *Enterobacteriaceae* are prevalent in the uterus and follicular fluid of females, although *Lactobacillus* species predominate in the vaginal canal and serve protective functions. Numerous bacteria, including *Corynebacterium, Staphylococcus, Lactobacillus,* and *Anaerococcus*, are found in the male reproductive system; some of these species are associated with the quality of sperm (Boldyreva et al. 2017). Biofilms and site-specific communities that affect immunological responses and fertility are present in both systems. The human microbiome operates as a symbiotic system, governed by relationships of mutualism and commensalism. In healthy states, the human microbiome functions symbiotically through mutualism and commensalism. In health, it aids digestion, immune regulation, vitamin synthesis, and pathogen resistance via metabolites like short-chain fatty acids. Disruption of gut microbial homeostasis referred to as dysbiosis is associated with a spectrum of diseases such as IBD, metabolic syndromes, and neurodevelopmental or neurodegenerative disorders (Nysten and Van Dijck 2023) (Table 1.2). The microbiota also interacts continuously with the external environment such as soil, water, food, animals, and other humans forming a dynamic microbial interface that impacts population health (Galvani et al. 2016). Understanding both its taxonomy (microbiota) and function (microbiome) is key for advancing diagnostics, therapeutics, and preventive medicine.

Table 1.2 Host–microbe interactions in the human body

Category	Microbe (examples)	Location in human body	Role/impact
Mutualism	*Bacteroides thetaiotaomicron*	Gut microbiota	Breaks down complex polysaccharides; provides nutrients to host
	Lactobacillus spp.	Vagina, gut	Produces lactic acid → protects against pathogens by lowering pH
	Escherichia coli (nonpathogenic strains)	Large intestine	Produces vitamin K and B vitamins; benefits from nutrients
	Faecalibacterium prausnitzii	Gut	Produces butyrate → anti-inflammatory, supports colon health
Commensalism	*Staphylococcus epidermidis*	Skin	Harmless resident; may inhibit pathogens by competing for nutrients
	Corynebacterium spp.	Skin, nasal passages	Resides without harming host
	Veillonella spp.	Oral cavity, gut	Consumes lactate made by other microbes; neutral to host
	Propionibacterium acnes (*Cutibacterium acnes*)	Skin (sebaceous glands)	Lives on skin oils; mostly harmless, sometimes linked to acne
	Neisseria lactamica	Nasopharynx	Harmless; may protect against pathogenic *N. meningitidis.*
	Streptococcus salivarius	Oral cavity	Normal oral flora; helps maintain microbial balance
Parasitism	*Trichomonas vaginalis* (protozoan)	Urogenital tract	Feeds on host tissues; causes irritation
	Entamoeba histolytica	Intestine	Invades gut wall; causes amoebiasis
	Giardia lamblia	Small intestine	Attaches to intestinal wall → diarrhea, malabsorption

(continued)

Table 1.2 (continued)

Category	Microbe (examples)	Location in human body	Role/impact
Pathogenic	*Mycobacterium tuberculosis*	Lungs	Causes tuberculosis
	Salmonella enterica	Intestine	Causes gastroenteritis and typhoid fever
	Streptococcus pyogenes	Throat, skin	Causes strep throat, scarlet fever, necrotizing fasciitis
	Helicobacter pylori	Stomach lining	Causes gastritis, ulcers, linked to gastric cancer
	Vibrio cholerae	Intestine	Produces cholera toxin → severe diarrhea, dehydration
	Neisseria gonorrhoeae	Urogenital tract	Causes gonorrhea
	Treponema pallidum	Multiple tissues (systemic)	Causes syphilis
	Clostridium tetani	Wounds, anaerobic sites	Produces tetanospasmin toxin → tetanus
	Clostridium botulinum	Intestine/ wounds	Produces botulinum toxin → botulism
	Yersinia pestis	Blood, lymph nodes	Causes plague
	Klebsiella pneumoniae	Lungs, urinary tract	Causes pneumonia, UTIs, sepsis in immunocompromised
Opportunistic	*Candida albicans* (fungus)	Mouth, vagina, gut	Overgrowth causes thrush or candidiasis
	Clostridium difficile	Gut (after antibiotics)	Overgrowth causes pseudomembranous colitis
	Pseudomonas aeruginosa	Lungs, wounds	Opportunistic infections, especially in cystic fibrosis patients
	Aspergillus fumigatus (fungus)	Lungs	Causes aspergillosis in immunocompromised hosts
Symbionts	*Akkermansia muciniphila*	Gut mucus layer	Degrades mucin, supports metabolic health
	Prevotella spp.	Gut	Contributes to carbohydrate metabolism
	Ruminococcus spp.	Gut	Breaks down dietary fibers.
	Methanobrevibacter smithii (archaea)	Gut	Uses H_2 → methanogenesis, helps gut fermentation balance

1.4 A Healthy Human Microbiome

Defining a "healthy" human microbiome is a controversial and complex issue because it is not characterized by a single configuration but is understood as a dynamic and multifaceted concept. This challenge arises from a number of factors:

the substantial interindividual variation of microbiomes, the dynamic equilibrium over time of disease and health states and large influence of various environmental forces on microbial populations. A 'healthy' microbiome represents a range of compositions rather than a fixed ideal composition. Its definition should arise from detailed population-scale studies that reflect diverse representations and incorporate intrinsic microbiome health features such as diversity and resilience along with the absence of overt disease (Joos et al. 2025). The intricate ecosystem of the human gastrointestinal (GI) tract maintains a sensitive equilibrium between beneficial and potentially pathogenic bacteria, which critically impacts human health. A well-balanced GI microbiome is marked by considerably higher biodiversity and a rich abundance of specific beneficial phyla and genera. For example, a healthy adult colon is usually enriched with an abundance of microbes, with dominant bacterial phyla being *Bacteroidetes, Firmicutes, Verrucomicrobia, Proteobacteria*, and *Actinobacteria*. Certain genera such as *Bifidobacterium* and *Bacteroides* species are the dominant members in infants who were breastfed, making significant contributions to the early-life microbiome. In adults, stable core microbes such as *Eubacterium, Bacteroides, Faecalibacterium, Ruminococcus, Alistipes, Clostridium, Blautia,* and *Roseburi* are always present and comprise a stable microbial core. Conversely, changes in the GI microbiome have been associated with the development and progression of several chronic diseases, reflecting the presence or overgrowth of 'bad' bacteria (Fig. 1.11). Reduced microbial diversity is linked to various disease states. Some examples of bacterial changes signaling danger are the increased frequency of *Coriobacteriaceae* in patients with colorectal cancer. A significant rise in mucus-degrading bacteria, with higher abundance of *Ruminococcus gnavus* in mucosal samples, has also been found in inflammatory bowel disease (IBD) patients. These changes from a normally healthy, diverse microbiome to one of dominance by some "bad" or opportunistic bacteria can interfere with normal gut function and are involved in the cause and progression of many diseases (Hall et al. 2017). It has been found in studies that prebiotics, probiotics, and synbiotics aid in promoting balanced gut microbiota by inducing a healthy and diverse microbial community (Aggarwal et al. 2022). Several important factors facilitate the preservation of homeostasis, while various natural and artificial factors could disrupt it, ultimately leading to a pathological state (Ahrodia et al. 2022).

1.5 The Human Microbiome as a Force in Promoting Health, Homeostasis, and Disease Prevention

The human microbiome is a dynamic and intricate ecosystem living in harmony with host cells, playing a substantial role in physiological homeostasis, a balanced internal environment essential for health and wellness (Belkaid and Harrison 2017). When balanced, microbial communities aid in immune function, nutrient processing and resistance to pathogen invasion ultimately, avoiding illness. This balance is impacted by a number of natural factors like lifestyle, diet, age, hormonal

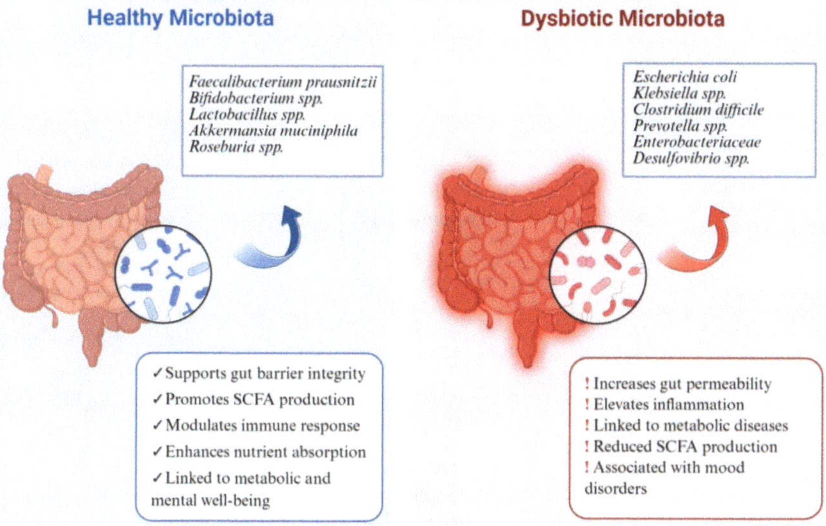

Fig. 1.11 Healthy and dysbiotic gut microbiota comparison. This diagram provides the structural and functional variations between healthy gut microbiota and dysbiosis. The healthy microbiota is dominated by beneficial taxa like *Lactobacillus* spp., *Bifidobacterium* spp., *Faecalibacterium prausnitzii, Akkermansia muciniphila and Roseburia* spp. which play roles like immune system modulation, maintaining gut barrier function, improved nutrient uptake, and production of short-chain fatty acids (SCFAs), such as butyrate, acetate, and propionate. These SCFAs are crucial for intestinal homeostasis and are linked with beneficial metabolic and neurobehavioral conditions. Dysbiotic microbiota reveal an abundance of potentially pathogenic taxa such as *Escherichia coli, Klebsiella* spp., *Clostridium difficile, Prevotella* spp., *Enterobacteriaceae, and Desulfovibrio* spp., which are associated with chronic inflammation, augmented intestinal permeability, diminished SCFA production and elevated risk of metabolic and neuropsychiatric disturbances

fluctuations, genetics and underlying disease. Disruption of microbial homeostasis, or dysbiosis, can also occur from these normal transitions but also by artificial means (Fig. 1.12). Antibiotics are a well-documented example, but other medications such as antipsychotics, nonsteroidal anti-inflammatory agents, calcium-channel blockers, antimetabolites, antivirals, metformin, proton-pump inhibitors, and antiseptics also change gut microbiota composition to a considerable extent and decrease microbial diversity (Vila et al. 2020). Such alterations can compromise gut barrier function, enhance inflammation, and enhance susceptibility to infections and long-term diseases. In an effort to reinstate microbial balance, probiotics and prebiotics have been investigated for their therapeutic and preventive role. Prebiotics are compounds such as indigestible carbohydrates from fibrous foods that supply nutrients to beneficial gut bacteria. Commensal bacteria fermentation of prebiotics leads to short-chain fatty acids (SCFAs) production like butyrate, which aid in gut health by strengthening the intestinal barrier, enhancing mucin secretion and enhancing

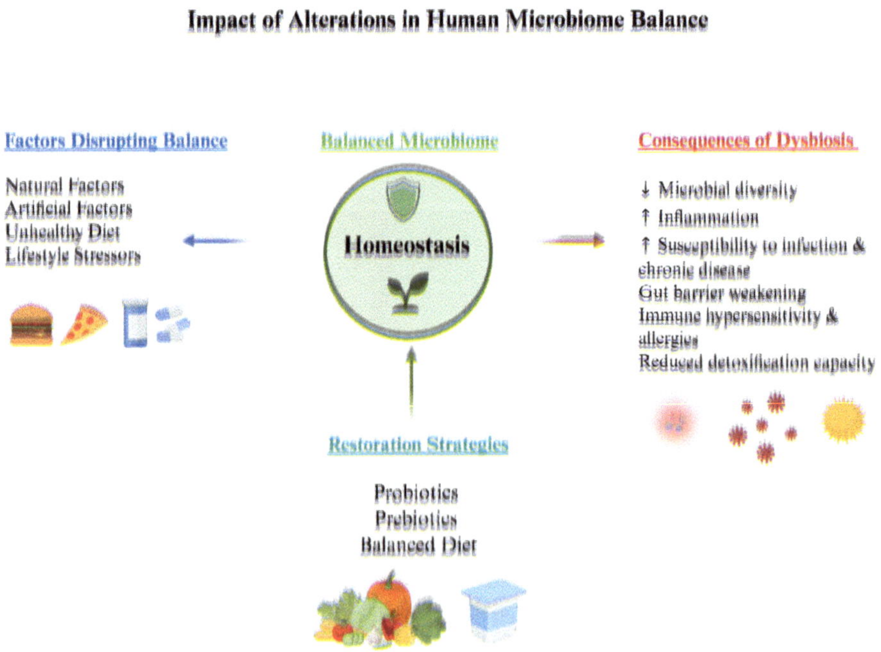

Fig. 1.12 Impact of alterations in human microbiome balance. The diagram depicts the dynamic equilibrium of the human microbiome, describing significant factors that disrupt its homeostasis and cause dysbiosis. It enumerates the resulting health effects, like systemic inflammation and shows strategies of restoring this balance in the form of probiotics, prebiotics and a good diet

overall intestinal integrity. Probiotics, being live normal flora, have immunomodulatory effects by inducing host defense. They increase the production of immunoglobulin A (IgA), defensins, and other antimicrobial peptides, thus inhibiting pathogenic organisms. These actions are mediated by signaling pathways that modulate cytokine production and inflammatory reactions (Bron et al. 2012). Dietary structure is another important determinant of microbial diversity. Diets with high fats and low fiber are negative, frequently depleting microbial richness within the gastrointestinal tract (Singh et al. 2017). Certain dietary structures promote particular microbial taxa: for instance, Proteobacteria thrive with diets high in simple sugars, *Prevotella copri* and *Xylanobacteria* with high-energy carbohydrates, and Bacteroides with protein-rich conditions. Surprisingly, even short-term diet modulation can induce significant, though temporary, changes in microbial structure, demonstrating the microbiome's responsiveness to nutritional inputs. In addition to their recognized role in digestion and immune response, human microbiome communities also engage with wider physiological and environmental systems. During infancy, the establishment of microbial communities is crucial for shaping neonatal immune responses and maintaining inflammatory balance. These have been found to be higher in children who are likely to develop allergies but show lower abundance of protective microbes including *Bifidobacterium adolescentis*, *B.*

bifidum, and *Lactobacillus* species. Such microbial imbalances can cause impairment of immune tolerance and favor hypersensitivity, contributing towards the susceptibility to allergic disorders. This imbalance in the microbiota can lead them to immune hypersensitivity (Sjögren et al. 2009). Research also emphasizes the significance of the microbiome in detoxification. Microbes in the gut, urinary tract, and other niche microbiota help break down toxic xenobiotics like heavy metals, pesticides, polycyclic aromatic hydrocarbons, ochratoxins, and plastic monomers (Claus et al. 2016). Upon renal filtration, some of these toxins are further metabolized by urinary microbiota, thereby pointing to an integrated detoxification process. Within the female reproductive tract, resident microbial populations develop protective immunity by releasing inhibitory substances, cytokines, and antimicrobial compounds. Such functions preserve vaginal health and avoid infection (Li et al. 2020). Overall, the microbiome is central to digestive and immune health as well as systemic detoxification and prevention of disease. Maintenance of its balance via proper lifestyle, nutrition and therapeutic management is critical in preserving overall human health.

1.6 Factors Influencing Human Microbiome

The variety and complexity of microbiome increase quickly in the first few years of life, but it becomes fairly constant once we are adults. Even though our microbiome is fairly constant, things like food, big changes in diet and medicines can change it. These things affect the makeup, activity, and microbiota stability (Sommer et al. 2017). When the balance is maintained, it helps control different diseases. Eating healthy and having this good balance helps keep people healthy. Taking a lot of antibiotics can upset the balance of bacteria in your gut and lead to diseases throughout the body (Santacroce et al. 2021). Many studies of different groups of people have shown that the bacteria in our guts are very helpful for healthy people. These studies also show that it is important to understand how these bacteria are structured and what affects them, like food, age, where you live, diseases and medicines (Fig. 1.13) (Rowland et al. 2018; Wang et al. 2017).

1.6.1 Mode of Delivery

The microbiome of an infant is deeply influenced by a variety of interlinked factors, among which the mode of birth is revealed to be an important determinant, whose impacts are maintained across the entire first year of life. Infants who are born through cesarean section have unique microbial populations and retarded immune system activation in relation to vaginally delivered infants, influencing key processes such as immune tolerance. The microbiota of infants delivered vaginally closely resembles the vaginal flora of mother, characterized by the dominance of *Lactobacillus* species. Apart from birth mode, the research brings out the prominence of diet, feeding practices (breast-fed vs. bottle-fed) and exposure to the environment in

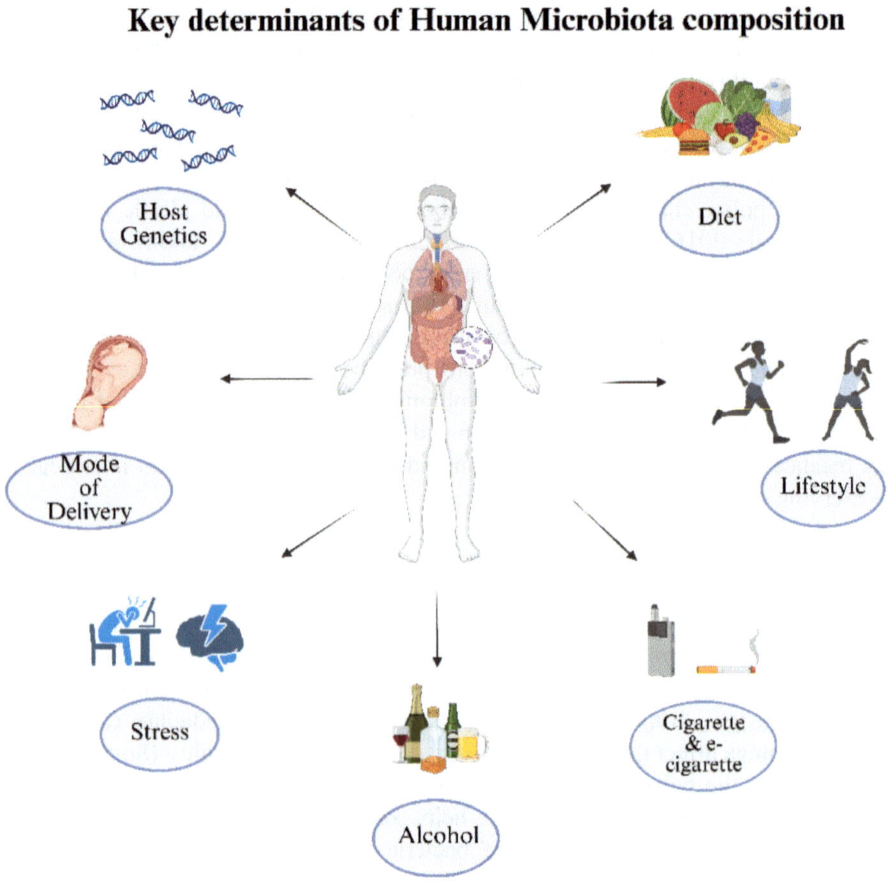

Fig. 1.13 Schematic representation of major factors modulating the human gut microbiome. Both extrinsic and intrinsic factors influence microbial diversity, metabolic activity, and immunological interactions. These factors determine the gut microbial community from infancy to adulthood and thereby have an impact on health outcomes during the entire lifespan

defining the emerging microbiome. The initial gut microbiome of breastfed babies is also dominated by *Bifidobacteria* and *Lactobacillus*, since breast milk is rich in these beneficial bacteria by nature (Ahrodia et al. 2022). In addition, the research comprehensively illustrates how these early microbial variations affect the occurrence and longevity of antimicrobial resistance genes. Even antibiotic regimes given to mothers tend to have an effect on the infant microbiome (Busi et al. 2021). Together, these determinants shape the distinctive course of the microbiome development of each infant, with enduring consequences for immune development and the landscape of antimicrobial resistance.

1.6.2 Host Genetics

Environmental determinants, such as diet, antibiotics, and lifestyle, are well-known modulators of the microbiome but a more valued aspect is the contribution of host genetics. Individual variations in genetics affect microbial diversity, composition, and functional potential, especially in early life. It is well established that host genetics significantly contribute to microbiome constitution in the initial months of life, while environmental impacts start to predominate from an age of approximately 12 months (Fig. 1.14). Information from the Human Microbiome Project (HMP) integrating host genomic data with microbial abundance profiles has shown correlations between host genetic markers and microbial diversity at different body sites. Genetic similarity among hosts, measured in terms of genome-wide identity-by-state (IBS), is negatively correlated with beta diversity (difference between

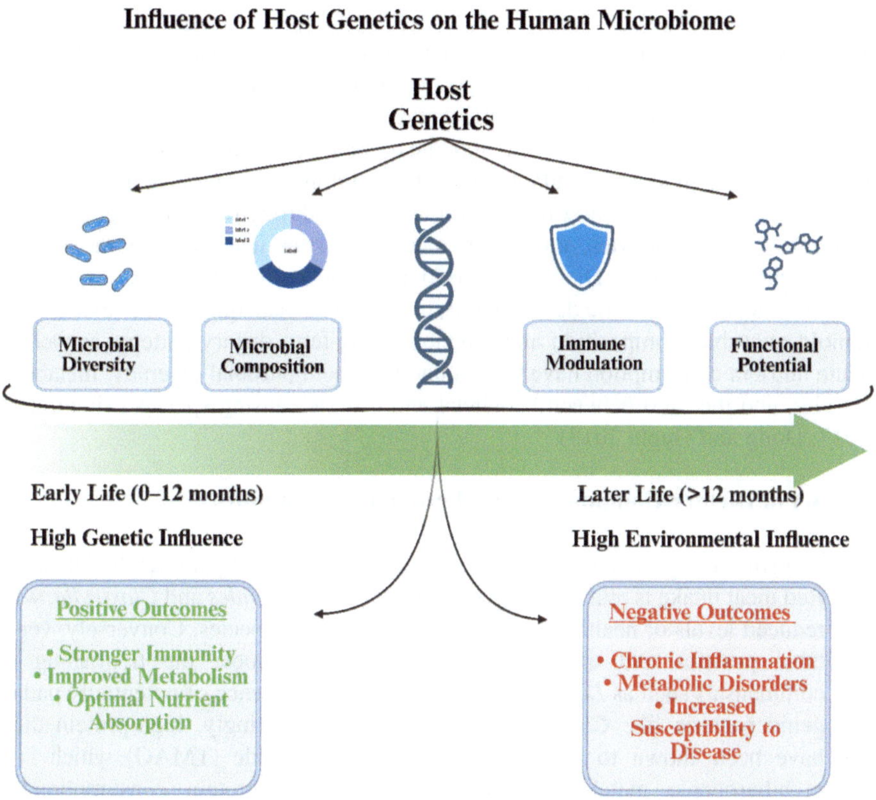

Fig. 1.14 The impact of host genetics on the human microbiome. This figure illustrates how the host's genetic predisposition and changing environmental conditions (from early to adult life) determine the microbial qualities of the gut, such as its diversity, structure, functional capacity, and immune-modulating profile. These microbial statuses are subsequently illustrated as having a direct association with favorable health state or an elevated risk of disease

individuals) in sites such as stool and the oral cavity, so genetically similar individuals will have more similar microbiomes (Blekhman et al. 2015). Twin research indicates monozygotic twins always have more comparable gut microbiota compared to dizygotic twins, which supports the genetic influence (Goodrich et al. 2014). Secretory IgA deficiency, usually genetic, impacts microbial containment, making one more prone to colonization by certain microbes. Genetic effects on the microbiome also vary with context, including diet, history of infection, and local environment and can represent evolutionary adaptations (Quintana-Murci and Clark 2013). Overall, host genetics determines immunity, metabolic activity, and balance of microbes and is a determinant of dysbiosis-related disease risk. Clarifying these interactions will be key to the development of microbiome-targeted approaches for genetically at-risk individuals and precision medicine (Blekhman et al. 2015).

1.6.3 Diet

The human gut microbiota is a densely populated ecosystem made up of trillions of microbes responsible for playing a large role in host health and preventing disease. Primarily dwelling in the distal gut, this gut microbiota affects several physiological processes, like nutrient metabolism, immune homeostasis, and synthesis of bioactive metabolites like short-chain fatty acids (SCFAs). Of these, propionate, butyrate, and acetate are vital in sustaining intestinal homeostasis through functioning as energy substrates for colonocytes, enhancing mucosal barrier integrity and modulating inflammatory networks. Diet serves as one of the strongest shapers of the gut microbiome; unlike genetic determinants, dietary intake can quickly and extensively remold microbial composition and function. Long-term dietary patterns as well as acute nutrient consumption have been found to affect microbial diversity, metabolic activity, and the ratio between beneficial and disease-causing species (Singh et al. 2017; Dong and Gupta 2018).

Impact of Dietary Components on Microbial Communities

(a) *Proteins*: The protein source and type may greatly impact gut microbial balance. Red meat intake is related to higher numbers of *Bacteroides* and *Clostridia*, with reduced levels of health-promoting *Bifidobacterium* species. Conversely, vegetable proteins and dairy proteins like whey promote the flourishing of commensals such as *Lactobacillus* and lower the presence of potentially pathogenic bacteria like *Clostridium perfringens*. Interestingly, high-protein diets have been shown to increase trimethylamine-N-oxide (TMAO) which is a proatherogenic metabolite, implicating excessive protein consumption in cardiovascular risk.

(b) *Fats*: The quality and amount of fat in diet also change microbiota profiles. Saturated and trans-fat diets, which are characteristic of western diets, are found to elevate the count of bile-tolerant anaerobes like *Bilophila and Bacteroides*. On the other hand, unsaturated fats, especially from fish oils, correlate with higher levels of beneficial *Actinobacteria* and *Akkermansia muciniphila*. These modifications can impact inflammatory signaling by Toll-like receptors (TLRs) and other immune mechanisms.

(c) *Carbohydrates*: Simple sugars (e.g., fructose, glucose, sucrose) and available starches regulate microbial communities by enhancing fermentable substrates for certain bacterial populations. High sugar consumption can result in decreased microbial diversity and potentially enhance dysbiosis. Artificial sweeteners like saccharin are associated with unfavorable microbiome alterations and glucose intolerance. Conversely, nondigestible carbohydrates, dietary fiber, and resistant starch are prebiotics, which favor the growth of beneficial species like *Bifidobacteria*, *Roseburia*, *Lactobacillus*, and *Eubacterium rectale*, while promoting SCFA production and enhancing metabolic well-being.

(d) *Probiotics, prebiotics, synbiotics and fermented foods*: Probiotics are live microorganisms that, when administered in sufficient quantities, confer benefits to the host. Fermented foods such as kefir, kimchi, and yogurt are typical dietary sources of probiotics, frequently containing strains of *Bifidobacterium*, *Lactobacillus*, and *Streptococcus thermophilus*. These bacteria have been shown to enhance gut barrier function, affect immune system modulation, and promote microbial diversity (Wastyk et al. 2021). Prebiotics, such as inulin, galacto-oligosaccharides, fructo-oligosaccharides, and resistant starches, are dietary nondigestible components that specifically stimulate beneficial microbiota. They enhance the proliferation of SCFA-producing microbes like *Bifidobacterium* and *Faecalibacterium prausnitzii*, enhancing metabolic and inflammatory profiles (Gibson et al. 2017). Synbiotics integrate probiotics and prebiotics within the same product, enhancing strains' colonization and longevity and optimizing SCFA production. Frequent intake of probiotic-containing fermented food, together with prebiotic fibers from fruit, vegetables, legumes, and whole grains or specific synbiotic supplementation, is linked with a more robust and functionally rich gut microbiome that underlies long-term gastrointestinal and systemic health.

(e) *Polyphenols*: These plant bioactive compounds, found in abundant amounts in fruits, vegetables, tea, and wine, possess antimicrobial and anti-inflammatory activities. Polyphenols selectively increase the commensal bacteria *Bifidobacterium* and *Lactobacillus* while suppressing opportunistic pathogens including *Clostridium* spp., thus making contribution to metabolic and immune control (Singh et al. 2017).

Influence of Different Dietary Patterns has Distinctive Influences on Microbiota Composition

(a) *Western diets,* defined by high levels of intake in saturated fats and animal protein with low fiber intake, are all connected with decreased microbial diversity and elevated loads of proinflammatory bacterial taxa. The diet also contributes to the production of carcinogenic nitrosamines and higher risk of metabolic and inflammatory disorders (Singh et al. 2017).

(b) *Mediterranean diets,* high in monounsaturated fats, polyphenols, complex carbohydrates, and fiber, are positively correlated with microbial diversity and anti-inflammatory taxa. Enhanced SCFA production, better lipid profiles, and decreased markers of inflammation have been reported in those consuming this diet (Singh et al. 2017).

(c) *Gluten-free diets* can accidentally decrease numbers of beneficial microbial taxa such as *Bifidobacterium* and *Lactobacillus* and support the proliferation of *E. coli* and other pathogenic microorganisms possibly through decreased prebiotic consumption (Singh et al. 2017).

Diet is the chief determinant of gut microbial ecology, influencing the balance between disease and health. Certain nutrients and dietary habits can promote microbial diversity and functional integrity or, alternatively, induce dysbiosis and inflammatory conditions (Table 1.3). As the connections among diet, microbiota, and chronic disease are elucidated, increasing potential exists for individualized nutrition interventions to shape the microbiome for preventive and therapeutic purposes. Emerging progress in microbiome science will increasingly refine our comprehension of how dietary components define host–microbe interactions and translate this to precision medicine and public health nutrition (Singh et al. 2017; Dong and Gupta 2018).

1.6.4 Drugs

Most widely used medications, ranging from over-the-counter analgesics to chronic prescription drugs, have the ability to affect the gut microbiome in a manner that is unintended compared to their primary function. Increasing research has shown that drugs beyond antibiotics can alter gut microbiota, behavior, and neurophysiology (Clarke et al. 2014a, b). The diverse gut microbiota in the large Belgian Flemish Gut Flora Project study was shown to be affected by hormones, osmotic laxatives, antimicrobials, antidepressants, antihistamines, and inflammatory bowel disease medications (Falony et al. 2016). Although antibiotics are most prominently known to decrease microbial diversity and allow for overgrowth of pathogenic species such as *Clostridioides difficile* (Nguyen et al. 2015), numerous nonantibiotic medications also influence the microbiome. Proton-pump inhibitors enhance upper-GI bacteria like *Streptococcus*; metformin tends to enrich short-chain fatty acid producers and *Akkermansia muciniphila* but decrease diversity temporarily (Wu et al. 2017); and long-term opioids decrease *Lactobacillus* and *Bifidobacterium* and increase inflammatory taxa, compromising gut barrier function (Thomas et al. 2022). These drug-induced changes not only impact health consequences but also modify the efficacy and side effects of the drugs themselves, making it critical to take gut–drug interactions into account in both practice and research.

Table 1.3 Impacts of different types of diet on gut microbiome

Dietary component	Representative food sources	Positive impact on gut microbiota	Negative impact on gut microbiota	References
Animal-based protein	Red meat, poultry, eggs	None strongly established as positive in general population	Increases abundance of *Bacteroides* and *Clostridia*, decreases *Bifidobacterium* species; associated with elevated levels of trimethylamine-N-oxide, a cardiovascular risk factor	Di Rosa et al. (2023)
Plant-based protein	Legumes, soy products, nuts, seeds	Promotes growth of *Lactobacillus* species and reduces potentially pathogenic bacteria such as *Clostridium perfringens*	Limited evidence of negative effects	Singh et al. (2017)
Saturated and trans fats	Processed meats, butter, pastries, fried foods	None strongly established as positive	Increases bile-tolerant anaerobes like *Bilophila and Bacteroides*; linked to reduced microbial diversity and higher inflammatory potential	Singh et al. (2017)
Unsaturated fats	Olive oil, nuts, seeds, fatty fish	Increases beneficial bacteria such as *Akkermansia muciniphila* and *Actinobacteria*; supports anti-inflammatory microbial profile	None strongly established as negative in moderate consumption	Ghosh et al. (2020)
Simple sugars	Table sugar, sweetened beverages, pastries	None strongly established as positive	Decreases microbial diversity, promotes dysbiosis, and may encourage growth of opportunistic pathogens	Singh et al. (2017) and Suez et al. (2014)
Dietary fiber and resistant starch	Whole grains, fruits, vegetables, legumes	Promotes growth of *Bifidobacterium*, *lactobacillus*, *Roseburia,* and *Eubacterium rectale*; increases	None strongly established as negative	Makki et al. (2018) and Singh et al. (2017)

(continued)

Table 1.3 (continued)

Dietary component	Representative food sources	Positive impact on gut microbiota	Negative impact on gut microbiota	References
		production of SCFAs such as butyrate, propionate, and acetate		
Probiotics	Yogurt, kefir, fermented vegetables	Increases beneficial bacteria including *Lactobacillus*, *Bifidobacterium*, and *Streptococcus* species; supports immune regulation and gut barrier integrity	May cause transient gastrointestinal discomfort in sensitive individuals	Wastyk et al. (2021)
Polyphenols	Berries, tea, cocoa, red wine	Selectively increases *Bifidobacterium* and *Lactobacillus*, reduces pathogenic bacteria such as *Clostridium* species; supports anti-inflammatory balance	None strongly established as negative in moderate consumption	Singh et al. (2017)
Western diet	High intake of processed meats, refined sugars, saturated fats	None strongly established as positive	Associated with reduced microbial diversity, increased proinflammatory taxa, and elevated production of harmful metabolites	Severino et al. (2024)
Mediterranean diet	High consumption of fish, olive oil, vegetables, fruits, legumes, whole grains	Enhances microbial diversity, increases anti-inflammatory taxa, promotes short-chain fatty acid production, and supports metabolic health	None strongly established as negative	Ghosh et al. (2020)

1.6.5 Lifestyle

A lifestyle which involves regular physical training, especially moderate aerobic and resistance exercise, is significant in determining the composition and function of the human gut microbiota (Fig. 1.15). Comparative research among athletes and non-athlete controls has revealed prominent microbial diversity and abundance differences, with athletes typically having higher microbial richness, more

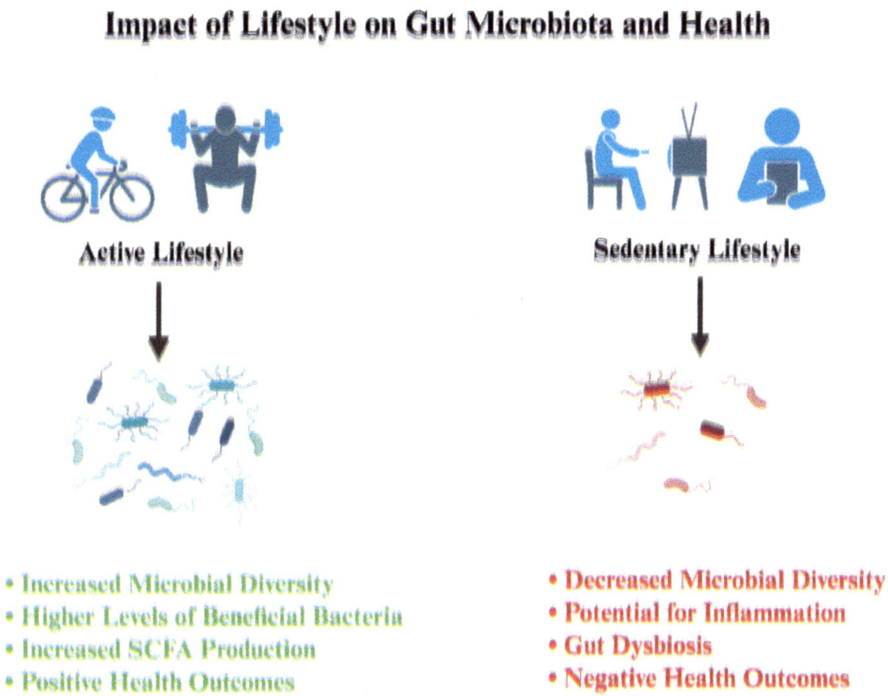

Fig. 1.15 The influence of lifestyle on gut microbiota and health. The diagram compares the consequences of an active lifestyle, which supports microbial diversity and health-promoting bacteria to a sedentary lifestyle, which causes gut dysbiosis and an inflammatory risk. The resulting microbiota condition is demonstrated to be directly associated with overall health consequences

Firmicutes levels and health-promoting taxa like *Faecalibacterium*, together with fewer Bacteroidetes (Matsumoto et al. 2008; Mailing et al. 2019). The differences are also affected by the pattern of diet common in athletes, e.g., increased protein consumption, which in turn can affect gut microbial composition. Exercise has been also shown to be related with increased production of short-chain fatty acids (SCFAs), mainly butyrate, linked with better gut health and systemic effects (Matsumoto et al. 2008). In older men and women, consistent resistance training is particularly helpful for the maintenance of muscle mass, strength, and independence. Combined with sufficient protein consumption, the advantage is optimized, although exercise by itself is still the strongest determinant of musculoskeletal health. Resistance and endurance training enhance mitochondrial quality, which is associated with improved metabolic health and healthy aging. In addition to musculoskeletal changes, exercise has a direct impact on gut microbial stability and diversity. Moderate exercise increases positive microbial levels and stimulates greater SCFA production, upholding a stable gut environment. Controlled trials have shown that lean and obese subjects can both achieve enhanced microbial diversity with regular aerobic exercise, although the magnitude of change may be influenced by metabolic status. Notably, exercise dose is also crucial, in while moderate exercise is helpful, excessive endurance training can weaken gut barrier function, lower microbial diversity, and enhance inflammation. The inclusion of daily, moderate-intensity exercise in lifestyles cultivates a better gut microbiome, improves functional ability, and promotes overall well-being. This holistic impact from exercise presents itself as a useful non-pharmacological intervention for healthy aging and long-term health preservation (Pedroza Matute and Iyavoo 2023).

1.6.6 Cigarette and E-cigarette

Cigarette and e-cigarette smoking are associated with decreased gut microbial diversity and a potential causative factor for multiple systemic diseases; however, significant compositional variations are often observed between smokers and non-smokers, despite the dissimilar changes in microbial phyla and genera across research. This inconsistency can be due to inter-individual heterogeneity, variations in smoking strength, or coexistent lifestyle factors. The underlying processes are not yet understood; the enormous diversity of cigarette smoke toxicants such as aldehydes, polycyclic aromatic hydrocarbons, benzene, nitrosamines, and heavy metals can directly affect gut ecology. These chemicals may change gut pH, interfere with the production of short-chain fatty acids or function as selective antimicrobials, allowing some taxa of bacteria to thrive while inhibiting others. Nicotine exposure was linked to higher relative abundances of Proteobacteria and Bacteroidetes and lower abundances of Actinobacteria and Firmicutes, a change indicating a proinflammatory gut microbiota and a decrease in SCFA-producing bacteria like Lachnospiraceae (Fan et al. 2025), which may compromise immune function, metabolic homeostasis, and mucosal integrity (Pedroza Matute and Iyavoo 2023). There is considerable evidence that cigarette and e-cigarette consumption adversely affects the gut microbial richness and composition, while patterns differ based on exposure details and subject-related factors.

1.6.7 Alcohol

Alcohol drinking is a well-recognized modulator of the gut microbiome. Chronic alcohol use or addiction is associated with gut dysbiosis and contributes profoundly toward alcoholic liver disease development and progression (Fig. 1.16). Dysbiosis in Bacteroidetes and Enterococcaceae is often seen in these patients, showing evidence for an association between microbial dysbiosis and hepatic inflammation. Even acute alcohol exposure, as demonstrated in animal models, may stimulate bacterial overgrowth of Enterobacteriaceae specifically and trigger intestinal inflammation, indicating an immediate onset of microbial derangement after alcohol consumption. The key mechanism is presumed to be the increased acetate levels, a product of ethanol metabolism by host enzymes that can set the stage for dysbiosis (Pedroza Matute and Iyavoo 2023). Epidemiological data indicates that increased use of

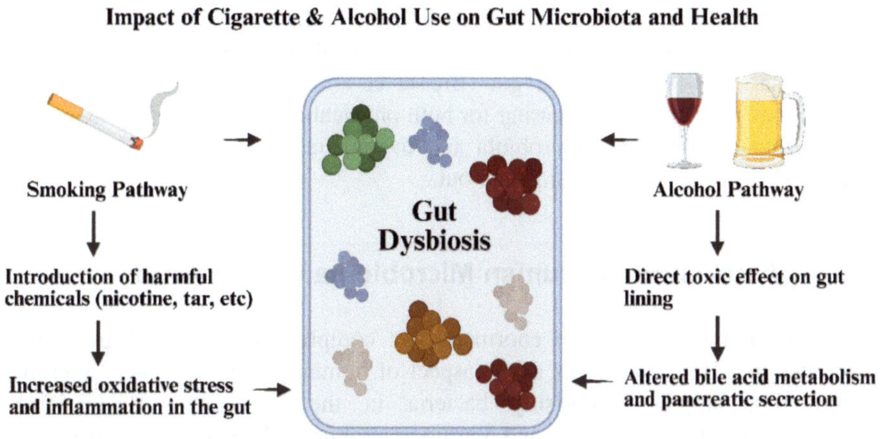

Fig. 1.16 Impact of cigarette/e-cigarette and alcohol consumption on gut microbiota and health outcomes. Both factors are associated with decreased microbial diversity and richness, altered bacterial composition, reduced SCFA production and increased proinflammatory taxa, contributing to dysbiosis, inflammation, metabolic disruption and heightened risk for chronic diseases

alcohol is linked with lower microbial richness, whereas organized detoxification or diminished use facilitates partial restoration of beneficial taxa within certain subjects (Koponen et al. 2025; Lee and Lee 2021). Total alcohol consumption is associated with a less rich gut microbiome with alterations towards taxa normally linked to inflammation and metabolic derangement alterations seen consistently within studies in humans with different levels of intake (Koponen et al. 2025).

1.6.8 Stress

Besides the physiological and behavioral consequences of alcohol and smoking, stress significantly affects the gut microbiota by interacting pathways that include the nervous system, hormonal signals, and immune mechanisms. The gut–brain axis represents a bidirectional communication network facilitated by endocrine signals, vagus nerve, immune elements, and microbial metabolites like serotonin and GABA, which may be disturbed under acute or chronic stress, leading to dysbiosis (Cussotto et al. 2018). Hormones such as estrogen and testosterone regulate microbial content and function of the intestinal barrier, contributing in part to observed gender differences in gut responses to stress, with women and men having different microbiota profiles and susceptibilities to stress (Santos-Marcos et al. 2023). Occupational stressors like excessive job demands, nonstandard work schedules and psychosocial stress also add to dysbiosis, connecting workplace settings with changes in microbial richness and diversity (Lopez-Santamarina et al. 2023). Chronic miscommunication between the brain and gut under these compounded effects could enhance susceptibility to IBS, anxiety, and depression. On the other hand, stress-reducing practices such as mindfulness meditation, yoga, and supportive workplace policies have been linked to enhanced SCFA production and decreased inflammation and this underscores the healing potential of environmental and mental well-being for the gut (Bressa et al. 2017). Acknowledging these connections underscores that caring for both physical and mental health is essential to ensuring balanced gut microbiota and overall health in order to live a more complete, better lives both inside and out.

1.7 Association of Human Microbiome with Disease

The human microbiota is an enormous and complex population of trillions of microbes that has surfaced as a key aspect of human physiology. This microbial community, largely comprising bacteria in the Firmicutes, Bacteroidetes, Actinobacteria, Proteobacteria, and Verrucomicrobia phyla, is founded early in life and perpetually molded by genetics, diet, environment, antibiotic exposure, and lifestyle. The past years, scientific evidence has recognized the gut microbiota as a metabolically active and dynamic entity. Instead of being a passive inhabitant, it has a central role in nutrient uptake, vitamin production, xenobiotic metabolism, ensuring the integrity of intestinal barrier and host immune system maturation. The microbes have intricated bidirectional communication with the host by means of immune, neural and endocrine pathways, together described as the gut-organ axes. A healthy microbial community or "eubiosis" is crucial to preserve homeostasis, while disturbances, termed "dysbiosis" may activate or worsen pathological processes (Fig. 1.17). Mounting evidence has implicated imbalances of gut microbiota in numerous pathological conditions ranging from inflammatory, metabolic, autoimmune, neurological to even oncological diseases. With growing insight, the gut microbiota is being increasingly understood not just as a diagnostic signature but as an emerging therapeutic target. In the following sections, the composition of microbiota in particular disease situations will be discussed in more detail to reveal their mechanistic relationships and significance for human health.

1.7.1 Microbiome and Allergies

Human microbiota, especially the gut and respiratory tract microbiota, is critical in the ontogeny of the immune system and modulation of allergic disorders. There is evidence to suggest a robust gut–lung axis, with a healthy gut microbiome influencing the health of respiratory mucosae and microbial imbalance reflecting enhanced susceptibility to respiratory diseases like asthma and allergic airway disease (Liu et al. 2025). The structure of microbiota varies among healthy and allergic subjects. Microbiota profiles among healthy subjects are characterized by increased microbial diversity and a higher representation of protective bacteria like *Bacteroidetes, Bifidobacterium, Akkermansia*, and *Faecalibacterium* that are linked with anti-inflammatory properties and lower allergy risk. Allergic subjects have lower microbial richness, decreased levels of the protective taxa, and a higher rate

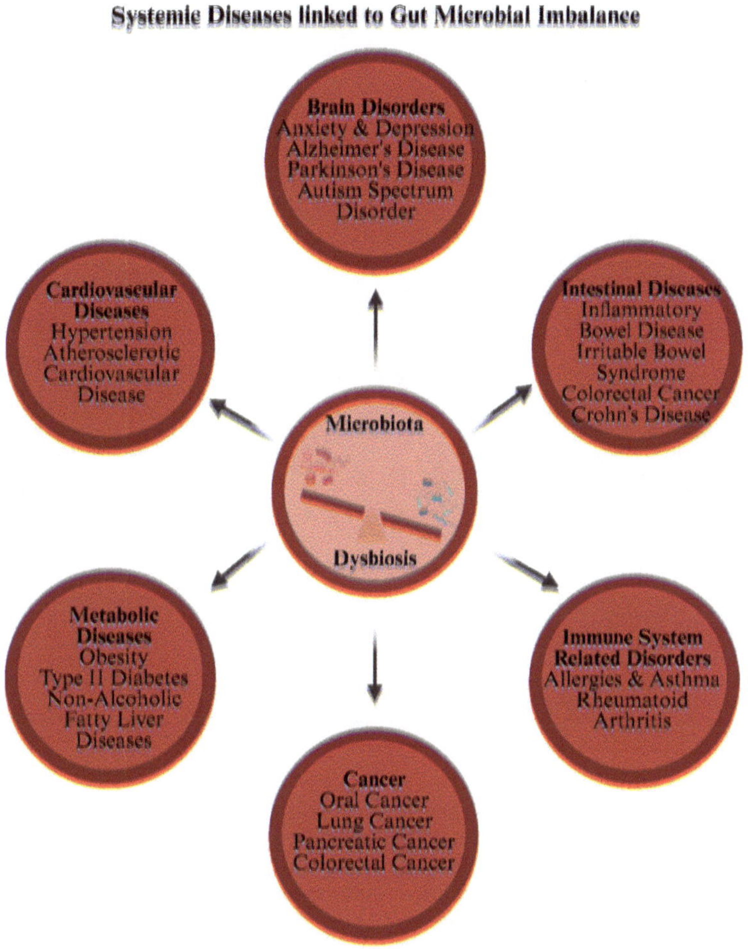

Fig. 1.17 Systemic effect of gut microbiota dysbiosis on human health. Disruption in composition and function of the gut microbiota, known as dysbiosis, is linked with a broad range of diseases involving many organ systems. They include brain disorders (e.g., anxiety, depression, neurodegenerative diseases), cardiovascular disease (e.g., hypertension, atherosclerosis), metabolic disorders (e.g., obesity, type II diabetes, NAFLD), immune disorders (e.g., asthma, allergies, rheumatoid arthritis), intestinal disorders (e.g., IBD, IBS, colorectal cancer), and cancers. The gut microbiota acts as a key controller of host homeostasis and its dysbiosis leads to systemic pathophysiology by immune, metabolic, endocrine, and neural mechanisms

of dysbiotic communities (Fig. 1.18). Mode of delivery is an important determinant; cesarean-section (CS) born children lack exposure to maternal vaginal flora and acquire gut colonization that is abnormal. Scientific studies have shown that CS-born individuals have decreased levels of Bacteroidetes, a family of beneficial bacteria with anti-inflammatory function. Absence of this family of bacteria may lead to chronic inflammation and manifestations of allergy such as asthma (Lee-Sarwar et al. 2023). Early-life microbial richness is also implicated, with reduced abundance

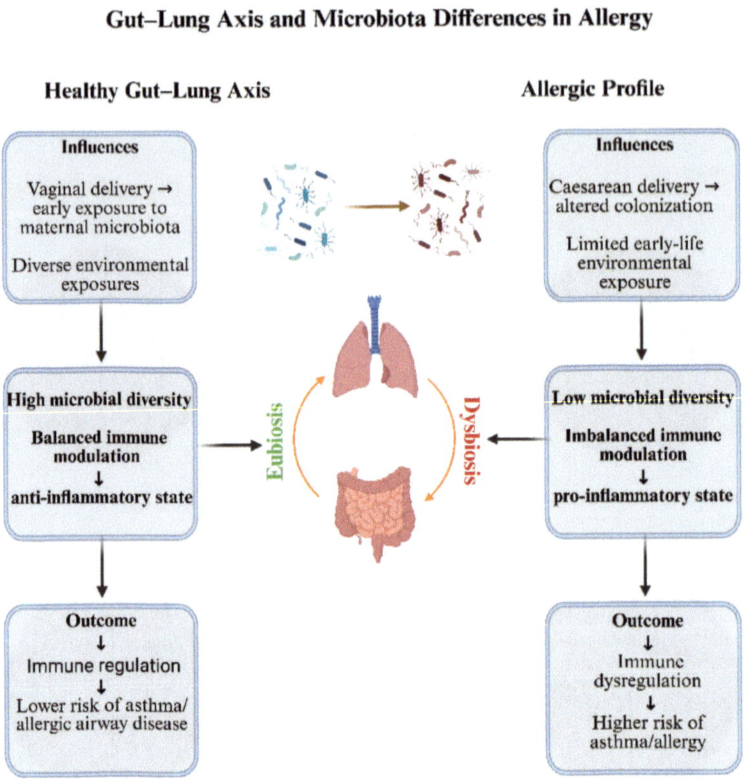

Fig. 1.18 The gut-lung axis and microbiota differences in allergy. This flowchart compares a healthy immune condition, characterized by a rich gut microbiota and early-life determinants such as vaginal delivery, with an allergic phenotype characterized by modified colonization. It illustrates how a switch from eubiosis to dysbiosis affects immune control and results in increased likelihood of asthma and other allergic illnesses

of *Bifidobacterium, Akkermansia,* and *Faecalibacterium* associated with polysensitization and enhanced risk of asthma at four years of age. Environmental exposure also influences microbial communities and allergy risk. Epidemiological research shows that those with diverse microbial communities to which they have been exposed, like farm kids are less prone to developing allergic airway diseases (Depner et al. 2020). In contrast, restricted exposure to microbial diversity in early life can cause heightened allergy risk and immune dysregulation.

1.7.2 Microbiome and Inflammatory Bowel Disease (IBD)

Inflammatory bowel disease (IBD) refers to a group of chronic immune-mediated diseases of the gastrointestinal tract, which mainly includes Crohn's disease (CD) and ulcerative colitis (UC). It is defined by chronic mucosal inflammation and is controlled by genetic, environmental, microbial, and immune factors (Lee and Chang 2021). Gut microbiota in IBD is significantly different from that of normal people. A consistent finding is decreased microbial richness (Matsuoka and Kanai 2015; Willing et al. 2010). Another significance is the reduction in members of phylum Firmicutes, including butyrate-producing bacteria like *Faecalibacterium prausnitzii* and *Roseburia* spp., which are normally found in high numbers in healthy intestines (Lee and Chang 2021). Conversely, the phylum Proteobacteria, most commonly harboring pro-inflammatory Gram-negative organisms, is generally over-abundant in IBD patients (Halfvarson et al. 2017). Some taxa like *Ruminococcus gnavus* are more prevalent in affected individuals and have been identified to be associated with inflammatory disease (Örtqvist et al. 2019). These shifts in composition also identify a gross dysbiosis in IBD versus healthy gut profiles. Overall, IBD correlates with decreased microbial diversity, loss of butyrate-producing benefiters, and increased abundance of potentially pathogenic taxa, making its microbiota profile distinct from that of healthy subjects. The question of whether these microbial alterations are a cause or effect of disease thus remains open despite considerable effort. Regardless, the association of changed microbiota with IBD is firmly established (Fig. 1.19).

1.7.3 Microbiome and Colorectal Cancer (CRC)

Colorectal cancer (CRC) is a significant public health issue and is the second most common cause of cancer-related death globally (Fitzmaurice et al. 2017). The

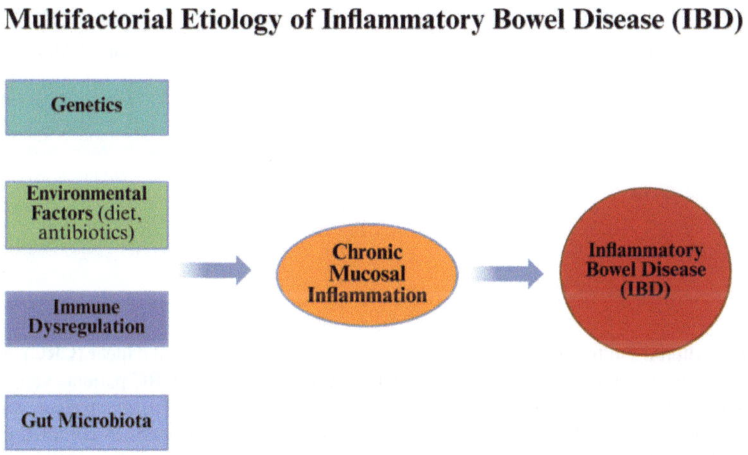

Fig. 1.19 Mechanistic overview of factors causative of inflammatory bowel disease (IBD). IBD pathogenesis arises from a multifactorial interplay between host genetics, environmental stimuli (i.e., diet, exposure to antibiotics), immune system dysregulation and changes in gut microbiota. These elements together contribute to chronic inflammation of the mucosa, finally resulting in IBD, which encompasses ulcerative colitis and Crohn's disease

etiology is multifactorial, with genetic factors, diet, environmental carcinogens, lifestyle and increasingly the gut microbiota (Arthur et al. 2012). Dysregulation of the colonic microbial ecosystem is highly linked with colorectal carcinogenesis, as illustrated by germ-free animal models indicating that the presence of microbiota is essential for CRC development (Arthur et al. 2012). Microbiota populations vary significantly between healthy humans and patients with CRC. One of the most ubiquitous findings is a high abundance of *Fusobacterium nucleatum* in colorectal tumor patients (Thomas et al. 2019; Kong et al. 2023; Cheng et al. 2020). Research indicates enrichment of *F. nucleatum, B. fragilis*, and other inflammatory taxa, in addition to the depletion of beneficial short-chain fatty acid producers like *Faecalibacterium prausnitzii* and *Roseburia* spp. Such microbial alterations can enhance CRC via inflammation, disruption of epithelial barrier function, and genotoxin production, thus enabling tumorigenesis (Modeel et al. 2025) (Fig. 1.20). Apart from oncogenesis, composition of gut microbiota determines the outcomes of cancer immunotherapy and highlights their dual role in disease progression and treatment (Bharti et al. 2023). Patients with CRC also have higher oral bacteria prevalence in the gut microbiome, indicating potential oral-gut microbial continuity (Thomas et al. 2019). Aside from *F. nucleatum*, the species *Bacteroides fragilis, Peptostreptococcus anaerobius,* and *Peptostreptococcus stomatis,* have been associated with CRC development (Cheng et al. 2020; Obón-Santacana et al. 2022). Diet plays an important role in determining microbial composition and CRC

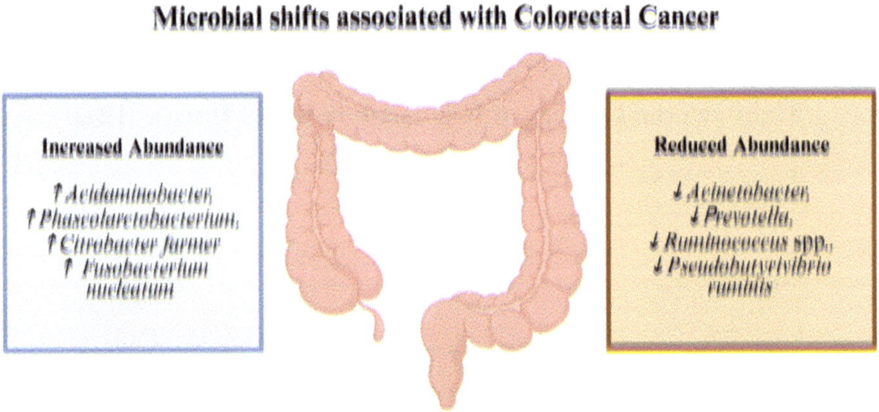

Fig. 1.20 Altered gut microbiota composition associated with colorectal cancer (CRC). Microbial profiling reveals a significant shift in the abundance of bacterial taxa in CRC patients versus healthy individuals. Tumor-promoting and proinflammatory species like *Fusobacterium nucleatum, Citrobacter farmer,* and *Acidaminobacter* are found at greater abundance. In contrast, protective commensals like *Ruminococcus* spp., *Pseudobutyrivibrio ruminis,* and *Prevotella,* which are typically associated with the production of short-chain fatty acid (SCFA) and maintenance of the gut integrity, are greatly reduced. Such microbiota changes could promote CRC development through modulation of inflammation, immune signaling and host-microbe interactions

risk. High-fiber, whole-grain diets correlate with a decreased risk of *F. nucleatum*-positive colon tumors (Mehta et al. 2017), whereas low-fiber, high-fat Western diets are likely to promote procarcinogenic microbial patterns. Metagenomics and metabolomics revealed cancer stage-dependent microbial signatures with potential for prognosis and early diagnosis (Yachida et al. 2019; Thomas et al. 2019; Yu et al. 2017). Cases of late-onset CRC (>50 years) tend to exhibit increased *F. nucleatum* levels and decreased short-chain fatty acids like butyrate (Kong et al. 2023). Beyond CRC, changed microbiota patterns like enhanced *Bacteroides massiliensis*, decreased *Eubacterium rectale*, and *Faecalibacterium prausnitzii* are detected in gastric and prostate cancers, where *F. nucleatum* and *Clostridium colicanis* can function as biomarkers (Sha et al. 2020). Longitudinal research is vital for further validation of cancer diagnosis and treatment using microbiome-based methods (Bultman 2014).

1.7.4 Microbiome and Cardiovascular Diseases

Cardiovascular diseases (CVDs) continue to be a leading factor of mortality across globe, with an increasing burden in economically developing nations despite the application of preventive and therapeutic interventions (Odutayo et al. 2016). Emerging evidence demonstrates a substantial correlation between gut microbiota composition and CVD risk, microbial disruption increasingly being recognized as an etiologic component for disease onset and aggravation (Jie et al. 2017). Significant alterations in gut microbiota composition are observed between healthy individuals and those with cardiovascular disease (Fig. 1.21). CVD patients have diminished microbial diversity coupled with augmented proportions of pro-inflammatory bacterial phyla like *Streptococcus, Enterobacteriaceae, Klebsiella*, and *Prevotella* (Jie et al. 2017). In coronary heart disease, alterations were observed with elevated abundance of Firmicutes and reduced abundance of Bacteroidetes, trends that are extensively associated with metabolic derangement (Ramírez-Macías et al. 2022). Alterations in the Firmicutes-to-Bacteroidetes ratio were also observed in human hypertensive subjects as well as in animal models and support the connection between dysbiosis and CVD (Silveira-Nunes et al. 2020). Animal studies also support this connection; for example, fecal microbiota transfer from hypertensive individuals to germ-free mice caused recipients to have an increased blood pressure, which suggests a causal role of the gut microbiota in cardiovascular physiology (Lin et al. 2024). In general, CVD-related microbiota profiles are distinguished by decreased diversity, enrichment of proinflammatory bacteria, and alterations in dominant phyla. These compositional changes distinguish the atherosclerotic from their healthy controls and suggest microbial imbalance as both a biomarker and a potential target for cardiovascular therapeutics (Barrington and Lusis 2017; Jonsson and Bäckhed 2017).

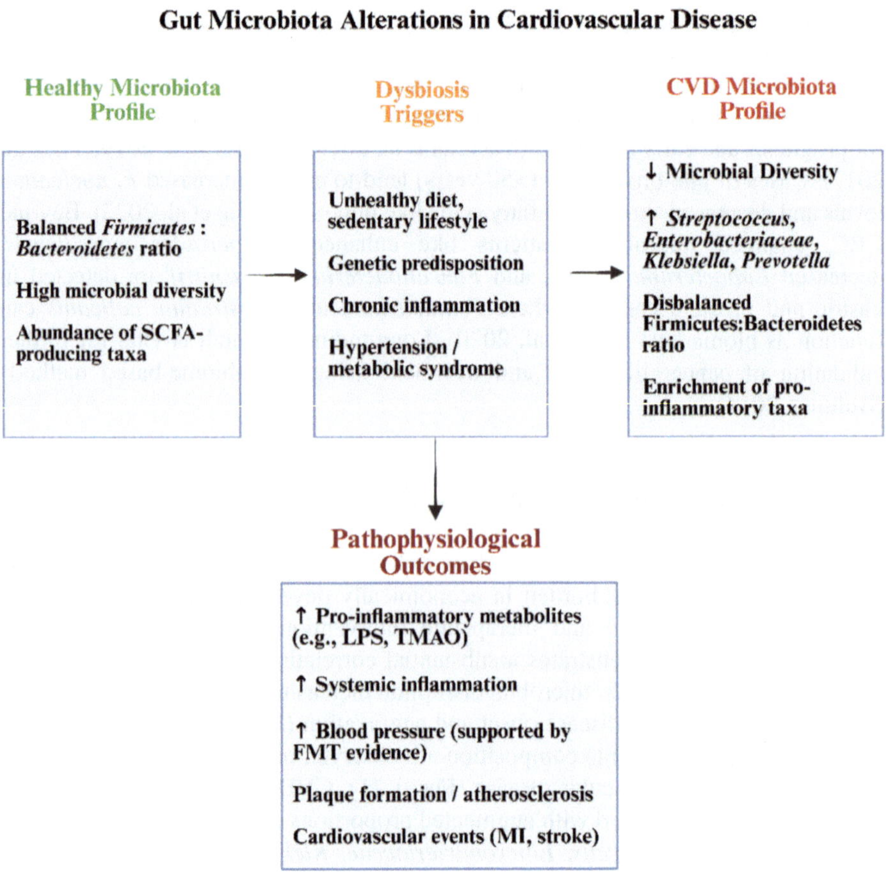

Fig. 1.21 Alterations of gut microbiota in cardiovascular disease. This figure shows how dysbiosis, initiated by the presence of an unhealthy diet, differs from a normal gut microbiota. The proinflammatory microbial signature leads to the development of proinflammatory metabolites, ultimately causing systemic inflammation, plaque formation, and cardiovascular disease development

1.7.5 Microbiota and Obesity

Obesity has become a significant global health issue, impacting over 650 million individuals worldwide, with prevalence rates steadily rising over the past four decades (Sørensen et al. 2022). Traditionally linked to increased caloric intake and reduced physical activity (Pascale et al. 2019), its development is now recognized as multifactorial, involving genetic, epigenetic, environmental, and microbial influences (Xia and Grant 2013). Growing evidence indicates a strong association between gut microbiota and host energy balance, with alterations in microbial populations potentially contributing to obesity risk (Bäckhed et al. 2004; Turnbaugh et al. 2006). Obesity is associated with distinct alterations in gut microbiota

Table 1.4 Key bacterial taxa associated with leanness or obesity

Category	Bacterial taxa	Abundance	Function	Source
Leanness-associated	*Faecalibacterium prausnitzii*	↓ Reduced	Butyrate producer; anti-inflammatory, maintains gut barrier integrity	Remely et al. (2014) and Gophna et al. (2017)
	Blautia	↓ Reduced	Negatively correlated with visceral fat accumulation	Ozato et al. (2019)
	Bacteroides thetaiotaomicron	↓ Reduced	Ferments glutamate; linked to lower serum glutamate and lean body mass	Liu et al. (2017)
Obesity-associated	*Firmicutes*	↑ Increased	Enhances energy harvest from diet	Ley et al. (2006) and Turnbaugh et al. (2009)
	Eubacterium ventriosum	↑ Increased	Produces SCFAs; enriched in obesity	Tims et al. (2013)
	Bacteroidetes	↓ Reduced	Often found in lower abundance; alters B:F ratio	Ley et al. (2005, 2006) and Turnbaugh et al. (2009)

compared to healthy states (Table 1.4). A prominent signature of obesity is a reduced Bacteroidetes-to-Firmicutes ratio, though findings are not entirely consistent across studies (Koliada et al. 2017; Schwiertz et al. 2010; Ley et al. 2005, 2006). Some studies report higher Firmicutes and lower Bacteroidetes in obese individuals, suggesting a higher efficiency for dietary energy utilization (Ley et al. 2005; Turnbaugh et al. 2009), whereas others find no significant correlation with Body Mass Index(BMI) (Tims et al. 2013). Despite such discrepancies, microbial richness and diversity are consistently lower in obesity (Cotillard et al. 2013). Specific taxa also vary with weight status. *Eubacterium ventriosum*, associated with short-chain fatty acid production, is increased in obesity (Tims et al. 2013), while butyrate-producer *Faecalibacterium prausnitzii* is linked to leanness (Remely et al. 2014; Gophna et al. 2017). *Bacteroides thetaiotaomicron* depleted in obesity, correlates inversely with serum glutamate (Liu et al. 2017). Christensenellaceae and *Blautia* show negative associations with obesity and visceral fat, respectively (Waters and Ley 2019; Ozato et al. 2019). Diet plays a central role, with fat-rich diets supporting obesogenic microbes and fiber-rich diets favoring leanness-associated species (Davis 2018; Bouter et al. 2017). These compositional differences highlight the gut microbiota's role as both a biomarker and potential target in obesity management.

1.7.6 Liver Diseases

Liver diseases like nonalcoholic fatty liver disease (NAFLD), nonalcoholic steatohepatitis (NASH), alcoholic liver disease, and cirrhosis are increasingly understood to be modulated by gut microbiota function and structure. Disruptions in the gut–liver axis, usually precipitated by dietary irregularities, alcohol consumption, and increased intestinal permeability, lead to the development and exacerbation of these liver diseases (Aqel and DiBaise 2015). NAFLD, the liver expression of metabolic syndrome, is currently the most prevalent chronic liver disease in Western nations (Henao-Mejia et al. 2012; Jiang et al. 2015). Gut microbial composition varies extensively between healthy subjects and liver patients. NAFLD patients typically show elevated counts of *Clostridium*, *Escherichia*, *Streptococcus*, *Lactobacillus*, and *Anaerobacter* and lower *Odoribacter*, *Oscillobacter*, and *Alistipes* (Jiang et al. 2015). In NASH, *Escherichia* is especially elevated, whereas children with steatosis or NASH present with higher *Dorea* and *Ruminococcus* and deplete *Oscillospira* (Del Chierico et al. 2017). *Klebsiella pneumoniae* with the ability to produce endogenous alcohol has also been found in NAFLD (Yuan et al. 2019). Multiomics analyses report reduced microbial gene richness, increased lipid processing capacity and higher endotoxin-producing Proteobacteria in hepatic steatosis (Hoyles et al. 2018). In alcoholic liver disease, there is also a similar dysbiosis profile with reduced Bacteroidaceae in humans (Mutlu et al. 2012). Chronic alcohol misuse is associated with an overgrowth of disease-causing Enterobacteriaceae (*E. coli, Klebsiella, Proteus, Enterobacter*) and Prevotellaceae (Bajaj et al. 2012; Yanfei et al. 2011). Overall, reduced microbial diversity, pathogenic taxon enrichment, and an increase towards endotoxin producers characterize liver-diseased subjects compared to their healthy counterparts (Fig. 1.22).

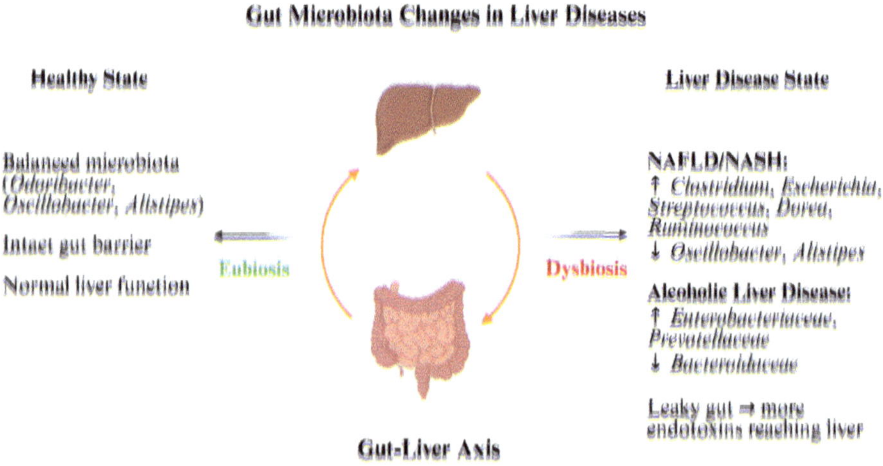

Fig. 1.22 Changes in gut microbiota in liver diseases. The figure compares a healthy eubiosis with a disease state of dysbiosis. It shows how certain microbial changes result in a weakened gut barrier and higher endotoxins, culminating in liver disease such as NAFLD and alcoholic liver disease

1.8 Emerging Tools and Challenges

Biotechnological advances are propelling microbiome science toward a new frontier in which microbes themselves become active agents of intervention, rather than passive objects of research. Foremost among these new tools are Clustered Regularly Interspaced Short Palindromic Repeats (CRISPR)-Cas 9 based gene editing technology, synthetic biology, and engineering of designed microbial consortiacombined to create the frontier of microbiome engineering. CRISPR/Cas systems are being reengineered to edit the genomes of resident gut microbiota to enable targeted targeting of genes for metabolite biosynthesis, immune interaction, or pathogenic virulence. In- situ editing approaches permit beneficial microbes to be active while inhibiting deleterious biochemical pathways. Bacterial genes native to the microbiome can be silenced or rewired in situ without altering microbiome composition, inhibiting disease promoting function or augmenting therapeutic function. Such approaches enjoy unprecedented specificity relative to broad spectrum antibiotics, permitting interventions as finely targeted as inhibition of synthesis of an asthma relevant metabolite in childhood gut microbiota (Münch et al. 2021). Synthetic biology to complement gene editing permits the construction of living therapeutics with programmed function. Engineered bacteria have been instructed to respond to environmental stimuli or host metabolites and provide targeted outputs such as anti-inflammatory cytokines or diagnostic reporters only under defined conditions. Such systems typically take advantage of genetically contained circuits or engineered auxotrophy to provide safety and control of the engineered microbe. For instance, microbial biosensors are being engineered that detect early gastrointestinal inflammation markers, releasing therapeutic molecules only on demand (Ezzamouri et al. 2021). A third frontier is designing engineered microbes consortia of live microbes that are assembled with customizable composition and function. These consortia can restore ecological balance more robustly than single-strain probiotics. They are being targeted in human medicine and agriculture (e.g., to suppress methane production in livestock). CRISPR mediated editing is central to evolving community structure, metabolic exchange, and stability in these consortia (Bai et al. 2023). Promising as these new technologies are, they also pose ethical, regulatory, and social concerns. Gene-editing of in- situ microbes meets with issues of informed consent, microbiome privacy, and microbial genetic information ownership. Privacy is at stake because microbiome profiles reveal intimate health or lifestyle data. Lack of validated legal frameworks for microbiome biobanks, data sharing, and therapeutic engineering risks abuse and dampens public trust. Potential risks of unanticipated effects such as horizontal gene transfer, ecological disruption, or biothreat generation require stringent biosafety and biosecurity protocols mimicking those in place for existing GMP and synthetic biology environments. Equally critical is consideration of gender-specific dynamics of microbiome interactions. Male and female microbiota differ in composition and metabolic output depending on hormonal, immune, and physiological determinants. Gender biology underlies gender-specific disease incidence patterns; type 1 diabetes and autoimmune diseases, for example, have female biases that correlate with gut microbial distinctions. But most studies are male-biased and few interventions are gender-specific. Closing the gap requires intentional inclusion of heterogeneous cohorts, stratified analysis, and intervention design on the basis of gender biology. Charting these frontiers responsibly requires the establishment of ethical guidelines, regulatory control, and gender-sensitive frameworks. Regulations need to care not only for organism safety but for data protection, transparency, long-term follow-up of human subjects, and control of microbiome engineering. Institutional review boards, bioethics committees, and regulatory agencies will need to modify procedures to evaluate risk, provide informed consent for microbiome modification, and balance innovation with caution.

1.9 Conclusions

The study of microbiomes has developed over the past century from the first microscopic observations of microorganisms to a complex multi-omics science that can determine the identification, role, and dynamics of microbial communities in a variety of hosts and environments. The realization that microbiomes are active contributors to health, illness, and ecological stability rather than passive inhabitants has revolutionized our knowledge of biology on all fronts. The composition and metabolic capacity of microbial consortia have been revealed in previously unheard-of detail thanks to developments in sequencing technology and complementing techniques including metagenomics, metatranscriptomics, metabolomics, and metaproteomics. By using these methods, it has become clear that a variety of factors, including host genetics, developmental stage, diet, environment, and human-induced stressors like antibiotics, pollution, and farming methods, influence microbiomes.

The understanding that dysbiosis, or imbalance within these communities, can have serious repercussions is equally crucial. Dysbiosis has been associated with neurological, metabolic, and inflammatory disorders in humans; growth, immunity, and productivity in animals; and illnesses and decreased ecosystem resilience in plants and aquatic systems. These results highlight how important the microbiome is in determining health in all biological systems.

The application of new tools and the integration of various analytical techniques continue to improve our capacity to relate microbial diversity to functional consequences as the field develops. From sustainable practices in agriculture, aquaculture, and environmental management to microbiome-based diagnostics and treatments in medicine, this expanding body of knowledge not only fortifies the conceptual underpinnings of microbiome science but also creates exciting opportunities for interventions. In the end, the study of microbiomes demonstrates the profound interdependence of all life, highlighting the fact that stable and healthy microbial relationships are essential for the well-being of both individual organisms and ecosystems.

References

Abu-Ali GS, Mehta RS, Lloyd-Price J, Mallick H, Branck T, Ivey KL, Drew DA, DuLong C, Rimm E, Izard J, Chan AT, Huttenhower C (2018) Metatranscriptome of human faecal microbial communities in a cohort of adult men. Nat Microbiol 3(3):356–366. https://doi.org/10.1038/s41564-017-0084-4

Aggarwal N, Kitano S, Puah GRY, Kittelmann S, Hwang IY, Chang MW (2022) Microbiome and human health: current understanding, engineering, and enabling technologies. Chem Rev 123(1):31–72. https://doi.org/10.1021/acs.chemrev.2c00431

Aharoni A, Goodacre R, Fernie AR (2023) Plant and microbial sciences as key drivers in the development of metabolomics research. Proc Natl Acad Sci 120(12):e2217383120. https://doi.org/10.1073/pnas.2217383120

Ahrodia T, Das S, Bakshi S, Das B (2022) Structure, functions, and diversity of the healthy human microbiome. Prog Mol Biol Transl Sci 191(1):53–82

Aqel B, DiBaise JK (2015) Role of the gut microbiome in nonalcoholic fatty liver disease. Nutr Clin Pract 30(6):780–786. https://doi.org/10.1177/0884533615605811

Arthur JC, Perez-Chanona E, Mühlbauer M, Tomkovich S, Uronis JM, Fan TJ, Campbell BJ, Abujamel T, Dogan B, Rogers AB, Rhodes JM, Stintzi A, Simpson KW, Hansen JJ, Keku TO, Fodor AA, Jobin C (2012) Intestinal inflammation targets cancer-inducing activity of the microbiota of microbes and cancer. Science 338(6103):120–123. https://doi.org/10.1126/science.1224820

Auchtung TA, Fofanova TY, Stewart CJ, Nash AK, Wong MC, Gesell JR, Auchtung JM, Ajami NJ, Petrosino JF (2018) Investigating colonization of the healthy adult gastrointestinal tract by fungi. mSphere 3(2):e00092–e00018. https://doi.org/10.1128/mSphere.00092-18

Bäckhed F, Ding H, Wang T, Hooper LV, Koh GY, Nagy A, Semenkovich CF, Gordon JI (2004) The gut microbiota as an environmental factor that regulates fat storage. Proc Natl Acad Sci 101 (44):15718–15723. https://doi.org/10.1073/pnas.0407076101

Bai X, Huang Z, Duraj-Thatte AM, Ebert MP, Zhang F, Burgermeister E et al (2023) Engineering the gut microbiome. Nat Rev Bioeng 1(9):665–679

Bajaj JS, Ridlon JM, Hylemon PB, Thacker LR, Heuman DM, Smith S, Sikaroodi M, Gillevet PM (2012) Linkage of gut microbiome with cognition in hepatic encephalopathy. Am J Physiol Gastrointest Liver Physiol 302(1):G168–G175. https://doi.org/10.1152/ajpgi.00190.2011

Barrington WT, Lusis AJ (2017) Association between the gut microbiome and atherosclerosis. Nat Rev Cardiol 14(12):699–700. https://doi.org/10.1038/nrcardio.2017.169

Bashiardes S, Zilberman-Schapira G, Elinav E (2016) Use of metatranscriptomics in microbiome research. Bioinform Biol Insights 10:19–25. https://doi.org/10.4137/BBI.S34610

Belkaid Y, Harrison OJ (2017) Homeostatic immunity and the microbiota. Immunity 46(4): 562–576. https://doi.org/10.1016/j.immuni.2017.04.008

Berche P (2012) Louis Pasteur, from crystals of life to vaccination. Clin Microbiol Infect 18:1–6. https://doi.org/10.1111/j.1469-0691.2012.03945.x

Berg G, Rybakova D, Fischer D, Cernava T, Vergès MCC, Charles T et al (2020) Microbiome definition revisited: old concepts and new challenges. Microbiome 8:1–22. https://doi.org/10.1186/s40168-020-00875-0

Bharti M, Modeel S, Yadav S, Yadav P, Siwach S, Dolkar P, Nagar S, Negi T, Negi RK (2023) Emerging role of gut microbiome in cancer immunotherapy. In: Sobti R, Kuhad RC, Lal R, Rishi P (eds) Role of microbes in sustainable development. Springer, pp 409–427. https://doi.org/10.1007/978-981-99-3126-2_18

Blekhman R, Goodrich JK, Huang K, Sun Q, Bukowski R, Bell JT, Spector TD, Keinan A, Ley RE, Gevers D, Clark AG (2015) Host genetic variation impacts microbiome composition across human body sites. Genome Biol 16(1). https://doi.org/10.1186/s13059-015-0759-1

Boldyreva MN, Chaplin AV, Rebrikov DV (2017) The human microbiome. Bull Russian State Med Univ 2:5–13. https://doi.org/10.24075/brsmu.2017-02-01

Bouchard C (1887) Leçons sur les auto-intoxications: dans les maladies. [Publisher], p 348

Bouter KE, van Raalte DH, Groen AK, Nieuwdorp M (2017) Role of the gut microbiome in the pathogenesis of obesity and obesity-related metabolic dysfunction. Gastroenterology 152 (7):1671–1678. https://doi.org/10.1053/j.gastro.2016.12.048

Bressa C, Bailén-Andrino M, Pérez-Santiago J, González-Soltero R, Pérez M, Montalvo-Lominchar MG et al (2017) Differences in gut microbiota profile between women with active lifestyle and sedentary women. PLoS One 12(2):e0171352. https://doi.org/10.1371/journal.pone.0171352

Bron PA, Van Baarlen P, Kleerebezem M (2012) Emerging molecular insights into the interaction between probiotics and the host intestinal mucosa. Nat Rev Microbiol 10(1):66–78. https://doi.org/10.1038/nrmicro2690

Bultman SJ (2014) Emerging roles of the microbiome in cancer. Carcinogenesis 35(2):249–255. https://doi.org/10.1093/carcin/bgt392

Busi SB, De Nies L, Habier J, Wampach L, Fritz JV, Heintz-Buschart A, May P, Halder R, De Beaufort C, Wilmes P (2021) Persistence of birth mode-dependent effects on gut microbiome composition, immune system stimulation and antimicrobial resistance during the first year of life. ISME Commun 1(1). https://doi.org/10.1038/s43705-021-00003-5

Cabrera JM, Monte DFM (2024) Microbiota [mī″-krō-bī′-ō-′tə], microbiome [mī″-krō-bī′-ōm]. Emerg Infect Dis 30(8):1663. https://doi.org/10.3201/eid3008.230677

Cavaillon JM, Legout S (2022) Louis Pasteur: between myth and reality. Biomolecules 12(4):596. https://doi.org/10.3390/biom12040596

Cheng Y, Ling Z, Li L (2020) The intestinal microbiota and colorectal cancer. Front Immunol 11. https://doi.org/10.3389/fimmu.2020.615056

Clarke SF, Murphy EF, O'Sullivan O, Lucey AJ, Humphreys M, Hogan A, Hayes P, O'Reilly M, Jeffery IB, Wood-Martin R, Kerins DM, Quigley E, Ross RP, O'Toole PW, Molloy MG, Falvey E, Shanahan F, Cotter PD (2014a) Exercise and associated dietary extremes impact on gut microbial diversity. Gut 63(12):1913–1920. https://doi.org/10.1136/gutjnl-2013-306541

Clarke G, Stilling RM, Kennedy PJ, Stanton C, Cryan JF, Dinan TG (2014b) Mini review: gut microbiota: the neglected endocrine organ. Mol Endocrinol 28(8):1221–1238. https://doi.org/10.1210/me.2014-1108

Claus SP, Guillou H, Ellero-Simatos S (2016) The gut microbiota: a major player in the toxicity of environmental pollutants? NPJ Biofilms Microb 2(1):16003. https://doi.org/10.1038/npjbiofilms.2016.3

Collard P (1976) The development of microbiology / Patrick Collard. Cambridge University Press, Cambridge; New York

Cotillard A, Kennedy SP, Kong LC, Prifti E, Pons N, Le Chatelier E, Almeida M, Quinquis B, Levenez F, Galleron N, Gougis S, Rizkalla S, Batto JM, Renault P, Doré J, Zucker JD, Clément K, Ehrlich SD, Blottière H, Leclerc M, Juste C, de Wouters T, Lepage P, Fouqueray C, Basdevant A, Henegar C, Godard C, Fondacci M, Rohia A, Hajduch F, Weissenbach J, Pelletier E, Le Paslier D, Gauchi JP, Gibrat JF, Loux V, Carré W, Maguin E van de Guchte M, Jamet A, Boumezbeur F, Layec S (2013) Dietary intervention impact on gut microbial gene richness. Nature 500(7464):585–588. https://doi.org/10.1038/nature12480

Cullender TC, Chassaing B, Janzon A, Kumar K, Müller CE, Werner JJ, Angenent LT, Bell ME, Hay AG, Peterson DA, Walter J, Vijay-Kumar M, Gewirtz AT, Ley RE (2013) Innate and adaptive immunity interact to quench microbiome flagellar motility in the gut. Cell Host Microbe 14(5):571–581. https://doi.org/10.1016/j.chom.2013.10.009

Cussotto S, Sandhu KV, Dinan TG, Cryan JF (2018) The neuroendocrinology of the microbiota-gut-brain axis: a behavioural perspective. Front Neuroendocrinol 51:80–101. https://doi.org/10.1016/j.yfrne.2018.04.002

Davis HC (2018) Can the gastrointestinal microbiota be modulated by dietary fibre to treat obesity? Ir J Med Sci 187(2):393–402. https://doi.org/10.1007/s11845-017-1686-9

Del Chierico F, Nobili V, Vernocchi P, Russo A, De Stefanis C, Gnani D, Furlanello C, Zandonà A, Paci P, Capuani G, Dallapiccola B, Miccheli A, Alisi A, Putignani L (2017) Gut microbiota profiling of pediatric nonalcoholic fatty liver disease and obese patients unveiled by an integrated meta-omics-based approach. Hepatology 65(2):451–464. https://doi.org/10.1002/hep.28572

DeLong EF (2021) Genome-enabled exploration of microbial ecology and evolution in the sea: a rising tide lifts all boats. Environ Microbiol 23(3):1301–1321. https://doi.org/10.1111/1462-2920.15403

Depner M, Taft DH, Kirjavainen PV, Kalanetra KM, Karvonen AM, Peschel S, Schmausser-Hechfellner E, Roduit C, Frei R, Lauener R, Divaret-Chauveau A, Dalphin JC, Riedler J, Roponen M, Kabesch M, Renz H, Pekkanen J, Farquharson FM, Louis P, Mills DA et al (2020) Maturation of the gut microbiome during the first year of life contributes to the protective farm effect on childhood asthma. Nat Med 26(11):1766–1775. https://doi.org/10.1038/s41591-020-1095-x

Di Rosa C, Di Francesco L, Spiezia C, Khazrai YM (2023) Effects of animal and vegetable proteins on gut microbiota in subjects with overweight or obesity. Nutrients 15(12):2675. https://doi.org/10.3390/nu15122675

Diakite A, Dubourg G, Dione N, Afouda P, Bellali S, Ngom II et al (2020) Optimization and standardization of the culturomics technique for human microbiome exploration. Sci Rep 10(1): 9674. https://doi.org/10.1038/s41598-020-66738-8

Dickson I (2017) Gut microbiota: culturomics: illuminating microbial dark matter. Nat Rev Gastroenterol Hepatol 14(1):3. https://doi.org/10.1038/nrgastro.2016.189

Dong TS, Gupta A (2018) Influence of early life, diet, and the environment on the microbiome. Clin Gastroenterol Hepatol 17(2):231–242. https://doi.org/10.1016/j.cgh.2018.08.067

Escherich T (1885) Die Darmbacterien des Neugeborenen und Säuglings. Fortschritte der Medicin 3(16 & 17):515–554

Escherich T (1886) Die darmbakterien des säuglings und ihre beziehungen zur physiologie der Verdauung. Enke

Ezzamouri B, Shoaie S, Ledesma-Amaro R (2021) Synergies of systems biology and synthetic biology in human microbiome studies. Front Microbiol 12:681982

Falony G, Joossens M, Vieira-Silva S, Wang J, Darzi Y, Faust K, Kurilshikov A, Bonder MJ, Valles-Colomer M, Vandeputte D, Tito RY, Chaffron S, Rymenans L, Verspecht C, De Sutter L, Lima-Mendez G, D'hoe K, Jonckheere K, Homola D, Garcia R, Tigchelaar EF, Eeckhaudt L, Fu J, Henckaerts L, Zhernakova A, Wijmenga C, Raes J (2016) Population level analysis of gut microbiome variation "normal" for the gut microbiota. Science 352(6285):560–564. https://doi.org/10.1126/science.aad3503

Fan J, Zeng F, Zhong H, Cai J, Shen W, Cheng C, He C, Liu Y, Zhou Y, Chen S, Zhu Y, Liu T, Zheng J, Wang L, Chen Y, Ma W, Zhou D (2025) Potential roles of cigarette smoking on gut microbiota profile among Chinese men. BMC Med 23(1):25. https://doi.org/10.1186/s12916-025-03852-2

Farré-Maduell E, Casals-Pascual C (2019) The origins of gut microbiome research in Europe: from Escherich to Nissle. Human Microbiome J 14:100065. https://doi.org/10.1016/j.humic.2019.100065

Fitzmaurice C, Christine A, Barber RM, Barregard L, Bhutta ZA, Brenner H, Dicker DJ, Chimed-Orchir O, Dandona R, Dandona L, Fleming T, Forouzanfar MH, Hancock J, Hay RJ, Hunter-Merrill R, Huynh C, Hosgood HD, Johnson CO, Jonas JB, Khubchandani J, Kumar GA, Kutz M, Lan Q, Larson HJ, Liang X, Lim SS, Lopez AD, MacIntyre MF, Marczak L, Marquez N, Mokdad AH, Pinho C, Pourmalek F, Salomon JA, Sanabria JR, Sandar L, Sartorius B, Schwartz SM, Shackelford KA, Shibuya K, Stanaway J, Steiner C, Sun J, Takahashi K, Vollset SE, Vos T, Wagner JA, Wang H, Westerman R, Zeeb H, Zoeckler L, Abd-Allah F, Ahmed MB, Alabed S, Alam NK, Aldhahri SF, Alem G, Alemayohu MA, Ali R, Al-Raddadi R, Amare A, Amoako Y, Artaman A, Asayesh H, Atnafu N, Awasthi A, Saleem HB, Barac A, Bedi N, Bensenor I, Berhane A, Bernabé E, Betsu B, Binagwaho A, Boneya D, Campos-Nonato I, Castañeda-Orjuela C, Catalá-López F, Chiang P, Chibueze C, Chitheer A, Choi JY, Cowie B, Damtew S, das Neves J, Dey S, Dharmaratne S, Dhillon P, Ding E, Driscoll T, Ekwueme D, Endries AY, Farvid M, Farzadfar F, Fernandes J, Fischer F, G/Hiwot TT, Gebru A, Gopalani S, Hailu A, Horino M, Horita N, Husseini A, Huybrechts I, Inoue M, Islami F, Jakovljevic M, James S, Javanbakht M, Jee SH, Kasaeian A, Kedir MS, Khader YS, Khang YH, Kim D, Leigh J, Linn S,

Lunevicius R, El Razek HMA, Malekzadeh R, Malta DC, Marcenes W, Markos D, Melaku YA, Meles KG, Mendoza W, Mengiste DT, Meretoja TJ, Miller TR, Mohammad KA, Mohammadi A, Mohammed S, Moradi-Lakeh M, Nagel G, Nand D, Le Nguyen Q, Nolte S, Ogbo FA, Oladimeji KE, Oren E, Pa M, Park EK, Pereira DM, Plass D, Qorbani M, Radfar A, Rafay A, Rahman M, Rana SM, Søreide K, Satpathy M, Sawhney M, Sepanlou SG, Shaikh MA, She J, Shiue I, Shore HR, Shrime MG, So S, Soneji S, Stathopoulou V, Stroumpoulis K, Sufiyan MB, Sykes BL. Tabarés-Seisdedos R, Tadese F, Tedla BA, Tessema GA, Thakur JS, Tran BX, Ukwaja KN, Uzochukwu BSC, Vlassov VV, Weiderpass E, Wubshet Terefe M, Yebyo HG, Yimam HH, Yonemoto N, Younis MZ, Yu C, Zaidi Z, Zaki MES, Zenebe ZM, Murray CJL, Naghavi M (2017) Global regional and national cancer incidence mortality years of life lost years lived with disability and disability-adjusted life-years for 32 cancer groups 1990 to 2015. JAMA Oncol 3(4):524. https://doi.org/10.1001/jamaoncol.2016.5688

Galvani AP, Bauch CT, Anand M, Singer BH, Levin SA (2016) Human–environment interactions in population and ecosystem health. Proc Natl Acad Sci 113(51):14502–14506. https://doi.org/10.1073/pnas.1618138113

Ghosh TS, Rampelli S, Jeffery IB, Santoro A, Neto M, Capri M, Giampieri E, Jennings A, Candela M, Turroni S, Zoetendal EG, Hermes GDA, Elodie C, Meunier N, Brugere CM, Pujos-Guillot E, Berendsen AM, De Groot LCPGM, Feskins EJM, O'Toole PW (2020) Mediterranean diet intervention alters the gut microbiome in older people reducing frailty and improving health status: the NU-AGE 1-year dietary intervention across five European countries. Gut 69(7):1218–1228. https://doi.org/10.1136/gutjnl-2019-319654

Gibson GR, Hutkins R, Sanders ME, Prescott SL, Reimer RA, Salminen SJ, Scott K, Stanton C, Swanson KS, Cani PD, Verbeke K, Reid G (2017) Expert consensus document: the international scientific Association for Probiotics and Prebiotics (ISAPP) consensus statement on the definition and scope of prebiotics. Nat Rev Gastroenterol Hepatol 14(8):491–502. https://doi.org/10.1038/nrgastro.2017.75

Gilbert JA, Blaser MJ, Caporaso JG, Jansson JK, Lynch SV, Knight R (2018) Current understanding of the human microbiome. Nat Med 24(4):392–400. https://doi.org/10.1038/nm.4517

Goodrich JK, Waters JL, Poole AC, Sutter JL, Koren O, Blekhman R, Beaumont M, Van Treuren W, Knight R, Bell JT, Spector TD, Clark AG, Ley RE (2014) Human genetics shape the gut microbiome. Cell 159(4):789–799. https://doi.org/10.1016/j.cell.2014.09.053

Gophna U, Konikoff T, Nielsen HB (2017) Oscillospira and related bacteria – from metagenomic species to metabolic features summary. Environ Microbiol 19(3):835–841. 10.1111/emi.2017.19.issue-3. https://doi.org/10.1111/1462-2920.13658

Guardino RF (2005) Early history of microbiology and microbiological methods. Encyclop Rapid Microbiol Methods 1:1–27

Hall BA, Moran H, Jenny Y, Ashley S, Xiaofang G, Timothy J, Georgia K. A, Tommi L, Nadine V, Robin F, Madeline W, Melissa B, John C, Hamed G, Dirk K, Ashwin N. G, Subra A, Eric S. K, Paul L, Hera B, Ramnik J. V, Curtis X, Huttenhower (2017) A novel Ruminococcus gnavus clade enriched in inflammatory bowel disease patients. Genome Med 9(1). https://doi.org/10.1186/s13073-017-0490-5

Halfvarson J, Brislawn CJ, Lamendella R, Vázquez-Baeza Y, Walters WA, Bramer LM, D'Amato M, Bonfiglio F, McDonald D, Gonzalez A, McClure EE, Dunklebarger MF, Knight R, Jansson JK (2017) Dynamics of the human gut microbiome in inflammatory bowel disease. Nat Microbiol 2(5). https://doi.org/10.1038/nmicrobiol.2017.4

Handelsman J, Rondon MR, Brady SF, Clardy J, Goodman RM (1998) Molecular biological access to the chemistry of unknown soil microbes: a new frontier for natural products. Chem Biol 5(10):R245–R249. https://doi.org/10.1016/s1074-5521(98)90108-9

Henao-Mejia J, Elinav E, Jin C, Hao L, Mehal WZ, Strowig T, Thaiss CA, Kau AL, Eisenbarth SC, Jurczak MJ, Camporez JP, Shulman GI, Gordon JI, Hoffman HM, Flavell RA (2012) Inflammasome-mediated dysbiosis regulates progression of NAFLD and obesity. Nature 482(7384):179–185. https://doi.org/10.1038/nature10809

Hofer U (2018) The majority is uncultured. Nat Rev Microbiol 16(12):716–717. https://doi.org/10.1038/s41579-018-0097-x

Höll D, Bossert LN (2022) Introducing the microbiome: interdisciplinary perspectives. Endeavour 46(1–2):100817. https://doi.org/10.1016/j.endeavour.2022.100817

Hoyles L, Fernández-Real JM, Federici M, Serino M, Abbott J, Charpentier J, Heymes C, Luque JL, Anthony E, Barton RH, Chilloux J, Myridakis A, Martinez-Gili L, Moreno-Navarrete JM, Benhamed F, Azalbert V, Blasco-Baque V, Puig J, Xifra G, Ricart W, Tomlinson C, Woodbridge M, Cardellini M, Davato F, Cardolini I, Porzio O, Gentileschi P, Lopez F, Foufelle F, Butcher SA, Holmes E, Nicholson JK, Postic C, Burcelin R, Dumas ME (2018) Molecular phenomics and metagenomics of hepatic steatosis in non-diabetic obese women. Nat Med 24 (7):1070–1080. https://doi.org/10.1038/s41591-018-0061-3

Hugon P, Lagier JC, Colson P, Bittar F, Raoult D (2017) Repertoire of human gut microbes. Microb Pathog 106:103–112. https://doi.org/10.1016/j.micpath.2016.06.020

The Human Microbiome Project Consortium (2012) Structure, function and diversity of the healthy human microbiome. Nature 486(7402):207–214. https://doi.org/10.1038/nature11234

Jiang W, Wu N, Wang X, Chi Y, Zhang Y, Qiu X, Hu Y, Li J, Liu Y (2015) Dysbiosis gut microbiota associated with inflammation and impaired mucosal immune function in intestine of humans with non-alcoholic fatty liver disease. Sci Rep 5(1). https://doi.org/10.1038/srep08096

Jie Z, Xia H, Zhong SL, Feng Q, Li S, Liang S, Zhong H, Liu Z, Gao Y, Zhao H, Zhang D, Su Z, Fang Z, Lan Z, Li J, Xiao L, Li J, Li R, Li X, Li F, Ren H, Huang Y, Peng Y, Li G, Wen B, Dong B, Chen JY, Geng QS, Zhang ZW, Yang H, Wang J, Wang J, Zhang X, Madsen L, Brix S, Ning G, Xu X, Liu X, Hou Y, Jia H, He K, Kristiansen K (2017) The gut microbiome in atherosclerotic cardiovascular disease. Nat Commun 8(1). https://doi.org/10.1038/s41467-017-00900-1

Jonsson AL, Bäckhed F (2017) Role of gut microbiota in atherosclerosis. Nat Rev Cardiol 14 (2):79–87. https://doi.org/10.1038/nrcardio.2016.183

Joos R, Boucher K, Lavelle A et al (2025) Examining the healthy human microbiome concept. Nat Rev Microbiol 23:192–205. https://doi.org/10.1038/s41579-024-01107-0

Joseph TA, Pe'er I (2021) An introduction to whole-metagenome shotgun sequencing studies. In: Deep sequencing data analysis, pp 107–122. https://doi.org/10.1007/978-1-0716-1103-6_6

Kahraman Ilıkkan Ö (2025) Lactobacillomics as a new notion in lactic acid bacteria research through omics integration. World J Microbiol Biotechnol 41(2):68. https://doi.org/10.3389/fcimb.2023.1196967

Koch R (1882) Die aetiologie der tuberculose. Berl Klin Wochenschr 15:221–236

Koliada A, Syzenko G, Moseiko V, Budovska L, Puchkov K, Perederiy V, Gavalko Y, Dorofeyev A, Romanenko M, Tkach S, Sineok L, Lushchak O, Vaiserman A (2017) Association between body mass index and Firmicutes/Bacteroidetes ratio in an adult Ukrainian population. BMC Microbiology 17(1). https://doi.org/10.1186/s12866-017-1027-1

Kolmeder CA, de Been M, Nikkilä J, Ritamo I, Mättö J, Valmu L, Salojärvi J, Palva A, Salonen A, de Vos WM (2012) Comparative metaproteomics and diversity analysis of human intestinal microbiota testifies for its temporal stability and expression of core functions. PLoS One 7(1): e29913. https://doi.org/10.1371/journal.pone.0029913

Kong C, Liang L, Liu G, Du L, Yang Y, Liu J, Shi D, Li X, Ma Y (2023) Integrated metagenomic and metabolomic analysis reveals distinct gut-microbiome-derived phenotypes in early-onset colorectal cancer. Gut 72(6):1129–1142. https://doi.org/10.1136/gutjnl-2022-327156

Koponen K, McDonald D, Jousilahti P, Meric G, Inouye M, Lahti L, Niiranen T, Männistö S, Havulinna A, Knight R, Salomaa V (2025) Associations of alcohol with the human gut microbiome and prospective health outcomes in the FINRISK 2002 cohort. Eur J Nutr 64(4): 153. https://doi.org/10.1007/s00394-025-03668-z

Lagier JC, Armougom F, Million M, Hugon P, Pagnier I, Robert C et al (2012) Microbial culturomics: paradigm shift in the human gut microbiome study. Clin Microbiol Infect 18 (12):1185–1193

Lagier JC, Dubourg G, Million M, Cadoret F, Bilen M, Fenollar F, Levasseur A, Rolain JM, Fournier PE, Raoult D (2018) Culturing the human microbiota and culturomics. Nat Rev Microbiol 16(9):540–550. https://doi.org/10.1038/s41579-018-0041-0

Lakhtakia R (2014) The legacy of Robert Koch: surmise, search, substantiate. Sultan Qaboos Univ Med J 14(1):e37–e41. https://doi.org/10.12816/0003334

Lee M, Chang EB (2021) Inflammatory bowel diseases (IBD) and the microbiome—searching the crime scene for clues. Gastroenterology 160(2):524–537. https://doi.org/10.1053/j.gastro.2020.09.056

Lee E, Lee J (2021) Impact of drinking alcohol on gut microbiota: recent perspectives on ethanol and alcoholic beverage. Cur Opin Food Sci 37:91–97. https://doi.org/10.1016/j.cofs.2020.10.001

Lee-Sarwar KA, Chen YC, Chen YY, Kozyrskyj AL, Mandhane PJ, Turvey SE, Subbarao P, Bisgaard H, Stokholm J, Chawes B, Sørensen SJ, Kelly RS, Lasky-Su J, Zeiger RS, O'Connor GT, Sandel MT, Bacharier LB, Beigelman A, Carey VJ, Harshfield BJ et al (2023) The maternal prenatal and offspring early-life gut microbiome of childhood asthma phenotypes. Allergy 78(2):418–428. https://doi.org/10.1111/all.15516

Leininger S, Urich T, Schloter M, Schwark L, Qi J, Nicol GW, Prosser JI, Schuster SC, Schleper C (2006) Archaea predominate among ammonia-oxidizing prokaryotes in soils. Nature 442(7104):806–809. https://doi.org/10.1038/nature04983

Ley RE, Bäckhed F, Turnbaugh P, Lozupone CA, Knight RD, Gordon JI (2005) Obesity alters gut microbial ecology. Proc Natl Acad Sci 102(31):11070–11075. https://doi.org/10.1073/pnas.0504978102

Ley RE, Turnbaugh PJ, Klein S, Gordon JI (2006) Microbial ecology: human gut microbes associated with obesity. Nature 444(7122):1022–1023. https://doi.org/10.1038/4441022a

Li H, Zang Y, Wang C, Li H, Fan A, Han C, Xue F (2020) The interaction between microorganisms metabolites and immune system in the female genital tract microenvironment. Front Cell Infect Microbiol 10. https://doi.org/10.3389/fcimb.2020.609488

Lindner BG, Gerhardt K, Feistel DJ, Rodriguez-R LM, Hatt JK, Konstantinidis KT (2024) A user's guide to the bioinformatic analysis of shotgun metagenomic sequence data for bacterial pathogen detection. Int J Food Microbiol 410:110488. https://doi.org/10.1016/j.ijfoodmicro.2023.110488

Lingyu L, Xu S, Cai M, Li S, Chen Y, Chen L, Lin Y (2024) Effects of fecal microbiota transfer on blood pressure in animal models: a systematic review and meta-analysis. PLOS ONE 19(4): e0300869. https://doi.org/10.1371/journal.pone.0300869

Linton DS (2010) "War Dysentery" and the limitations of German military hygiene during World War I. Bull Hist Med 84(4):607–639. https://doi.org/10.1353/bhm.2010.a408205

Liu Y, Zhou J, Omelchenko MV, Beliaev AS, Venkateswaran A, Stair J, Wu L, Thompson DK, Xu D, Rogozin IB, Gaidamakova EK, Zhai M, Makarova KS, Koonin EV, Daly MJ (2003) Transcriptome dynamics of Deinococcus radiodurans recovering from ionizing radiation. Proc Natl Acad Sci USA 100(7):4191–4196. https://doi.org/10.1073/pnas.0630387100

Liu R, Hong J, Xu X, Feng Q, Zhang D, Gu Y et al (2017) Gut microbiome and serum metabolome alterations in obesity and after weight-loss intervention. Nat Med 23(7):859–868. https://doi.org/10.1038/nm.4358

Liu Y, Zhou Y, Zhang H, Zhao K, Yang D (2025) Gut-lung Axis mediates asthma pathogenesis: roles of dietary patterns and their impact on the gut microbiota. Exp Mol Pathol 142:104964. https://doi.org/10.1016/j.yexmp.2025.104964

Lopez-Santamarina A, Mondragon ADC, Cardelle-Cobas A, Santos EM, Porto-Arias JJ, Cepeda A, Miranda JM (2023) Effects of unconventional work and shift work on the human gut microbiota and the potential of probiotics to restore Dysbiosis. Nutrients 15(13):3070. https://doi.org/10.3390/nu15133070

Lozupone CA, Stombaugh JI, Gordon JI, Jansson JK, Knight R (2012) Diversity, stability and resilience of the human gut microbiota. Nature 489(7415):220–230. https://doi.org/10.1038/nature11550

Mahajan M (2021) Etymologia: petri dish. Emerg Infect Dis 27(1):261. https://doi.org/10.3201/eid2701.ET2701

Mailing LJ, Allen JM, Buford TW, Fields CJ, Woods JA (2019) Exercise and the gut microbiome: a review of the evidence potential mechanisms and implications for human health. Exerc Sport Sci Rev 47(2):75–85. https://doi.org/10.1249/JES.0000000000000183

Makki K, Deehan EC, Walter J, Bäckhed F (2018) The impact of dietary fiber on gut microbiota in host health and disease. Cell Host Microbe 23(6):705–715. https://doi.org/10.1016/j.chom.2018.05.012

Mancini R, Nigro M, Ippolito G (2007) Lazzaro Spallanzani e la confutazione della teoria della generazione spontanea [Lazzaro Spallanzani and his refutation of the theory of spontaneous generation]. Infez Med 15(3):199–206

Matsuoka K, Kanai T (2015) The gut microbiota and inflammatory bowel disease. Semin Immunopathol 37(1):47–55. https://doi.org/10.1007/s00281-014-0454-4

Matsumoto M, Inoue R, Tsukahara T, Ushida K, Chiji H, Matsubara N et al (2008) Voluntary running exercise alters microbiota composition and increases n-butyrate concentration in the rat cecum. Biosci Biotechnol Biochem 72(2):572–576. https://doi.org/10.1271/bbb.70474

Mayers MD, Moon C, Stupp GS, Su AI, Wolan DW (2017) Quantitative metaproteomics and activity-based probe enrichment reveals significant alterations in protein expression from a mouse model of inflammatory bowel disease. J Proteome Res 16(2):1014–1026. https://doi.org/10.1021/acs.jproteome.6b00938

Mehta RS, Nishihara R, Cao Y, Song M, Mima K, Qian ZR, Nowak JA, Kosumi K, Hamada T, Masugi Y, Bullman S, Drew DA, Kostic AD, Fung TT, Garrett WS, Huttenhower C, Wu K, Meyerhardt JA, Zhang X, Willett WC, Giovannucci EL, Fuchs CS, Chan AT, Ogino S (2017) Association of dietary patterns with risk of colorectal cancer subtypes classified by Fusobacterium nucleatum in tumor tissue. JAMA Oncology 3(7):921. https://doi.org/10.1001/jamaoncol.2016.6374

Modeel S, Siwach S, Dolkar P, Chaurasia M, Yadav P, Atri A, Yadav A, Negi T, Negi RK (2025) Emerging risk factors and the role of gut microbiota in immunomodulation and therapeutic implications in colorectal cancer. Cancer Pathogen Ther 4(01):14–30. https://doi.org/10.1016/j.cpt.2025.06.007

Münch PC, Franzosa EA, Stecher B, McHardy AC, Huttenhower C (2021) Identification of natural CRISPR systems and targets in the human microbiome. Cell Host Microbe 29(1):94–106

Mutlu EA, Gillevet PM, Rangwala H, Sikaroodi M, Naqvi A, Engen PA, Kwasny M, Lau CK, Keshavarzian A (2012) Colonic microbiome is altered in alcoholism. Am J Physiol Gastrointest Liver Physiol 302(9):G966–G978. https://doi.org/10.1152/ajpgi.00380.2011

Nguyen TL, Vieira-Silva S, Liston A, Raes J (2015) How informative is the mouse for human gut microbiota research? Dis Model Mech 8(1):1–16. https://doi.org/10.1242/dmm.017400

Ning L, Zhou YL, Sun H, Zhang Y, Shen C, Wang Z et al (2023) Microbiome and metabolome features in inflammatory bowel disease via multi-omics integration analyses across cohorts. Nat Commun 14(1):7135

Nysten J, Van Dijck P (2023) Can we microbe-manage our vitamin acquisition for better health? PLoS Pathog 19(5):e1011361. https://doi.org/10.1371/journal.ppat.1011361

Obón-Santacana M, Mas-Lloret J, Bars-Cortina D, Criado-Mesas L, Carreras-Torres R, Díez-Villanueva A, Moratalla-Navarro F, Guinó E, Ibáñez-Sanz G, Rodríguez-Alonso L, Mulet-Margalef N, Mata A, García-Rodríguez A, Duell EJ, Pimenoff VN, Moreno V (2022) Meta-analysis and validation of a colorectal cancer risk prediction model using deep sequenced fecal metagenomes. Cancers 14(17):4214. https://doi.org/10.3390/cancers14174214

Odutayo A, Wong CX, Hsiao AJ, Hopewell S, Altman DG, Emdin CA (2016) Atrial fibrillation and risks of cardiovascular disease renal disease and death: systematic review and meta-analysis. BMJ:i4482. https://doi.org/10.1136/bmj.i4482

Ortega C, Anderson LN, Frando A, Sadler NC, Brown RW, Smith RD, Wright AT, Grundner C (2016) Systematic survey of serine hydrolase activity in Mycobacterium tuberculosis defines changes associated with persistence. Cell Chem Biol 23(2):290–298. https://doi.org/10.1016/j.chembiol.2016.01.003

Örtqvist AK, Lundholm C, Halfvarson J, Ludvigsson JF, Almqvist C (2019) Fetal and early life antibiotics exposure and very early onset inflammatory bowel disease: a population-based study. Gut 68(2):218–225. https://doi.org/10.1136/gutjnl-2017-314352

Ozato N, Saito S, Yamaguchi T, Katashima M, Tokuda I, Sawada K, Katsuragi Y, Kakuta M, Imoto S, Ihara K, Nakaji S (2019) Blautia genus associated with visceral fat accumulation in adults 20–76 years of age. NPJ Biofilms Microbiomes 5(1). https://doi.org/10.1038/s41522-019-0101-x

Parro V, Paz MM (2023) Metatranscriptome. In: Encyclopedia of astrobiology. Springer, Berlin/Heidelberg, pp 1868–1869. https://doi.org/10.1007/978-3-662-65093-6_5171

Parro V, Moreno-Paz M, González-Toril E (2007) Analysis of environmental transcriptomes by DNA microarrays. Environ Microbiol 9(2):453–464. https://doi.org/10.1111/j.1462-2920.2006.01162.x

Pascale A, Marchesi N, Govoni S, Coppola A, Gazzaruso C (2019) The role of gut microbiota in obesity diabetes mellitus and effect of metformin: new insights into old diseases. Curr Opin Pharmacol:491–495. https://doi.org/10.1016/j.coph.2019.03.011

Pasolli E, Asnicar F, Manara S, Zolfo M, Karcher N, Armanini F, Beghini F, Manghi P, Tett A, Ghensi P, Collado MC, Rice BL, DuLong C, Morgan XC, Golden CD, Quince C, Huttenhower C, Segata N (2019) Extensive unexplored human microbiome diversity revealed by over 150,000 genomes from metagenomes spanning age, geography, and lifestyle. Cell 176(3):649–662.e20. https://doi.org/10.1016/j.cell.2019.01.001

Pedroza Matute S, Iyavoo S (2023) Exploring the gut microbiota: lifestyle choices disease associations and personal genomics. Front Nutr 10. https://doi.org/10.3389/fnut.2023.1225120

Pinart M, Dötsch A, Schlicht K, Laudes M, Bouwman J, Forslund SK et al (2021) Gut microbiome composition in obese and non-obese persons: a systematic review and meta-analysis. Nutrients 14(1):12. https://doi.org/10.3390/nu14010012

Prescott SL (2017) History of medicine: origin of the term microbiome and why it matters. Human Microb J 4:24–25. https://doi.org/10.1016/j.humic.2017.05.004

Qin J, Li R, Raes J, Arumugam M, Burgdorf KS, Manichanh C et al (2010) A human gut microbial gene catalogue established by metagenomic sequencing. Nature 464(7285):59–65. https://doi.org/10.1038/nature08821

Quintana-Murci L, Clark AG (2013) Population genetic tools for dissecting innate immunity in humans. Nat Rev Immunol 13(4):280–293. https://doi.org/10.1038/nri3421

Ramírez-Macías I, Orenes-Piñero E, Camelo-Castillo A, Rivera-Caravaca JM, López-García C, Marín F (2022) Novel insights in the relationship of gut microbiota and coronary artery diseases. Crit Rev Food Sci Nutr 62(14):3738–3750. https://doi.org/10.1080/10408398.2020.1868397

Remely M, Hippe B, Geretschläger I, Stegmayer B, Haslberger AG (2014) Gut microbiota composition correlates with changes in body fat content due to weight loss. Benefic Microbes 6(4):431–439. https://doi.org/10.3920/BM2014.0106

Rinninella E, Raoul P, Cintoni M, Franceschi F, Miggiano GAD, Gasbarrini A, Mele MC (2019) What is the healthy gut microbiota composition? A changing ecosystem across age, environment, diet, and diseases. Microorganisms 7(1):14. https://doi.org/10.3390/microorganisms7010014

Rowland I, Gibson G, Heinken A, Scott K, Swann J, Thiele I, Tuohy K (2018) Gut microbiota functions: metabolism of nutrients and other food components. Eur J Nutr 57(1):1–24. https://doi.org/10.1007/s00394-017-1445-8

Santacroce L, Man A, Charitos IA, Haxhirexha K, Topi S (2021) Current knowledge about the connection between health status and gut microbiota from birth to elderly: a narrative review. Front Biosci 26:135–148. https://doi.org/10.52586/4930

Santacroce L, Passarelli PC, Azzolino D, Bottalico L, Charitos IA, Cazzolla AP et al (2023) Oral microbiota in human health and disease: a perspective. Exp Biol Med 248(15):1288–1301. https://doi.org/10.1177/15353702231187645

Santos-Marcos JA, Mora-Ortiz M, Tena-Sempere M, Lopez-Miranda J, Camargo A (2023) Interaction between gut microbiota and sex hormones and their relation to sexual dimorphism in metabolic diseases. Biol Sex Diff 14(1). https://doi.org/10.1186/s13293-023-00490-2

Savage DC (1977) Microbial ecology of the gastrointestinal tract. Ann Rev Microbiol 31(1): 107–133

Schena M, Shalon D, Davis RW, Brown PO (1995) Quantitative monitoring of gene expression patterns with a complementary DNA microarray. Science 270(5235):467–470. https://doi.org/10.1126/science.270.5235.467

Schwiertz A, Taras D, Schäfer K, Beijer S, Bos NA, Donus C, Hardt PD (2010) Microbiota and SCFA in lean and overweight healthy subjects. Obesity 18(1):190–195. https://doi.org/10.1038/oby.2009.167

Segal LN, Rom WN, Weiden MD (2014) Lung microbiome for clinicians. new discoveries about bugs in healthy and diseased lungs. Ann Am Thorac Soc 11(1):108–116. https://doi.org/10.1513/AnnalsATS.201310-339FR

Sender R, Fuchs S, Milo R (2016) Revised estimates for the number of human and bacterial cells in the body. PLoS Biol 14(8):e1002533. https://doi.org/10.1371/journal.pbio.1002533

Severino A, Tohumcu E, Tamai L, Dargenio P, Porcari S, Rondinella D, Venturini I, Maida M, Gasbarrini A, Cammarota G, Ianiro G (2024) The microbiome-driven impact of western diet in the development of noncommunicable chronic disorders. Best Pract Res Clin Gastroenterol 72: 101923. https://doi.org/10.1016/j.bpg.2024.101923

Severn MM, Horswill AR (2023) Staphylococcus epidermidis and its dual lifestyle in skin health and infection. Nat Rev Microbiol 21(2):97–111. https://doi.org/10.1038/s41579-022-00780-3

Sha S, Ni L, Stefil M, Dixon M, Mouraviev V (2020) The human gastrointestinal microbiota and prostate cancer development and treatment. Investig Clin Urol 61(Suppl 1):S43. https://doi.org/10.4111/icu.2020.61.S1.S43

Silveira-Nunes G, Durso DF, Jr LRAO, Cunha EHM, Maioli TU, Vieira AT, Speziali E, Corrêa-Oliveira R, Martins-Filho OA, Teixeira-Carvalho A, Franceschi C, Rampelli S, Turroni S, Brigidi P, Faria AMC (2020) Hypertension is associated with intestinal microbiota Dysbiosis and inflammation in a Brazilian population. Front Pharmacol 11. https://doi.org/10.3389/fphar.2020.00258

Singh RK, Chang H, Yan D, Lee KM, Ucmak D, Wong K, Abrouk M, Farahnik B, Nakamura M, Zhu TH, Bhutani T, Liao W (2017) Influence of diet on the gut microbiome and implications for human health. J Transl Med 15(1):73. https://doi.org/10.1186/s12967-017-1175-y

Sjögren YM, Jenmalm MC, Böttcher MF, Björkstén B, Sverremark-Ekström E (2009) Altered early infant gut microbiota in children developing allergy up to 5 years of age. Clin Exp Allergy 39 (4):518–526. https://doi.org/10.1111/j.1365-2222.2008.03156.x

Sommer F, Anderson JM, Bharti R, Raes J, Rosenstiel P (2017) The resilience of the intestinal microbiota influences health and disease. Nat Rev Microbiol 15(10):630–638

Sood U, Kumar R, Hira P (2021) Expanding culturomics from gut to extreme environmental settings. mSystems 6(4):10–1128

Sørensen TIA, Martinez AR, Jørgensen TSH, Eckel J, Clément K (2022) From obesity to diabetes. In: Epidemiology of obesity. Springer International Publishing, Cham, pp 3–27

Sravanthi K, Sattiraju KS, Paul S, Nihal NG, Salunkhe S, Mane SV (2024) Robert Koch: from anthrax to tuberculosis—a journey in medical science. Cureus 16(11):e72955. https://doi.org/10.7759/cureus.72955

Suez J, Korem T, Zeevi D, Zilberman-Schapira G, Thaiss CA, Maza O, Israeli D, Zmora N, Gilad S, Weinberger A, Kuperman Y, Harmelin A, Kolodkin-Gal I, Shapiro H, Halpern Z, Segal E, Elinav E (2014) Artificial sweeteners induce glucose intolerance by altering the gut microbiota. Nature 514(7521):181–186. https://doi.org/10.1038/nature13793

Thomas AM, Manghi P, Asnicar F, Pasolli E, Armanini F, Zolfo M, Beghini F, Manara S, Karcher N, Pozzi C, Gandini S, Serrano D, Tarallo S, Francavilla A, Gallo G, Trompetto M, Ferrero G, Mizutani S, Shiroma H, Shiba S, Shibata T, Yachida S, Yamada T, Wirbel J, Schrotz-King P, Ulrich CM, Brenner H, Arumugam M, Bork P, Zeller G, Cordero F, Dias-Neto E, Setubal JC, Tett A, Pardini B, Rescigno M, Waldron L, Naccarati A, Segata N (2019) Metagenomic analysis of colorectal cancer datasets identifies cross-cohort microbial diagnostic signatures and a link with choline degradation. Nat Med 25(4):667–678. https://doi.org/10.1038/s41591-019-0405-7

Thomas KR, Watt J, Wu CMJ, Akinrinoye A, Amjad S, Colvin L, Cowe R, Duncan SH, Russell WR, Forget P (2022) Pain and opioid-induced gut microbial dysbiosis. Biomedicine 10(8): 1815. https://doi.org/10.3390/biomedicines10081815

Tims S, Derom C, Jonkers DM, Vlietinck R, Saris WH, Kleerebezem M, de Vos WM, Zoetendal EG (2013) Microbiota conservation and BMI signatures in adult monozygotic twins. ISME J 7(4):707–717. https://doi.org/10.1038/ismej.2012.146

Tissier MH (1899) La réaction chromophile d'Escherich et le Bacterium coli. CR Société de Biologie 51:943–945

Turnbaugh PJ, Ley RE, Mahowald MA, Magrini V, Mardis ER, Gordon JI (2006) An obesity-associated gut microbiome with increased capacity for energy harvest. Nature 444(7122): 1027–1031. https://doi.org/10.1038/nature05414

Turnbaugh PJ, Hamady M, Yatsunenko T, Cantarel BL, Duncan A, Ley RE et al (2009) A core gut microbiome in obese and lean twins. Nature 457(7228):480–484. https://doi.org/10.1038/nature07540

Van Den Bossche T, Arntzen MØ, Becher D, Benndorf D, Eijsink VGH, Henry C, Jagtap PD, Jehmlich N, Juste C, Kunath BJ, Mesuere B, Muth T, Pope PB, Seifert J, Tanca A, Uzzau S, Wilmes P, Hettich RL, Armengaud J (2021) The metaproteomics initiative: a coordinated approach for propelling the functional characterization of microbiomes. Microbiome 9(1):243. https://doi.org/10.1186/s40168-021-01176-w

Vila AV, Collij V, Sanna S, Sinha T, Imhann F, Bourgonje AR, Mujagic Z, Jonkers DMAE, Masclee AAM, Fu J, Kurilshikov A, Wijmenga C, Zhernakova A, Weersma RK (2020) Impact of commonly used drugs on the composition and metabolic function of the gut microbiota. Nat Commun 11(1). https://doi.org/10.1038/s41467-019-14177-z

Wan X, Yang Q, Wang X, Bai Y, Liu Z (2023) Isolation and cultivation of human gut microorganisms: a review. Microorganisms 11(4):1080. https://doi.org/10.3390/microorganisms11041080

Wang B, Yao M, Lv L, Ling Z, Li L (2017) The human microbiota in health and disease. Engineering 3(1):71–82. https://doi.org/10.1016/J.ENG.2017.01.008

Wastyk HC, Fragiadakis GK, Perelman D, Dahan D, Merrill BD, Yu FB, Topf M, Gonzalez CG, Van Treuren W, Han S, Robinson JL, Elias JE, Sonnenburg ED, Gardner CD, Sonnenburg JL (2021) Gut-microbiota-targeted diets modulate human immune status. Cell 184(16):4137–4153. e14. https://doi.org/10.1016/j.cell.2021.06.019

Waters JL, Ley RE (2019) The human gut bacteria Christensenellaceae are widespread heritable and associated with health. BMC Biol 17(1). https://doi.org/10.1186/s12915-019-0699-4

Whipps JM, Lewis K, Cooke RC (1988) Mycoparasitism and plant disease control. In: Fungi in biological control systems, pp 161–187

Willing BP, Dicksved J, Halfvarson J, Andersson AF, Lucio M, Zheng Z, Järnerot G, Tysk C, Jansson JK, Engstrand L (2010) A pyrosequencing study in twins shows that gastrointestinal microbial profiles vary with inflammatory bowel disease phenotypes. Gastroenterology 139 (6):1844–1854.e1. https://doi.org/10.1053/j.gastro.2010.08.049

Wilmes P, Bond PL (2004) The application of two-dimensional polyacrylamide gel electrophoresis and downstream analyses to a mixed community of prokaryotic microorganisms. Environ Microbiol 6(9):911–920. https://doi.org/10.1111/j.1462-2920.2004.00687.x

Wilmes P, Heintz-Buschart A, Bond PL (2015) A decade of metaproteomics: where we stand and what the future holds. Proteomics 15(20):3409–3417. https://doi.org/10.1002/pmic.201500183

Wishart DS, Feunang YD, Marcu A, Guo AC, Liang K, Vázquez-Fresno R, Sajed T, Johnson D, Li C, Karu N, Sayeeda Z, Lo E, Assempour N, Berjanskii M, Singhal S, Arndt D, Liang Y, Badran H, Grant J, Serra-Cayuela A et al (2018) HMDB 4.0: the human metabolome database for 2018. Nucleic Acids Res 46(D1):D608–D617. https://doi.org/10.1093/nar/gkx1089

Wu H, Esteve E, Tremaroli V, Khan MT, Caesar R, Mannerås-Holm L, Ståhlman M, Olsson LM, Serino M, Planas-Fèlix M, Xifra G, Mercader JM, Torrents D, Burcelin R, Ricart W, Perkins R, Fernàndez-Real JM, Bäckhed F (2017) Metformin alters the gut microbiome of individuals with treatment-naive type 2 diabetes, contributing to the therapeutic effects of the drug. Nat Med 23(7):850–858. https://doi.org/10.1038/nm.4345

Xia Q, Grant SFA (2013) The genetics of human obesity. Ann N Y Acad Sci 1281:178–190. https://doi.org/10.1111/nyas.12020

Yanfei C, Fengling Y, Haifeng L, Baohong W, Yunbo L, Dajiang L, Yuezhu W, Baoli Z, Lanjuan L (2011) Characterization of fecal microbial communities in patients with liver cirrhosis. Hepatology 54(2):562–572. https://doi.org/10.1002/hep.24423

Yang SY, Han SM, Lee JY, Kim KS, Lee JE, Lee DW (2025) Advancing gut microbiome research: the shift from metagenomics to multi-omics and future perspectives. J Microbiol Biotechnol 35: e2412001. https://doi.org/10.4014/jmb.2412.12001

Yu J, Feng Q, Wong SH, Zhang D, Liang QY, Qin Y, Tang L, Zhao H, Stenvang J, Li Y, Wang X, Xu X, Chen N, Wu WKK, Al-Aama J, Nielsen HJ, Kiilerich P, Jensen BAH, Yau TO, Lan Z, Jia H, Li J, Xiao L, Lam TYT, Ng SC, Cheng AS-L, Wong VW-S, Chan FKL, Xu X, Yang H, Madsen L, Datz C, Tilg H, Wang J, Brünner N, Kristiansen K, Arumugam M, Sung JJ-Y, Wang J (2017) Metagenomic analysis of faecal microbiome as a tool towards targeted non-invasive biomarkers for colorectal cancer. Gut 66(1):70–78. https://doi.org/10.1136/gutjnl-2015-309800

Yuan J, Chen C, Cui J, Lu J, Yan C, Wei X, Zhao X, Li N, Li S, Xue G, Cheng W, Li B, Li H, Lin W, Tian C, Zhao J, Han J, An D, Zhang Q, Wei H, Zheng M, Ma X, Li W, Chen X, Zhang Z, Zeng H, Ying S, Wu J, Yang R, Liu D (2019) Fatty liver disease caused by high-alcohol-producing Klebsiella pneumoniae. Cell Metab 30(4):675–688.e7. https://doi.org/10.1016/j.cmet.2019.08.018

Zhang GF, Sadhukhan S, Tochtrop GP, Brunengraber H (2011) Metabolomics, pathway regulation, and pathway discovery. J Biol Chem 286(27):23631–23635. https://doi.org/10.1074/jbc.R110.171405

Zhu B, Wang X, Li L (2010) Human gut microbiome: the second genome of the human body. Protein Cell 1(8):718–725. https://doi.org/10.1007/s13238-010-0093-z

The Human Microbiota: Composition and Functions

2

Abstract

The human body is home to the intricate and ever-changing human microbiota. The gut microbiota prevents infections by controlling inflammatory reactions and fostering nutritional competition, in addition to being necessary for digestion, food absorption, energy harvesting, bile acid metabolism, and immune system development. Beyond the colon, endotoxins and microbial metabolites affect distant organs like the liver, influencing the growth of health or illness. Dysbiosis, or a disturbance of the microbial balance, can result in pathobiont overgrowth, immunological dysregulation, and heightened vulnerability to metabolic, allergy, and infectious disorders. With current clinical trials examining their ability to prevent and treat microbiota-associated illnesses, growing familiarity with the makeup and purpose of the microbiota has stoked interest in therapeutic approaches, including probiotics, prebiotics, and fecal microbiota transplantation.

Keywords

Human microbiota · Gut microbiome · Dysbiosis · Microbial composition · Immune regulation · Host-microbe interaction · Disease susceptibility · Microbial metabolism

2.1 Overview

Humans and their symbiotic bacteria coexist in a single unit, a metaorganism, in multicellular organisms. Health and disease in humans are largely determined by the over 100 trillion symbiotic microbes that reside in the human body. Others have even argued that the human microbiome, specifically the gut microbiome, can be considered an "essential organ." This is because the microbiome possesses enormous gene diversity, having about 150 times as many distinct genes as the genome of an

R. C. Sobti et al., *Microbiota Unveiled: Science, Health, and Future Therapies*,
https://doi.org/10.1007/978-981-95-9031-5_2

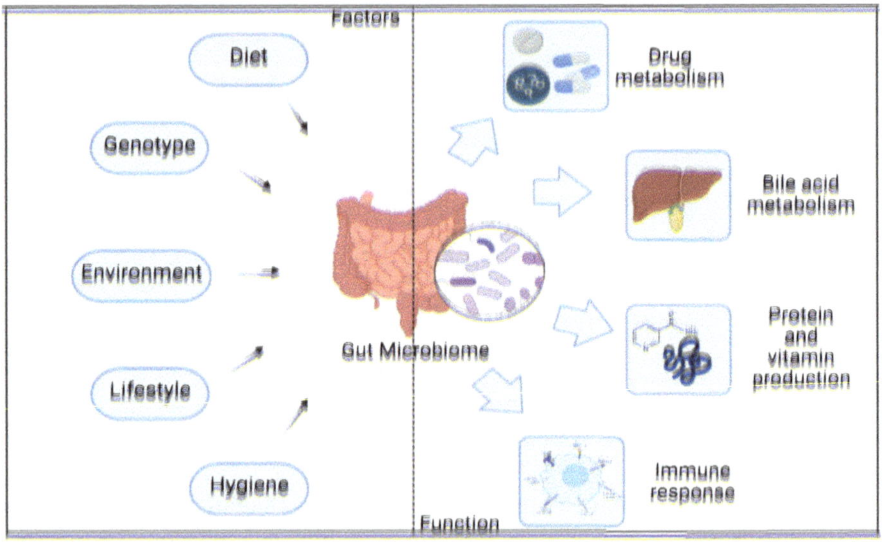

Fig. 2.1 Factors affecting gut microbiome and gut microbiome function: several factors affect the gut microbiome, like diet, host genome, host environment, host lifestyle, etc. Gut microbiome takes part in many functions including the production of protein and vitamins, immune response, production of secondary bile acids, and drug metabolism

individual human and more than 10-times more distinct genes as its host's genome. Each host has diverse microbial populations that are present across different regions in the body, which vary significantly in terms of function and composition. This microbial diversity is affected by many factors like their collective role, host genotype, hygiene, diet, environment, health status, and lifestyle (Fig. 2.1). The number of different microbes and their genetic diversity existing in a microbiome are regulated partly by the specific biology of the body site and partly by the local environment (Dekaboruah et al. 2020). Early-life experiences, including delivery mode, breastmilk or bottle-feed, gestational age (GA), and introduction to antibiotics are linked to the formation of the characteristic core microbiome, which starts early in life and is influenced by early life events as well as factors connected to the mother's pregnancy (Van den Elsen et al. 2019). These alterations may be a precursor to future lifetime disruptions in the core microbiome, which could predict senescence pathologies such as immunological senescence, inflammation, insulin resistance, mutation loading, and epigenetic changes. Age and frailty are also responsible for large alterations in the microbiome, which is found to alter along with an individual's life (O'Toole and Jeffery 2015). These microbes are found on the skin surface, as well as within the mouth, respiratory tract, urogenital system, and gastrointestinal (GI) tract.

Current studies have shown that the microbes, particularly the microbes residing inside gut, take part in fundamental biological functions of humans, including epithelial cell development, metabolism, and natural immunity. A wide variety of illnesses, such as the inflammatory bowel diseases (IBD), cardiovascular condition, metabolic diseases, allergic reactions, autism, cancer, and neurodegenerative conditions, can be impacted by dysbiosis of the gut, which is defined as disturbances in the relative abundances of the gut microbiome brought on by the commensal's community loss, decreased diversity, or growth of infectious populations. The microbiome formed during the first three years largely stays stable through adulthood. While genera like *Bifidobacterium, Roseburia, Ruminococcus, Faecalibacterium*, and *Dialister* are more prevalent in infants than adults, phylum abundances are relatively stable (Hollister et al. 2015). It is possible for taxonomic classification discovered in the infant's gut microbiota to persist as colonies in the microbiomes while growing up for years (Yee et al. 2019). Investigation of the connection between early childhood microbiome development and future illness and health is essential. Hence, understanding the composition, role, and function of these microbes is essential for developing innovative therapies and promoting overall host well-being.

2.2 How Is the Microbiota Acquired and Shaped?

The notion of a placental microbiota remains disputed, and the placenta is presently considered to be sterile. Thus, the gut microbiota is first seeded at birth. It should come as no surprise that the gut microbiota of newborns closely resembles that of the mother's stomach, vagina, oral cavity, and skin. The microbiota composition of the mother's and the infant's stool overlaps by around 72% between 2 and 5 days after birth, and by the age of month 4, most of the microbes that are not from the mother disappear from the infant's gut (Donald and Finlay 2023). These initial interactions are believed to predispose the immune system to maintain balance throughout the individual's life. Although the precise process by which infants accommodate microbial colonization is not yet fully established, factors like gestational age (GA), mode of birth, and type of milk used for feeding play a substantial role in the establishment of the commensals (Fig. 2.2) (Sarkar et al. 2021).

The composition of gut microbes is directly related to gestational age at delivery, indicating that these changes are driven by physiological variations in the intestinal and immune systems and that these variations last for up to a year. A recent study identified GA as a crucial factor influencing microbiota colonization in 45 preterm infants, though they eventually matched the microbiota of healthy term infants (Korpela et al. 2018). Another study analyzing 189 fecal samples from term infants and 65 from preterm infants confirmed the significance of GA and showed that many bacterial genera were absent in preterm children (Chernikova et al. 2018). Premature infants contain fewer species of gut bacteria, often lower counts of *Bifidobacteria*, and dominance of *Proteobacteria*. *Enterobacteriaceae* are also ubiquitous in the feces of preterm newborns.

The greatest determinant of the initial microbiota of neonates is the delivery method. Delivery mode has a long-lasting effect that can be months, if not longer, and affects normal metabolism, health, and susceptibility to certain diseases. The

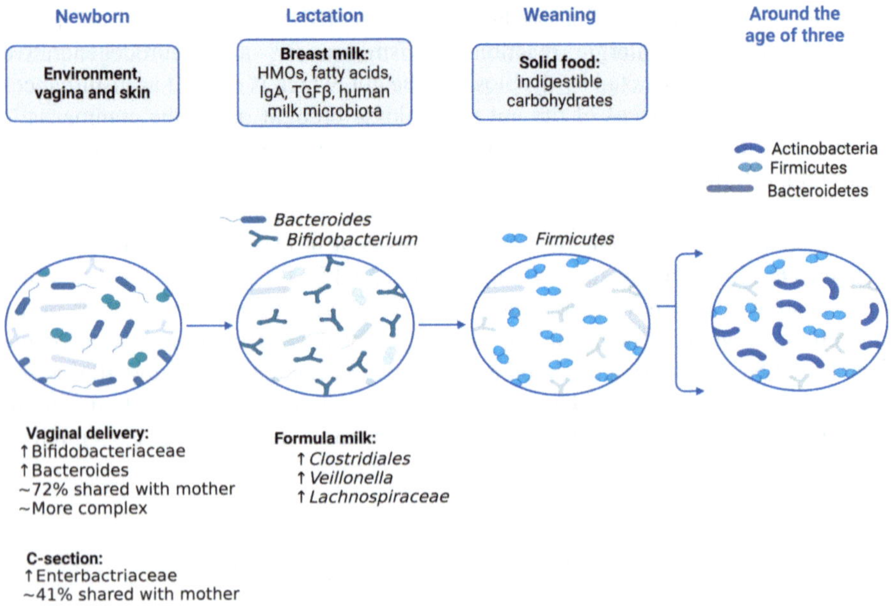

Fig. 2.2 Microbiota in an infant shaped the healthy gut microbiome depending on the various factors like the delivery mode, then by the feeding, i.e., lactation, and later changes when introduced to solid food. The microbiota of an infant then takes the shape of the adult when it is around the age of three

initial contact of the neonate with the mother's vaginal and gastrointestinal (GI) tract microbiota after a vaginal delivery determines the first gut microbiota. Children born through caesarean section (CS) gain microbiota from their surroundings, including the mother's skin and the hospital environment, as CS delivery prevents exposure to vaginal and intestinal microbials (Sarkar et al. 2021). 41% of the gut microorganisms in babies born via caesarean section are shared with the mother, compared to 72% in babies born vaginally. While *Bacteroides and Bifidobacteriaceae* are decreased in infants born via CS, the Enterobacteriaceae family is found in abundance. Infants born through vaginal delivery have an extra diverse gut microbiota at birth, featuring a greater presence of maternal gut and vaginal bacteria. In the first 24 hours after delivery, the gut microbiota is simple and varies widely from baby to baby. Aerobic bacteria and facultative anaerobes, such as Enterobacteriaceae, are among the first to colonize. Within a few days, the gut turns anaerobic as the oxygen is consumed by aerobes and facultative anaerobes and proliferate, supporting organisms from the genera *Bacteroides, Clostridium, and Bifidobacterium*, that are commonly present in the human gut (Donald and Finlay 2023). The gut microbiota starts to be dominated by *Bifidobacterium and Bacteroides*, and the primary factor influencing the composition of the microbiome shifts from birth mode to feeding mode.

A newborn's first encounter with food comes in the form of lactation. The newborn's diet also determines his or her gut microbiome. The milk and colostrum contain living microorganisms, metabolites, cytokines, immune cells, and secretory

IgA (sIgA), all of which define the wet-nursed infant's microbiota and host defense mechanisms triggered by these microorganisms. Obligate anaerobes from the genera *Bacteroides and Bifidobacterium* dominate the gut microbiota during the first weeks after birth, with feeding mode becoming the main factor shaping the microbiota instead of birth mode. Human milk oligosaccharides are the food source for microorganisms in the genus *Bifidobacterium*, which are constantly prevalent in breastfed infants. During the first few months, some Proteobacteria and bacteria produce lactic acid. Breastfed infants also exhibit delayed gut microbiota composition maturation and decreased gut microbiota diversity, and nursing cessation causes a sharp rise in complexity. Conversely, formula-fed babies have early microbial diversification, lower Bifidobacteria levels, and higher *Clostridiales, Veillonella*, and *Lachnospiraceae*. Infants who are fed formula are also more prone to opportunistic pathogens (Donald and Finlay 2023).

Other than the microbiota, maternal IgA complexes with nutritional antigens of the microbiota, suppressing immune activation and microbial adhesion (Fig. 2.3). Thus, the newborn's microbiota is not only seeded through its derivation from the mother's vaginal or epidermal origin, but also maintained by the sIgA secreted by the mother's microbiota. Furthermore, the sIgA also provides the baby with significant protection against infection before the immune system reaches maturity (Pietrasanta et al. 2024). The developing immune system possesses characteristics of unbalanced growth of T and B cells, with regulatory response bias and minimal secretion of inflammatory cytokines. Hence, the tolerogenic environment of early mammalian life and the underdeveloped condition of the neonatal immune defense allow the microbiota from the mother, its surroundings, and the breastmilk to also be tolerated by the system (Kalbermatter et al. 2021).

Once delivered, environmental factors, for example, hospitalization, diet, and intake of antibiotics, further influence how the infant's immune system and microbiota develop (Parizadeh and Arrieta 2023). In line with this, the use of antibiotics brings about extensive alterations in the gut microbiota of newborns. The microbiome is impacted more severely if the antibiotics are administered earlier in juvenility. Infants aged 3 months have a higher relative abundance of *Clostridium* and *Enterococcus,* and a lower *Bacteroides* and *Parabacteroides* abundance when their mothers receive antibiotics during or before labor. When infants are exposed to antibiotics, the quantity of *Bifidobacteria* rapidly declines, and *Proteobacteria* takes its place. In addition, the problem is compounded by the fact that when preterm infants receive more frequent antibiotics, especially in early life (Donald and Finlay 2023). Specifically, antibiotic therapy changes the gut microbiome's α-diversity and community composition, or ecology, when consumed for long period in both adults and children. It also results in reduced production of IgG and antibacterial components, rendering newborns vulnerable to infection. Immune-mediated diseases like diabetes and asthma are likely to be caused by such conditions, as the metabolism of short-chain fatty acids (SCFAs) is likely to be changed and correlated with an upsurge in inflammatory cytokines (Francino 2016). Compared to controls, the antibiotic-treated neonates had high levels of *Klebsiella* and *Enterococcus* species and low level of *Bifidobacterium* species. Penicillin + gentamicin has the

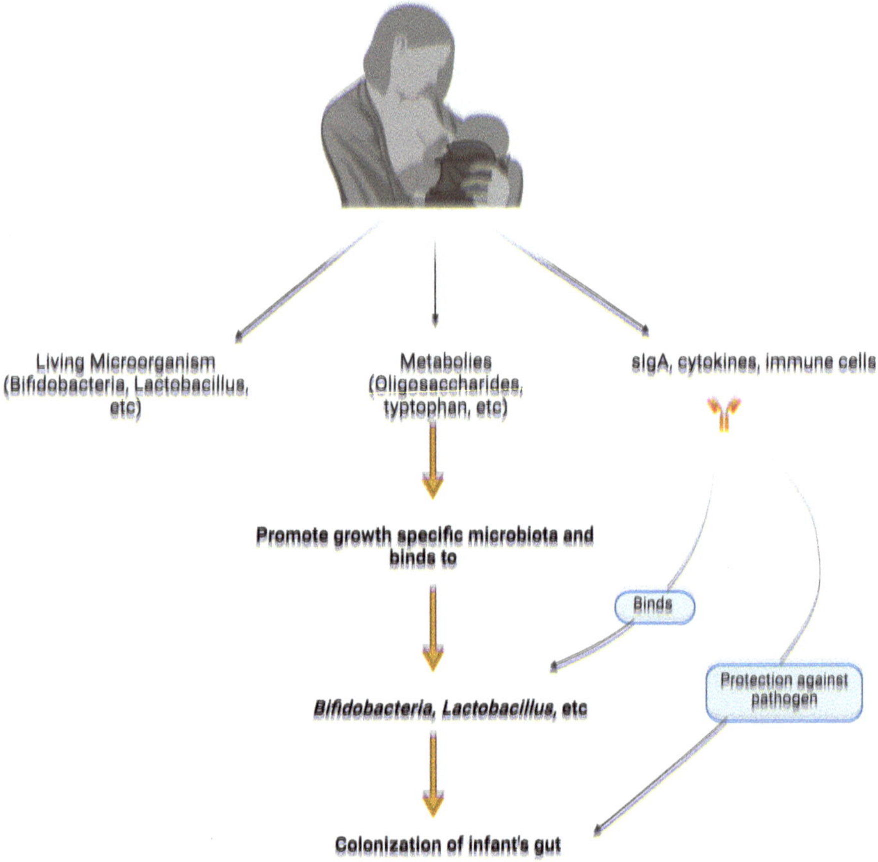

Fig. 2.3 Breastfeeding modulates the colonization of the infant's gut microbiota

least impact on the antimicrobial resistance gene profile and the composition of the microbial community, while amoxicillin + cefotaxime has the greatest effects. According to these findings, the selection of empirical antibiotics has an impact on unfavorable ecological side effects (Reyman et al. 2022).

An infant's surroundings greatly affect the development of their gut microbiota as well. The development of allergy illnesses has been linked to alterations in gut microbiota that are generally influenced by dust exposure during infancy, siblings, pets, and a rural family setting during early ages. The precise species causing this occurrence is not well characterized. Bifidobacterial dominance and low diversity characterize a healthy gut microbiota in the initial few months after birth, and exposure to a wider variety of microorganisms appears to be crucial for immunological maturation and gut microbiota health. At around the age of three, their microbial structure is comparable to that of adults and is predominantly populated with the *Actinobacteria, Firmicutes, and Bacteroidetes* phyla (Rinninella et al. 2019). The structure of human microbiota is then influenced by changes in the immune system, environmental factors, hygiene, and diet, etc., during the aging process as discussed in Chap. 1. Specifically, old individuals have more *Clostridium* and *Proteobacteria* and less *Bifidobacteria* (Ma et al. 2024).

2.3 Major Human Microbial Communities

The human body hosts diverse microbial populations that develop from infancy, childhood, and change throughout life with lifestyle and health conditions (Ronan et al. 2021). Microbiome composition and abundance vary significantly across different body sites (Dekaboruah et al. 2020), as the lower respiratory tract has fewer populations than the upper regions (Bishop et al. 2024) (Table 2.1). The areas like the stomach and duodenum inside the GI tract have a number of microbes, while the caecum and colon have a more dense microbial composition (Belizário and Napolitano 2015). The maximum number of microbiotas inside the human body is made up of *Bacteroidetes, Firmicutes, Actinobacteria, and Proteobacteria*. Individual variations in bacterial species are mainly affected by different body habitats. The

Table 2.1 Human body microbiota

Human body sites		Major bacterial genera
Skin		*Corynebacterium, Propionibacterium, Staphylococcus, and Streptococcus*
Oral	Saliva	*Streptococcus, Veillonella, and Prevotella*
	Soft tissue	*Eubacterium, Rothia, and Streptococcus salivarus*
	Supragingival region	*Actinomyces and Corynebacterium*
	Subgingival area	*Proteobacteria, Actinobacteria, Fusobacteria, Spirochaetes*
Respiratory tract	Upper respiratory tract	*Corynebacterium, Propionibacterium, Staphylococcus, Streptocouus, and Moraxella*
	Lower respiratory tract	*Prevotella, Veillonella, Streptococcus, and Tropheryma*
Gut		*Bifidobacterium, Streptococcus, Lactobacillus, Bacteroides, Clostridium, Escherichia, and Ruminococcus*
Urogenital tract	Female urethra	*Enterobacteriaceae, Escherichia coli, Fusobacterium spp, Corynebacterium* spp., *Veillonella spp, Lactobacillus* spp..
	Male urethera	*Staphylococci, Corynebacterium* spp., *Streptococcus viridans, Mollicute*
	Vagina	*Lactobacillus* spp., *Clostridium* spp., *Staphylococcus* spp., *Enterococcus* spp., *Porphyromonas* spp., *Bifidobacterium* spp., *Propionibacterium* spp., *Streptococcus* spp., *Corynebacterium* spp., *Bacteroides* spp., *Fusobacterium* spp., *Veillonella* spp., *Prevotella spp*

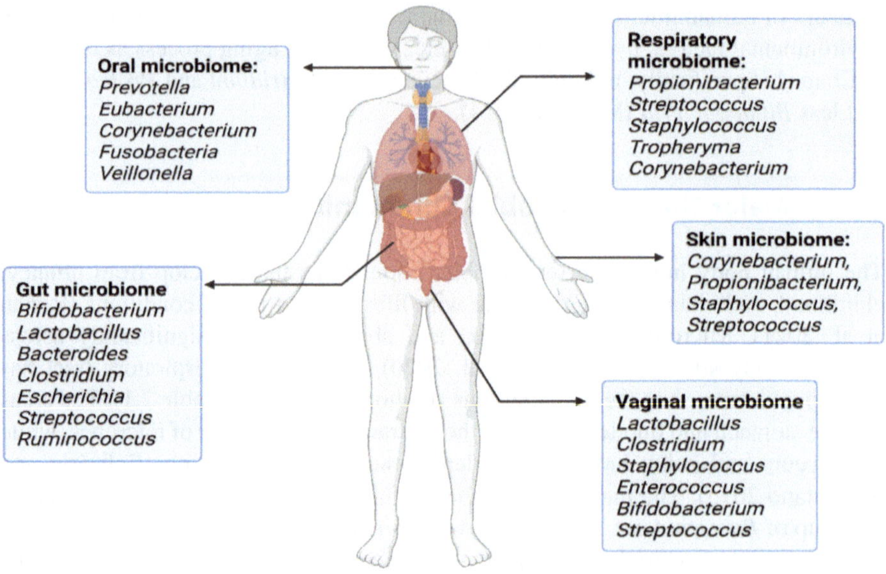

Fig. 2.4 Major human microbiome composition in different anatomical sites

composition of microbes varies dramatically depending on humidity at different locations on the body, predominantly including *Firmicutes, Actinobacteria, and Proteobacteria*, while *Bacteroidetes* are less common depending on skin site (Wang et al. 2017). In the GI tract, Proteobacteria, Firmicutes, and Bacteroidetes dominate, whereas *Firmicutes* are present in the majority in the vagina (Fig. 2.4) (Han et al. 2024). The gut microbiome in infants varies remarkably based on the timing of bacterial species acquisition.

The gastrointestinal tract of infants is initially occupied by facultative bacteria like *Staphylococcus spp, Enterococcus spp, and E. coli* (Rodríguez et al. 2015). After birth, when oxygen is available, these bacteria create an anaerobic habitat by using the available oxygen inside the gut. This anaerobic shift, when combined with breast milk present, favors the development of anaerobic bacteria like *Bifidobacterium and Bacteroides* as some oligosaccharides were present inside milk. The gastrointestinal tract also contains rare, phylogenetically diverse taxa that may play key roles in physiological functions and diseases (Bhute et al. 2017).

By the age of one year, the core human microbiome is developed, but it continues evolving throughout due to various factors (Dekaboruah et al. 2020). Many microbes present on different sites on human are well acquired to the specific habitat of where they are present, like *Candida* species are acidophiles and are found in vagina's acidic habitat (Dhakar and Pandey 2016). Also, *Propionibacterium* species which are lipophilic are present on sebaceous site on skin and *Corynebacterium and Staphylococcus* species are present on feet and elbow bends as they grow in a humid environment (Byrd et al. 2018). It implies that organisms which are found in different locations in human body corresponding to similar condition in their natural habitats therefore have similar adaptation mechanism.

2.3.1 Skin Microbiome

The skin protects the human cells by serving as a barrier between the environment and the human cells and consists of the dermis and the epidermis. Various microorganisms are present on the skin surface like bacteria, viruses, and fungi (Fig. 2.5). *Malassezia* spp. is one of the fungi species distributed along the skin surface. The microbial communities on the skin are mainly represented by bacteria from many phyla, like *Proteobacteria, Firmicutes, and Actinobacteria* (Egert and Simmering 2016). The key genera from these phyla residing on human skin are *Streptococcus, Staphylococcus, Propionibacterium, and Corynebacterium* (Lunjani et al. 2019). These bacteria are distributed along the skin based on their environmental needs; for example, the anaerobic bacterial species *Propionibacterium* are found in sebaceous glands because of the anaerobic environment. The skin is considered to be ranks fourth for the highest bacterial count.

Before birth, the skin of the baby is sterile inside the uterus, so the microbes start to colonize the skin just after birth until they reach an equilibrium. The baby skin microbiota is affected by many factors, including the delivery type. The baby born

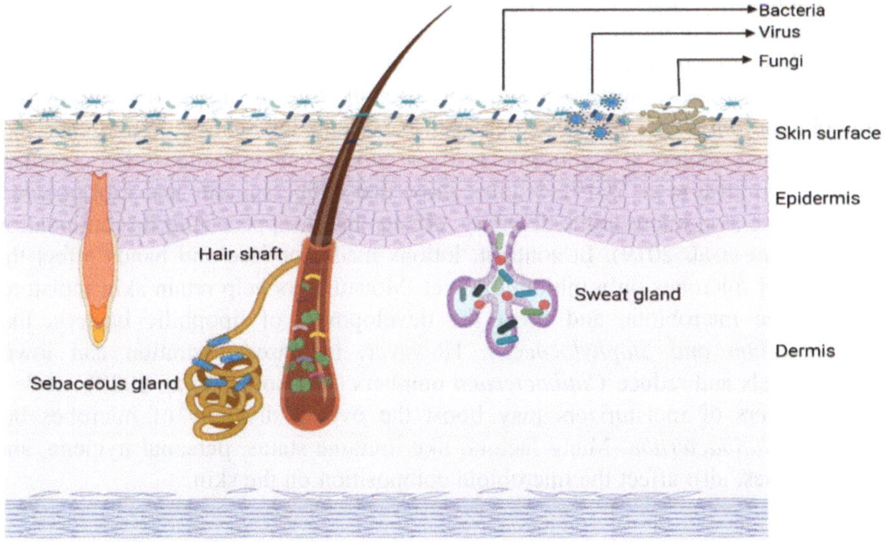

Fig. 2.5 Skin surface, glands, and hair follicles are covered with various microorganism like bacteria, virus, and fungi. Skin surface contains many microorganisms like *Staphylococcus* spp. and Proteobacteria (Bacteria), *Malassezia* spp. (Fungi) and virus live in both bacteria cells and freely on surface. Sebaceous gland is associated with lipophilic bacteria like *Propionibacterium* whereas *Cutibacterium* and *Corynebacterium* are present in axilla in sweat gland giving sweat its odor

from the birth canal naturally has the skin microbiota resembling that of their mother's vaginal flora, mainly represented by *Prevotella and Lactobacillus* species. While the baby born via C-section has the skin microbiota resembling that of their mother's skin, mainly represented by species like *Propionibacterium, Corynebacterium, and Staphylococcus* (Zhang et al. 2021). The skin has different microbes on their surface, which are distributed unevenly. Different studies have been conducted to identify these bacteria; one of these studies found that the skin on the front of the body is colonized by *Proteobacteria, Corynebacterium, and Propionibacterium*, while the skin on the back is colonized by *Propionibacterium, Corynebacterium, and Staphylococcus* (Egert and Simmering 2016). Skin microbiota is also found to be different between females and males. Male and female skin are different physiologically because of skin pH, perspiration rate, and hormonal difference. Fierer et al. 2008 found significant differences in hand microbiota, with women showing greater alpha diversity, while men had higher abundances of *Corynebacteriuma and Cutibacterium*. Conversely, women had more *Lactobacillaceae, Moraxellaceae, Enterobacteriales, and Pseudomonadaceae*. Oh et al. 2012 found no sex-associated differences in microbiota among individuals aged 2 to 40. Callewaert et al. 2013 distinguished two main microbiota groups in the axillary vault, with females clustering with *Staphylococcus* and males with *Corynebacterium*. Other studies by Prohic et al. 2014 and Leung et al. 2015 observed no significant effects of sex on certain microorganisms, although distinct patterns emerged in specific age groups and skin sites, such as males having higher species richness and unique *Corynebacterium* species. Overall, sex differences in skin microbiota appear to vary based on age, skin location, and specific taxa. This difference is because of many factors like sebum production, hormones, and makeup use (Robert et al. 2022). Cosmetics are used to increase the quality of skin and slow aging, diversifying the microbiome on skin with its regular use. Cosmetic's active ingredients like N-acetylglucosamine, a hyaluronic acid precursor, can influence microorganism growth (Dimitriu et al. 2019). Studies show that antiperspirants and foot powders can enhance skin microbiome diversity, effects that disappear after discontinuation (Bouslimani et al. 2019). In contrast, lotions made for face and hands affect the diversity of microbes on a minimum level. Moisturizers help retain skin moisture, support the microbiota, and favor the development of lipophilic bacteria like *Cutibacterium and Staphylococcus*. However, increased hydration can lower sebum levels and reduce *Cutibacterium* numbers (Moskovicz et al. 2020), while a few numbers of moisturizers may boost the overall diversity of microbes but decrease *Cutibacterium*. Many factors, like immune status, personal hygiene, and skin diseases, also affect the microbiota composition on the skin.

2.3.2 Oral Microbiota

The oral cavity is home to both hard and mucous surfaces (i.e., lips, teeth, cheeks, tongue, and palate), which are constantly moistened by saliva and gingival fluid, which nourishes the diverse microbiome. The oral microbiota contains more than

700 microbial species and is estimated to be the second most complex microbiota niche after the gut (Zhang et al. 2018; Koliarakis et al. 2019). Numerous studies have been done to classify the bacterial phyla present in the mouth, with fewer studies suggesting that the key phyla representing the oral microbiome are *Synergistetes, Proteobacteria, Firmicutes, Bacteroidetes, and Actinobacteria* as key phyla, while others highlight *Spirochaetes and Fusobacteria* as key phyla (Siqueira and Rôças 2017).

The oral microbiota consists of various species, either common or site-specific. The common microbiota is represented with *Granulicatella, Streptococcus, Veillonella, Gemmella, and Fusobacterium*, while others like *Corynebacterium, Neisseria, Prevotella, Pasterutella, and Bacteroidetes* are site-specific (Fig. 2.6) (Li et al. 2022). The oral cavity contains many parts, such as soft tissues (tongue, palate, cheek), saliva, and hard tissues (teeth), each supporting different microbial populations (Zhang et al. 2018). Saliva mainly hosts *Prevotella, Veillonella, and Streptococcus*, while soft tissues are home to *Eubacterium, Rothia, and Streptococcus salivarius*. Teeth harbor *Actinomyces and Corynebacterium* in the supragingival area, and anaerobic species in the subgingival region such as *Bacteroidetes, Proteobacteria, Actinobacteria, Fusobacteria, and Spirochaetes* genera.

Aggregatibacter actinomycetemcomitans and *Porphyromonas gingivalis* are found to be predominant in the gingival sulcus in the first month of pregnancy while *Veillonella parvula, Staphylococcus oralis, Fusobacterium nucleatum naviforme, Capnocytophaga sputigena, Leptotrichia buccalis, Prevotella intermedia, Staphylococcus aureus,* and *Streptococcus sanguinis*, and are present

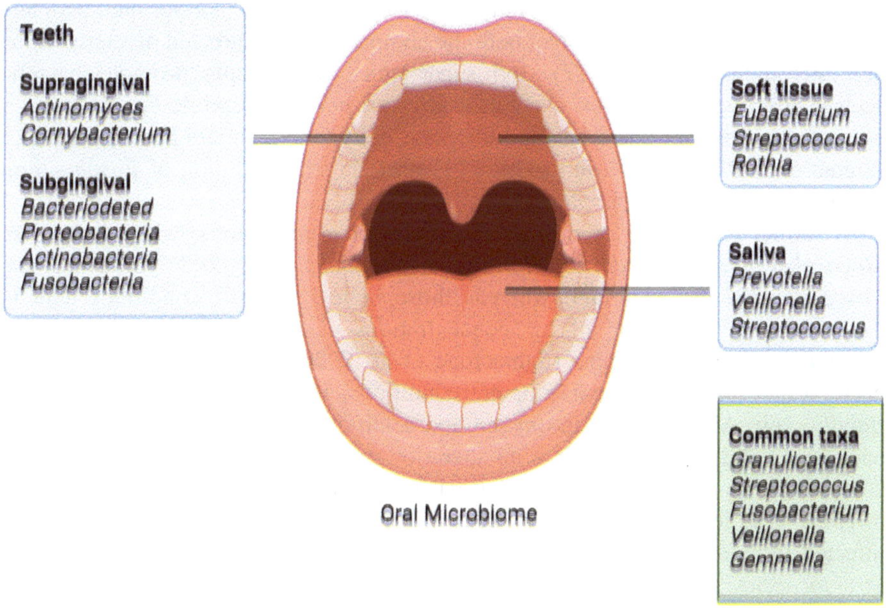

Fig. 2.6 Major human oral microbiome

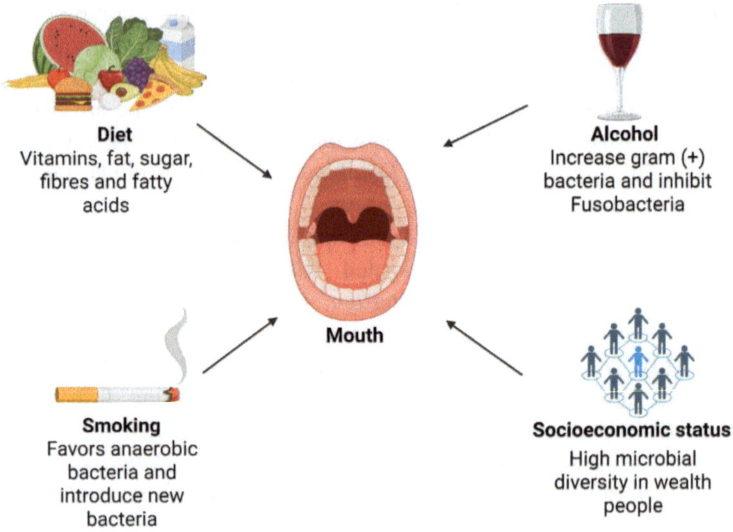

Fig. 2.7 Factors affecting the oral microbiome in humans

more in the last trimester (Saadaoui et al. 2021). During early childhood, the oral microbiota are predominant by *Streptococcus* spp. which protects against pathogenic strains by producing bacteriocins.

Factors like age, diet, alcohol use, smoking, social status, and genetics significantly affect microbial composition (Fig. 2.7). Nutrient which affects the most are vitamins, fats, and sugars. Fibers, fatty acids (medium-chain, monounsaturated, and polyunsaturated fatty acids) affect both the community structure and diversity of the oral microbiota (Li et al. 2022). High sugar intake, for example, promotes bacteria such as *Streptococcus mutans and Fusobacterium nucleatum*, while *Fusobacteria* is promoted by vitamin c and fats and fiber and dairy help maintain bacterial balance (Sultan et al. 2018). Increase in Gram (+) bacteria and inhibition of *Fusobacteria* activity are observed in people consuming alcohol.

Wealthier individuals often have a more diverse microbiome with species like *Megasphaera micronuciformis, Rothia mucilaginosa, Tannerella forsythia, and Fusobacterium periodontium*, whereas those in less well-off conditions tend to have lower diversity and higher concentrations of *Aggregatibacter segnis and Achromobacter xylosoxidans* (Santacroce et al. 2020). Smoking introduces new bacteria and alters oral environmental conditions, favoring *Clostridium* spp. *and Bacillus* spp. Also decrease in oxygen and acidification of saliva favors anaerobic bacteria. Oral dysbiosis, which is an alteration in microbial communities, has been associated with diseases; oral candidiasis is linked with increased *Lactobacillus* and *Streptococcus* levels (Lamont et al. 2018). Human immunodeficiency virus (HIV)-infected individuals tend to show lower microbial diversity and higher levels of *Streptococcus, Veillonella, and Rothia* in their mouth (Huang and Segal 2020).

2.3.3 Gut Microbiota

The gut microbiota is the most comprehensively researched, encompassing about 70% of the total microbiota (Thursby and Juge 2017). Its composition and relative abundance vary along the GI tract (Pascale et al. 2018) due to different immunological, nutritional, and chemical factors in the digestive tract (Thursby and Juge 2017). The gut microbiota and the host work together in a way that benefits each other. This mutualism is formed by a set of interactions that include signals related to metabolism, immune responses, and hormones. Microbes in the gut produce substances that serve as signals affecting how the body's immune and nervous systems function. This can create connections between the gut and other organs in the body. The microbiota starts colonizing in the gut a few moments after birth and is initially influenced by early life events (Mogoş et al. 2025), including delivery method and breastfeeding. The digestive tract is first colonized with facultative anaerobes before shifting to anaerobic species. While throughout the GI tract, the largest concentration of bacteria is found in the colon (Rosenberg 2024), primarily from the *Firmicutes and Bacteroidetes* phyla present in majority (Pascale et al. 2018). Key genera representing gut microbiota include *Streptococcus, Escherichia, Clostridium, Bacteroides, Lactobacillus, Bifidobacterium, and Ruminococcus.*

Various factors influence gut microbiota composition, including the geography (Kostic et al. 2013), host genome, psychological stress, circadian rhythms, diet, and physical activity (Smith et al. 2019; Karl et al. 2018). Amongst these, diet plays the most vital role in determining the gut microbiome. It profoundly affects the microbiota's composition and diversity (Rinninella et al. 2023) through dietary patterns, macronutrients, bioactive components, and the use of probiotics, which enhance host health (Kolodziejczyk et al. 2019). Studies show that microbes inside the gut react to fasting and malnutrition (Paukkonen et al. 2024). Research on the effects of malnutrition has mainly been conducted on lab animals due to ethical constraints. For instance, nutrient deficiency over several weeks diversifies the microbial composition in many animals, like fish and mice, but decreases in geckos, with no change in quails. This variability complicates the study of nutrient deficiency in humans, who experience it only in specific cases. Malnutrition is a leading cause of child mortality, and while foods that are rich in nutrient aid recovery, children often struggle to regain body mass completely due to their developing microbiomes. The early growth of the microbes in the gut is significant, as their composition evolves with age and dietary changes (Derrien et al. 2019).

The human gut microbiome has increasingly shifted from descriptive studies to translational applications with direct clinical relevance. A major focus has been on oncology, where accumulating evidence indicates that the gut microbiome modulates responses to immune checkpoint inhibitors, leading to clinical trials exploring fecal microbiota transplantation (FMT), live biotherapeutics, probiotics, and dietary interventions as adjuvant strategies to improve therapeutic efficacy and mitigate toxicity (Jian and Wei 2025; Shaikh et al. 2025). At the same time, precision nutrition has advanced considerably, with randomized controlled trials demonstrating that microbiome-informed, personalized diets outperform standard dietary recommendations in improving cardiometabolic risk and alleviating symptoms in irritable bowel syndrome patients (Sanna Nybacka et al. 2022; Martínez-González et al. 2025). While FMT remains well established for recurrent *Clostridioides difficile* infection, it is now being tested in ulcerative colitis and supportive cancer care, with new emphasis on donor standardization, safety monitoring, and microbial engraftment tracking to ensure predictable outcomes (Porcari et al. 2023; Allegretti et al. 2025). Methodological innovations, such as genome-resolved metagenomics, strain-level analysis, and integration with metabolomics and host multiomics, are enabling causal insights into microbe–metabolite–immune interactions underlying disease phenotypes (Sarkar et al. 2025). Beyond bacterial communities, recent large-scale studies have expanded catalogs of the gut virome; notably, a 2024 resource of the early-life gut virome assembled over 160,000 nonredundant viral sequences, providing insights into how viral communities shape microbiome succession and immune development (Liu et al. 2024). Taken together, these advances illustrate a paradigm shift in gut microbiome science toward precision interventions, where dietary modulation, microbial therapies, and metabolite-based strategies are now being prospectively tested in rigorous clinical trials.

2.3.4 Respiratory Microbiome

The human respiratory system consists of the respiratory tract, which is divided into two zones, and its colonization with microbes starts in utero. The conducting portion is represented by the nose, pharynx, larynx, trachea, and bronchi, and it is rich in microorganisms, while the respiratory portion is typically sterile (Zhao et al. 2023). The upper respiratory path, consisting of the pharynx and nasal passages, transports air to the lower respiratory path (trachea, bronchi, and lungs) and is directly exposed to the environment. Despite mucosal immune defenses, various microbial species coexist and affect human health (Lee et al. 2019). The nasal cavity hosts microbes like *Corynebacterium* spp., *Dolosigranulum* spp., *and Moraxella* spp., while the *Streptococcus spp* is found to be predominant in the oropharynx. This microbiota can help prevent pathogen entry from the lower respiratory pathway. Dysbiosis in the upper airway is linked to inflammation and obstruction in diseases such as asthma, cystic fibrosis, COPD, and idiopathic pulmonary fibrosis. Monitoring changes in the upper airway microbiome could provide insights into lung disease. The lower airway is generally considered an unfavorable environment for microbial survival, resembling a desert due to limited nutrients and high oxygen levels. The lower airway was once considered sterile, but recent research reveals the presence of many microbial species (Santacroce et al. 2020), primarily anaerobes like Gram-negative *Prevotella and Veillonella* species, as well as Gram-positive *Coprococcus and Dorea* species (Tong et al. 2019), in healthy lungs. Two models describe the lower airway microbiome: the adapted island model, which highlights the influence of microbial immigration from the upper respiratory path and the decrease of microbial population from upper to lower respiratory path. The second model is the neutral model, which suggests that various microbial communities are formed by random population dynamics. In summary, the healthy human lower airway contains a wide microbial composition immigrated from the upper airway and oral cavity. The respiratory tract is commonly represented by *Pseudomonas aeruginosa, Mycoplasma pneumoniae, Neisseria meningitidis, Klebsiella* spp., *Bordetella pertussis, Mycobacterium tuberculosis, Haemophilus* influenzae, *Staphylococcus aureus, Corynebacterium diphtheriae, Streptococcus pneumoniae, Streptococcus viridans,* and *Streptococcus pyogenes* (Kelly et al. 2016). Additionally, viruses (like Adenovirus and Influenza virus) and fungi (such as *Aspergillus and Candida* species) also associate with the respiratory tract (Dickson et al. 2016).

Recent large-scale studies show that when microbial diversity in the nose and throat decreases, people are more likely to develop illnesses such as asthma, influenza, Respiratory Syncytial Virus (RSV), or pneumonia, and that these patterns can differ with age, geography, and even the part of the airway that is sampled (Koceva et al. 2025). Other findings have shown that the microbiome of the nasopharynx continues to change well into young adulthood, usually stabilizing between ages 15 and 24, and that beneficial bacteria like *Dolosigranulum pigrum* and *Corynebacterium* appear to help guard against respiratory infections (Hurst et al. 2022). Environmental influences are pollution, climate conditions, and indoor air quality as key factors shaping the composition of the respiratory microbiome (Wang et al. 2025). Moving beyond just identifying microbes, workers are now combining microbiome data with host gene expression, and this approach has already been shown to improve the accuracy of diagnosing lower respiratory tract infections (Zhao et al. 2025). Together, these findings highlight the respiratory microbiome as a sensitive, evolving ecosystem that not only reflects health status but could soon become a target for personalized diagnostics and treatments.

2.3.5 Urogenital Microbiome

The urinary system of human is represented by the ureters, paired kidneys, urethra, and bladder. The anatomy and microbial composition of the urogenital system differ significantly between the sexes (Neugent et al. 2020). The study of urogenital microbiota of healthy individuals has helped us to understand urinary tract infections (UTIs) and their recurrence mechanisms (Neugent et al. 2020). Suprapubic aspiration is used to collect urine samples, which are analyzed and used to demonstrate the bladder microbiome and its bacterial composition closely resemble that obtained through transurethral catheterization. Significant similarities between female vaginal and urinary microbiomes suggest a possible connection between the two. Research by Thomas-White and colleagues identified shared strains such as *Lactobacillus crispatus and Escherichia coli* in paired vaginal and bladder samples from women with urinary incontinence (Thomas-White et al. 2018). In females, the kidneys and ureters are typically sterile, but the microorganisms are colonized in the urethra. The urethra is colonized by *Lactobacillus* spp., *Bacteroides* spp., *coagulase-negative staphylococci, Burkholderia* spp., *Enterobacteriaceae (Enterobacter, Proteus),*

Corynebacterium spp., *Escherichia coli, Gram-positive anaerobic cocci including Sarcina, Coprococcus, Ruminococcus, Micromonas, Peptoniphilus, Anaerococcus, Peptostreptococcus, Peptococcus*, etc. (Sheerin 2011). The urethral composition of microbes is less diverse in males, primarily featuring coagulase-negative *staphylococci and Corynebacterium* spp.

The vaginal microbiota plays an important part in many functions inside the human body like nutrition, physiology, and immunity, existing in a mutualistic relationship but also has a few disease-causing opportunistic pathogens. Vaginal microbiome is primarily comprised of beneficial bacteria like *Lactobacillus* spp., *Mycoplasma* spp., *Prevotella* spp., *Fusobacterium* spp., *Veillonella* spp., *Clostridium* spp., *Porphyromonas* spp., *Propionibacterium* spp., *Corynebacterium* spp., *Streptococcus* spp., *Bifidobacterium* spp., *Enterococcus* spp., *Staphylococcus* spp. (Dhakar and Pandey 2016). The microbes play a crucial part in providing colonization resistance against infections, like uropathogens in UTIs. Urothelial glycosaminoglycans (GAGs) are believed to be important for the urinary microbiota. A recent study on postmenopausal women found that urinary chondroitin sulfate was significantly elevated during active UTIs and negatively correlated with the abundance of specific bacteria (Neugent et al. 2023). This suggests that active UTIs damage the GAG layer and alter the urinary microbiome, potentially affecting functions like antibiotic resistance and metabolism. A 2019 study showed that metagenomics can accurately detect antibiotic resistance associated with urosepsis (Barraud et al. 2019), identifying resistance to multiple antibiotic classes. Additionally, a history of recurrent UTIs may enrich the urinary microbiome with antibiotic resistance genes, leaving a lasting impact even when no active infection is present.

The human urogenital microbiome has moved beyond the earlier assumption of a sterile urinary tract, showing that both men and women harbor diverse microbial communities that play important roles in health. Longitudinal studies in women undergoing pelvic floor surgery indicate that while surgical intervention temporarily disturbs the microbiome, recovery typically occurs within several weeks, although postmenopausal women demonstrate greater microbial diversity and less stability than premenopausal women (Kim et al. 2023). Advances in sequencing technologies, including nanopore-based approaches, have confirmed significant overlap between the urinary and vaginal microbiomes, with both niches frequently dominated by *Lactobacillus* spp. or *Gardnerella* spp. (Ferneyhough et al. 2025). At the same time, mounting evidence supports the existence of a resident "urobiome," in which species such as *Escherichia coli*, *Enterococcus faecalis*, and *Aerococcus urinae* may persist without causing disease, suggesting a homeostatic role in urinary health (Wu et al. 2025). Dysbiosis in these microbial communities is increasingly linked to oncological outcomes, with reviews noting potential roles of the urogenital microbiome in bladder, kidney, cervical, and prostate cancers through pathways involving hormone regulation, chronic inflammation, and microbial metabolites (Chen et al. 2025). Emerging clinical data also connect shifts in the urinary microbiome with male urogenital disorders such as benign prostatic hyperplasia and lower urinary tract symptoms, where organisms like *E. coli* and *Burkholderia* may drive chronic inflammation and disease progression (Kim and Jung 2021).

Together, these findings suggest that therapeutic targeting of the urogenital microbiome could become a viable strategy for improving genitourinary health in both sexes.

2.4 The Microbiota-Gut-Brain Axis

Our understanding of the vital roles of the gut–brain axis in sustaining homeostasis was strengthened following the initiation of brain imaging techniques in the 1980s (Cryan et al. 2019). Preclinical sign suggests that it can utilize these processes, part of the brain–gut–microbiota axis, to modulate behavior, brain development, and function. The gut microbiota affects the brain via immunological, neurological, and endocrine pathways, making it a crucial regulator of the gut-brain axis. It achieves this by modifying neurotransmitters, like gamma-aminobutyric acid (GABA) and serotonin (5-HT), and neurotransmitter precursors such as tryptophan (Fig. 2.8). Furthermore, the gut microbiota also modulates the brain by regulating vital proteins and metabolites associated with neuropeptides and gut hormone secretion (Agirman and Hsiao 2021). The neurological contact between the gut bacteria and the central nervous system is mediated by the vagus nerve and spinal afferent pathways. The gut and brain can communicate directly and reciprocally thanks to the gastrointestinal

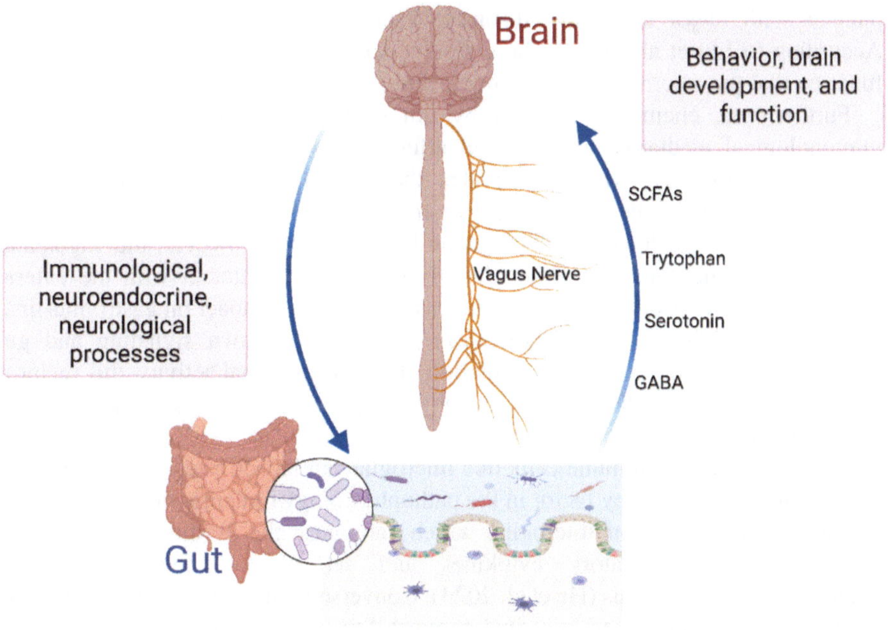

Fig. 2.8 The microbiota–gut–brain axis. The microbiota affects the brain via immunological, neurological, and endocrine pathways through the gamma-aminobutyric acid (GABA) and serotonin, metabolites like tryptophan, and SCFAs

branches of the vagus nerve. Since vagus nerve stimulation is linked to serotonin (5-HT) release because it directly projects to 5-HT neuronal centers, fluctuations in the gut microbiota must preserve vagal function to cause depression-like behaviors in mice (Siopi et al. 2023).

Additionally, gut bacteria trigger the cytokines release which then simultaneously affect immunological signaling to the brain received from the gut. Proinflammatory cytokines like interleukin-1β (IL-1β), interleukin-6 (IL-6), interleukin-4 (IL-4), and tumor necrosis factor-α (TNF-α) can be released in response to an unbalanced gut flora. These cytokines can damage intestinal epithelial cells' tight junctions, jeopardize the blood-brain and gut vascular barriers, and impact brain function (Agirman et al. 2021). These complex relationships of the gut, gut microbiota, and brain highlight how psychological stress and other stressors affect the balance of the gut microbiota.

Several animal models of depression caused by diverse techniques have shown changes in gut flora. For example, dysbiosis of the phylum Firmicutes was observed in the depressive-like macaque model, which was associated with depression-like symptoms (Zheng et al. 2021), and gut microbiota diversity and richness were changed in the chronic unpredictable mild stress mouse model (Li et al. 2019). Furthermore, rodents that get a transplant of "depression-associated microbiota" may exhibit depressive-like symptoms. For instance, fecal matter transplantation (FMT) from healthy mice reduced depression-like behaviors in recipient mice. But when germ-free (GF) or pseudo-germ-free mice were infected with FMT from patients with major depressive disorder, it promoted depression-like behaviors. According to Liu et al. (2023), these results demonstrate the critical and complex function that gut microbiota and its microecology play in the onset of depression.

Furthermore, chemicals released by gut bacteria, regulatory hormones, and immunological mediators affect spinal afferent nerves, enteric neurons, and the vagus nerve directly, which innervates the gut. By flowing through sensory pathways to areas of the brain related to emotion, bodily sensation, eating habits, and cognition, these local signals regulate brain activity. However, the vagus and spinal efferent neurons that either directly or indirectly interact with the enteric nervous system to project to the gut mucosa also have an impact on gastrointestinal homeostasis (Agirman and Hsiao 2021). Through top-down signaling and gut microbiota-related biochemical modulation of enteric neuronal activity, this reciprocal connection eventually impacts gut physiology, local immunological function, and gut microbiota composition.

Moreover, central immune cells like microglia can be activated by variations in the gut microbiome. A key factor in the pathophysiology of depression is thought to be the overstimulation of microglia, which can cause the production of copious quantities of inflammatory cytokines and set off a chain reaction of neuroinflammatory signals (He et al. 2024). Conversely, individuals with depression frequently exhibit amplified intestinal permeability, also recognized as "leaky gut." Systemic inflammatory reactions are exacerbated by the translocation of bacterial metabolites, cellular components, and even entire bacteria into the systemic circulation due to the weakened gut barrier. The pathophysiology of depression is strongly linked to the reuptake, production, metabolism, and receptor expression of neurotransmitters, all of which can be affected by these inflammatory processes (Singh et al. 2024).

Scientists uncovered what might be considered a true "sixth sense" residing in our gut a real-time neural pathway in which the gut communicates with the brain to regulate appetite. Specialized cells in the colon, called neuropods, were found to detect a bacterial protein known as flagellin through the immune receptor TLR5, and relay that information directly to the brain via the vagus nerve. In mouse experiments, flagellin triggered a sense of fullness and curbed feeding, whereas animals lacking TLR5 failed to register the signal, overate, and gained weight (Liu et al. 2025). This finding redefines how we think about gut–brain communication—no longer is it only about hormonal or immune signaling, but rather a sensory-like neural route that responds instantly to what is happening in the gut. It opens up compelling possibilities for treating obesity and other disorders by targeting this neural pathway.

2.5 Role in Metabolism, Digestion, and Immune Function

The human gut microbiome is very complex and consists of many microbial species; many of these bacteria are not being cultured. Approximately 160 species are typically found in an individual's gut (Rajilić-Stojanović and De Vos 2014). The gut microbiome comprises around three million genes, significantly exceeding the human genome. This diverse microbial community plays a crucial role in metabolism by contributing enzymes for breaking down polyphenols, polysaccharides, and synthesizing vitamins functions not encoded by human genes, as microbiota have an extensive metabolic repertoire complementing mammalian enzyme activity in gut microbiota and liver (Nicholson et al. 2012). Studies comparing microbiota of patients with healthy individuals indicate that the microbiota is linked to various gastrointestinal diseases, with emerging evidence suggesting its involvement in diabetes and obesity as well (Marchesi et al. 2016).

2.5.1 Digestion

The human microbiome is an important part of our body as it takes part in many metabolic activities essential for gut and liver function, influencing health by altering the biochemical profile. Research highlights the significance of gut microbiota in immunity and their role in metabolic pathways to metabolize dietary components (Cardona and Roman 2022) and phytochemicals, particularly polyphenols (Rowland et al. 2018). Studies show that gut microbes rapidly adapt to significant changes in diet, such as shifting between plant and meat-based diets or altering fiber intake, with changes in composition observed within 1–2 days. This adaptability may reflect the various and competitive nature of the microbial composition, aiding in the protein and peptide degradation through fermentation. Bacteria in the upper digestive tract are dependent on undigested dietary substances for survival in the large intestine. Saccharolytic fermentation produces beneficial metabolites, but with limited sugar, bacteria use different energy resource, leading to the production of potentially harmful metabolites. Key products of dietary carbohydrate fermentation include short-chain fatty acids and gases.

Macfarlane and Cummings (1986) discovered that the gut microbiome efficiently breaks down dietary and host proteins with proteolytic power into peptides, amino acids, and gases like ammonia and Carbon dioxide. They identified *Propionibacterium and Bacteroides* as the main proteolytic species in fecal samples, with other contributors including *Bacillus and clostridia* and species. Gibson et al. (1989) later noted that fecal microbiota exhibited different proteolytic activity than that in the ileum, effectively degrading bovine serum albumin despite lower overall activity. In 1996, Macfarlane team explored amino acid fermentation processes in gut bacteria. Recently, it was discovered that tyrosine and phenylalanine (aromatic amino acids) can be fermented into phenolic metabolites by bacteria, including species of *Clostridium barlettii, Eubacterium hallii, and Bacteroides* which are also produced from plant polyphenol breakdown.

2.5.2 Metabolism

For over four decades, research using both conventional and germfree rodents has demonstrated that colonic microbiota can synthesize important vitamins, mainly those from the B and K groups, including folates, thiamine, riboflavin, and nicotinic acid. Gut microbiome synthesizes micronutrients such as vitamins, play an important part in both microbial and host metabolism. Bacteria like *Bacteroides fragilis* and *Eubacterium lentum* produce vitamin K2, which helps reduce vascular calcification and lower cholesterol, thereby decreasing the risk of cardiovascular disorders. Additionally, gut microbiota provides important vitamins B, particularly B5 and B12, which are key coenzymes in various biochemical processes essential for nervous system function (Wan et al. 2022). Deficiencies in these vitamins are related to gastrointestinal issues, insomnia, and neuropsychological disorders (Gominak 2016). However, the connection between the loss of vitamin-producing gut bacteria and disease onset remains unclear. Studies have demonstrated that germ-free rats lacking vitamin K supplementation develop low prothrombin levels, while conventional rats do not. Human subjects on restricted vitamin K diets showed no deficiency, but those treated with antibiotics had reduced plasma prothrombin levels. Recent metagenomic sequencing revealed that certain gut microbial sequences are enriched for pathways involved in vitamin synthesis. Le Blanc et al. (2013) explored the gut microbial sequences and their metabolic potential and discovered that their gut microbiomes can produce many biosynthetic pathways. The gut bacterial genome was analyzed and found that 256 of these bacteria have genomes for biosynthetic pathways for eight B vitamins, and found varying degrees of potential producers among different bacterial phyla. Bacteroidetes had the highest number of predicted pathways for vitamin B, with over 90% being potential producers for all B vitamins except B12.

Carbohydrates and simple sugars are key components that gut microbiota metabolize. Microbial species, especially Bacteroidetes and Firmicutes, ferment indigestible fibers to produce SCFAs, such as acetate, propionate, and butyrate, along with other byproducts. These SCFAs are essential for human health, typically found in a 3:1:1 to 10:2:1 molar ratio in feces (Rowland et al. 2018). Butyrate acts as the key foundation of energy for colon cells and has anti-cancer capability by promoting apoptosis in colon cancer cells and regulating gene expression (Wang et al. 2019). Propionate is vital for liver epithelial cells and gluconeogenesis, while acetate supports the development of beneficial bacteria like *Faecalibacterium prausnitzii.*

The gut microbes also take part in enhancing the metabolism of micronutrients, amino acids, and polysaccharides, thereby improving host metabolic efficiency and energy harvesting. Experiments with germ-free (GF) mice discovered that an additional 10–30% food is needed with 40% less epididymal fat to maintain body mass compared to mice with normal microbiota. Gut microbiota ferment soluble dietary fiber into SCFAs, which act as an additional energy substrate, contributing 10% of daily energy needs. These SCFAs account for 70% of ATP production in the colon, with butyrate as the main source for colonocytes, and it supports mitochondrial respiration and ATP synthesis in GF mice. Additionally, SCFAs activate the G-protein-coupled receptor 41 (GPR41) in gut epithelial cells, regulating energy homeostasis through leptin production, which affects various physiological functions including energy metabolism and appetite, highlighting the interaction between host and microbiota (GPR41 regulation).

Bile acids are key trans-genomic metabolites that are formed by the interaction of host genome and gut microbiome. A bidirectional relationship is established: gut microbiome can modify the properties and structure of bile acids, while it exerts antimicrobial effects on the microbiota and can exert selection pressure in gut microbiota community structure (Rowland et al. 2018). For instance, deoxycholic acid, a secondary bile acid derived from bacteria, is ten times more potent than its precursor, cholic acid (CA), due to its stronger detergent properties (Kurdi et al. 2006). Many bacterial species in the gut of human contains bile salt hydrolase, an enzyme that hydrolyzes the amide bond in bile acids. BSH genes are found in prominent gut bacterial genera such as *Bacteroides, Listeria, Lactobacillus, and Clostridium*, allowing them to hydrolyze both glyco and tauro-conjugates.

The gut microbiome also enhances the host's capacity to metabolize various dietary substrates, particularly carbohydrates, proteins, and vitamins (Fig. 2.9). The microbiota's flexible metabolism adapts to the availability of these substances, with well-studied pathways for processing them. Notably, the metabolism of phytochemicals, especially polyphenols, varies widely because of different composition of microbes in individuals, impacting health effects. Additionally, the dynamic interactions between microbial metabolites and the host complicate the measurement of metabolite concentrations over time. Utilizing stable isotope probes along with mathematical modeling may provide more insights in this area.

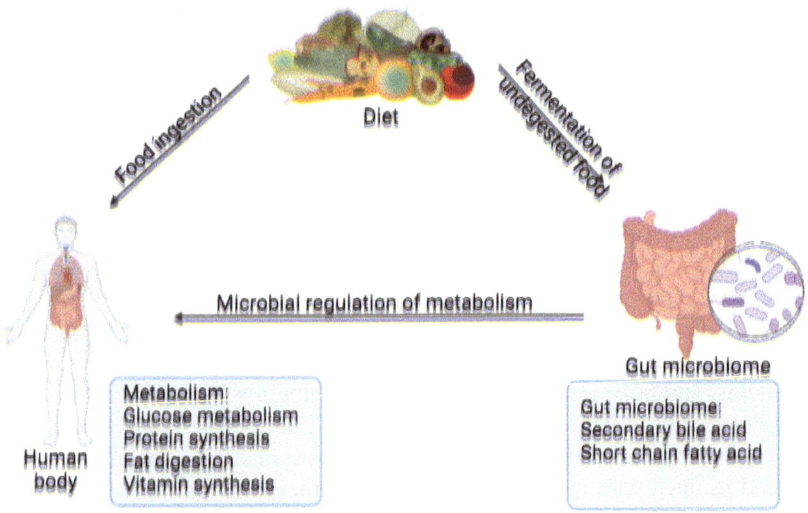

Fig. 2.9 Host metabolism regulated by gut microbiota: gut microbiota releases microbial metabolites like secondary bile acid and short-chain fatty acid which in turn influence key metabolic processes like glucose metabolism, protein and vitamin synthesis via metabolically active hormones

2.5.3 Immune System

The adaptive and innate immune responses, which form the human immune system, have been found to have extensive interaction with microbiota. The innate immune system is essential for maintaining a homeostatic state because it eradicates harmful microbes and alters the response of adaptive immunity to microbiota. Early exposure to a wide variety of microorganisms is crucial for the immune system's healthy development because it activates immune regulatory pathways, especially those that include Toll-like receptors (TLRs), which promote a balanced immunological response. It is believed that this mechanism encourages the production of regulatory T cells (Treg), which reduce overactive immune responses and may lower the risk of developing autoimmune and allergy illnesses by producing anti-inflammatory cytokines, including TGF-β and IL-10 (Ma et al. 2024).

Autophagy, Toll-like receptor 5 (TLR5), inflammasomes, and secretory IgA (sIgA) are some of the mechanisms through which these effects are created. Commensal gut bacteria like *B. fragilis* are an intricate part of the development of innate immunity as well as the recognition of pathogens by triggering reactive oxygen species production, apoptosis, and Toll-like receptor (TLR) signaling (Vasquez Ayala et al. 2023). Commensal bacteria and slgA can bind and become complexed, presenting the bacterial components to tolerogenic dendritic cells in a selective manner. Since slgA is an anti-inflammatory substance, it inhibits the inflammatory reaction that the organs' thriving bacterial load would cause. However, microbiota dysbiosis can cause unchecked microbial development by interfering with the sIgA defense (Fig. 2.10). Hapfelmeier and his colleagues used a reversible microbial

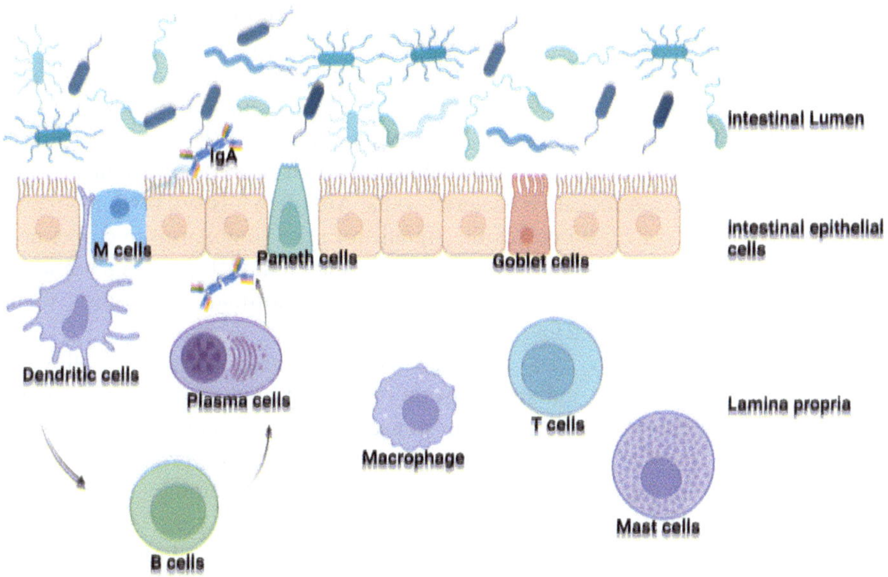

Fig. 2.10 The gut microbiota and host immune system

colonization system to demonstrate that germ-free (GF) mice exhibited a microbiota-specific sIgA response. The adaptive immune system's response is another essential component in preserving a healthy microbiota and balanced immune system (Hapfelmeier et al. 2010).

The gut microbiota controls the immune response in one way by modulating the migration of neutrophils, which in turn controls T cell differentiation into different types, such as regulatory T cells and helper T cells (Th1, Th2, and Th17) (Owaga et al. 2015). When the immune system matures, anomalies in microbiota development can lead to autoimmune illness and aberrant immunological tolerance. Besides, the many compounds produced by the microbiota can trigger an immunological response, trigger inflammation, or produce long-term tissue damage (Hou et al. 2022). The adaptive immune response is significantly shaped by the growth and differentiation of B and T cells as well as the establishment of immunological tolerance to the microbiota. The bacterium and CD4 + T cells react quite differently, which causes the cells to differentiate into many subsets and subsequently produce cytokines that encourage inflammation, such as interleukin IL-17A and interferon-γ (Hou et al. 2022).

Early colonization by these commensal bacteria can have significant, lasting impacts on the capacity on inflammatory disease susceptibility. Intestinal epithelial cell (IEC) growth, GI mucosal angiogenesis development, and GI tract lymphoid structure formation, like Peyer's patches and solitary lymphoid follicles, can all be influenced by commensal bacteria (Belkaid and Hand 2014; Borbet et al. 2023). Gut-associated lymphoid tissue (GALT) and gut microbial communicate by regulating inflammation by the early gut microbiota and influence the way adaptive immunity is educated (Belkaid and Hand 2014; Borbet et al. 2023).

The detection of conserved molecular patterns associated with microorganisms is the key channel of communication among the microbiota and the host. More specifically, epithelial cells of the gut become resistant to TLR stimulation subsequently due to an early reaction to microbial ligands such as lipopolysaccharides, the outer membrane of Gram-negative bacterial walls. Although it is still not known how the innate immune system processes the signal from the commensal microbe, new evidence reveals that the secretion of enzymes that remodel the epigenome by epithelial cells is needed for commensal-dependent intestinal homeostasis to be orchestrated (Sarkar et al. 2021).

Enteric microorganisms can stimulate Gram-negative bacterial walls and macromolecules, as well as antigens outside of the gut epithelium, establishing a strong correlation between the immune response and the gut microbial community. The major constituent of the bacterial flagellum, flagellin, the primary structural protein of bacterial flagella, has a significant role in maintaining gut epithelial integrity through TLR5 recognition on B cells and CD4+ T cells. Secreted IgA neutralizes pathogens, and within GALT including plasma cells, lymphocytes, and Peyer's patches microbiota help mature the immune system. Mucosal antibodies captured by CD11+ dendritic cells in Peyer's patches often bind luminal microbes. CD8+ T lymphocytes, abundant in the intraepithelial space, rely on the microbiota for functional maintenance. The reduced number of intestinal CD8+ T cells in GF mice supports this observation (Fig. 2.10) (Hou et al. 2022).

Besides, commensals facilitate the establishment of the immune system postnatally, which represses them. Experiments on GF animals animals reared without living microbes demonstrated that the microbiota is accountable for the development of secondary lymphoid structures. This is most clearly seen in the gut, as GF mice possess reduced Peyer's patches and reduced $CD4^+$ T cells and IgA-secreting plasma cells. Tertiary lymphoid structures in the colon, like cryptopatches or solitary lymphoid follicles, are induced by postnatal commensal contact. Furthermore, commensals enhance intestinal barrier integrity by stimulating angiogenesis and epithelial cell maturation (Fig. 2.11) (Belkaid and Hand 2014). Maintaining the neonatal immune–microbiota balance is critical for healthy immune development, achieved through a homeostatic, long-term host–microbe relationship that shapes lifelong health outcomes.

Research has also shown that commensal bacteria improve the function of the epithelial barrier, which protects against pathogen invasion. It has also been established that GF mice and mice with deficiencies in certain proteins, including Nod2 and MyD88, had defective antimicrobial peptide production, weakening barrier function, and allowing bacterial translocation (Vaishnava et al. 2008). These peptides are vital to limiting bacterial colonization and intestinal disease. MyD88 or Paneth cell-deficient mice have increased loads of mucosal bacteria associated to wild-type mice and contain more pathogenic bacteria at the epithelium lining of the small intestine (Park and Im 2020). The gut microbiota is also known to secrete interleukin 22 (IL-22), which activates the secretion of RegIIIγ and serves to

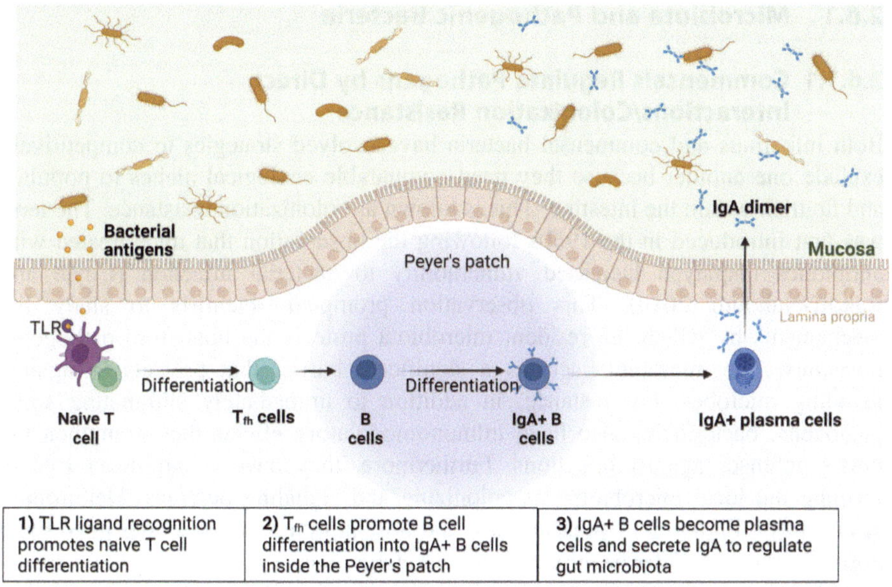

Fig. 2.11 IgA induction in Peyer's patches and regulation of gut microbiota. Diagram illustrating how IgA-mediated mucosal immunity in the small intestine is triggered by commensal-derived antigens. The three main steps shown are summarized in the boxes beneath the diagram: (1) TLR-dependent priming of T cells, (2) Tfh-driven class switching of B cells in Peyer's patches, and (3) plasma cell production of IgA and control by mucosal microbes. IgA stands for immunoglobulin A, TLR for toll-like receptor, Tfh for T follicular helper cell, and GALT for gut-associated lymphoid tissue (Peyer's patch is displayed)

protect against intestinal infection, for example, by *Citrobacter rodentium*, a murine bacterium that models human pathogens (Qiu et al. 2012). Also, the gut microbiota is crucial in ensuring homeostatic pro-IL-1β levels within resident macrophages to prime them for a quick response to infections (Franchi et al. 2012). The process is essential for the recruitment of neutrophils and clearing pathogens. Through MyD88-dependent signaling pathways, the gut microbiota maintains the integrity of the intestinal barrier and promotes antimicrobial defense, according to a 2024 study. Commensal disruption in MyD88 deficient animals resulted in decreased antimicrobial peptide, increased mucosal permeability, and enhanced pathogen translocation, demonstrating the critical function of the MyD88 microbiota axis in infection resistance and barrier maintenance (Zhou et al. 2023).

2.6 Microbiota and Disease Control

As our understanding of microbiota deepens, there has been growing interest in the potential to treat disease through targeted microbial manipulation. The human gut is to be modified to prevent or cure the resultant disorders since it is involved in a myriad of physiological processes. Hence, a number of clinical trials are now being undertaken to explore this possibility.

2.6.1 Microbiota and Pathogenic Bacteria

2.6.1.1 Commensals Regulate Pathogens by Direct Interactions/Colonization Resistance

Both infectious and commensal bacteria have evolved strategies to competitively exclude one another because they need comparable ecological niches to populate and flourish within the intestine. This is known as colonization resistance. The term was first introduced in the 1950s following the observation that mice treated with antibiotics displayed increased vulnerability to bacterial infection (Coyte and Rakoff-Nahoum 2019). This observation prompted scientists to study the mechanisms by which the resident microbiota protects the host from pathogenic invasions. The microbiota acts as a significant barrier that protects us against invading microbes. For instance, in addition to immediately eliminating some pathogens, bacteriocins also have immunomodulatory effects that strengthen the host's defenses against infections. Furthermore, they have a significant role in forming the local microbiome as colonizing and signaling peptides (Heilbronner et al. 2021). Figure 2.12 presents how the host immune response and microbiota normally interact under healthy and disease status.

Skin is yet another body organ that is predominantly colonized by the microbiota. The immune system's reaction to skin infections depends heavily on the commensal bacterial populations. According to Ridaura et al., the genus *Corynebacterium* can stimulate IL-17A γδT cells in the dermis, which attract immune cells, and enhance IL-23 signaling (Ridaura et al. 2018). It has been demonstrated that *Cutibacterium acnes*, which is linked to healthy skin, triggers the secretion of extracellular traps from certain Th17 subsets at the species level (Thiboutot and Nelson 2021). Coagulase-negative *Staphylococcus epidermidis*, applied to mouse skin, demonstrated improved innate protection against *Candida albicans* by upregulating Th17 immune mediators, including S100A8 and S100A9. *Staphylococci* (CoNS) such as *Staphylococcus cohnii* stimulate the host steroid system and encourage immunosuppression (Agak et al. 2021). Commensal bacteria are also necessary for the regulation of surface pH and the development of the epidermal barrier (Harris-Tryon and Grice 2022; Uberoi et al. 2021). For example, a lipase secreted by *Cutibacterium acnes* transforms triacylglycerols in sebum into propionic acid, which helps to acidify the skin's surface and prevents *S. aureus* from growing (Danby and Cork 2018). *Staphylococcus epidermidis* stimulates the generation of skin ceramides, while *Roseomonas mucosa* secretes glycerophospholipids that enhance cholinergic activation via TNFR2 signaling, hence promoting host epithelium healing (Myles et al. 2020; Zheng et al. 2022). Notably, commensals directly provide colonization resistance by secreting certain chemicals that stop dangerous bacteria, including *S. aureus*, from growing or producing virulence factors.

The vaginal microbiota inhibits the colonization of invasive pathogens; hence, it is essential in preventing vulvovaginal candidiasis, urinary tract infections (UTI), and sexually transmitted diseases (STDs). Most of the population of the vagina is comprised of *Lactobacillus* species, which secrete bacteriocins and lactic acid, among other bactericidal and virucidal compounds. Consequently,

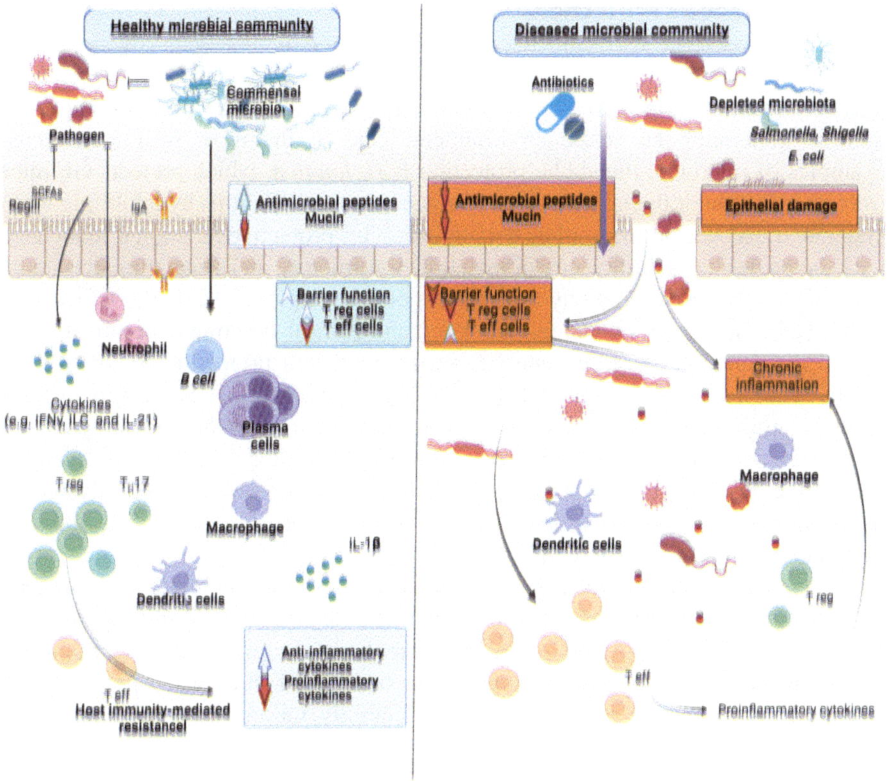

Fig. 2.12 By creating bacteriocins and short-chain fatty acids, reducing oxygen levels, and modifying pathogen pathogenicity, the resident commensal microbiota in a healthy gut occupy colonization niches and inhibit the proliferation of pathogens and pathobionts. By boosting the mucus barrier, triggering antimicrobial proteins (including RegIIIβ/γ), and controlling IgA secretion, they improve host defenses. Commensals also encourage TH17 and innate lymphoid cell activity, which results in the generation of cytokines (such IL-22) for microbial control, and prime macrophages to activate IL-1β during infection, attracting neutrophils. When antibiotics or environmental variables disrupt colonization resistance, pathogenic infections (like Salmonella and S. flexneri) and pathobiont overgrowth (like C. difficile and VRE) can occur, leading to systemic illness

vaginal *Lactobacilli* are viewed as a protective element against sexually transmitted illnesses, such as human papillomavirus infection, herpes simplex virus infection, and human immunodeficiency virus (HIV). It has been observed that females with unusual Gram-positive stain smears have a reduced rate of chlamydial genital infections. Likewise, healthy vaginal flora can also decrease vaginal pH to prevent the colonization of urinary tract pathogens, which need neutral pH to thrive (Kazlauskaitė et al. 2025).

2.6.1.2 Metabolic Protection

Commensal bacteria also influence the harmfulness of pathogens by producing specific metabolites (Fig. 2.13). Some *Bifidobacterium* species, like *Bifidobacterium longum* subsp. *longum* JCM 1217 T, inhibits pathogenic *Escherichia coli* colonization by generating an acidic environment. One of the most well-documented examples of microbial metabolic protection is *B. longum*, which protects GF mice from an infection of the fatal *E coli*, *B. longum* accomplishes this by inhibiting toxin systemic translocation and epithelial apoptosis. This protective function is associated with the bacterium's proficiency in metabolizing fructose and generating acetate by particular ABC-type carbohydrate transporters (e.g., COG1879, COG1172, COG1129). Genome sequencing, knockout, and gain-of-function experiments confirmed the function of these genes in acetate-mediated protection (Fukuda et al. 2011). The intestinal commensal *E. coli* Nissle and some other pathogens are strongly inhibited by the acetate, propionate, and butyrate at the physiological pH of the cecum and ascending colon. The anaerobic gut commensals were considerably less affected by this suppression. Following antibiotic-induced dysbiosis, *Bacteroides thetaiotaomicron* and *Bifidobacterium adolescent* have been linked to microbiota recovery (Chang et al. 2024).

By competing for nutrients, commensal bacteria can also inhibit the colonization of intestinal pathogens. Pathogens such as *Salmonella Typhimurium* and *K. pneumoniae* are deprived of their metabolic niche when diverse microbial communities effectively deplete sugars and other vital nutrients (Spragge et al. 2023). Similarly, by controlling gluconate availability, tailored commensal consortia prevent Enterobacteriaceae from expanding in mice and restore colonization resistance. Further demonstrating how food constraint underlies colonization resistance,

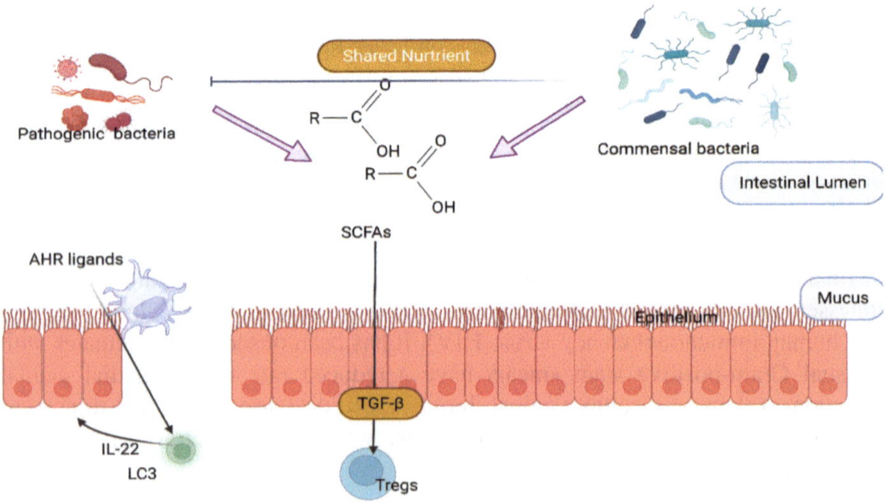

Fig. 2.13 Metabolic protection: the host commensal bacteria compete with the pathogens for the metabolites, which indirectly leads to the death of pathogens

Lactobacilli inhibited pathogen development in a mouse model of *C. jejuni* infection by reducing essential amino acids, particularly cysteine.

2.6.1.3 Anti-inflammatory and Barrier Function

In changing the host's environment, commensals can also limit pathogen infection and make it harder for pathogens to colonize. For example, propionate inhibits the development of many bacteria, but it has also been shown to promote adhesion and biofilm formation in adherent-invasive *E. coli* (AIEC), resulting in a variety of toxicities (Ormsby et al. 2020). In a related study, propionate was found to increase the expression of several significant pathogenic factors in this *E. coli*, including proteins involved in cell wall biosynthesis and oxidative stress defense, porins, and transcription factors (Pobeguts et al. 2020). The intricacy of the process by which SCFAs regulate bacterial virulency is demonstrated by the fact that, despite promoting AIEC virulence, propionate stimulation also causes an inflammatory response in the gut, which inhibits the growth of AIEC adhesion and invasion (Pace et al. 2021). Acetate, for instance, has been demonstrated in earlier studies to function as an environmental cue that enhances the expression of Salmonella invasion genes. Additionally, acetate and yeast extract can work together to increase Salmonella's invasiveness by activating the SPI-1 gene expression (Hamed et al. 2019). Apart from its impact on *Salmonella* virulence, another study discovered that acetate increased the adhesion and motility of enteropathogenic *E. coli* grown at varying pH levels by activating the flagella gene and locus of enterocyte effacement (LEE) expression (Yang et al. 2018).

2.6.2 Microbiota in Allergy Prevention

Microbial dysbiosis also plays a critical role in the advancement of allergy disorders through mechanisms that increase inflammation and decrease immunological tolerance, and new research highlights the gut microbiome as a potential goal for treatment. Allergy disorders can result from dysbiosis, which has been connected to alterations in the microbiome diversity that impact host immunological responses (Sutanto et al. 2025). Using information from the American Gut Project, a study examined the relationship between adult allergies and the composition of the gut microbiota. An analysis of 1879 volunteers showed that allergies were quite prevalent, and that the gut microbiota richness of allergic people was much lower than that of their non-allergic counterparts. Additionally, participants with allergies showed notable fluctuations in the makeup of the microbial community, as evidenced by decreased levels of *Clostridiales* taxa and increased levels of *Bacteroidales* (Yamaguchi et al. 2023). Food allergy is also often associated with atopic dermatitis, a recurrent and chronic skin condition that appears in early infancy, youth, or adolescence. It might go away as kids become older, but in adults, other atopic conditions like asthma or allergic rhinitis might surface. Atopic dermatitis is frequently linked to upregulated serum IgE levels and with at least one atopic condition, for instance, food allergies, allergic rhinitis, or asthma. According to epidemiologic

Table 2.2 Some bacteria's contributions to the pathophysiology of allergic disease

Good bacteria	Bad bacteria
Lactobacillus rhamnosus GG: Improves immunological lenience and lowers the prevalence of childhood atopic dermatitis	*Staphylococcus aureus:* Worsens the skin disorders like eczema
Bifidobacterium longum: Improves the function of the intestinal barrier and regulates the immune system, may lower the chance of developing allergy disorders	*Escherichia coli:* Inflammation and emergence of allergy disorders
Lactobacillus paracasei: Helps people with allergic rhinitis by reducing their symptoms	*Clostridium difficile:* Increased risk of developing allergy disorders, and is associated with disturbances in the gut flora
Bifidobacterium breve: Preventive function against allergic skin disorders by lessening the severity of atopic eczema in babies	*Klebsiella pneumoniae:* Pathophysiology of some allergic diseases, including asthma
Faecalibacterium prausnitzii: Produces butyrate and other SCFAs, which have anti-inflammatory and potentially protective effects against allergic disorders	*Helicobacter pylori*: Occasionally been linked to an elevated risk of rhinitis and allergic asthma

evidence, diseases such as asthma that are linked with dysregulated barrier responses may be predisposed to by changes in mothers' or infants' microbiota (Van den Elsen et al. 2019) (Table 2.2).

2.6.3 Role in Prevention of Other Diseases

2.6.3.1 Microbiota and Liver Diseases

The continuous exposure to gut-derived factors from the liver, the microbes, and microbial components. The intestinal microbiota synthesizes ammonia, ethanol, and acetaldehyde and can influence liver function by modulating liver metabolism or endotoxin release (Giannelli et al. 2014). The VSL#3 probiotic has been demonstrated in a clinical trial to reduce the severity of liver disease and hospitalizations in patients with cirrhosis (Dhiman et al. 2014). Moreover, rats administered *Lactobacillus* salivarius LI01 or *Pediococcus pentosaceus* LI05 show recovery from acute liver injury caused by D-galactosamine (Lv et al. 2014). A diet high in soluble dietary fiber reduces alcohol-induced liver damage in rats and boosts the number of *Bacteroides acidifaciens*. By producing unconjugated bile acids that activate the intestine farnesoid X receptor and its downstream target, fibroblast growth factor-15, *B. acidifaciens* therapy alone reduces liver damage. To provide enough glutamate for ammonia detoxification through the glutamine synthesis route, FGF15 stimulates hepatocyte production of ornithine aminotransferase, which helps the liver convert stored ornithine into glutamate. All of these results point to a possible treatment approach for Alcohol-related liver disease that includes *B. acidifaciens* and dietary fiber supplements.

2.6.3.2 Microbiota and Cancer

Current research has outlined the roles of the microbiota in the carcinogenesis process and the stages of treatment, even though genetics and environmental factors continue to be the primary causes of cancer and tumor formation. Alterations in the ratio of pathogenic microorganisms in mixed microbial systems might trigger or promote cancer-causing mechanisms such as inflammation, altered DNA methylation, metabolic issues, and immune system disarray (Garg et al. 2023). Additionally, microbial communication with the tumor microenvironment and host immune system has a significant impact on cancer and treatment outcomes. Because it identifies new strategies for cancer prevention, detection, and treatment, describing the processes governing microbiota-cancer interactions has thus become a critical area of research. [Explained in Detail in the Chapter: Microbiota and Cancer].

2.7 Microbiota Modulation

As knowledge of the connection involving the host and its microbiome and a range of diseases has evolved, there has been a lot of interest in leveraging these data to anticipate or cure disease.

Certain microbial enrichments have been pinpointed in some research as potential biomarkers by some studies; however, there is no specificity, as they are observed in other research. In addition, it has also been found that the ratio of *Bifidobacteria/Enterobacteriaceae* (B/E) is also considered an indicator of the perfect microbial balance in humans and that it expresses the resistance of the bowel to microbial colonization. People with healthy microbiomes possess a B/E ratio that is greater than 1, whereas cirrhotic patients and those who are infected by the avian influenza virus H7N9 have a B/E ratio much lower than 1 (Lu et al. 2014). Innumerable experiments on preventing and treating diseases through the intervention of the microbiome have been carried out, and some treatments have been efficaciously applied clinically (Wang et al. 2017).

H7N9 patients have been found to recover faster when probiotics are given. Further, modulation of these microbiomes is also important in treating GI disorders like necrotizing enterocolitis, inflammatory bowel disease, and antibiotic-associated diarrhea. For example, it was found that when consortia of 17 *Clostridium* strains from humans were orally administered to adult mice, it decreased the symptoms of models of allergic diarrhea and colitis (Lu et al. 2014). Additionally, several clinical findings have established the remarkable effectiveness of probiotics and their products in managing allergic conditions, exclusively in infants (West et al. 2016). The use of microbiota in treating other diseases and its usage is discussed in detail in subsequent chapters.

2.8 Conclusion

The human body hosts a diverse microbiota that exceeds its number of genes. This microbiota resides all over the body, inside the body in the urinogenital, the respiratory tract, the oral cavity, and especially in the gut. It develops from birth and changes throughout life based on various factors. From energy harvesting and nutrition digestion to immune system maturation and pathogen defense, these microbial communities support vital physiological functions. By fermenting food fibers, producing short-chain fatty acids, and controlling bile acid metabolism, the gut microbiota in particular plays a critical role in regulating host metabolism. New research shows that it has an impact outside of the gastrointestinal system, where endotoxins and microbial metabolites affect liver function and the course of disease. Studies highlight that competition for limited nutrients in the intestine influences how well commensals can inhibit pathogen growth. Furthermore, inflammatory responses of these commensals significantly affect the growth of pathogens. The microbiota in the gut maintains a fragile balance; when disrupted, pathobiont overgrowth and dysbiosis occur, resulting in disease and immune issues. Identifying natural "competitors" that suppress pathogen growth could lead to new treatment strategies for intestinal diseases. Besides infectious diseases, microbiota also help prevent allergies and other non-communicable diseases, as discussed in the upcoming chapters.

As our understanding of microbiota has expanded, so too has interest in the potential of using microbial manipulation to treat illnesses. Modification of the human gut is expected to prevent or treat the associated disorders because it is engaged in a wide variety of physiological functions. Consequently, an increasing number of clinical trials are being conducted to look at this potential.

References

Agak GW, Mouton A, Teles RM, Weston T, Morselli M, Andrade PR et al (2021) Extracellular traps released by antimicrobial T H 17 cells contribute to host defense. J Clin Invest 131(2)

Agirman G, Hsiao EY (2021) SnapShot: the microbiota-gut-brain axis. Cell 184(9):2524–2524

Agirman G, Yu KB, Hsiao EY (2021) Signaling inflammation across the gut-brain axis. Science 374(6571):1087–1092

Allegretti JR, Fischer M, Kelly CR (2025) Diagnostic uncertainties and delayed treatment protocols impact efficacy of FMT for Clostridioides difficile infection. Clin Infect Dis 81(3):e60–e61. https://doi.org/10.1093/cid/ciaf140

Barraud O, Ravry C, François B, Daix T, Ploy MC, Vignon P (2019) Shotgun metagenomics for microbiome and resistome detection in septic patients with urinary tract infection. Int J Antimicrob Agents 54(6):803–808

Belizário JE, Napolitano M (2015) Human microbiomes and their roles in dysbiosis, common diseases, and novel therapeutic approaches. Front Microbiol 6:151578

Belkaid Y, Hand TW (2014) Role of the microbiota in immunity and inflammation. Cell 157(1): 121–141

Bhute SS, Suryavanshi MV, Joshi SM, Yajnik CS, Shouche YS, Ghaskadbi SS (2017) Gut microbial diversity assessment of Indian type-2-diabetics reveals alterations in eubacteria, archaea, and eukaryotes. Front Microbiol 8:214

Bishop RC, Migliorisi A, Holmes JR, Kemper AM, Band M, Austin S et al (2024) Microbial populations vary between the upper and lower respiratory tract, but not within biogeographic regions of the lung of healthy horses. J Equine Vet Sci 140:105141

Borbet TC, Pawline MB, Li J, Ho ML, Yin YS, Zhang X et al (2023) Disruption of the early-life microbiota alters Peyer's patch development and germinal center formation in gastrointestinal-associated lymphoid tissue. Iscience 26(6)

Bouslimani A, da Silva R, Kosciolek T, Janssen S, Callewaert C, Amir A, Dorrestein K, Melnik AV, Zaramela LS, Kim JN et al (2019) The impact of skin care products on skin chemistry and microbiome dynamics. BMC Biol 17:47. https://doi.org/10.1186/s12915-019-0660-6

Byrd AL, Belkaid Y, Segre JA (2018) The human skin microbiome. Nat Rev Microbiol 16(3): 143–155

Callewaert C, Kerckhof F-M, Granitsiotis MS, Van Gele M, Van de Wiele T, Boon N (2013) Characterization of Staphylococcus and Corynebacterium clusters in the human axillary region. PLoS ONE 8:e70538

Cardona D, Roman P (2022) New perspectives in health: gut microbiota. Int J Environ Res Public Health 19:5828. https://doi.org/10.3390/ijerph19105828

Chang KC, Nagarajan N, Gan YH (2024) Short-chain fatty acids of various lengths differentially inhibit Klebsiella pneumoniae and Enterobacteriaceae species. Msphere 9(2):e00781–e00723

Chen A, Teng C, Wei J, Wu X, Zhang H, Chen P et al (2025) Gut microbial dysbiosis exacerbates long-term cognitive impairments by promoting intestinal dysfunction and neuroinflammation following neonatal hypoxia-ischemia. Gut Microbes 17(1):2471015. https://doi.org/10.1080/19490976.2025.2471015

Chernikova DA, Madan JC, Housman ML, Zain-Ul-Abideen M, Lundgren SN, Morrison HG et al (2018) The premature infant gut microbiome during the first 6 weeks of life differs based on gestational maturity at birth. Pediatr Res 84(1):71–79

Coyte KZ, Rakoff-Nahoum S (2019) Understanding competition and cooperation within the mammalian gut microbiome. Curr Biol 29(11):R538–R544

Cryan JF, O'Riordan KJ, Cowan CS, Sandhu KV, Bastiaanssen TF, Boehme M et al (2019) The microbiota-gut-brain axis. Physiol Rev 99:1877

Danby SG, Cork MJ (2018) pH in atopic dermatitis. Curr Probl Dermatol 54:95–107

Dekaboruah E, Suryavanshi MV, Chettri D, Verma AK (2020) Human microbiome: an academic update on human body site specific surveillance and its possible role. Arch Microbiol 202(8): 2147–2167

Derrien M, Alvarez AS, De Vos WM (2019) The gut microbiota in the first decade of life. Trends Microbiol 27:997–1010. https://doi.org/10.1016/j.tim.2019.08.001

Dhakar K, Pandey A (2016) Wide pH range tolerance in extremophiles: towards understanding an important phenomenon for future biotechnology. Appl Microbiol Biotechnol 100(6):2499–2510

Dhiman RK, Rana B, Agrawal S, Garg A, Chopra M, Thumburu KK et al (2014) Probiotic VSL# 3 reduces liver disease severity and hospitalization in patients with cirrhosis: a randomized, controlled trial. Gastroenterology 147(6):1327–1337

Dickson RP, Erb-Downward JR, Martinez FJ, Huffnagle GB (2016) The microbiome and the respiratory tract. Annu Rev Physiol 78(1):481–504

Dimitriu PA, Iker B, Malik K, Leung H, Mohn WW, Hillebrand GG (2019) New insights into the intrinsic and extrinsic factors that shape the human skin microbiome. MBio 10:e00839–19. https://doi.org/10.1128/mBio.00839-19

Donald K, Finlay BB (2023) Early-life interactions between the microbiota and immune system: impact on immune system development and atopic disease. Nat Rev Immunol 23(11):735–748

Ferneyhough B, Roddis M, Millington S, Quirk J, Clements C, West S et al (2025) A highly accurate nanopore-based sequencing workflow for culture and PCR-free microbial metagenomic profiling of urogenital samples. BMC Urol 25(1):41. https://doi.org/10.1186/s12894-025-01723-9

Egert M, Simmering R (2016) The microbiota of the human skin. In: Microbiota of the human body: implications in health and disease, pp 61–81

Fierer N, Hamady M, Lauber CL, Knight R (2008) The influence of sex, handedness, and washing on the diversity of hand surface bacteria. Proc Natl Acad Sci 105(46):17994–17999

Franchi L, Kamada N, Nakamura Y, Burberry A, Kuffa P, Suzuki S et al (2012) NLRC4-driven production of IL-1β discriminates between pathogenic and commensal bacteria and promotes host intestinal defense. Nat Immunol 13(5):449–456

Francino M (2016) Antibiotics and the human gut microbiome: dysbioses and accumulation of resistances. Front Microbiol 6:164577. https://doi.org/10.3389/fmicb.2015.01543

Fukuda S, Toh H, Hase K, Oshima K, Nakanishi Y, Yoshimura K et al (2011) Bifidobacteria can protect from enteropathogenic infection through production of acetate. Nature 469(7331): 543–547. https://doi.org/10.1038/nature09646

Garg S, Sharma N, Bharmjeet, Das A (2023) Unraveling the intricate relationship: influence of microbiome on the host immune system in carcinogenesis. Cancer Reports 6(11):e1892. https://doi.org/10.1002/cnr2.1892

Giannelli V, Di Gregorio V, Iebba V, Giusto M, Schippa S, Merli M, Thalheimer U (2014) Microbiota and the gut-liver axis: bacterial translocation, inflammation and infection in cirrhosis. World J Gastroenterol: WJG 20(45):16795

Gibson SA, McFarlan C, Hay S, MacFarlane GT (1989) Significance of microflora in proteolysis in the colon. Appl Environ Microbiol 55:679–683

Gominak S (2016) Vitamin D deficiency changes the intestinal microbiome reducing B vitamin production in the gut. The resulting lack of pantothenic acid adversely affects the immune system, producing a "pro-inflammatory" state associated with atherosclerosis and autoimmunity. Med Hypotheses 94:103–107. https://doi.org/10.1016/j.mehy.2016.07.007

Hamed S, Wang X, Shawky RM, Emara M, Aldridge PD, Rao CV (2019) Synergistic action of SPI-1 gene expression in salmonella enterica serovar typhimurium through transcriptional crosstalk with the flagellar system. BMC Microbiol 19(1):211

Han M, Wang N, Han W, Liu X, Sun T, Xu J (2024) Specific vaginal and gut microbiome and the anti-tumor effect of butyrate in cervical cancer women. Transl Oncol 44:101902

Hapfelmeier S, Lawson MA, Slack E, Kirundi JK, Stoel M, Heikenwalder M et al (2010) Reversible microbial colonization of germ-free mice reveals the dynamics of IgA immune responses. Science 328(5986):1705–1709

Harris-Tryon TA, Grice EA (2022) Microbiota and maintenance of skin barrier function. Science 376(6596):940–945

He H, He H, Mo L, Yuan Q, Xiao C, Ma Q et al (2024) Gut microbiota regulate stress resistance by influencing microglia-neuron interactions in the hippocampus. Brain Behavior Immun Health 36:100729

Heilbronner S, Krismer B, Brötz-Oesterhelt H, Peschel A (2021) The microbiome-shaping roles of bacteriocins. Nat Rev Microbiol 19(11):726–739

Hollister EB, Riehle K, Luna RA, Weidler EM, Rubio-Gonzales M, Mistretta TA et al (2015) Structure and function of the healthy pre-adolescent pediatric gut microbiome. Microbiome 3:1–13

Hou K, Wu ZX, Chen XY, Wang JQ, Zhang D, Xiao C et al (2022) Microbiota in health and diseases. Signal Transduct Target Ther 7(1):135

Huang YJ, Segal LN (2020) Looking higher: Is it prime time for the oral–lung axis in HIV-related lung disease? Am J Respir Crit Care Med 201(4):402–403

Hurst JH, McCumber AW, Aquino JN, Rodriguez J, Heston SM, Lugo DJ et al (2022) Age-related changes in the nasopharyngeal microbiome are associated with severe acute respiratory syndrome coronavirus 2 (SARS-CoV-2) infection and symptoms among children, adolescents, and young adults. Clin Infect Dis 75(1):e928–e937. https://doi.org/10.1093/cid/ciac184

Jian J, Wei J (2025) Ferroptosis: a new pathway in the interaction between gut microbiota and multiple sclerosis. Front Biosci Landmark 30(1):26265. https://doi.org/10.31083/FBL26265

Kalbermatter C, Fernandez Trigo N, Christensen S, Ganal-Vonarburg SC (2021) Maternal microbiota, early life colonization and breast milk drive immune development in the newborn. Front Immunol 12:683022

Karl JP, Hatch AM, Arcidiacono SM, Pearce SC, Pantoja-Feliciano IG, Doherty LA, Soares JW (2018) Effects of psychological, environmental and physical stressors on the gut microbiota. Front Microbiol 9:2013

Kazlauskaitė J, Žukienė G, Rudaitis V, Bartkevičienė D (2025) The vaginal microbiota, human papillomavirus, and cervical dysplasia—a review. Medicina 61(5):847

Kelly BJ, Imai I, Bittinger K, Laughlin A, Fuchs BD, Bushman FD, Collman RG (2016) Composition and dynamics of the respiratory tract microbiome in intubated patients. Microbiome 4(1):7

Kim MS, Jung SI (2021) The urinary tract microbiome in male genitourinary diseases: focusing on benign prostate hyperplasia and lower urinary tract symptoms. Int Neurourol J 25(1):3. https://doi.org/10.5213/inj.2040174.087

Kim M, Lee S, Kim HS, Kwon MY, Yoo J, Kim MJ (2023) Microbiome alterations in women with pelvic organ prolapse and after anatomical restorative interventions. Sci Rep 13(1):17547. https://doi.org/10.1038/s41598-023-44988-6

Koceva H, Amiratashani M, Akbarimoghaddam P, Hoffmann B, Zhurgenbayeva G, Gresnigt MS et al (2025) Deciphering respiratory viral infections by harnessing organ-on-chip technology to explore the gut–lung axis. Open Biol 15(3). https://doi.org/10.1098/rsob.240231

Koliarakis I, Messaritakis I, Nikolouzakis TK, Hamilos G, Souglakos J, Tsiaoussis J (2019) Oral bacteria and intestinal dysbiosis in colorectal cancer. Int J Mol Sci 20(17):4146

Kolodziejczyk AA, Zheng D, Elinav E (2019) Diet–microbiota interactions and personalized nutrition. Nat Rev Microbiol 17(12):742–753

Korpela K, Blakstad EW, Moltu SJ, Strømmen K, Nakstad B, Rønnestad AE et al (2018) Intestinal microbiota development and gestational age in preterm neonates. Sci Rep 8(1):2453

Kostic AD, Howitt MR, Garrett WS (2013) Exploring host–microbiota interactions in animal models and humans. Genes Dev 27(7):701–718

Kurdi P, Kawanishi K, Mizutani K, Yokota A (2006) Mechanism of growth inhibition by free bile acids in lactobacilli and bifidobacteria. J Bacteriol 188(5):1979–1986

Lamont RJ, Koo H, Hajishengallis G (2018) The oral microbiota: dynamic communities and host interactions. Nat Rev Microbiol 16(12):745–759

Le Blanc JG, Milani C, de Giori GS et al (2013) Bacteria as vitamin suppliers to their host: a gut microbiota perspective. Curr Opin Biotechnol 24:160–168

Lee J-J, Kim S-H, Lee M-J, Kim B-K, Song W-J, Park H-W et al (2019) Different upper airway microbiome and their functional genes associated with asthma in young adults and elderly individuals. Allergy 74:709–719

Leung MHY, Wilkins D, Lee PKH (2015) Insights into the pan-microbiome: skin microbial communities of Chinese individuals differ from other racial groups. Sci Rep 5:11845

Li N, Wang Q, Wang Y, Sun A, Lin Y, Jin Y, Li X (2019) Fecal microbiota transplantation from chronic unpredictable mild stress mice donors affects anxiety-like and depression-like behavior in recipient mice via the gut microbiota-inflammation-brain axis. Stress 22(5):592–602

Li X, Liu Y, Yang X, Li C, Song Z (2022) The oral microbiota: community composition, influencing factors, pathogenesis, and interventions. Front Microbiol 13:895537

Liu L, Wang H, Chen X, Zhang Y, Zhang H, Xie P (2023) Gut microbiota and its metabolites in depression: from pathogenesis to treatment. EBioMedicine 90:104527

Liu Y, Huang Q, Zhuang Z, Yang H, Gou X, Xu T et al (2024) Gut virome alterations in patients with chronic obstructive pulmonary disease. Microbiol Spectrum 12(7):e04287–e04223. https://doi.org/10.1128/spectrum.04287-23

Liu WW, Reicher N, Alway E, Rupprecht LE, Weng P, Schaefgen C et al (2025) A gut sense for a microbial pattern regulates feeding. Nature 645(8081):729–736. https://doi.org/10.1038/s41586-025-09301-7

Lu H, Zhang C, Qian G, Hu X, Zhang H, Chen C et al (2014) An analysis of microbiota-targeted therapies in patients with avian influenza virus subtype H7N9 infection. BMC Infect Dis 14:1–12

Lunjani N, Hlela C, O'Mahony L (2019) Microbiome and skin biology. Curr Opin Allergy Clin Immunol 19(4):328–333

Lv LX, Hu XJ, Qian GR, Zhang H, Lu HF, Zheng BW et al (2014) Administration of Lactobacillus salivarius LI01 or Pediococcus pentosaceus LI05 improves acute liver injury induced by D-galactosamine in rats. Appl Microbiol Biotechnol 98(12):5619–5632. https://doi.org/10.1007/s00253-014-5638-2

Ma Z, Zuo T, Frey N, Rangrez AY (2024) A systematic framework for understanding the microbiome in human health and disease: from basic principles to clinical translation. Signal Transduct Target Ther 9(1):237

Macfarlane GT, Cummings JH, Allison C (1986) Protein degradation by human intestinal bacteria. Microbiology 132:1647–1656

Marchesi JR, Adams DH, Fava F, Hermes GD, Hirschfield GM, Hold G et al (2016) The gut microbiota and host health: a new clinical frontier. Gut 65(2):330–339

Martínez-González MA, Planes FJ, Ruiz-Canela M, Toledo E, Estruch R, Salas-Salvado J et al (2025) Recent advances in precision nutrition and cardiometabolic diseases. Revista Española de Cardiología (Engl Ed) 78(3):263–271. https://doi.org/10.1016/j.rec.2024.09.003

Mogoş GFR, Manciulea M, Enache RM, Pavelescu LA, Popescu OA, Cretoiu SM, Marinescu I (2025) Intestinal microbiota in early life: latest findings regarding the role of probiotics as a treatment approach for Dysbiosis. Nutrients 17(13):2071

Moskovicz V, Gross A, Mizrahi B (2020) Extrinsic factors shaping the skin microbiome. Microorganisms 8:1023. https://doi.org/10.3390/microorganisms8071023

Myles IA, Castillo CR, Barbian KD, Kanakabandi K, Virtaneva K, Fitzmeyer E et al (2020) Therapeutic responses to Roseomonas mucosa in atopic dermatitis may involve lipid-mediated TNF-related epithelial repair. Sci Transl Med 12(560):eaaz8631

Neugent ML, Hulyalkar NV, Nguyen VH, Zimmern PE, De Nisco NJ (2020) Advances in understanding the human urinary microbiome and its potential role in urinary tract infection. MBio 11(2):10–1128

Neugent ML, Hulyalkar NV, Kumar A, Xing C, Zimmern PE, Shulaev V, De Nisco NJ (2023) Urinary glycosaminoglycans are associated with recurrent UTI and urobiome ecology in postmenopausal women. ACS infectious diseases 9(4):1022–1032

Nicholson JK, Holmes E, Kinross J, Burcelin R, Gibson G, Jia W, Pettersson S (2012) Host-gut microbiota metabolic interactions. Science 336(6086):1262–1267

O'Toole PW, Jeffery IB (2015) Gut microbiota and aging. Science 350(6265):1214–1215

Oh J, Conlan S, Polley EC, Segre JA, Kong HH (2012) Shifts in human skin and nares microbiota of healthy children and adults. Genome Med 4:77

Ormsby MJ, Johnson SA, Carpena N, Meikle LM, Goldstone RJ, McIntosh A et al (2020) Propionic acid promotes the virulent phenotype of Crohn's disease-associated adherent-invasive Escherichia coli. Cell Rep 30(7):2297–2305

Owaga E, Hsieh RH, Mugendi B, Masuku S, Shih CK, Chang JS (2015) Th17 cells as potential probiotic therapeutic targets in inflammatory bowel diseases. Int J Mol Sci 16(9):20841–20858

Pace F, Rudolph SE, Chen Y, Bao B, Kaplan DL, Watnick PI (2021) The short-chain fatty acids propionate and butyrate augment adherent-invasive Escherichia coli virulence but repress inflammation in a human intestinal enteroid model of infection. Microbiol Spectrum 9(2): e01369–e01321

Parizadeh M, Arrieta MC (2023) The global human gut microbiome: genes, lifestyles, and diet. Trends Mol Med 29(10):789–801

Park JC, Im SH (2020) Of men in mice: the development and application of a humanized gnotobiotic mouse model for microbiome therapeutics. Exp Mol Med 52(9):1383–1396

Pascale A, Marchesi N, Marelli C, Coppola A, Luzi L, Govoni S et al (2018) Microbiota and metabolic diseases. Endocrine 61(3):357–371

Paukkonen I, Törrönen EN, Lok J, Schwab U, El-Nezami H (2024) The impact of intermittent fasting on gut microbiota: A systematic review of human studies. Front Nutr 11:1342787

Pietrasanta C, Carlosama C, Lizier M, Fornasa G, Jost TR, Carloni S et al (2024) Prenatal antibiotics reduce breast milk IgA and induce dysbiosis in mouse offspring, increasing neonatal susceptibility to bacterial sepsis. Cell Host Microbe 32(12):2178–2194

Pobeguts OV, Ladygina VG, Evsyutina DV, Eremeev AV, Zubov AI, Matyushkina DS et al (2020) Propionate induces virulent properties of Crohn's disease-associated Escherichia coli. Front Microbiol 11:1460

Porcari S, Severino A, Rondinella D, Bibbo S, Quaranta G, Masucci L et al (2023) Fecal microbiota transplantation for recurrent Clostridioides difficile infection in patients with concurrent ulcerative colitis. J Autoimmun 141:103033. https://doi.org/10.1016/j.jaut.2023.103033

Prohic A, Simic D, Sadikovic TJ, Krupalija-Fazlic M (2014) Distribution of Malassezia species on healthy human skin in Bosnia and Herzegovina: correlation with body part, age and gender. Iran J Microbiol 6:253–262

Qiu J, Heller JJ, Guo X, Chen ZME, Fish K, Fu YX, Zhou L (2012) The aryl hydrocarbon receptor regulates gut immunity through modulation of innate lymphoid cells. Immunity 36(1):92–104

Rajilić-Stojanović M, De Vos WM (2014) The first 1000 cultured species of the human gastrointestinal microbiota. FEMS Microbiol Rev 38(5):996–1047

Reyman M, Van Houten MA, Watson RL, Chu MLJ, Arp K, De Waal WJ et al (2022) Effects of early-life antibiotics on the developing infant gut microbiome and resistome: a randomized trial. Nat Commun 13(1):893. https://doi.org/10.1038/s41467-022-28525-z

Ridaura VK, Bouladoux N, Claesen J, Chen YE, Byrd AL, Constantinides MG et al (2018) Contextual control of skin immunity and inflammation by Corynebacterium. J Exp Med 215(3):785–799

Rinninella E, Raoul P, Cintoni M, Franceschi F, Miggiano GAD, Gasbarrini A, Mele MC (2019) What is the healthy gut microbiota composition? A changing ecosystem across age, environment, diet, and diseases. Microorganisms 7(1):14

Rinninella E, Tohumcu E, Raoul P, Fiorani M, Cintoni M, Mele MC et al (2023) The role of diet in shaping human gut microbiota. Best Pract Res Clin Gastroenterol 62:101828

Robert C, Cascella F, Mellai M, Barizzone N, Mignone F, Massa N et al (2022) Influence of sex on the microbiota of the human face. Microorganisms 10(12):2470

Rodríguez JM, Murphy K, Stanton C, Ross RP, Kober OI, Juge N, Marchesi JR (2015) The composition of the gut microbiota throughout life, with an emphasis on early life. Microb Ecol Health Dis 26(1):26050. https://doi.org/10.3402/mehd.v26.26050

Ronan V, Yeasin R, Claud EC (2021) Childhood development and the microbiome—the intestinal microbiota in maintenance of health and development of disease during childhood development. Gastroenterology 160(2):495–506

Rosenberg E (2024) Diversity of bacteria within the human gut and its contribution to the functional unity of holobionts. NPJ Biofilms Microb 10(1):134

Rowland I, Gibson G, Heinken A, Scott K, Swann J, Thiele I et al (2018) Gut microbiota functions: metabolism of nutrients and other food components. Eur J Nutr 57:1–24. https://doi.org/10.1007/s00394-017-1445-8

Saadaoui M, Singh P, Al Khodor S (2021) Oral microbiome and pregnancy: a bidirectional relationship. J Reprod Immunol 145:103293

Sanna Nybacka RD, Törnblom H, Simren M, Störsrud S (2022) The role of carbohydrates in irritable bowel syndrome: protocol for a randomized controlled trial comparing three different treatment options. JMIR Res Protoc 11(1):e31413

Santacroce L, Charitos IA, Ballini A, Inchingolo F, Luperto P, De Nitto E et al (2020) The human respiratory system and its microbiome at a glimpse. Biology (Basel) 9:318

Sarkar A, Yoo JY, Valeria Ozorio Dutra S, Morgan KH, Groer M (2021) The association between early-life gut microbiota and long-term health and diseases. J Clin Med 10(3):459. https://doi.org/10.3390/jcm10030459

Sarkar D, Roy P, Saha S (2025) Meta-analysis of faecal microbiome studies followed by machine learning to identify intestinal disease-specific taxonomic signatures. Microb Pathog:108221. https://doi.org/10.1016/j.micpath.2025.108221

Shaikh RG, Dey A, Singh VP, Khandagle A, Naik S, Hasan A, Singh VP Sr (2025) Understanding the impact of the gut microbiome on mental health: a systematic review. Cureus 17(1). https://doi.org/10.7759/cureus.78100

Sheerin NS (2011) Urinary tract infection. Medicine 39(7):384–389. https://doi.org/10.1016/j.mpmed.2011.04.003
Singh J, Singh A, Biswal S, Zomuansangi R, Lalbiaktluangi C, Singh BP et al (2024) Microbiota-brain axis: exploring the role of gut microbiota in psychiatric disorders-a comprehensive review. Asian J Psychiatr 97:104068
Siopi E, Galerne M, Rivagorda M, Saha S, Moigneu C, Moriceau S et al (2023) Gut microbiota changes require vagus nerve integrity to promote depressive-like behaviors in mice. Mol Psychiatr 28(7):3002–3012
Siqueira JF, Rôças IN (2017) The oral microbiota in health and disease: an overview of molecular findings. Methods Mol Biol 1537:127–138
Smith RP, Easson C, Lyle SM, Kapoor R, Donnelly CP, Davidson EJ, Parikh E, Lopez JV, Tartar JL (2019) Gut microbiome diversity is associated with sleep physiology in humans. PLoS One 14:e0222394
Spragge F, Bakkeren E, Jahn MT, BN Araujo E, Pearson CF, Wang X et al (2023) Microbiome diversity protects against pathogens by nutrient blocking. Science 382(6676):eadj3502. https://doi.org/10.1126/science.adj3502
Sultan AS, Kong EF, Rizk AM, Jabra-Rizk MA (2018) The oral microbiome: a lesson in coexistence. PLoS Pathog 14:e1006719
Sutanto H, Elisa E, Rachma B, Fetarayani D (2025) Gut microbiome modulation in allergy treatment: the role of fecal microbiota transplantation. Am J Med 138:769
Thiboutot DM, Nelson AM (2021) Keeping the peace: commensal Cutibacterium acnes trains CD4+ T H 17 cells to trap and kill. J Clin Invest 131(2)
Thomas-White K, Forster SC, Kumar N, Van Kuiken M, Putonti C, Stares MD et al (2018) Culturing of female bladder bacteria reveals an interconnected urogenital microbiota. Nat Commun 9:1557
Thursby E, Juge N (2017) Introduction to the human gut microbiota. Biochem J 474:1823–1836
Tong X, Su F, Xu X, Xu H, Yang T, Xu Q et al (2019) Alterations to the lung microbiome in idiopathic pulmonary fibrosis patients. Front Cell Infect Microbiol 9:149
Uberoi A, Bartow-McKenney C, Zheng Q, Flowers L, Campbell A, Knight SA et al (2021) Commensal microbiota regulates skin barrier function and repair via signaling through the aryl hydrocarbon receptor. Cell Host Microbe 29(8):1235–1248
Vaishnava S, Behrendt CL, Ismail AS, Eckmann L, Hooper LV (2008) Paneth cells directly sense gut commensals and maintain homeostasis at the intestinal host-microbial interface. Proc Natl Acad Sci 105(52):20858–20863
Van den Elsen LW, Garssen J, Burcelin R, Verhasselt V (2019) Shaping the gut microbiota by breastfeeding: the gateway to allergy prevention? Front Pediatr 7:47. https://doi.org/10.3389/fped.2019.00047
Vasquez Ayala A, Hsu CY, Oles RE, Matsuo K, Loomis LR, Buzun E et al (2023) Commensal bacteria promote type I interferon signaling to maintain immune tolerance in mice. J Exp Med 221(1):e20230063
Wan Z, Zheng J, Zhu Z, Sang L, Zhu J, Luo S et al (2022) Intermediate role of gut microbiota in vitamin B nutrition and its influences on human health. Front Nutr 9:1031502
Wang B, Yao M, Lv L, Ling Z, Li L (2017) The human microbiota in health and disease. Engineering 3(1):71–82. https://doi.org/10.1016/J.ENG.2017.01.008
Wang M, Wichienchot S, He X, Fu X, Huang Q, Zhang B (2019) In vitro colonic fermentation of dietary fibers: fermentation rate, short-chain fatty acid production and changes in microbiota. Trends Food Sci Technol 88:1–9. https://doi.org/10.1016/j.tifs.2019.03.005
Wang X, Bi L, Li J, Liu D, Li W, Wang Z (2025) Dynamic balance of the lung microbiome in health and respiratory diseases. Chin Med J:10–1097. https://doi.org/10.1097/CM9.0000000000003712
West CE, Jenmalm MC, Kozyrskyj AL, Prescott SL (2016) Probiotics for treatment and primary prevention of allergic diseases and asthma: looking back and moving forward. Expert Rev Clin Immunol 12(6):625–639

Wu H, Forslund S, Wang Z, Zhao G (2025) Human gut microbiome researches over the last decade: current challenges and future directions. Phenomics 5(1):1–7. https://doi.org/10.1007/s43657-023-00131-z

Yamaguchi T, Nomura A, Matsubara A, Hisada T, Tamada Y, Mikami T, Ishida M (2023) Effect of gut microbial composition and diversity on major inhaled allergen sensitization and onset of allergic rhinitis. Allergol Int 72(1):135–142. https://doi.org/10.1016/j.alit.2022.06.005

Yang F, Yang L, Chang Z, Chang L, Yang B (2018) Regulation of virulence and motility by acetate in enteropathogenic Escherichia coli. Int J Med Microbiol 308(7):840–847

Yee AL, Miller E, Dishaw LJ, Gordon JM, Ji M, Dutra S et al (2019) Longitudinal microbiome composition and stability correlate with increased weight and length of very-low-birth-weight infants. MSystems 4(1):10–1128

Zhang Y, Wang X, Li H, Ni C, Du Z, Yan F (2018) Human oral microbiota and its modulation for oral health. Biomed Pharmacother 99:883–893

Zhang C, Li L, Jin B, Xu X, Zuo X, Li Y, Li Z (2021) The effects of delivery mode on the gut microbiota and health: state of art. Front Microbiol 12:724449

Zhao L, Luo JL, Ali MK, Spiekerkoetter E, Nicolls MR (2023) The human respiratory microbiome: current understandings and future directions. Am J Respir Cell Mol Biol 68(3):245–255

Zhao X, Zhu X, Wang J, Ye C, Zhao S (2025) The epidemiological analysis of respiratory virus infections in children in Hangzhou from 2019 to 2023. Virus Res 355:199558. https://doi.org/10.1016/j.virusres.2025.199558

Zheng P, Wu J, Zhang H, Perry SW, Yin B, Tan X et al (2021) The gut microbiome modulates gut–brain axis glycerophospholipid metabolism in a region-specific manner in a nonhuman primate model of depression. Mol Psychiatry 26(6):2380–2392

Zheng Y, Hunt RL, Villaruz AE, Fisher EL, Liu R, Liu Q et al (2022) Commensal Staphylococcus epidermidis contributes to skin barrier homeostasis by generating protective ceramides. Cell Host Microbe 30(3):301–313

Zhou F, Zhang GD, Tan Y, Hu SA, Tang Q, Pei G (2023) NOD-like receptors mediate homeostatic intestinal epithelial barrier function: promising therapeutic targets for inflammatory bowel disease. Ther Adv Gastroenterol 16:17562848231176889

The Gut-Cancer Connection and Microbial Insights

3

Abstract

The gut microbiota is a significant factor in cancer development and systemic health, affecting tumorigenesis and treatment outcomes. This chapter analyzes the dual function of microbiota in cancer. Some microorganisms can effectively enhance immune responses and suppress tumor growth whereas others may lead to chronic inflammation, produce carcinogenic metabolites, alter immune responses and affect gene expression thereby promoting the development of malignancies such as colorectal, pancreatic, liver, breast and lung cancers. The gut microbiota affects the effectiveness and toxicity of cancer treatments including chemotherapy and immunotherapy, positioning it as both a biomarker and a therapeutic target. Innovative strategies such as fecal microbiota transplantation, engineered bacteria for targeted tumor therapy, precision probiotics, synbiotics and microbiome diagnostics are being increasingly investigated to enhance treatment outcomes and minimize adverse effects. Microbial metabolites including short chain fatty acids and indoles highlight the potential for microbiota to modulate cancer pathways. Despite significant progress, challenges remain including inadequate large scale clinical validation, limited understanding of mechanisms, substantial inter individual variability and a lack of standardized methodologies. Future directions emphasize precision microbiome engineering, artificial intelligence-driven predictive tools and multiomics strategies to develop personalized microbiome guided interventions. This chapter underscores the transformative potential of microbiota-based strategies in cancer prevention and therapy highlighting the necessity for additional mechanistic research, standardized clinical frameworks and interdisciplinary collaboration to bridge current gaps in science and translation.

R. C. Sobti et al., *Microbiota Unveiled: Science, Health, and Future Therapies*,
https://doi.org/10.1007/978-981-95-9031-5_3

Keywords

Gut microbiome · Dysbiosis · Tumorigenesis · Chemotherapy · Immunotherapy · Fecal microbiota transplantation · engineered bacteria · Probiotics · Synbiotics · Microbial metabolites

3.1 Overview of Cancer and Its Global Impact

Cancer is a collection of over 100 different malignancies that are characterized by the uncontrolled growth of aberrant cells. Such cells have the ability to invade adjacent tissues and in a majority of instances to travel to distant organs via the bloodstream or lymphatic metastasis. Cancer development is a multistep process that includes genetic and epigenetic mutations that interfere with the regulation of cell growth, differentiation and death (Hanahan and Weinberg 2011). Cancer can effectively occur in nearly any organ in the body and has a wide array of histologic and molecular features. It is produced by a plethora of risk factors such as environmental factors (tobacco smoke, radiation, and carcinogenic chemicals), infectious agents (human papillomavirus (HPV), hepatitis B and C, and *Helicobacter pylori*), lifestyle factors (diet, alcohol intake, inactivity, and obesity), and inherited germ cell mutations (Fig. 3.1). Cancer continues to be a leading cause of morbidity and death worldwide. Based on the GLOBOCAN 2020 estimates by the International Agency for Research on Cancer (IARC), there were an estimated 19.3 million new cases and nearly ten million cancer-related deaths worldwide in 2020 (Sung et al. 2021). Top 5 most common cancers were: breast (11.7%), lung (11.4%), colorectal (10.0%), prostate (7.3%), and stomach (5.6%) cancers. Lung cancer is the top cause of cancer death with 1.8 million fatalities (18%) in 2020. The burden of cancer will continue to elevate exponentially over the next few decades with population growth, aging, and increased risk factor exposure due to urbanization and socioeconomic development. In 2040, the burden of cancer worldwide is likely to be 28.4 million cases annually, i.e., a 47% hike over the year 2020.

3.1.1 Leading Types of Cancer Worldwide

Lung cancer is the leading cause of cancer-related death. Breast cancer is the most common cancer overall, but colorectal, prostate, stomach, and liver cancers are among the most frequently diagnosed cancers (Sung et al. 2021). About 22% of cancer deaths are caused by a combination of alcohol use, poor diet, inactivity, environmental pollution, and tobacco use. Cancer has a huge economic impact. The cost of cancer worldwide, including medical bills and lost productivity, was $1.16 trillion in 2010. The price has increased ever since. This economic strain is particularly felt by developing countries with limited resources. International programs like the World Health Organization (WHO) Global Action Plan for the Prevention and Control of Non communicable Diseases 2013–2020 support integrated cancer

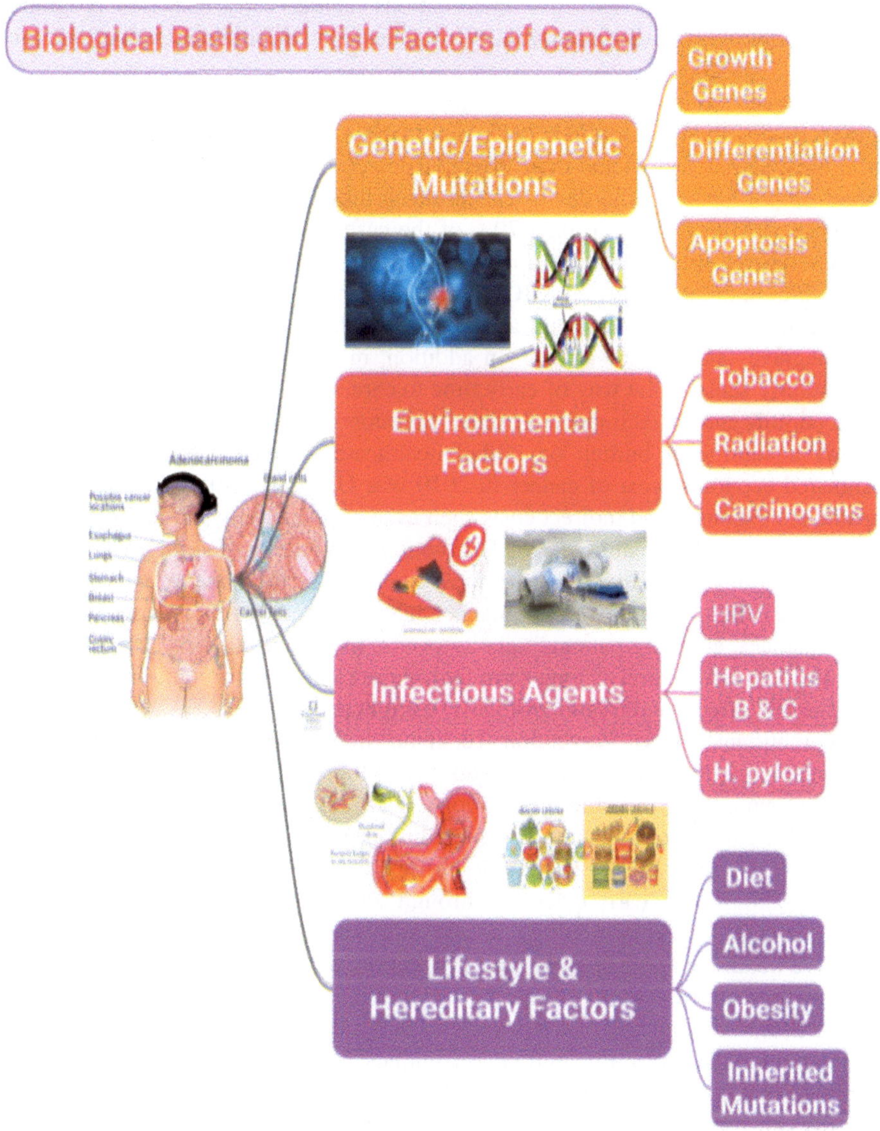

Fig. 3.1 Biological basis and risk factors of cancer: illustrates different risk factors responsible for causing cancer in different organs. The multistep process by which cancer develops is fueled by genetic and epigenetic mutations that interfere with the regular control of cell division, growth, and programmed cell death. Several risk factors contribute to the development of cancer such as environmental exposures (radiation, tobacco smoke, and chemical carcinogens), infectious agents (*Helicobacter pylori*, hepatitis B and C viruses, and human papillomavirus), lifestyle-related factors (diet, alcohol consumption, physical inactivity, and obesity), and inherited germline mutations. These factors work in concert to encourage malignant transformation, growth, and metastasis

prevention and control. Effective interventions include early detection programs, tobacco control laws, vaccination (HPV vaccines), and affordable treatment options (Atun et al. 2015).

Disparities in incidence and mortality across regions: cancer burden is not uniformly distributed worldwide. Incidence is more common in high-income nations with lifestyle risk factors and more effective detection, whereas a disproportionately larger proportion of the burden of cancer death falls upon low- and middle-income countries (LMICs). In more than 70% of the cases of cancer death, this occurs in LMICs where early diagnosis, appropriate treatment, and palliative care are not accessible. In the majority of LMICs, there are no national programs for controlling cancer, there is limited health infrastructure, and there are economic constraints to late diagnosis and worse survival. In addition, misinformation stigma and culture consistently stand in the way of early time to care.

Economic and social impact of cancer: the economic cost of cancer is staggering, posing a significant burden on global healthcare systems and economies. The estimated economic cost of cancer in 2010 was US$1.16 trillion, including direct healthcare expenses and indirect costs such as lost productivity and income attributable to illness and premature death. This burden continues to rise each year, placing immense strain on individuals, families, and healthcare systems worldwide. In addition to the economic impact, cancer has profound psychological and social consequences, affecting patients' quality of life, mental health, and social functioning. Many patients survive long term and require ongoing care and monitoring to manage late effects and treatment-related complications.

International Cancer Prevention Techniques: A comprehensive and equitable approach is required to control and reduce the global burden of cancer. The World Health Organization and other international health organizations emphasize a range of effective cancer control strategies, including primary prevention (such as tobacco control, vaccination against oncogenic viruses like HPV and HBV, and promotion of healthy lifestyles), early detection through screening programs (breast, cervical, and colorectal cancer screening), timely and accurate diagnosis, and access to effective treatment and palliative care. In addition, public health policies, awareness campaigns, and reduction of environmental and occupational risk factors play a crucial role in minimizing cancer incidence and mortality worldwide.

International Cancer Prevention Techniques: A comprehensive and equitable approach is required to reduce the global burden of cancer. These strategies can be broadly categorized into four key areas: (1) Primary Prevention, (2) Early Detection and Screening, (3) Accessibility of Diagnostic and Curative Care, and (4) Palliative Care.

1. Primary Prevention: This approach focuses on reducing exposure to major risk factors such as unhealthy diets, alcohol consumption, tobacco use, and infections. Preventive measures include lifestyle modifications; vaccination programs such as human papillomavirus (HPV) and Hepatitis B virus (HBV) vaccines, and public health initiatives aimed at reducing cancer incidence.

2. Early Detection and Screening: This involves implementing systematic screening programs for high-incidence cancers such as colorectal, breast, and cervical cancer to enable early diagnosis and improve treatment outcomes.
3. Accessibility of Diagnostic and Curative Care: The initiative emphasizes expanding access to essential healthcare services, including imaging, pathology, surgery, radiotherapy, and chemotherapy, particularly in low-resource and underserved regions.
4. Palliative Care: This focuses on providing supportive and end-of-life care, including pain management and psychological support, to improve the quality of life for patients with advanced cancer.

In addition, international collaboration, increased research funding, and the strengthening of healthcare and educational systems are crucial to reducing global disparities and improving cancer outcomes worldwide.

3.1.2 Traditional Cancer Therapies and Their Limitations

Conventional cancer therapies like radiation therapy, chemotherapy, and surgery have long been the mainstay of cancer treatment. Even so, the survival rates of many cancers have significantly increased thanks to these modalities. In addition, they have serious disadvantages that have prompted research into more specialized and customized treatment modalities (Table 3.1).

- Surgery: It is often the first line of treatment for solid tumors in their early stages. It has the advantage of immediate tumor removal or debulking, which in some cases may be curative. However, surgery is limited by factors such as tumor location, metastatic spread, and the risk of recurrence (Edge and Compton 2010). Moreover, surgical procedures are often associated with significant morbidity, lengthy recovery periods, and can be highly invasive.
- Radiation therapy: Radiation therapy uses high-energy ionizing radiation to destroy cancer cells by damaging their DNA. Although radiation therapy is beneficial (for localized cancers), it can effectively damage nearby healthy tissues, leading to adverse effects like fatigue, skin irritation, and chronic organ failure (Delaney et al. 2005). Furthermore, some tumors might develop a resistance to radiation which would reduce the efficacy of treatment.
- Chemotherapy: Chemotherapy uses systemic administration of cytotoxic drugs to target rapidly dividing cells. It has several serious disadvantages especially "non-specificity" which can effectively damage healthy cells that are proliferating such as bone marrow, the gastrointestinal tract, and hair follicles even though it is still the standard of care for many cancers. In addition to causing serious side effects like immunosuppression, nausea, alopecia, and neuropathy, drug resistance is caused by genetic changes in cancer cells (Longley and Johnston 2005).

Table 3.1 Evaluation of new and conventional cancer treatments

Type of therapy	Explanation	Benefits	Restrictions	References
Surgery	Tumors are physically removed	Cures localized tumors	Risk of complications; invasive; limited use in metastatic disease	Edge and Compton (2010)
Radiation therapy	Ionizing radiation is used to destroy cancer cells	Noninvasive and effective for localized cancers	Radio-resistance; harm to nearby healthy tissues	Delaney et al. (2005)
Chemotherapy	Use of cytotoxic medications systemically	Widely used; capable of treating metastatic or widespread cancer	Severe side effects, drug resistance, and non-specific toxicity	Longley and Johnston (2005)
Targeted therapy	Medications that target particular cancer cell molecular pathways	Fewer adverse effects and high specificity	Resistance development; restricted to known target cancers	Sawyers (2004)
Immunotherapy	Uses the immune system to combat cancer	Long-lasting effects; useful for certain advanced cancers	Adverse events linked to the immune system; varying effectiveness	Postow et al. (2015)
Hormonal therapy	Slows the growth of hormone-sensitive cancer by altering hormone levels	Beneficial for prostate and breast cancers	Long-term resistance; adverse effects of hormonal fluctuations	Jordan (2003)
Gene/cell therapy	Alters cells or genes to combat cancer	Individualized care with the possibility of a sustained remission	Expensive, scarce, and prone to severe reactions	Maude et al. (2014)
Nanomedicine	Employs nanoparticles to deliver drugs precisely	Decreased systemic toxicity and improved drug delivery	Research is ongoing; scalability and regulatory issues exist	Bobo et al. (2016)

3.1.2.1 Broad Restrictions of Traditional Methods

Traditional treatments usually ignore tumor heterogeneity, minimal residual disease, and metastatic progression, all of which are significant contributors to treatment failure and cancer recurrence (Hanahan and Weinberg 2011). Moreover, these approaches typically follow a one size fits all paradigms which are increasingly recognized as inadequate for the complex biology of cancer. These limitations have led to the development of novel therapies such as gene-based interventions, immunotherapy, and targeted molecular therapies that aim to enhance treatment accuracy and lower systemic toxicity.

Novel and emerging treatment methods: The development of innovative therapeutic approaches targeted at increasing efficacy, lowering toxicity, and conquering resistance has accelerated due to the shortcomings of conventional cancer treatments. By applying knowledge from genomics, immunology, and molecular biology, these innovative treatments offer more individualized and focused cancer treatment.

- Targeted Therapy: Molecular Pathway Inhibition: Targeted therapy aims to interfere with specific molecular pathways that regulate the growth, survival, and proliferation of cancer cells. Unlike conventional chemotherapy, which affects both rapidly dividing healthy and cancerous cells, targeted therapies act selectively on cancer-specific molecules, thereby reducing damage to normal tissues and minimizing side effects (Sawyers 2004).
  Examples include monoclonal antibodies such as trastuzumab, which targets HER2-positive breast cancer, and tyrosine kinase inhibitors such as imatinib, used in chronic myeloid leukemia. These therapies work by blocking signaling pathways essential for tumor progression. However, challenges such as tumor heterogeneity, development of drug resistance due to secondary mutations, and limited efficacy in some patients remain significant concerns.
- Immunotherapy: Harnessing the Immune System: Immunotherapy aims to harness the patient's immune system to recognize and eliminate cancer cells. Key strategies include immune checkpoint inhibitors (e.g., anti-PD-1 and anti-CTLA-4 therapies), chimeric antigen receptor (CAR) T-cell therapy, which is particularly effective in hematological malignancies, cancer vaccines, and cytokine-based therapies. Immunotherapy has demonstrated durable responses in certain advanced cancers, including melanoma and lung cancer. However, it can be associated with immune-related adverse events and is not universally effective across all tumor types (Postow et al. 2015).
- Hormonal Therapy for Hormone-Dependent Cancers: Hormonal therapy is particularly effective for hormone-sensitive cancers such as breast and prostate cancer, where it works by blocking or reducing the body's hormone levels to slow or inhibit tumor growth. Common approaches include androgen deprivation therapy, aromatase inhibitors, and selective estrogen receptor modulators such as tamoxifen (Jordan 2003). However, the development of treatment resistance and disease relapse remains a significant clinical challenge.
- Gene and Cell-Based Therapies: Gene therapy aims to treat cancer by correcting or modifying genetic abnormalities that drive tumor development. This can be achieved through the delivery of tumor suppressor genes, inhibition of oncogenes using small interfering RNA (siRNA), or precise genome editing using CRISPR-Cas9 technology. These approaches work by restoring normal cellular function or preventing uncontrolled cell proliferation.
  Cell-based therapies, such as chimeric antigen receptor (CAR) T-cell therapy and other engineered immune cells, improve the body's immune response by

specifically targeting and killing cancer cells. These therapies have shown remarkable success, particularly in hematological malignancies. However, their broader application is limited by challenges such as high cost, complex manufacturing processes, potential toxicity (cytokine release syndrome), and issues related to scalability and accessibility (Maude et al. 2014).

- Nanomedicine and Precision Drug Delivery: Nanotechnology enables targeted drug delivery by improving drug solubility, stability, and bioavailability while allowing controlled and sustained release of therapeutic agents. Nanoparticles can preferentially accumulate in tumor tissues due to the enhanced permeability and retention (EPR) effect, which arises from the leaky vasculature and poor lymphatic drainage characteristic of tumors. This targeted accumulation increases drug concentration at the tumor site while minimizing exposure to healthy tissues, thereby reducing systemic toxicity and improving therapeutic efficacy (Bobo et al. 2016).

3.1.2.2 Gut Health and Its Impact on Cancer

This chapter aims to address new research showing the role of gut health (composition and function of the gut microbiome) in influencing the risk and progression of cancer like 1. Dysbiosis and carcinogenesis: Research suggests that an imbalance in the gut's microbial community may contribute to oncogenesis by triggering the release of genotoxin causing chronic inflammation and altering the immune system (Garrett 2015) 2. Microbial species linked to cancer: By promoting tumorigenesis and inhibiting immune cell function, certain bacterial species such as *Fusobacterium nucleatum* have been linked to colorectal cancer (Kostic et al. 2013) 3. Microbial metabolites and tumor modulation: Depending on the circumstances surrounding their production and the host environment, microbial metabolites such as short-chain fatty acids can effectively have both procarcinogenic and beneficial effects (Louis et al. 2014) 4. Interactions between microbiome and therapy (Immunotherapy, Chemotherapy): There is proof that the gut microbiota serves a pivotal role in a crucial role in the development of cancer as well as its treatment, and it also regulates the effectiveness and toxicity of immunotherapy and chemotherapy (Gopalakrishnan et al. 2018). The current paper reviews the literature in an effort to highlight the complex connections between gut health and cancer biology as well as the potential of microbiome-directed cancer prevention and treatment strategies (Fig. 3.2).

3.2 Human Microbiome Developing Functions in Health and Illness

We now understand that the human microbiome serves a pivotal role in a significant role in preserving physiological homeostasis and influencing the etiology of disease. These microbial communities especially those in the gastrointestinal tract have a complex impact on metabolic, neuroendocrine, and host immunity systems. Next-generation sequencing technology advancements have made it possible to profile

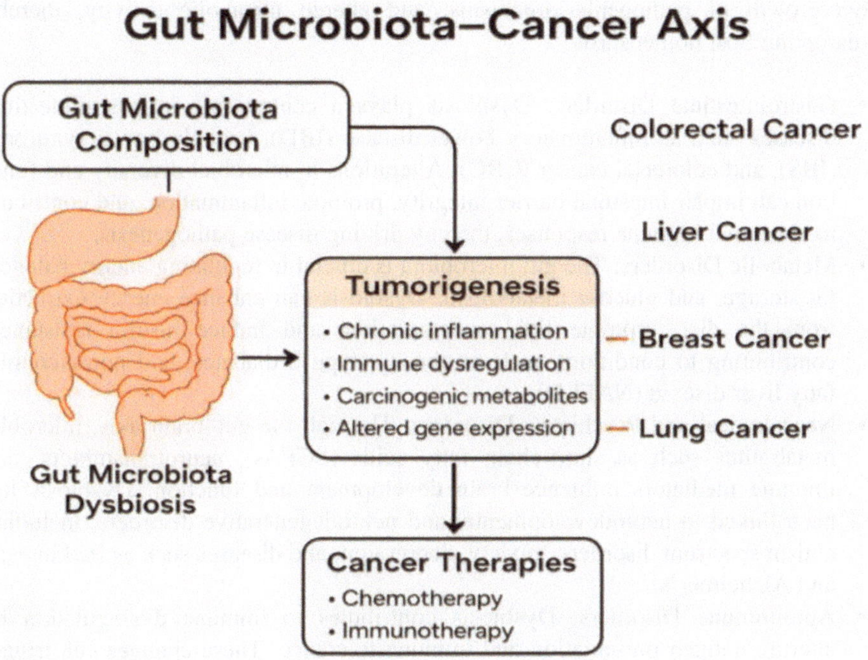

Fig. 3.2 Role of gut microbiota in cancer development and treatment

these intricate microbial ecosystems in great detail which has demonstrated a significant influence on both health and disease.

3.2.1 Physiological Homeostasis and the Microbiome

The digestion of complex polysaccharides, the production of vitamins (K and B), maintenance of the integrity of the gut barrier, and immune modulation are all facilitated by a balanced microbiome. By producing short-chain fatty acids (butyrate, propionate, and acetate), the gut microbiota maintains immunological tolerance and provides guidance to the host immune system during early development. These metabolites promote the health of epithelial cells and have anti-inflammatory properties.

3.2.2 Dysbiosis and Its Links to Chronic Diseases

Dysbiosis, defined as an imbalance in the composition and function of the gut microbiota, has been increasingly associated with the development and progression of several chronic diseases. This imbalance can lead to loss of beneficial microbes, overgrowth of pathogenic organisms, and altered metabolic activity, thereby disrupting host homeostasis.

- Gastrointestinal Disorders: Dysbiosis plays a central role in gastrointestinal diseases such as inflammatory bowel disease (IBD), irritable bowel syndrome (IBS), and colorectal cancer (CRC). Alterations in microbial diversity and function can impair intestinal barrier integrity, promote inflammation, and contribute to abnormal immune responses, thereby driving disease pathogenesis.
- Metabolic Disorders: The gut microbiota is crucial in regulating energy balance, fat storage, and glucose metabolism. Dysbiosis can enhance energy extraction from the diet, promote lipid accumulation, and induce insulin resistance, contributing to conditions such as obesity, type 2 diabetes, and non-alcoholic fatty liver disease (NAFLD).
- Neurological and Psychiatric Disorders: Through the gut–brain axis, microbial metabolites such as short-chain fatty acids (SCFAs), neurotransmitters, and immune mediators influence brain development and function. Dysbiosis has been linked to neurodevelopmental and neurodegenerative disorders, including autism spectrum disorders, anxiety, depression, and diseases such as Parkinson's and Alzheimer's.
- Autoimmune Disorders: Dysbiosis contributes to immune dysregulation by altering antigen presentation and immune tolerance. These changes can trigger inappropriate immune responses against self-antigens, playing a significant role in autoimmune diseases such as multiple sclerosis, rheumatoid arthritis, and systemic lupus erythematosus.

3.2.3 Therapeutic Modulation of the Microbiome (Probiotics, Prebiotics, and Fecal Microbiota Transplantation)

Therapeutic modulation of the microbiome has emerged as a promising strategy for disease prevention and treatment. Approaches such as probiotics, prebiotics, synbiotics, dietary interventions, and fecal microbiota transplantation (FMT) aim to restore microbial balance and improve host health. As the role of these microbiome modulators has been discussed in detail in a previous chapter, only a brief overview is provided here to avoid redundancy. Additionally, advances in personalized medicine are enabling the development of microbiome-based therapies tailored to individual microbial profiles.

3.3 Increasing Awareness of the Gut Microbiome's Role in Health

3.3.1 Gut Microbiome Diversity and Composition

The human gastrointestinal tract is home to a highly diverse community of trillions of microorganisms called "gut microbiome." Among them, bacterial species serves a pivotal role in crucial role in most of the bodily function. Advances in sequencing

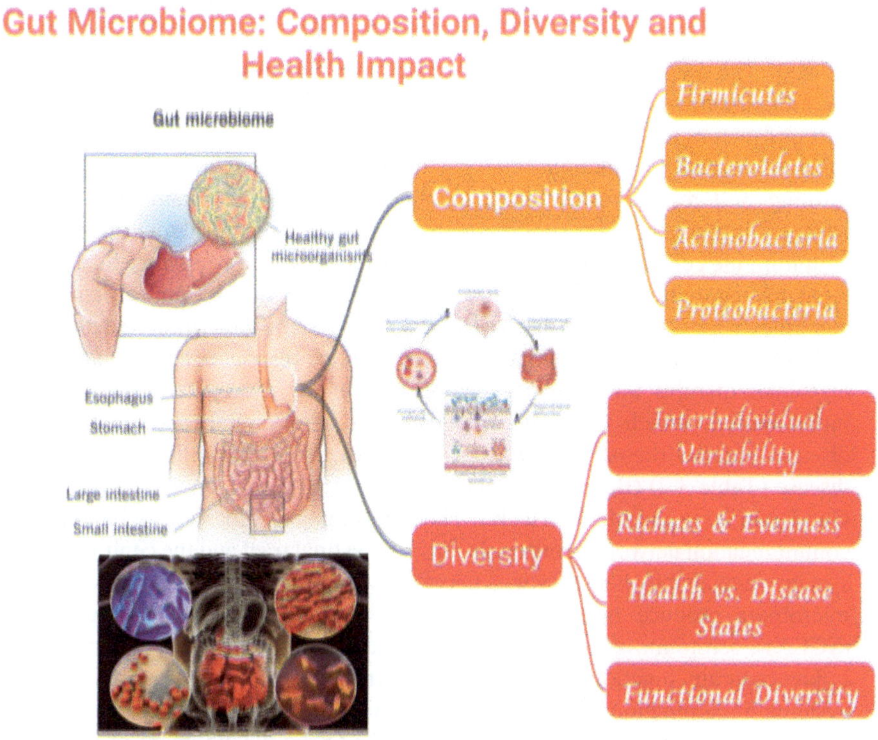

Fig. 3.3 The composition, diversity, and health impact of the gut microbiome

methods have expanded our understanding of this community and demonstrated the critical role the gut microbiota serves a pivotal role in in both health as well as disease when dysregulation takes place. Bacteria from the phyla Firmicutes, Bacteroidetes, Actinobacteria, and Proteobacteria make up the majority of the gut microbiome. Every individual has a unique microbiota that is impacted by lifestyle, environmental, dietary, and genetic factors. Understanding the functional capacity and microbial richness will facilitate one better understand how it affects health (Fig. 3.3) (Turnbaugh et al. 2007).

3.3.2 Functions of the Gut Microbiome

Figure 3.3 illustrates the main bacterial phyla of the gut microbiome (Proteobacteria, Actinobacteria, Bacteroidetes, and Firmicutes), their role in maintaining homeostasis, regulating immunity, and metabolizing nutrients. It also represents microbial diversity, which includes interindividual variability, species richness/evenness, functional diversity, and distinctions between health and disease states as one of the main factors influencing the microbiome's overall impact on human health.

1. Digestion and Nutrient Metabolism: When gut bacteria ferment indigestible dietary fibers, short-chain fatty acids (acetate, butyrate, and propionate) are created. The SCFAs produce energy and have anti-inflammatory qualities. In addition, the microbiota produces vitamins (K and B), which are essential for gut barrier function and nutrient metabolism (Koh et al. 2016).
2. Immune System Regulation: The gut microbiome plays an important role in forming and regulating the host immune system by maintaining a balance between immune tolerance and immune activation. Commensal microbes interact with immune cells through pattern recognition receptors, promoting the development of regulatory T cells (Tregs) and the production of anti-inflammatory cytokines such as interleukin-10 (IL-10). At the same time, they enhance host defense by stimulating innate and adaptive immune responses against pathogens. Dysbiosis disrupts this balance, leading to immune dysregulation and chronic inflammation, which has been implicated in autoimmune and inflammatory diseases such as rheumatoid arthritis, type 1 diabetes, and inflammatory bowel disease (Table 3.2) (Belkaid and Hand 2014).
3. Gut–Brain Axis and Mental Health: The gut–brain axis represents a bidirectional communication network between the gastrointestinal tract and the central nervous system, mediated by neural (vagus nerve), endocrine, and immune pathways. Gut microbiota produce neuroactive compounds such as short-chain fatty acids (SCFAs), serotonin, gamma-aminobutyric acid (GABA), and dopamine, which influence brain function and behavior. Microbial metabolites can also change systemic inflammation and the integrity of the blood-brain barrier. Dysbiosis can disrupt these signaling pathways, contributing to the development of neuropsychiatric conditions such as depression, anxiety, autism spectrum disorders, and neurodegenerative diseases (Table 3.2) (Cryan and Dinan 2012).
4. Obesity and Metabolic Health: The gut microbiota plays a significant role in metabolic regulation by influencing energy extraction from the diet, lipid metabolism, and insulin sensitivity. Certain microbial communities enhance the breakdown of complex polysaccharides, increasing caloric availability and promoting fat storage. Dysbiosis is associated with altered production of metabolites such as short-chain fatty acids and bile acids, which affect metabolic signaling pathways and glucose homeostasis. Despite the association of an increased Firmicutes-to-Bacteroidetes ratio with obesity, studies yield inconsistent findings. Overall, gut microbiota contribute to the pathogenesis of metabolic disorders, including obesity, insulin resistance, and type 2 diabetes (Fig. 3.3) (Turnbaugh et al. 2006).

3.3.3 Therapeutic Strategies and Microbiome-Based Personalized Medicine

Probiotics, prebiotics, and fecal microbiota transplantation are therapeutic strategies that target the gut microbiome. FMT is being researched for use with a number of conditions and has been found to be effective in treating recurrent *Clostridium difficile* infections. In addition, both prebiotics and probiotics have the potential to

Table 3.2 Comprehensive function of the gut microbiota in the development and treatment of cancer

Mechanism	Description	Key microbial players	Associated cancers	References
Chronic inflammation and dysbiosis	Chronic inflammation and a cancer-promoting tumor microenvironment are fostered by persistent immune activation caused by microbial imbalance	*Escherichia coli* and *Fusobacterium nucleatum*	Pancreatic, liver and colorectal	Garrett (2015) and Zitvogel et al. (2016)
Carcinogenic metabolite production	Some bacteria produce genotoxic metabolites that damage DNA such as nitrosamines, hydrogen sulfide, and secondary bile acids	Bacteroides species and *Clostridium* species	Gastric and colorectal	Louis et al. (2014) and Schwabe and Jobin (2013a)
Modulation of the immune system	The gut microbiota affects how the body reacts to checkpoint inhibitors such as PD-1/PD-L1 and regulates host immunity	*Bifidobacterium* species and *Akkermansia muciniphila*	Renal cell carcinoma, lung cancer and melanoma	Gopalakrishnan et al. (2018) and Routy et al. (2018)
Changes in epigenetics	Metabolites produced by bacteria change the expression of genes linked to cancer by influencing DNA methylation and histone modifications	*Lactobacillus* species and *Roseburia* species	Breast, colorectal	Johnson et al. (2015)
Drug metabolism in chemotherapy	The effectiveness and toxicity profiles of chemotherapy medications are impacted by the way gut microbes alter their metabolism	*E. coli* and *Enterococcus faecalis*	Breast, colorectal	Alexander et al. (2017)
Intervention in therapy	Immunotherapy results can be improved by techniques like probiotic	*Lactobacillus* species, donor diverse microbiota	Melanoma and colorectal	Davar et al. (2021) and Matson et al. (2018)

(continued)

Table 3.2 (continued)

Mechanism	Description	Key microbial players	Associated cancers	References
	administration and fecal microbiota transplantation.			
Prognostic and diagnostic biomarkers	Emerging noninvasive methods for early cancer detection and risk assessment include microbiome signatures	*Peptostreptococcus* species and *Fusobacterium nucleatum*	Pancreatic and colorectal	Zeller et al. (2014) and Ren et al. (2017)

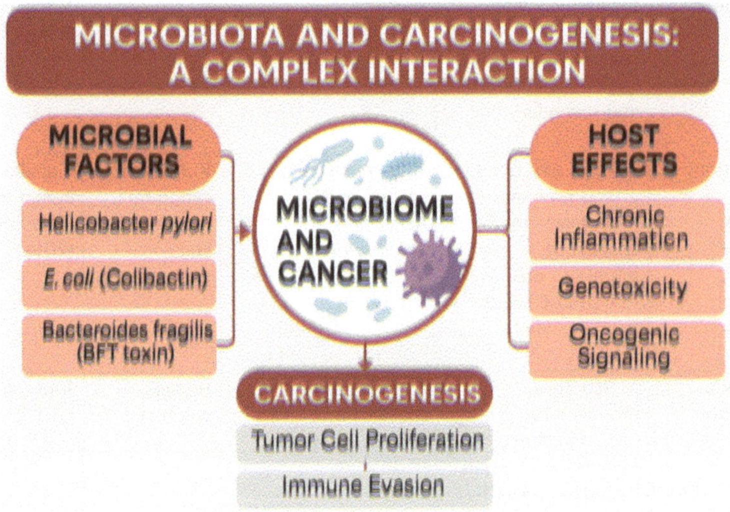

Fig. 3.4 Microbiota and carcinogenesis highlight the main ways that the gut microbiota influences the onset and spread of cancer. A protumorigenic environment is fostered by chronic inflammation such as that caused by *Helicobacter pylori* infection. Certain bacteria elevate genotoxicity by causing DNA damage and mutagenesis such as *Escherichia coli* (colibactin) and Bacteroides fragilis (BFT toxin). In addition, microbial metabolites affect oncogenic signaling pathways such as NF-κB and Wnt/β-catenin which promote aberrant cell survival, proliferation, and tumor initiation

alter gut health (Table 3.2) (Smits et al. 2013). Moreover, personal microbiome profiles are being targeted by personalized nutrition and medicine. The composition of the microbial community can effectively predict the pharmacologic and dietary response, resulting in customized health plans (Zeevi et al. 2015). With effects on metabolism, immunity, mental health, and digestion, the gut microbiota is an essential component of total health. Microbiome therapeutics and diagnostics hold the potential to transform medicine as the field develops. To fully utilize this dynamic system's therapeutic potential, more research is required (Table 3.2).

3.4 Microbial Insights in Oncology

3.4.1 Carcinogenesis and Microbiota

Recent advances in microbiome research have significantly transformed our understanding of cancer biology, with important implications for its diagnosis, prevention, and treatment. The gut microbiota plays a critical role in carcinogenesis by influencing inflammation, genomic stability, immune responses, and cellular signaling pathways.

One of the primary mechanisms linking microbiota to cancer is chronic inflammation, which is often driven by pathogenic microorganisms such as *Helicobacter pylori*. Persistent inflammation promotes the production of pro-inflammatory cytokines and reactive oxygen species, leading to DNA damage and increased risk of malignant transformation.

In addition, certain microbes exert genotoxic effects through the production of harmful metabolites. For example, *Bacteroides fragilis* produces *B. fragilis* toxin (BFT), while some strains of *Escherichia coli* produce colibactin; both are capable of inducing DNA damage and promoting mutagenesis.

Furthermore, the microbiota can influence oncogenic signaling pathways by modulating key molecular mechanisms such as nuclear factor-kappa B (NF-κB) and Wnt/β-catenin pathways. These alterations affect cell proliferation, apoptosis, and survival, thereby contributing to tumor initiation and progression. Overall, the complex interactions between the microbiota and host cellular processes highlight its critical role in carcinogenesis (Fig. 3.4; Table 3.2).

3.4.2 Microbial Signatures as Cancer Biomarkers

The microbiome has emerged as a promising biomarker for the early detection of cancer, as distinct microbial signatures have been associated with different tumor types. In colorectal cancer, patients often exhibit an increased abundance of *Fusobacterium nucleatum* along with a reduction in beneficial bacterial groups such as Firmicutes, which may contribute to tumor development and progression. Similarly, in pancreatic and liver cancers, alterations in both oral and gut microbiota have been identified, suggesting that microbial dysbiosis could serve as a non-invasive diagnostic marker for hepatobiliary malignancies. These findings highlight the potential of microbiome-based profiling as a tool for early cancer detection and risk assessment.

3.4.3 Microbiota and Cancer Therapy Outcomes

The microbiome plays a significant role in influencing both the efficacy and toxicity of cancer therapies, thereby offering promising opportunities for microbiota-driven precision oncology. In the context of immunotherapy, the gut microbiota has been shown to modulate host responses to immune checkpoint inhibitors such as anti-PD-1/PD-L1 and anti-CTLA-4 therapies. For instance, an increased abundance of beneficial microbes such as *Bifidobacterium* species, *Faecalibacterium prausnitzii*, and *Akkermansia muciniphila* has been associated with improved therapeutic responses in cancers such as melanoma and lung cancer.

Similarly, the microbiome also affects the efficacy and toxicity of chemotherapeutic agents by influencing their metabolism. Certain gut bacteria can enhance the immunogenic effects of drugs like cyclophosphamide, while others may inactivate drugs such as irinotecan, thereby altering treatment outcomes and side-effect profiles.

Furthermore, the microbiome contributes to radiotherapy tolerance by regulating radiation-induced inflammation and facilitating tissue repair mechanisms. These interactions highlight the importance of the microbiome in optimizing cancer treatment strategies and reducing therapy-related toxicity (Table 3.2).

3.4.4 Modulation of the Microbiome in Cancer Therapy

Modifying the microbiome is increasingly being explored as a complementary approach in cancer treatment, particularly to enhance therapeutic efficacy and reduce treatment-related toxicity. Fecal microbiota transplantation (FMT) is a new strategy that aims to restore beneficial microbial communities and has shown promise in improving responses to immunotherapy.

In addition, approaches such as probiotics and prebiotics may support gut health and mitigate treatment-associated side effects; however, as these have been discussed in detail in previous chapters, only a brief mention is included here.

Conversely, the use of antibiotics in oncology requires careful consideration, as they may disrupt beneficial microbiota and negatively impact treatment outcomes, including responses to immunotherapy. Therefore, their use should be judicious and closely monitored (Table 3.2)

3.5 Overview of the Microbiome and Cancer

3.5.1 The Human Microbiome

Composition and scope: human microbiome describes the combined genomic makeup of all microorganisms (bacteria, fungi, viruses, archaea, and protozoa) that live in different parts of the human body, including skin, gut, oral cavity, and respiratory and urogenital tracts (Clemente et al. 2012; Grice and Segre 2011;

Sekirov et al. 2010). Host development, physiology, immunity, and even psychological functions like mood and cognition are all significantly impacted by this dynamic and complex ecosystem (Marchesi and Ravel 2015; Clemente et al. 2012). While microbiota refers specifically to the microbial community found in a particular environment, microbiome refers to both the microorganisms and their genetic material (Turnbaugh et al. 2007). Microbial cells are thought to outnumber human cells by a factor of ten and their genes are far more varied and functional than those of the human genome as per the Human Microbiome Project Consortium (2012). A critically important effort to map the human microbiome and comprehend its role in health and disease was the Human Microbiome Project (HMP), which was started by the NIH in 2008. Genetics, diet, lifestyle, environment, and medical interventions all have an impact on the microbiome (Schwabe and Jobin 2013b). Bacteria are the most prevalent of its essential components particularly in the GI tract, where they support the production of vitamins like B12, the digestion of complex carbohydrates and the preservation of mucosal integrity (Ley et al. 2006; Marchesi and Ravel 2015). Actinobacteria, Proteobacteria, Bacteroides, and Firmicutes are common bacterial phyla (Clemente et al. 2012). Archaea (*Methanobrevibacter smithii*) facilitate the production of methane during fermentation (Duller and Moissl-Eichinger 2024). When dysbiosis occurs, infections can effectively be caused by Candida species and other fungi that are more prevalent in the gut, skin, and oral cavity (Grice and Segre 2011). The viral component, which includes bacteriophages and human viruses, influences microbial balance and promotes horizontal gene transfer, which impacts microbial evolution and host interactions (Duan et al. 2022). Lastly, although they receive less attention, protozoa are present in the gut and may encourage microbial diversity while preventing the growth of pathogenic microorganisms (Chabé et al. 2017).

3.5.2 Ecological Roles of the Microbiome in Host Physiology

To maintain host homeostasis, microorganisms in the human microbiome engage in a range of ecological interactions including commensalism, competition, and mutualism. (Belkaid and Hand 2014) revealed that the human microbiome is a dynamic and complex ecosystem. Microbes in symbiotic relationships benefit from their host's nutrients and environment while carrying out essential tasks like vitamin synthesis and pathogen defense (Kamada et al. 2013). Dysbiosis may result from disturbances in this tranquil community. For example, obesity has been linked to an excess of some pathogenic Firmicutes while diseases such as inflammatory bowel disease (IBD) are linked to a decrease in microbial diversity (Hou et al. 2022). The immune system and the microbiome work together to control inflammation, preserve immunological tolerance, and prevent the spread of infections. The microbiome affects both innate and adaptive immune responses. The innate immune system is the first line of defense. It uses immune cells like neutrophils, dendritic cells, and macrophages, as well as physical barriers like the skin and mucosa and pattern recognition receptor to identify pathogen-associated molecular patterns and initiate inflammation (Iwasaki and Medzhitov 2015). The adaptive immune system which uses T and B cells to identify microbial antigens and initiate specific responses like antibody production and immunological memory complements this by offering targeted and durable protection. Collectively, these connections show how the microbiome shapes and influences the immune landscape, promoting general health and resilience.

3.5.3 The Microbiome and the Immune System

3.5.3.1 Function in the Development of the Immune System

The human microbiome has a substantial impact on the maturation and development of the innate and adaptive immune systems particularly during the early years of life. Immune programming is significantly impacted by microbial colonization during the neonatal period, a critical time that establishes the foundation for long-term health outcomes. Microorganism-derived metabolites and structural components are crucial immunological educators that guide immune responses through a number of channels. To maintain immunological tolerance and prevent autoimmunity, commensal gut bacteria generate short-chain fatty acids that encourage the differentiation of regulatory T cells (Tregs) (Round and Mazmanian 2009). Furthermore, by controlling the maturation and activation of these immune populations and upholding a delicate balance between immune stimulation and tolerance, the microbiome has a substantial impact on the gut-associated lymphoid tissue which is home to a significant portion of the body's immune cells. In addition to immune regulation, the microbiota maintains the integrity of the mucosal barrier by promoting the production of mucus and antimicrobial peptides, which strengthen epithelial defenses and aid in preventing colonization and invasion by pathogens (Belkaid and Hand 2014). The microbiome fosters immune education and builds lifelong immune resilience through these complex interactions.

3.5.3.2 Microbial Immune Modulation Mechanisms

A variety of signaling pathways allow microbes to communicate with the immune system (Fig. 3.5). Among the most crucial mechanisms are:

1. Toll-like receptors (TLRs) and pathogen recognition: These pattern recognition receptors (PRRs) trigger inflammation when they detect microbial substances like lipopolysaccharides (LPS) or peptidoglycan. TLRs, which are expressed on immune cells such as dendritic cells and macrophages, are essential for pathogen detection and the initiation of immune responses.
2. Antigen presentation and adaptive immunity: When dendritic cells and macrophages expose T cells to microbial antigens, the adaptive immune system is triggered. This process is necessary for the development of immunity against specific pathogens.
3. Microbial metabolites: Immune modulation depends on short-chain fatty acids (SCFAs) and other bacterial metabolites. Dietary fibers are fermented by the gut

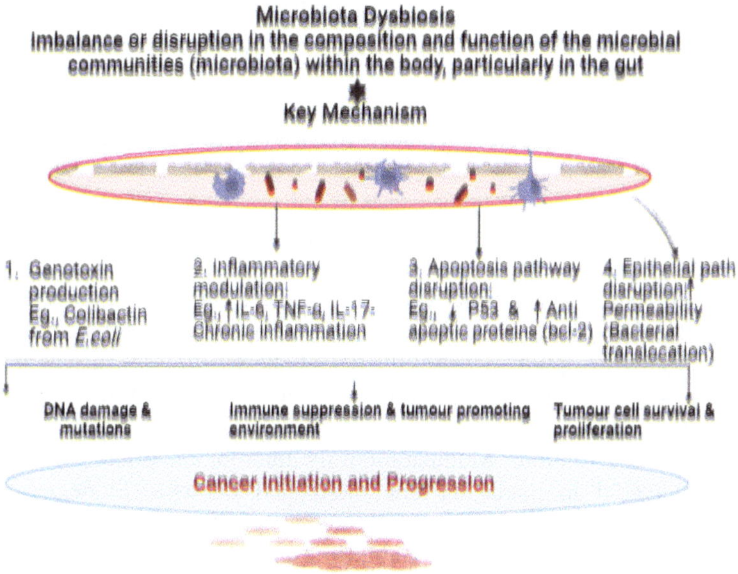

Fig. 3.5 The mechanistic relationship between cancer development and microbiota illustrates various ways that dysregulated microbiota contribute to the development of cancer such as (1) Microbial genotoxins (colibactin) are produced and causing DNA damage and mutagenesis (2) Disruption of host apoptotic pathways which is characterized by upregulation of antiapoptotic proteins (Bcl-2) and inhibition of tumor suppressors(p53) (3) Modulation of inflammatory responses through cytokines (IL-6, TNF-α & IL-17) which promote a tumor supportive microenvironment and (4) Compromise of epithelial barrier integrity which results in increased intestinal permeability and microbial translocation. All of these changes contribute to the development, spread, and immune evasion of cancer

microbiota to produce SCFAs such as butyrate, which have anti-inflammatory properties. These metabolites facilitate regulatory T cell (Treg) differentiation, which is necessary for immune tolerance and the prevention of inflammatory diseases (Arpaia et al. 2013).

4. Immuno-modulatory molecules: Certain microorganisms produce chemicals that directly affect immune responses. For example, *Bifidobacterium* species are known to produce bioactive compounds that elevate the synthesis of anti-inflammatory cytokines thereby promoting a state of immune tolerance.

3.5.3.3 Dysbiosis and Immune-Related Diseases

Dysbiosis, characterized by an imbalance in the gut microbiota, can significantly impact immune function and is associated with a wide range of diseases. Reduced microbial diversity may lead to immune dysregulation, resulting in increased susceptibility to infections, autoimmune disorders, and chronic inflammatory conditions. For example, diseases such as rheumatoid arthritis, asthma, and inflammatory bowel disease have been strongly linked to alterations in gut microbial composition (Sartor 2008).

1. Inflammatory and autoimmune diseases: Chronic low-grade inflammation resulting from gut microbiota dysbiosis plays a key role in the development and progression of autoimmune and inflammatory diseases such as Crohn's disease. Dysbiosis disrupts immune homeostasis by enhancing pro-inflammatory cytokine production and impairing regulatory immune mechanisms, leading to persistent immune activation and tissue damage (Sartor 2008).
2. Immune tolerance and allergic reactions: Reduced microbial exposure, particularly during early life, has been associated with impaired development of immune tolerance. This disruption increases the risk of allergic diseases such as asthma, as the immune system becomes more prone to hypersensitivity reactions (García-Rivero 2020).
3. Therapeutic interventions: Modulation of the microbiome offers promising therapeutic potential in managing immune-related diseases. Strategies such as probiotics, prebiotics, and fecal microbiota transplantation (FMT) aim to restore microbial balance and regulate immune responses. Probiotics and prebiotics promote the growth of beneficial microorganisms, enhance immune function, and reduce inflammation. FMT, which involves the transfer of microbiota from a healthy donor to a patient, has been shown to be highly effective in treating Clostridium difficile infections. Ongoing research is exploring its potential applications in conditions such as inflammatory bowel disease and cancer (Sekirov et al. 2010; Kelly et al. 2016; Wu et al. 2024).

3.5.4 The Microbiome and Cancer

3.5.4.1 Microbiome in Cancer Initiation

The human microbiome plays a significant role in influencing various physiological processes, including cancer development. Increasing evidence suggests that the microbiota contributes to cancer initiation, progression, and metastasis through multiple mechanisms, including direct interactions with host cells, modulation of immune responses, and regulation of inflammatory pathways. Cancer initiation typically begins with genetic mutations in normal cells that lead to uncontrolled proliferation, and the microbiome can facilitate this process through several interconnected mechanisms (Fig. 3.5).

1. Chronic inflammation and tumorigenesis: A dysbiotic microbiome can promote chronic inflammation, which is a well-established risk factor for cancer initiation. Persistent inflammation leads to the release of pro-inflammatory cytokines such as tumor necrosis factor-alpha (TNF-α) and interleukin-6 (IL-6), which can induce DNA damage, stimulate cell proliferation, and impair immune surveillance. Certain pathogenic bacteria, such as *Fusobacterium nucleatum* in colorectal cancer, have been shown to sustain inflammatory responses that create a

tumor-promoting microenvironment (Coussens and Werb 2002; Lahouty et al. 2025).

2. Carcinogenic metabolites and DNA damage: Some microbial species produce carcinogenic metabolites that directly damage DNA or promote mutagenesis. These include secondary bile acids (deoxycholic acid and lithocholic acid), hydrogen sulfide (H_2S), N-nitroso compounds (nitrosamines), and genotoxins such as colibactin. For instance, *Bacteroides fragilis* produces *B. fragilis* toxin (BFT), which can damage colonic epithelial DNA and contribute to colorectal cancer development. Additionally, metabolites such as butyrate produced by *Clostridium* species may exert both protective and harmful effects depending on their concentration and the surrounding microbial context (Koh et al. 2016). These metabolites promote genomic instability and increase the risk of malignant transformation.
3. Microbial DNA in tumor tissue: The presence of microbial DNA, particularly from pathogenic bacteria, has been detected in various tumor tissues, suggesting a direct association between microbes and cancer development. A well-established example is *Helicobacter pylori*, which is strongly linked to gastric cancer. This bacterium contributes to carcinogenesis by inducing chronic inflammation and generating reactive oxygen species (ROS) that cause DNA damage and genomic instability.

3.5.4.2 Role of Microbiota in Development and Spread of Cancer

The microbiome can effectively support tumor growth in a number of ways once cancer has begun such as

1. Immunosuppression by tumor-associated microbes: Certain bacteria can effectively impair the host's defenses, making tumors invisible. For example, *Fusobacterium nucleatum* has been associated with immune suppression in colorectal cancer by drawing myeloid-derived suppressor cells which inhibit T cell activation and encourage tumor growth.
2. Alteration of the tumor microenvironment: The microbiome can effectively influence the tumor microenvironment by altering the extracellular matrix, cytokine production, and local immune cell populations. Microbial metabolites such as SCFAs can effectively promote tumor growth by regulating the ratio of proinflammatory to anti-inflammatory cytokines. Dysbiosis may lead to a shift toward a more proinflammatory Tumor Microenvironment (TME) that encourages the development and spread of cancer cells.
3. Angiogenesis and metastatic spread: The formation of new blood vessels or angiogenesis is influenced by microbes. For example, it has been shown that Bacteroides fragilis and *Fusobacterium nucleatum* influence the immune system and promote blood vessel formation both of which aid in the spread of colon cancer. Microbe-induced TME modulation and immune suppression allow cancer cells to spread to far-off places.

3.5.4.3 Microbiome in Cancer Suppression

It is intriguing to learn that in certain situations the microbiota may also have a protective function and aid in the suppression of cancer (Table 3.2).

1. Anticancer immune activation: A diverse and stable gut microbiome can enhance the immune system's ability to recognize and eliminate cancer cells by activating antitumor immune cells such as CD8+ T cells and natural killer (NK) cells, while certain probiotics, particularly from the *Lactobacillus* genus, promote cytokine production such as interferon-gamma (IFN-γ), thereby supporting an anticancer immune environment.
2. Metabolic modulation: Microorganisms produce butyrate and other metabolites that can either promote or inhibit tumor growth. In colorectal cancer, butyrate produced by beneficial gut bacteria inhibits histone deacetylases (HDACs), leading to cancer cell apoptosis and suppression of tumor growth. Additionally, short-chain fatty acids (SCFAs) promote regulatory T cell (Treg) differentiation, thereby maintaining immune tolerance and preventing chronic inflammation that may contribute to cancer development (Arpaia et al. 2013).
3. Direct inhibition of tumor growth by microbial metabolites: Some microbes produce bioactive substances that can suppress tumor growth and prevent cancer progression. For example, certain Lactobacillus species produce metabolites that inhibit the proliferation of tumor cells in cancers such as breast and liver cancer and induce apoptosis.

3.5.4.4 Therapeutic Potential of the Microbiome in Cancer

Given the connection between the microbiome and cancer, several treatment options could be available (Table 3.2) such as: probiotics and prebiotics in cancer prevention and management: these can effectively facilitate restoring the microbial balance and strengthen the body's anti-cancer defenses. Probiotics have been shown to have anticancer effects by altering immune responses, inhibiting the growth of tumor cells and reducing inflammation (Sekirov et al. 2010). Dietary fibers and other prebiotics can effectively support the growth of beneficial bacteria and the synthesis of protective metabolites such as SCFAs. Synergy between immunotherapy and fecal microbiota transplantation: FMT has been investigated as a therapeutic strategy to alter the microbiome and influence the results of cancer treatment. According to research, by boosting immune responses, changing the microbiome may make cancer treatments like immune checkpoint inhibitors more effective (Gopalakrishnan et al. 2018). Engineered bacteria and targeted microbial therapy: creating targeted therapies using particular microbial strains could offer a fresh method of treating cancer. For instance, engineered bacteria can effectively be used to target the tumor with an immune response or to deliver anticancer drugs straight to the tumor site (Zhan et al. 2019).

3.6 Gut Microbiome: The Central to Health

3.6.1 Gut Microbial Diversity and Genetic Potential

The gut microbiome is a complex and dynamic community of trillions of microorganisms that primarily live in the human host's gastrointestinal tract with the colon having the highest concentration. This microbial community is comprised of a diverse array of bacteria alongside archaea, viruses (bacteriophages), fungi, and protozoa. In aggregate, the collective microbe population is more numerous than host cells and carries a vast genetic repertoire dubbed the microbiome with more than 150-fold more genes than contained within the genome of the human host (Qin et al. 2010). Such genes provide valuable metabolic functions that are not possessed by the human host itself such as the breakdown of polysaccharides by fermentation, production of vitamins (B12 and K), production of short-chain fatty acids, and detoxification of xenobiotics. The gut microbiota also maintains a critical function in developing and modulating the host's immunity, differentiating between pathogenic and commensal microorganisms and ensuring maintenance of immune homeostasis (Belkaid and Hand 2014). In a healthy host, the gut microbiome is typified by a diverse and relatively consistent microbial community, predominated by the bacterial phyla Firmicutes, Bacteroidetes, Actinobacteria, and Proteobacteria.

3.6.2 Microbiome Influencing Factors

The gut microbiome is a dynamic and adaptive ecosystem whose function and composition are strongly influenced by environmental and lifestyle factors upon which diet, antibiotics, and lifestyle are paramount. Amongst them, diet is generally regarded to be the strongest and quickest to change determinant of microbiome composition. The phyla of predominant gut bacteria can effectively be determined by long-term dietary patterns. For example, a diet high in plants and fiber promotes microbial diversity and the growth of good bacteria such as *Lactobacillus* and *Bifidobacterium* which produce short-chain fatty acids that restore health, such as butyrate, acetate, and propionate. These metabolites give colonocytes energy, regulate inflammation, and preserve the integrity of the intestinal barrier (Koh et al. 2016). On the other hand, diets high in saturated fat, added sugar and highly processed foods typical of the Western dietary pattern are linked to a decline in microbial diversity and a rise in proinflammatory bacteria such as Proteobacteria species and *Bilophila wadsworthia* which can effectively cause inflammatory diseases, obesity and metabolic syndrome (David et al. 2014; De Filippo et al. 2010). The ability of this variety of dietary interventions to produce noticeable changes in the composition and function of the microbiome within a few weeks is evidence of the microbiota's receptiveness to dietary feedback. Antibiotics have a strong effect on the gut microbiota and can effectively disrupt it profoundly and permanently. Despite being essential for managing bacterial infections, broad-spectrum antibiotics indiscriminately eliminate both commensal and pathogenic bacteria. It may lead to a loss of beneficial species, an overabundance of opportunistic pathogens such as *Clostridium difficile* or dysbiosis or decreased microbial diversity. Frequent or early exposure to antibiotics has been linked to an increased risk of developing chronic conditions later in life, including inflammatory bowel disease, asthma, allergies, and obesity (Dethlefsen and Relman 2011; Francino 2015). When antibiotics are stopped, some microbial communities recover, but others experience long-lasting changes, especially when disruption takes place during critical developmental windows like infancy. The gut microbiota is significantly influenced by lifestyle factors in addition to diet and antibiotics. For instance, there is a synergistic relationship between physical exercise and microbial health, as exercise has been associated with increased microbial richness and enrichment of bacteria that produce SCFA (Clarke et al. 2014). Another critically important modulator is psychological stress, which affects the composition of the gut microbiota through the gut-brain axis, a network of intercommunication that includes a two-way dialogue made up of immunologic, endocrine, and neural signaling. Changes in neurotransmitter and cortisol levels during psychological stress can effectively directly impact gut permeability and alter the makeup of the microbial community which may result in the emergence of disorders like irritable bowel syndrome (Foster et al. 2017). Research is still ongoing, but factors like smoking, drinking alcohol, getting poor sleep, and being around pollutants have also been shown to affect the gut microbiota in both positive and negative ways. Furthermore, early-life factors that are crucial for initial gut colonization and long-term gut microbiota stability include breastfeeding or formula feeding, the mode of delivery (vaginal versus cesarean section) and early-life microbial and antigenic exposure (Dominguez-Bello et al. 2010). When taken as a whole, these traits suggest that the gut microbiota is flexible and that both intervention and disruption are possible. Developing strategies to enhance microbial health and prevent disease requires an understanding of the relationship between the microbiome and lifestyle, antibiotics, and diet.

3.6.3 Substantial Functions of the Gut Microbiome

1. Digestion and short-chain fatty acid production: The gut microbiome functions are to play several central roles that are essential to ensuring overall human health like digestion, regulation of the immune system, and biosynthesis of essential compounds. One of its most basic functions is in the process of digestion and nutrient metabolism. The presence of microbial enzymes permits the degradation of indigestible dietary fibers such as complex carbohydrates by humans. Fermentation of this yields the production of short-chain fatty acids (acetate, propionate, and butyrate) which are beneficial energy sources to colonocytes and also have anti-inflammatory and immuno-modulatory activities (Koh et al. 2016).
2. Vitamin biosynthesis (B12, K, folate, and biotin): In addition to its metabolic functions, the microbiome contributes to the production of essential nutrients

such as vitamin K, biotin, folate, and several B vitamins, which are absorbed by the host and play key roles in vital metabolic processes (LeBlanc et al. 2013).

3. Immune regulation and epithelial barrier protection: The gut microbiota has also crucial role in the development and regulation of the immune system. Commensal microorganisms since early life contribute to the development of the innate and adaptive immunity by favoring tolerance and dampening over activation that will result in autoimmune disease (Belkaid and Hand 2014). The microbiome also protects against pathogenic microorganisms by inhabiting ecological niches, secreting antimicrobial substances, and eliciting protective immunity. In addition, gut microbiota contributes to the protection of intestinal integrity and minimizing induction of systemic inflammation by endotoxin and microbial translocation. Such related functions point to the gut microbiome as a determinant of systemic health and disease resilience.

3.7 Dysbiosis and Systemic Diseases

3.7.1 Chronic Inflammation and Cancer Precursor

Chronic inflammation has a crucial function in the genesis and development of a wide array of cancers. In contrast to acute inflammation which is a short-lived and self-limiting response to infection or injury aimed at restoring tissue homeostasis, chronic inflammation is a chronic and dysregulated process that has the potential to cause damage to DNA, promote cell proliferation, and suppress apoptosis each of them contributing to the process of oncogenesis. Inflammatory cells like neutrophils and macrophages invade tissue and release a rich cocktail of ROS and RNS that are able to cause mutation in oncogenes and tumor suppressor genes (Coussens and Werb 2002). Chronic inflammation also activates pathways like NF-κB and STAT3 that promote the overexpression of proinflammatory cytokines like IL-6 and TNF-α, creating a tumor-promoting microenvironment, tumor growth, angiogenesis, and immune evasion (Grivennikov et al. 2010). There have been several epidemiological and clinical studies that have confirmed firm associations between chronic inflammatory conditions and the development of cancer. For example, patients with chronic inflammatory bowel conditions like ulcerative colitis and Crohn's disease have a risk of developing colorectal cancer (Eaden et al. 2001). Chronic *Helicobacter pylori* infection that causes gastritis has been associated with gastric carcinoma, and chronic infection with hepatitis B and C has been a well-documented precursor to hepatocellular carcinoma (El-Serag 2012). These observations point to a requirement of anti-inflammatory measures and early screening in patients with chronic inflammatory conditions to curb the development of cancer.

3.7.2 Gut Dysbiosis and Systemic Inflammation

Imbalance in gut microbiota composition and function is a central player in initiating and sustaining inflammation in the system. The gut microbiota ensures intestinal homeostasis by facilitating digestion, modulating immunity, and strengthening the gut barrier. But when there is dysbiosis (caused by poor diet, overuse of antibiotics, stress, or infection), the protective equilibrium gets upset. This can effectively break down the integrity of the intestinal epithelial lining commonly known as leaky gut and let microbial metabolites, toxins (lipo-polysaccharides/ LPS), and undigested food components leak into the bloodstream (Cani et al. 2007).

3.7.3 Immune Modulation by Dysbiosis

These foreign substances incite immune responses and generate pro-inflammatory cytokines (TNF-α, IL-1β, and IL-6) by activating immune cells' pattern recognition receptors or PRRs (toll-like receptors or TLRs) (Belkaid and Hand 2014). Numerous non-communicable diseases including type-2 diabetes, obesity, cardiovascular disease and some types of cancer, are associated with low-grade systemic inflammation, which is caused by chronic immune system activation brought on by persistent microbial translocation (Tilg and Kaser 2011). In addition, gut dysbiosis can effectively change the proportion of pro-inflammatory immune cells to anti-inflammatory immune cells, which can effectively worsen systemic inflammation. For instance, it can effectively reduce regulatory T cells (Tregs) and boost Th17 responses (Round and Mazmanian 2009). The complex interplay between the gut microbiota and the host immune system highlights the significance of microbial balance in maintaining systemic immune homeostasis and preventing chronic disease.

3.7.4 Inflammatory Pathway Bacterial Participants

A few bacterial species in the gut microbiota have been directly linked to the activation of inflammatory pathways causing both localized and systemic inflammations. Notably, certain pathobiont microorganisms that are normally benign but can effectively become harmful in particular circumstances have a big impact on immune response regulation. *Escherichia coli* is a well-known example, especially the adherent invasive *E. coli* (AIEC) strain that has been linked to Crohn's disease. When AIEC infiltrates intestinal epithelial cells and macrophages, it triggers nuclear factor-kappa B (NF-κB), a central transcription factor in the pro-inflammatory signaling cascade which increases the production of cytokines such as TNF-α and IL-8 (Darfeuille-Michaud et al. 2004). Enterococcus faecalis is another pro-inflammatory bacterium that has been demonstrated to generate extracellular superoxide. Conversely, the depletion of anti-inflammatory commensals like *Faecalibacterium prausnitzii* is commonly noted in inflammatory conditions. *F. prausnitzii* generates short-chain fatty acids such as butyrate which function as an energy source for colonocytes inhibiting NF-κB activation and facilitate the differentiation of regulatory T cells thereby sustaining intestinal and systemic immune tolerance (Sokol et al. 2008). *Ruminococcus gnavus* and *Bilophila wadsworthia* have been linked to higher levels of endotoxins like lipopolysaccharide in people who are obese or have metabolic syndrome. LPS can effectively move into the bloodstream and activate toll-like receptor 4 (TLR4) signaling, which makes systemic inflammation worse (Devkota et al. 2012). These instances highlight the dualistic character of the gut microbiome, demonstrating how crucial it is for health to have a balanced population of microorganisms. While some bacteria support immune balance, others disrupt it by interacting with particular inflammatory pathways.

3.8 Microbiological Signs of Various Cancer Types

3.8.1 Gut Microbiota and Colorectal Cancer

1. Microbial dysbiosis in colorectal cancer (CRC): Microbial dysbiosis, defined as an imbalance in gut microbiota composition, has been strongly associated with colorectal cancer. This condition is characterized by a reduction in beneficial bacteria and an increase in pathogenic species, creating a pro-tumorigenic environment. For instance, patients with CRC often exhibit decreased levels of *Firmicutes* and increased levels of *Bacteroidetes* (Marchesi et al. 2016).
2. Inflammation and immune modulation: Chronic inflammation driven by dysbiotic microbial communities plays a central role in CRC development. Bacteria such as *Fusobacterium nucleatum*, *Bacteroides fragilis*, and *Enterococcus faecalis* contribute to an inflammatory microenvironment by producing pro-inflammatory metabolites, disrupting epithelial barrier integrity, and inducing DNA damage in colon cells. This persistent inflammation increases cellular susceptibility to malignant transformation. Moreover, dysregulation of immune responses by the gut microbiota is a key factor in CRC pathogenesis (Belkaid and Hand 2014).
3. Carcinogenic metabolites: In addition to inflammation, certain gut microbes produce carcinogenic metabolites that directly contribute to tumor development. For example, species of *Clostridium* and *Bacteroides* generate secondary bile acids and other toxic compounds that can damage epithelial cells and induce DNA mutations. These genotoxic effects promote genomic instability and increase the risk of colorectal cancer (Lahouty et al. 2025).

3.8.2 CRC-Associated Microbes

Fusobacterium nucleatum has been connected to colorectal cancer as it can effectively elude the immune system and induce inflammation. Elevated *F. nucleatum* levels are linked to metastases and a poor prognosis in patients with colorectal cancer. The bacterium can effectively change immune responses by decreasing

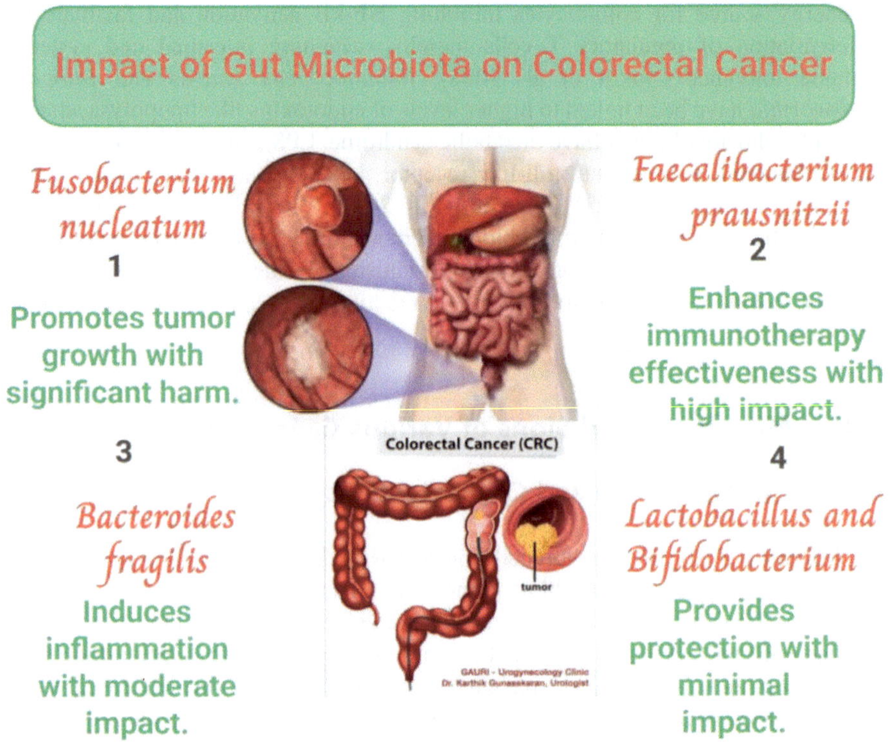

Fig. 3.6 Microbes associated with colorectal cancer and treatments based on the microbiome

T-cell activity and speeding up tumor growth (Fig. 3.6) (Bullman et al. 2017; Ou et al. 2022). *Bacteroides fragilis*: *Bacteroides fragilis* and its toxin (BFT) have been implicated in colorectal cancer. BFT induces inflammation and encourages the growth of epithelial cells, both of which can effectively lead to the formation of tumors. It has been shown that a toxin generated by Bacteroides fragilis directly promotes the growth of colon cancer (Fig. 3.6). *Escherichia coli*: Certain strains of *Escherichia coli* particularly those that produce the genotoxin colibactin have been connected to colorectal cancer. According to Liguori et al., colibactin contributes to the spread of cancer by causing mutations in the DNA of the colon epithelial cells (Fig. 3.6).

Figure 3.6 illustrates critically important microbiological contributors to colorectal cancer. By avoiding immune surveillance, decreasing T-cell activity, and promoting metastasis, *Fusobacterium nucleatum* accelerates the growth of tumors. Tumorigenesis is promoted by Bacteroides fragilis and its toxin (BFT), which causes inflammation and epithelial proliferation. Genotoxic *Escherichia coli* strains that produce colibactin damage colon epithelial cells' DNA and cause mutations. *Faecalibacterium prausnitzii* enhances response to immune checkpoint inhibitors (e.g., PD-1 blockers).

3.8.3 Microbiome and CRC Treatment

The effectiveness of immunotherapies for colorectal cancer is significantly influenced by the gut microbiota. Research demonstrates that a balanced and healthy microbiome can effectively enhance the immune system's capacity to combat cancer especially when combined with immune checkpoint inhibitors (PD-1 blockers). Gopalakrishnan et al. (2018) found that certain microbial profiles such as those with a high concentration of *Faecalibacterium prausnitzii* are associated with better immunotherapy outcomes. For patients with colorectal cancer who want to restore microbial balance, fecal microbiota transplantation has become a potentially effective treatment option. It has been demonstrated that fecal microbiota transplantation from healthy donors can effectively restore gut homeostasis and elevate the effectiveness of cancer treatments. Probiotics and prebiotics by changing the gut microbiota can effectively also enhance gut health, lower inflammation, and possibly lower the risk of colorectal cancer. When paired with prebiotics like dietary fiber, probiotic strains like *Lactobacillus* and *Bifidobacterium* can effectively support a healthy microbiota and offer protection against colorectal cancer. To completely comprehend their protective effects more research is needed.

3.8.4 Head and Neck Cancers (HNC) and Oral Microbiota

The importance of the oral microbiota in the initiation and spread of head and neck cancers (HNC) is becoming more widely acknowledged. Cancer pathophysiology, treatment plans and early detection can effectively all benefit from an understanding of the interactions between host cells and oral microbiota.

1. Microbial dysbiosis: It is believed that the development of HNC is influenced by an imbalance in microbial populations. Research has indicated that an increased risk of developing HNC is frequently linked to a higher abundance of pathogenic bacteria including *Treponema denticola, Porphyromonas gingivalis,* and *Fusobacterium nucleatum*. Nonetheless, it has been observed that HNC patients have lower levels of *Lactobacillus* and other helpful bacteria.
2. Inflammation and carcinogenesis: One known risk factor for HNC is inflammation, which is significantly influenced by the oral microbiota. Pro-inflammatory cytokines and oxidative stress can effectively result from inflammatory responses brought on by oral pathogens especially those that cause periodontitis. These elements encourage tumor growth, immunological evasion, and DNA damage (Irfan et al. 2020). It has been demonstrated that pathogens like *Fusobacterium nucleatum* interact with host immune cells to inhibit the immune response against tumors (Yan et al. 2025). Carcinogenic metabolites are produced by certain oral bacteria. *Fusobacterium nucleatum* for instance has been linked to the epithelial to mesenchymal transition and the encouragement of metastasis in head and neck squamous cell carcinoma. According to Zhang et al. (2020), these bacteria have

the ability to start inflammatory processes that promote tumor growth and resistance to treatment.

3. Carcinogenic mechanisms of oral pathogens: *Gingivalis porphyromonas* causes periodontal disease and has been found to be associated with an increased risk of HNC. Among the many virulence factors the bacterium produces are gingipains which inhibit the death of epithelial cells and encourage invasion and metastasis. In addition, these elements support immunological suppression and inflammation. Another well-known bacterium in HNC is *Fusobacterium nucleatum*. The epithelial to mesenchymal transition which enables cancer cells to multiply and become more invasive is known to be facilitated by interactions between immune cells and cancer cells. Furthermore, *F. nucleatum* has also been linked to lowering the effectiveness of cancer treatments (Zhang et al. 2020). *Treponema denticola* has also been connected to an elevated risk of HNC and is associated with periodontal disease. It promotes immunological evasion and inflammation, i.e., two factors that contribute to the development of head and neck cancers. The regulation of local macrophage and T-cell populations is one immune response that the oral microbiota influences. The immune system's capacity to react to cancer cells can effectively be enhanced or diminished by specific bacteria. For example, it has been demonstrated that *Fusobacterium nucleatum* reduces T-cell activity enabling tumors to elude immune surveillance (Fig. 3.7).

3.8.5 HPV-Associated HNC and Microbiome

One of the main risk factors for HNC especially oropharyngeal cancers is the human papillomavirus (HPV). By altering the immune response or the viral persistence in the host, the oral microbiota may have an impact on HPV-related carcinogenesis. According to studies, some oral bacteria may facilitate HPVs enter host cells or enhance their capacity to endure in the oral cavity's epithelial layers (Zhang et al. 2020). The composition of the oral microbiome may have an impact on how well immunotherapy treats HNC. The immune system's response to therapies like checkpoint inhibitors may be enhanced by a balanced, healthy microbiome. However, microbial dysbiosis may hinder the body's ability to mount a successful immune response. Although FMT has primarily been studied in relation to gastrointestinal cancers, it has been proposed as a potential treatment for patients with HNC. Among other cancer treatments, reestablishing a healthy microbiome may enhance the effectiveness of immunotherapy and chemotherapy. Probiotics like *Lactobacillus* and *Bifidobacterium* strains may facilitate control of the oral microbiome and prevent HNC by promoting a balanced population of bacteria, reducing inflammation and strengthening immune responses. Further research is required to confirm their efficacy in preventing cancer (Zhang et al. 2020).

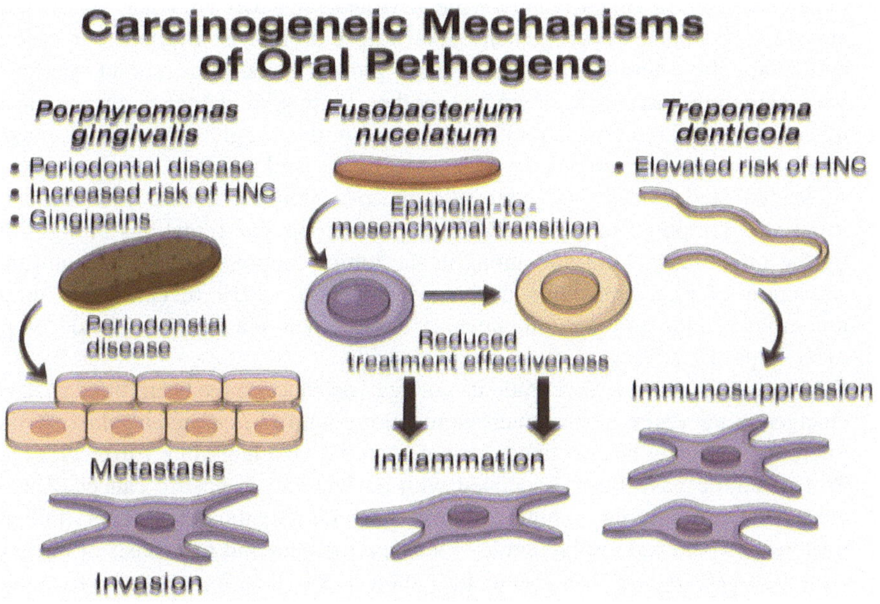

Fig. 3.7 Carcinogenic mechanisms of oral pathogens: illustrates oral bacteria linked to carcinogenesis. *Porphyromonas gingivalis* raises the risk of HNC by encouraging tissue destruction, invasion and metastasis. It also contributes to periodontal disease through virulence factors like gingipains. The epithelial to mesenchymal transition mediated by *Fusobacterium nucleatum* results in increased tumor growth, decreased treatment efficacy and chronic inflammation. By altering host responses and inducing immunosuppression, *Treponema denticola* is associated with an increased risk of HNC which promotes the development of cancer. These pathogens work together to cause periodontal disease, inflammation, invasion, and immune evasion all of which contribute to the development and spread of cancer

3.8.6 *Helicobacter pylori* and Microbes in Gastric Cancer

1. *Helicobacter pylori*'s function in gastric carcinogenesis: One of the most common and deadly cancers in the world is gastric cancer (GC). *Helicobacter pylori* and the human microbiome are strongly associated with gastric cancer and *H. pylori*, i.e., Gram negative bacterium is one of the main causes of gastric cancer.
2. Genetic and molecular alterations: *H. pylori* adheres to the stomach's epithelial cells after entering through the mucous layer. By generating urease enzyme, it neutralizes stomach acid and endure stomach's acidic environment. Over time, the bacterium can effectively cause chronic inflammation and damage to the stomach mucosa, which can effectively result in the development of gastric cancer. Chronic inflammation brought on by an *H. pylori* infection precedes carcinogenesis. While growth factors encourage genetic mutations and obstruct the normal turnover of epithelial cells, inflammation results in the release of proinflammatory cytokines (Kusters et al. 2006). Vacuolating cytotoxin-A

(VacA) and cytotoxin-associated gene A (CagA) are two virulence factors that elevate *Helicobacter* pathogenicity. The CagA protein is injected into gastric epithelial cells where it inhibits apoptosis, promotes cell division and interferes with signaling pathways. VacA is another toxin that damages the stomach mucosa by causing cellular vacuolization and immunological evasion. *H. pylori* infection is the primary cause of gastric ulcers and is also thought to be a substantial risk factor for noncardia gastric adenocarcinoma and mucosa-associated lymphoid tissue (MALT) lymphoma. Genetic mutations brought on by the infection include alterations in the tumor suppressor gene p53 and the activation of oncogenes like c-Myc. Zhang et al. 2019 revealed that these mutations and the ongoing inflammatory environment play a role in the development of gastric cancer from ulcers and gastritis.

3. Dysbiosis beyond *H. pylori*: Patients with gastric cancer have been found to have changes in the entire gastric microbiome along with *Helicobacter pylori*. While microbial diversity has declined, certain bacterial taxa including *Firmicutes* and Proteobacteria have been associated with an increase in gastric cancer (Zhang et al. 2020). By altering immune responses and affecting the inflammatory milieu, dysbiosis is believed to play a role in the development and metastasis of tumors. Although *H. pylori* has been the subject of the most research, other microorganisms such as *Escherichia coli*, *Streptococcus,* and *Clostridium* species have also been linked to the development of gastric cancer. According to Arthur et al., some strains of *Escherichia coli* that produce the genotoxin-colibactin have been linked to DNA damage in gastric epithelial cells and may play a role in the development of cancer.
   Immune system modification by *H. pylori.*
4. Immune system modulation: Both innate and adaptive immune responses are involved in the stomach mucosa's chronic inflammatory response brought on by *H. pylori.* To avoid immune detection, the bacteria can effectively control immune cells like T cells, dendritic cells, and macrophages. Because it modifies the local cytokine milieu which inhibits the activation of efficient immune responses the bacteria can effectively survive for decades. Cancer eventually develops as a result of this ongoing inflammation (Reyes 2023). *H. pylori* has also been found to elevate the number of regulatory T cells (Tregs) in the stomach mucosa. By obstructing antitumor immune responses, these immune cells can effectively aid in the growth of tumors and are essential for suppressing immune responses.
5. Strategies for prevention and treatment: Eliminating *Helicobacter pylori* has been demonstrated to dramatically lower the risk of gastric cancer especially if the infection is discovered early. Antibiotics that target *H. pylori* can effectively successfully eradicate the bacteria and lessen the chronic inflammation linked to the development of gastric cancer. According to studies, early eradication can effectively stop gastritis and peptic ulcers from developing into cancer (Ford et al. 2020). However, eliminating *H. pylori* completely is not always easy. Its efficacy may be restricted by re-infection, antibiotic resistance, and possible long term adverse effects. Furthermore, eliminating *H. pylori* may not always completely

eliminate the risk of developing cancer particularly in those who already have advanced precancerous lesions (Yan et al. 2025). Therefore, two dietary supplements that have been suggested as a preventative measure against stomach cancer and as an additional treatment option for *H. pylori* infections are prebiotics and probiotics. Probiotics (*Bifidobacterium* and *Lactobacillus*) have demonstrated promise in regulating the stomach microbiome and lessening the negative consequences of *H. pylori* infection. Prebiotics may also facilitate maintaining a healthy gastric microbiome and prevent dysbiosis by promoting the growth of beneficial bacteria. Vaccination development is also advance therapeutic strategies where research is being done on *H. pylori* vaccines to prevent infections before they cause cancer. More research is required to create safe and effective vaccines that can effectively be used widely even though early vaccine trials have shown promise.

3.9 Emerging Microbiome and Cancer

Knowledge about various cancers has led to new discoveries that are altering our comprehension, approach, and strategy for these diseases. The impact of the tumor environment, treatment resistance, and covert metabolic strategies that contribute to their survival are some of the emerging themes that these cancers share despite their wildly disparate behaviors and prognoses (Table 3.3).

3.9.1 Breast Cancer

The fact that breast cancer is actually a group of illnesses that are classified according to molecular types like HER2 expression or hormone receptor status is becoming more and more evident. Triple negative breast cancer (TNBC) which is resistant to hormone therapy is among the most difficult. Recent research indicates that the immune system is crucial in this case; tumors that have a higher level of immune cell infiltration may react better to immunotherapy. When combined with chemotherapy, PD-1/PD-L1 immune checkpoint inhibitors are showing promise in TNBC. Furthermore, real-time detection of resistance mutations and tracking of cancer progression are being accomplished through the use of tools such as liquid biopsies (circulating tumor DNA). Because pancreatic cancer frequently goes undiagnosed until it is too late, treatment is still difficult. However, current research focuses on the thick fibrous tissue that surrounds the tumor and protects it from drugs and immune cells. They are testing treatments that either attack this stroma or starve the cancer by denying it the nutrients it needs. Once believed to be incurable, the KRAS mutation which is highly prevalent in pancreatic tumors is now more successfully targeted.

Table 3.3 Certain cancers associated with gut health

Type of Cancer	Changes in the gut microbiome	Important microbiological species	Suggested mechanism	References
Colorectal cancer (CRC)	Proinflammatory and genotoxic bacteria are more prevalent, while beneficial microbes are less prevalent	*Bacteroides fragilis*, *Escherichia coli*, and *Fusobacterium nucleatum*	Toxins (colibactin), inflammation and immune response modulation can all cause DNA damage	Kostic et al. (2012)
Liver cancer (HCC)	Hepatic inflammation and elevated LPS production are caused by dysbiosis	*Escherichia coli* and *Enterococcus faecalis*	Bacterial products are transported to the liver through the portal vein, resulting in persistent inflammation	Yu and Schwabe (2017)
Cancer of the pancreas	Changes in gut and pancreatic microbial signatures; some bacteria encourage the growth of tumors.	*Proteobacteria*, *Malassezia species*, and *Fusobacterium species*	Oncogenic pathway activation, inflammation and immune suppression	Pushalkar et al. (2018)
Gastric cancer	Strong correlation with *Helicobacter pylori* infection; results are exacerbated by gut dysbiosis	*Lactobacillus* species and *Helicobacter pylori*	Prolonged infection causes oncogenic mutations, inflammation, and damage to the epithelium	Plummer et al. (2015)
Cancer of the esophagus	Barrett's esophagus and cancer risk are linked to altered gut and esophageal microbiomes	*Fusobacterium* species and *Prevotella* species	Bile acid metabolism, inflammation, and epithelial cell proliferation modulation	Blackett et al. (2013)

3.9.2 Breast and Lung Cancers

Direct microbial colonization of the primary tumor site is not a feature of breast or lung cancers, in contrast to colorectal, liver, or pancreatic cancers. Nonetheless, a growing amount of evidence suggests that the gut microbiota can effectively significantly and indirectly affect these cancers through systemic inflammation, immune modulation and hormonal metabolism. It has been demonstrated that the estrobolome- group of bacterial genes that can effectively metabolize estrogens is how the gut microbiota controls estrogen metabolism in breast cancer. Increased recirculation of active estrogens, which are known to encourage the growth of estrogen receptor-positive breast cancer cells can effectively result from dysbiosis in this system (Plottel and Blaser 2011). In addition, a decreased diversity of gut microbes has been associated with impaired anti-tumor immunity especially in triple-negative breast cancer, increased levels of proinflammatory cytokines (TNF-α, IL-6) and altered immune surveillance (Goedert et al. 2015). The gut-lung axis a two-way immunological pathway is also critically important in lung cancer. By changing systemic immune responses gut dysbiosis can effectively affect the risk and course of lung cancer. The immune landscape of the lung can effectively be shaped by metabolites derived from microorganisms, such as short-chain fatty acids which can effectively influence T regulatory cell function, dendritic cell maturation, and cytokine production (Zitvogel et al. 2015a). Furthermore, research on mice has shown that imbalances in the oral or intestinal microbiota can effectively reduce the effectiveness of immune checkpoint inhibitors such as anti-PD-1 and anti-CTLA-4 treatments which are used to treat nonsmall cell lung cancer (Routy et al. 2018). In preclinical models, it has been demonstrated that restoring microbial balance via probiotics or fecal microbiota transplantation (FMT) restores immunotherapy responsiveness. The gut microbiota functions as a distant yet potent regulator in both the cancers influencing immune tone and systemic inflammation to either aid in the removal of tumors or foster an immunosuppressive environment that is conducive to the growth of cancer. This demonstrates how microbiome-targeted therapies can effectively be used to treat cancers that are not related to the gastrointestinal tract (Table 3.3).

3.9.3 Prostate Cancer

While the majority of prostate cancers grow slowly, some become aggressive and are resistant to hormone therapy. Our knowledge of how genetic mutations like BRCA1/2 render tumors susceptible to PARP inhibitors like olaparib and rucaparib has significantly changed (Thapa et al. 2024). PSMA targeted therapy is an interesting use of radioactive medications that selectively target and kill prostate cancer cells (Longoria et al. 2024). Because of this, the pattern for all three cancers is evident; research is delving deeper into the biology underlying each tumor type, diagnostics are improving and becoming earlier and treatments are becoming more intelligent. Personalized cancer treatment is now a reality.

3.9.4 Pancreatic Cancer

Despite the pancreas's long-standing reputation as a sterile organ, new research over the past ten years indicates that oral and gut-derived microbial communities may have an impact on the development of pancreatic cancer especially pancreatic ductal adenocarcinoma (PDAC) the most prevalent and deadly type of the disease. The presence of microbial DNA and live bacteria in pancreatic tumor tissues suggests that microorganisms from the gut or oral cavity may move through the bloodstream, bile duct, or intestinal permeability (Pushalkar et al. 2018). These microorganisms are not passive bystanders rather they seem to influence immunological responses, alter inflammation and shape the tumor microenvironment all of which contribute to carcinogenesis and treatment resistance. Immune suppression is one of the main mechanisms. Toll-like receptors (TLRs) on immune and cancer cells can effectively be activated by bacteria in the tumor microenvironment which can effectively lead to chronic inflammation and reduce the cytotoxic T-cell response that is necessary for tumor control. *Malassezia* species which are typically found in the gut have been shown to use complement activation to speed up the growth of pancreatic tumors (Aykut et al. 2019). Fan et al. (2018) revealed that there is a connection between oral health and PDAC because a higher abundance of the periodontal pathogen Porphyromonas gingivalis in the oral microbiota has been linked to an increased risk of pancreatic cancer. Therapeutic results also seem to be influenced by the microbiome. Certain intratumoral bacteria can effectively inactivate chemotherapeutic agents such as gemcitabine through enzymatic degradation (Geller et al. 2017). Targeting the microbiome to enhance treatment response and lower drug resistance has gained attention as a result of these findings. Although research on the connection between the microbiome and pancreatic cancer is ongoing, mounting evidence points to the possibility that microbial factors play a significant role in the pathophysiology of PDAC. The potential of microbiome-based biomarkers for early detection and microbiota modulation (through probiotics, antibiotics or fecal transplants) as supplements to conventional therapy is being investigated in ongoing studies (Table 3.3).

3.9.5 Liver Cancer

It is becoming more widely acknowledged that gut microbiota affects liver cancer especially hepatocellular carcinoma (HCC) through the gut-liver axis a two-way communication system between the liver and the gastrointestinal tract via the portal vein which transports nutrients, blood, and microbial products straight from the intestines to the liver. By filtering and detoxifying bacterial components like lipopolysaccharide and preserving immunological tolerance liver serves a pivotal role in a vital immune role in healthy people. However, intestinal permeability also known as leaky gut can effectively be increased and the intestinal barrier disrupted by gut dysbiosis which is frequently brought on by a high-fat diet, alcohol consumption or chronic liver disease. This allows bacteria, endotoxins and microbial metabolites to move into the portal circulation. Hepatic inflammation is caused by this microbial translocation, primarily through activation of the toll-like receptor (TLR) on hepatocytes and Kupffer cells. TLR4 is particularly significant because it reacts to LPS from Gram (-ve) bacteria by starting a chain reaction that activates NF-κB, encourages the synthesis of proinflammatory cytokines (TNF-α and IL-6) and triggers fibrogenesis by activating hepatic stellate cells (Dapito et al. 2012). The development of HCC is considerably increased by chronic inflammation and fibrosis particularly in patients with viral hepatitis, nonalcoholic steatohepatitis (NASH) or non-alcoholic fatty liver disease (NAFLD). The presence of specific gut microbes or their byproducts is crucial for the progression of HCC as evidenced by the lower incidence of liver tumors in mice raised in germ-free environments or treated with gut-sterilizing antibiotics in experimental models. Liver carcinogenesis is also influenced by bile acid metabolism which is partially controlled by the gut microbiota. In the liver, dysregulated bile acid composition can effectively activate receptors like FXR and TGR5 which can effectively affect oncogenic signaling, inflammation and cell proliferation (Yoshimoto et al. 2013). As a result, the gut-liver axis is a promising target for microbiome-based interventions in HCC prevention and therapy since it is a crucial pathway through which the microbiota influences liver cancer risk (Table 3.3). The increasing amount of research on the human microbiome has transformed our knowledge of the patho-physiology of cancer by demonstrating the close relationship between gut health and a variety of cancers both inside and outside the gastrointestinal tract. With bacteria like Bacteroides fragilis and *Fusobacterium nucleatum* clearly involved in tumor initiation and immune evasion through inflammation and disruption of epithelial barriers, colorectal cancer offers the most direct and convincing evidence (Kostic et al. 2013; Rubinstein et al. 2013; Wu et al. 2009). Dapito et al. (2012) revealed that liver cancer illustrates how the gut-liver axis facilitates the translocation of microbial metabolites such as lipopolysaccharide which feed fibrogenesis and chronic inflammation two substantial predisposing factors to hepatocellular carcinoma. Research suggests that certain microorganisms colonize the microenvironment surrounding the pancreatic tumor and either deactivate chemotherapy drugs or suppress anti-tumor immunity. The impact on pancreatic cancer is more recent and complex (Pushalkar et al. 2018; Geller et al. 2017; Aykut et al. 2019). However, the gut microbiota influences immunological responses, inflammation and hormonal activity through the gut-lung axis and estrogen metabolism which have indirect systemic effects in lung and breast cancers (Plottel and Blaser 2011; Goedert et al. 2015; Zitvogel et al. 2015b; Routy et al. 2018). These microbiota-driven pathways can effectively have a substantial impact on the growth of tumors, the response to immune checkpoint therapy and the overall course of cancer. Together, these findings demonstrate that gut dysbiosis is often not merely a side effect of cancer or its treatment but often contributes to the onset and progression of disease. Probiotics, dietary changes, fecal microbiota transplantation and microbiota-targeted drugs and other microbiome-based cancer prevention, diagnostic, and treatment strategies have exciting prospects as a result of this growing understanding. As research progresses, a more integrated oncology approach that takes microbiome health into account may become more central to personalized cancer care (Table 3.3).

3.10 Gut Microbiota and Cancer Interaction

The development and risk of cancer are influenced by the gut microbiota in a complex and dynamic way. Its interactions with host physiology can effectively be advantageous or harmful depending on the microbial composition, host genetics, environmental factors, and disease context. Three primary ways that gut microbe's impact carcinogenesis are through immune system modulation, the production of carcinogenic or anti-cancer metabolites, and modifications in the body's reaction to cancer treatments.

3.10.1 Microbial Metabolites

Dual role in Carcinogenesis and Cancer Prevention: short-chain fatty acid (butyrate) produced by commensal gut bacteria has been demonstrated to serve two purposes. Even though they can effectively lower inflammation and promote normal epithelial turnover, some may in some circumstances elevate tumorigenesis (Louis et al. 2014). Secondary bile acids produced by microbial metabolism can effectively also cause cancer particularly liver and colorectal cancers (Ridlon et al. 2014). Therefore, gut microbes can effectively produce both carcinogenic as well as anti-carcinogenic compounds depending upon the microbial species and host environment. For example, in a dysbiotic gut environment some bacterial species can effectively produce secondary bile acids like deoxycholic acid (DCA) which have been connected to oxidative stress, DNA damage and the promotion of colorectal and liver cancer (Yoshimoto et al. 2013). Similarly, N-nitroso compounds (NOCs) and hydrogen sulfide which are genotoxic and can effectively harm the intestinal epithelium can effectively be produced by gut bacteria that break down red and processed meat (Louis et al. 2014). Conversely, good bacteria like Lactobacillus and Bifidobacterium generate short-chain fatty acids especially butyrate which have anti-inflammatory and anti-cancer properties. Butyrate maintains the integrity of the gut barrier, inhibits histone deacetylases, and promotes apoptosis in colon cancer cells (Canani et al. 2011). The importance of microbial balance in preventing cancer is highlighted by these two roles of microbial metabolites.

3.10.2 Gut Bacteria's Modification of the Immune System

The microbiota can effectively influence both systemic and local immune responses. The tumor immune microenvironment and the results of immunotherapy are impacted by certain bacterial species such as *Bacteroides fragilis* which change T-cell activity and cytokine expression (Vetizou et al. 2015). Therefore, the gut microbiota is a central regulator of the immune system and its effects on immune homeostasis can effectively directly affect the development and metastasis of cancer. Toll-like receptors (TLRs) and NOD-like receptors (NLRs) are two examples of pattern recognition receptors (PRRs) that microbes use to interact with immune cells and start pro or anti-inflammatory signaling cascades. Prolonged activation of these pathways particularly through Gram negative bacteria's lipopolysaccharides can effectively result in protumorigenic immune responses and chronic inflammation (Garrett 2015). The balance of T-helper cells, regulatory T cells, and cytotoxic T lymphocytes is also influenced by the microbiota. For instance, Th17 cells which are implicated in inflammation and have been linked to colorectal and liver cancer can effectively be encouraged to differentiate by segmented filamentous bacteria (Wu et al. 2009). On the other hand, commensal bacteria such as *Bacteroides fragilis* have the ability to control the growth of Tregs, which can effectively facilitate to preserve immunological tolerance or in certain situations inhibit anti-tumor immunity within the tumor microenvironment (Mazmanian et al. 2005). Some gut microbes like colibactin also produce genotoxins that can effectively damage host DNA directly and result in cancerous changes. Chronic microbially driven inflammation also contributes to oncogenic transformation through pathways involving NF-κB and STAT3 activation (Arthur et al. 2012).

3.10.3 Microbiota Influence on Cancer Treatments

The finding that the gut microbiota influences the effectiveness and toxicity of cancer treatments such as chemotherapy, radiation, and immunotherapy is one of the most exciting advances in cancer research. Chemotherapeutic medications are less effective when they are metabolized or broken down by certain bacterial species. For example, using an enzyme called cytidine deaminase, Gamma-proteobacteria have been demonstrated to inactivate gemcitabine a first-line chemotherapy treatment for pancreatic cancer (Geller et al. 2017). In immunotherapy, the gut microbiota is particularly significant. The effectiveness of immune checkpoint inhibitors such as anti-PD-1 and anti-CTLA-4 antibodies has been shown to be influenced by the composition of microorganisms. Studies show that patients respond better to ICIs when they have higher levels of beneficial bacteria including *Bifidobacterium, Faecalibacterium prausnitzii,* and *Akkermansia muciniphila* (Routy et al. 2018; Matson et al. 2018). These microbes enhance antigen presentation, boost T-cell activation, and reduce immunosuppressive cells in the tumor microenvironment. The fact that using antibiotics prior to or during immunotherapy can effectively decrease treatment efficacy by disrupting the gut microbiota further emphasizes the need for a balanced microbial ecosystem during cancer therapy (Derosa et al. 2018). When taken as a whole, these mechanisms highlight the gut microbiota's role as an active regulator of immunological surveillance, carcinogenesis and therapeutic response. Comprehending these microbial host interactions creates new opportunities for personalized medicine wherein dietary interventions, fecal microbiota transplantation (FMT), probiotics and prebiotics can effectively be used in conjunction with traditional cancer treatments to enhance results (Table 3.3).

3.11 Therapeutic Outcomes and Microbiome Modulation

3.11.1 Microbiome and Immunotherapy

The gut microbiota, which has emerged as a central modulator of host immunity, has a substantial impact on the efficacy of cancer immuno-therapies, particularly immune checkpoint inhibitors (ICIs). Through the promotion of dendritic cell maturation and T cell infiltration into tumors, commensal bacteria such as Bifidobacterium longum and *Akkermansia muciniphila* have been demonstrated to enhance the efficacy of anti-PD-1/PD-L1 and anti-CTLA-4 treatments (Routy et al. 2018; Gopalakrishnan et al. 2018). Conversely, dysbiosis is associated with a poor response to ICIs. Microbial composition is one possible predictive biomarker for treatment response. For example, responders to PD-1 blockade often exhibit distinct microbial signatures, suggesting that microbiome screening or modification may enhance outcomes (Matson et al. 2018). This makes it possible to develop personalized immunotherapy plans that consider the distinct microbial profile of every patient (Fig. 3.7).

3.11.2 Microbiota and Chemotherapy

Microbiota affects the outcomes of chemotherapy by affecting drug metabolism, bioavailability, and toxicity. Certain bacteria have the ability to enzymatically activate or deactivate chemotherapy agents. Some bacteria, such as *Enterococcus hirae* and *Barnesiella intestinihominis* have been linked to enhanced cyclophosphamide efficacy, while others can effectively worsen side effects or break down medications (Viaud et al. 2013). In addition, concurrent use of chemotherapy and antibiotics may change the gut microbiota and reduce the treatment's efficacy. The importance of preserving microbiome integrity during cancer treatment has been highlighted by numerous clinical studies that have connected the use of broad-spectrum antibiotics to worse outcomes in patients undergoing immunotherapy or chemotherapy (Fig. 3.7) (Derosa et al. 2018).

3.11.3 Fecal Microbiota Transplantation in Cancer Therapy

It has been shown that giving patients fecal material from healthy donors enhances immunotherapy response and brings back the diversity of the microbiome. Because FMT can effectively re-sensitize patients who are immunotherapy-resistant to PD-1 inhibitors, early clinical trials suggest that it could be used as a companion therapy (Baruch et al. 2021; Davar et al. 2021). Ongoing trials are evaluating the safety, efficacy, and donor selection criteria for FMT in oncology. FMT may become more well-liked as the field advances as an adjuvant treatment for microbiome modulation in cancer care, especially in relation to gastrointestinal cancers and immunotherapy (Fig. 3.8).

3.12 Microbes as Direct Therapeutics

3.12.1 Drug-Delivering Bacteria

Genetically altered bacterial strains transport medicinal substances straight to the sites of illness particularly in cases of cancer, infections or inflammation. This method of drug delivery is novel. These bacteria can effectively be genetically altered to produce and release therapeutic proteins, enzymes, or drugs in response to specific environmental stimuli. Since they can effectively be genetically altered to enter particular tissues or tumor microenvironments, a number of bacteria including

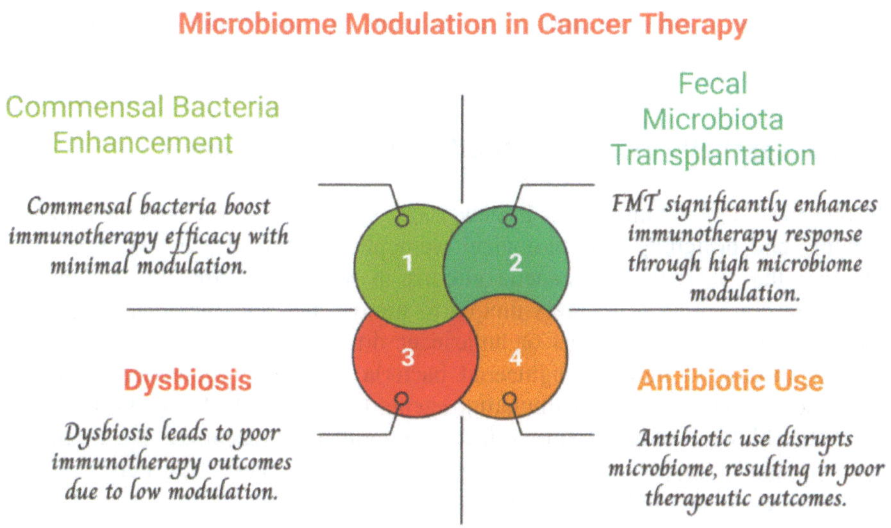

Fig. 3.8 Microbiome modulation in Cancer Therapy illustrates how the gut microbiota affects the results of cancer treatment (1) In immunotherapy, commensal bacteria like *Bifidobacterium longum* and *Akkermansia muciniphila* enhance the effectiveness of immune checkpoint inhibitors by encouraging dendritic cell maturation and T-cell infiltration (2) Dysbiosis or the use of antibiotics can effectively hinder treatment and elevate toxicity, microbial species such as *Enterococcus hirae* and *Barnesiella intestinihominis* enhance drug efficacy during chemotherapy. (3) Restoring microbial diversity, re-sensitizing patients to PD-1 inhibitors and demonstrating potential as an adjuvant therapy in oncology are all benefits of fecal microbiota transplantation

Salmonella, Lactococcus and *E. coli* are frequently used in drug delivery as an efficient way to deliver targeted therapy. By acting as carriers for targeted drug delivery, the bacteria can effectively lessen the negative effects of systemic therapies (Zhang et al. 2020; Zalatan et al. 2024). Microbial-based systems, for example, have been thoroughly investigated for tumor-specific drug delivery to target solid tumors with encouraging results.

3.12.2 Oncolytic Bacterial Treatments

Oncolytic bacterial therapies provide an alternative to conventional cancer treatments by using bacteria that selectively target and destroy tumor cells. Because they can effectively grow in hypoxic areas of tumors, *Clostridium* species like *Clostridium novyi* can effectively be used to target tumor cells. Furthermore, some *Salmonella* species have been altered to invade and kill tumor cells. These bacteria have the ability to either lyse tumor cells directly or produce toxins that cause the immune system to identify and eliminate the cancerous cells. The premise behind these treatments is that the bacteria kill tumor cells while protecting healthy tissues by selectively infecting them. This approach is becoming more popular in clinical trials because it may be able to circumvent the drawbacks of conventional treatments.

3.12.3 Synthetic Biology Methods in Cancer Treatment

In order to interact with tumors in new ways, synthetic biology approaches entail creating and designing new biological elements, instruments, and systems. This includes the emergence of genetic circuits in microorganisms that are able to recognize conditions or markers unique to tumors and start therapeutic reactions like the synthesis of prodrugs or anticancer drugs. For instance, in response to signals specific to tumors, engineered bacteria may be programmed to release cytokines or other immune-stimulating agents (Zhang et al. 2020). Furthermore, synthetic biology reduces side effects and enhances accuracy by producing microorganisms capable of synthesizing specific drugs. These tactics could lead to more successful and individualized cancer treatments.

3.13 Nutrition, Lifestyle, and Microbiota in Cancer Prevention

3.13.1 Diet and Microbiota Modulation

Nutrition and lifestyle choices have a big influence on the gut microbiota which influences inflammation, immunity, and carcinogenesis. Better metabolic health, a lower risk of cancer and a better response to treatment are all associated with a balanced, diverse microbiome that is full of beneficial bacteria and microbial metabolites. Therefore, promoting gut microbial health through diet, exercise, stress management and other modifiable behaviors is a powerful cancer prevention strategy. A diet rich in dietary fiber, fermented foods and polyphenols promotes microbial diversity and the growth of beneficial bacteria that produce anti-inflammatory and anti-cancer metabolites like short-chain fatty acids (SCFAs). Fiber: prebiotic fibers, which are present in whole plant foods such as fruits, vegetables, legumes and whole grains, promote the growth of *Lactobacillus* and *Bifidobacterium* species which produce SCFAs like butyrate which have been shown to inhibit the growth of cancer cells and promote the death of colon cancer cells (Canani et al. 2011; Louis et al. 2014). Fermented foods: Probiotics found in yogurt, kefir, sauerkraut, kimchi and kombucha naturally bring live, healthy bacteria into the digestive system. According to Marco et al. (2017), these can effectively enhance gut barrier function, lower inflammation, and restore microbial balance. Polyphenols: berries, green tea, olive oil, dark chocolate, and red wine are examples of plant compounds called polyphenols. These compounds are broken down by gut bacteria into bioactive metabolites that have anti-inflammatory and antioxidant qualities. In addition, polyphenols inhibit pathogens and encourage the growth of helpful microorganisms (Cardona et al. 2013).

3.13.2 Lifestyle Factors and Microbial Health

Stress, sleep, and exercise: in addition to diet, lifestyle choices have a big impact on the makeup and function of gut microbes (Fig. 3.8). Exercise: frequent exercise enhances mucosal immunity, increases butyrate-producing bacteria, and diversifies the gut microbiota. It has been demonstrated that athletes' microbiomes are more robust and diverse than those of sedentary people (Clarke et al. 2014). Sleep: poor sleep quality and disturbed circadian rhythms can effectively cause systemic inflammation, microbial dysbiosis and increased gut permeability all of which can effectively raise the risk of cancer (Benedict et al. 2016). Stress management: Through the gut-brain axis long-term psychological stress changes the gut microbiota increasing pathogenic bacteria and decreasing commensals. This dysbiosis brought on by stress can effectively worsen inflammation and interfere with immune regulation (Foster et al. 2017).

3.13.3 Probiotics, Prebiotics, and Cancer Risk Reduction

Reduced incidence of colorectal and other inflammation-driven cancers has been associated with adopting a microbiota-friendly diet and taking probiotics and prebiotics. Restoring microbial balance, decreasing the activity of bacteria that cause cancer, and boosting anti-tumor immune responses are all possible with probiotics, i.e., live beneficial bacteria and prebiotics, i.e., nondigestible fibers that feed probiotics (Rafter 2003). Prebiotic supplements (inulin and fructo-oligosaccharides) may lessen tumor burden, elevate the production of SCFAs, and enhance the results of chemotherapy as per the clinical and animal research (Gibson et al. 2017). A lower incidence of cancer, better gut health, and better regulation of systemic inflammation have all been linked to long-term adherence to Mediterranean-style diets, which are rich in fiber, polyphenols, and healthy fats. Therefore, people can effectively influence their gut microbiota to promote immunological function, decrease inflammation, and lower the risk of cancer by changing their diet and lifestyle. A comprehensive approach to supporting microbial health and cancer prevention includes incorporating foods high in fiber, fermented products, items rich in polyphenols, regular exercise, getting enough sleep, and managing stress effectively. As part of an all-encompassing prevention and treatment paradigm, personalized nutrition and microbiome-based interventions may become more prevalent in cancer care in the future (Fig. 3.9).

3.14 Current Research and Potential Directions

1. Microbiome-based biomarkers and diagnostics: microbial signatures are gaining popularity as noninvasive biomarkers for cancer detection, prognosis and response to treatment. Research indicates that individuals with colorectal cancer (CRC) have distinct microbial profiles, which include increased levels of pro-inflammatory

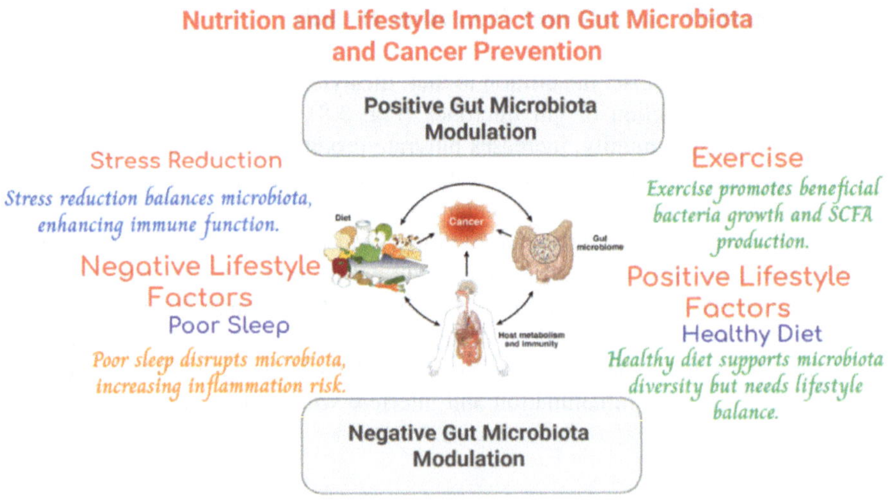

Fig. 3.9 Gut microbiota composition and activity modulated by diet and lifestyle factors. This figure illustrates how gut microbiota composition and activity are modulated by diet and lifestyle factors leading to enhanced immune regulation, reduced inflammation, and lower cancer risk. Beneficial microorganisms that generate anti-inflammatory metabolites like SCFAs are supported by fiber, fermented foods, and polyphenols. In addition to promoting microbial diversity and stability, exercise, enough sleep and stress management also lower the risk of carcinogenesis

bacteria like *Fusobacterium nucleatum* and *Peptostreptococcus anaerobius* (Yu et al. 2017). These microbial alterations can effectively be detected by stool-based DNA sequencing which may be a helpful tool for the early detection of colorectal cancer and other related disorders. In addition, the sensitivity and specificity of cancer screening are improved by integrating microbial data with host transcriptomics and metabolomics. For example, a microbiome-based machine learning model demonstrated over 90% accuracy in differentiating between individuals with colorectal cancer and healthy individuals (Zeller et al. 2014). 2. Fecal microbiota transplantation: This technique involves transferring gut microbes from a healthy donor to a patient and has shown promise in treating *Clostridium difficile* infections and is presently being researched in oncology. The goal of re-establishing a balanced microbial community is to enhance immunotherapy outcomes and strengthen immune responses. Recent clinical trials have demonstrated that FMT from immunotherapy-responsive patients can effectively convert nonresponders into responders in metastatic melanoma suggesting that gut microbes may be causally involved in controlling the efficacy of immune checkpoint inhibitors (Baruch et al. 2021; Davar et al. 2021). Scientists are being encouraged by these positive results to look into FMT in other cancers such as lung, liver and pancreatic cancers. Standardizing donor selection, screening and delivery methods is still very difficult. 3. Personalized medicine based on gut microbiome profiling: the future of cancer treatment is precision oncology where treatments are tailored to each patient's distinct microbiome profile in addition to genetic mutations. Synbiotic or targeted probiotic therapy, which aims to boost populations of beneficial microorganisms; dietary modifications that support microbial strains and boost immunological surveillance or lower inflammation and microbiome-informed immunotherapy which uses the gut flora composition to predict which patients will benefit from ICIs are examples of personalized treatments. Microbiome profiling may also have an impact on drug choice and dosage. For instance, certain microbes have the ability to metabolize chemotherapeutic agents which may affect their toxicity and bioavailability (Geller et al. 2017). Therefore, understanding a patient's microbial landscape may facilitate optimize treatment strategies and minimize adverse effects. Numerous biotech companies and academic collaborations are currently developing microbiome panels and AI-powered platforms in an attempt to bring microbial data closer to real-world applications. The gut microbiota therefore represents a frontier in cancer research since it offers new tools for diagnosis, prognosis, and treatment. Although there are still challenges to be addressed such as interindividual variability, standardization of microbiome interventions, and ethical concerns research is progressing toward personalized, microbiome-informed cancer treatment. Long-term investment in this field could revolutionize oncology by incorporating microbial insights into every stage of cancer treatment.

3.15 Challenges and Limitations

1. Variations in the microbiome among individuals: The human microbiome's interindividual variability is one of the main obstacles to the use of microbial therapies. The safety and efficacy of microbial therapies can effectively be significantly impacted by individual differences in the microorganisms present in each person's gut and other body parts (Zhang et al. 2020). People may react differently to oncolytic treatments or modified bacteria as a result of this variability. For instance, the ability of bacteria to colonize and efficiently deliver therapeutic agents may be impacted by genetic and environmental factors including diet, medications, and underlying medical conditions (Zhang et al. 2020). To maximize treatment results, individualized strategies that consider a patient's microbiome might be required.
2. Concerns about microbial treatments' safety and ethics: When using engineered microbes for therapeutic purposes, safety is still the top priority. Release of genetically modified organisms into the environment or human body can effectively result in the emergence of pathogenic strains, horizontal gene transfer or unintended genetic changes. In addition, engineered microbes may cause dysbiosis or unexpected immunological reactions both of which could be harmful. Because of the potential for unexpected ecological disruption and the long-term effects of introducing live bacteria into the microbiome the use of live

bacteria in humans also presents ethical concerns. Before such treatments are widely used, regulatory agencies must thoroughly evaluate their safety and ethical implications.

3. Barriers to translation and regulation: One substantial barrier to the clinical application of microbial therapies is the continuous evolution of regulatory frameworks. The special difficulties presented by microbial treatments might not be adequately addressed by the laws currently controlling biologics and genetically modified organisms. Concerns about the creation, description, and application of genetically modified bacteria in medicine are covered in this article. Comprehensive regulations that strike a balance between possible advantages and safety concerns must be developed by regulatory bodies like the U.S. Food and Drug Administration (FDA) and European Medicines Agency (EMA). Regulatory, financial, and logistical challenges also cause many promising therapies to stall in preclinical or early phase clinical trial stages because it is typically expensive and time consuming to convert laboratory-based research into clinical applications. To overcome these obstacles and expedite the implementation of microbial therapies in clinical settings more adaptable and efficient regulatory procedures must be created.
4. Prospects for the future: 1. Personalized cancer treatments based on microbiomes: the results of cancer treatment could be greatly enhanced by personalized cancer therapies that take into account a patient's microbiome profile. According to recent research, the microbiome's makeup may affect how well cancer treatments like immunotherapy and chemotherapy work (Zhang et al. 2020). Consequently, there is increasing interest in customizing cancer treatments according to a patient's microbiome. To enhance therapeutic responses such as boosting the efficacy of immune checkpoint inhibitors or enhancing chemotherapy tolerance, this may entail altering or enhancing the microbiome (Zhang et al. 2020). By enabling more targeted and less harmful treatments, personalized microbiome-based approaches are expected to transform cancer care by improving efficacy and lowering side effects. 2. Combining metagenomics and metabolomics: By combining omics technologies like metagenomics and meta-bolomics we can effectively gain a better understanding of how the microbiome influences the development of cancer and how it responds to treatment. While metabolomics studies the metabolites these microbes produce which can effectively have a substantial impact on tumor progression and treatment response, meta-genomics allows for a thorough examination of the microbial communities in the human body. Finding microbial biomarkers for early cancer detection and treatment outcome monitoring is made possible by combining these omics techniques with cancer research. It is anticipated that this multiomics approach will enhance the accuracy of cancer treatments and open up new therapeutic intervention opportunities. Future cancer prevention and treatment may benefit from the integration of artificial intelligence and microbiome data. Large volumes of data including microbiome sequencing, clinical patient data and treatment outcomes can effectively be processed and analyzed by artificial intelligence to find patterns that conventional approaches might miss. More individualized therapeutic

approaches may be made possible by training AI algorithms to anticipate how changes in the microbiome might impact the progression of cancer or the effectiveness of treatment. In addition, AI can effectively be used to predict a patient's reaction to a particular treatment or find new microbial-based biomarkers for early cancer detection which could result in the creation of more specialized and efficient cancer treatments (Zhang et al. 2020). 3. Cancer prevention strategies that take the microbiome into account: The creation of microbiome-aware cancer prevention techniques is being fueled by our expanding knowledge of the role the microbiota serves a pivotal role in in the development and metastasis of cancer. These tactics include probiotics, diet, and microbiome-targeting therapies and concentrate on altering the microbiome to lower the risk of cancer. For example, certain gut microbiota may elevate the risk of colorectal cancer, while others have been linked to a decreased risk. It might be possible to lower the risk of cancer or slow its progression by treating dysbiosis or preserving a healthy microbiome. Potential preventive treatments that offer a more proactive approach to cancer prevention include probiotics, prebiotics, and dietary modifications that impact the gut microbiota.

5. Conclusion: Microbial therapeutics are becoming more widely acknowledged as effective instruments in the fight against cancer because they provide innovative ways to both directly treat the illness and enhance conventional treatments. Microbes have great potential in the treatment of cancer from oncolytic bacterial therapies that specifically target and kill cancer cells to engineered bacteria that deliver medications directly to tumor sites. These microbes are highly adaptable to various cancer types and can effectively interact with tumors in novel ways due to the incorporation of synthetic biology. Furthermore, the development of microbiome-aware strategies has improved the effectiveness of chemotherapy and immunotherapy because it is now clearer how the microbiome affects immune responses. These developments point to a move toward more accurate, customized, and effective cancer therapies that enhance therapeutic results and reduce side effects. Multidisciplinary research is essential to the efficacy of microbial-based cancer treatments. The development of these treatments requires cooperation between microbiologists, oncologists, geneticists, and bioengineers. To overcome present obstacles, more investigation into the role of the microbiome in the development of cancer and the development of safer, more efficient engineered microbes will be required. Furthermore, to maximize the potential of these treatments, technologies like metagenomics, metabolomics, and artificial intelligence will need to be integrated. This will enable more accurate tumor targeting and better patient outcomes. Despite persistent safety and regulatory concerns, microbial therapeutics have enormous translational potential. These therapies will probably be incorporated into conventional cancer treatment regimens as preclinical and clinical research advances.

References

Alexander JL, Wilson ID, Teare J, Marchesi JR, Nicholson JK, Kinross JM (2017) Gut microbiota modulation of chemotherapy efficacy and toxicity. Nat Rev Gastroenterol Hepatol 14(6): 356–365

Arpaia N, Campbell C, Fan X et al (2013) Metabolites produced by commensal bacteria regulate the differentiation of regulatory T cells. Nature 504(7480):451–455. https://doi.org/10.1038/nature12726

Arthur JC, Perez-Chanona E, Mühlbauer M et al (2012) Intestinal inflammation targets cancer-inducing activity of the microbiota. Science 338(6103):120–123. https://doi.org/10.1126/science.1224820

Atun R, Jaffray DA, Barton MB et al (2015) Expanding global access to radiotherapy. Lancet Oncol 16(10):1153–1186. https://doi.org/10.1016/S1470-2045(15)00222-3

Aykut B, Pushalkar S, Chen R et al (2019) The fungal mycobiome promotes pancreatic oncogenesis via activation of MBL. Nature 574(7777):264–267. https://doi.org/10.1038/s41586-019-1608-2

Baruch EN, Youngster I, Ben-Betzalel G et al (2021) Fecal microbiota transplant promotes response in immunotherapy-refractory melanoma patients. Science 371(6529):602–609. https://doi.org/10.1126/science..abb5920

Belkaid Y, Hand TW (2014) Role of the microbiota in immunity and inflammation. Cell 157(1): 121–133. https://doi.org/10.1016/j.cell.2014.03.011

Benedict C, Vogel H, Jonas W et al (2016) Gut microbiota and glucometabolic alterations in response to recurrent partial sleep deprivation in normal-weight young individuals. Mol Metab 5(12):1175–1186. https://doi.org/10.1016/j.molmet.2016.10.003

Blackett KL, Siddhi SS, Cleary S, Steed H, Mille.r MH, Macfarlane S et al (2013 Jun) Oesophageal bacterial biofilm changes in gastro-oesophageal reflux disease, Barrett's and oesophageal carcinoma: association or causality? Aliment Pharmacol Ther 37(11):1084–1092. https://doi.org/10.1111/apt.12317

Bobo D, Robinson KJ, Islam J, Thurecht KJ, Corrie SR (2016) Nanoparticle-based medicines: a review of FDA-approved materials and clinical trials to date. Pharm Res 33(10):2373–2387. https://doi.org/10.1007/s11095-016-1958-5

Bullman S, Pedamallu CS, Sicinska E et al (2017) Analysis of Fusobacterium nucleatum in colorectal carcinoma tissue. Science 358(6369):1443–1448

Canani RB et al (2011) Potential beneficial effects of butyrate in intestinal and extraintestinal diseases. World J Gastroenterol 17(12):1519–1528. https://doi.org/10.3748/wjg.v17.i12.1519

Cani PD et al (2007) Metabolic endotoxemia initiates obesity and insulin resistance. Diabetes 56(7): 1761–1772. https://doi.org/10.2337/db06-1491

Cardona F, Andrés-Lacueva C, Tulipani S, Tinahones FJ, Queipo-Ortuño MI (2013) Benefits of polyphenols on gut microbiota and implications in human health. J Nutr Biochem 24(8): 1415–1422. https://doi.org/10.1016/j.jnutbio.2013.05.001

Chabé M, Lokmer A, Ségurel L (2017) Gut protozoa: friends or foes of the human gut microbiota? Trends Parasitol 33(12):925–934. https://doi.org/10.1016/j.pt.2017.08.005. Epub 2017 Sep 1

Clarke SF, Murphy EF, O'Sullivan O et al (2014) Exercise and associated dietary extremes impact on gut microbial diversity. Gut 63(12):1913–1920. https://doi.org/10.1136/gutjnl-2013-306541

Clemente JC, Ursell LK, Parfrey LW, Knight R (2012) The impact of the gut microbiota on human health: an integrative view. Cell 148(6):1258–1270. https://doi.org/10.1016/j.cell.2012.01.035

Coussens LM, Werb Z (2002) Inflammation and cancer. Nature 420(6917):860–867. https://doi.org/10.1038/nature01322

Cryan JF, Dinan TG (2012) Mind-altering microorganisms: the impact of the gut microbiota on brain and behaviour. Nat Rev Neurosci 13(10):701–712. https://doi.org/10.1038/nrn3346

Dapito DH, Mencin A, Gwak GY et al (2012) Promotion of hepatocellular carcinoma by the intestinal microbiota and TLR4. Cancer Cell 21(4):504–516. https://doi.org/10.1016/j.ccr.2012.02.007

Darfeuille-Michaud A et al (2004) High prevalence of adherent-invasive Escherichia coli associated with ileal mucosa in Crohn's disease. Gastroenterology 127(2):412–421. https://doi.org/10.1053/j.gastro.2004.04.061

Davar D, Dzutsev AK, McCulloch JA et al (2021) Fecal microbiota transplant overcomes resistance to anti–PD-1 therapy in melanoma patients. Science 371(6529):595–602. https://doi.org/10.1126/science.abf3363

David LA et al (2014) Diet rapidly and reproducibly alters the human gut microbiome. Nature 505(7484):559–563. https://doi.org/10.1038/nature12820

De Filippo C et al (2010) Impact of diet in shaping gut microbiota revealed by a comparative study in children from Europe and rural Africa. Proc Natl Acad Sci USA 107(33):14691–14696. https://doi.org/10.1073/pnas.1005963107

Delaney G, Jacob S, Featherstone C, Barton M (2005) The role of radiotherapy in cancer treatment: estimating optimal utilization from a review of evidence-based clinical guidelines. Cancer 104(6):1129–1137. https://doi.org/10.1002/cncr.21324

Derosa L, Routy B, Kroemer G, Zitvogel L (2018) The intestinal microbiota determines the clinical efficacy of immune checkpoint blockers targeting the PD-1/PD-L1 axis. Onco Targets Ther 7(6):e1434468. https://doi.org/10.1080/2162402X.2018.1434468

Derosa L, Routy B, Fidelle M et al (2018a) Gut bacteria composition drives primary resistance to cancer immunotherapy in renal cell carcinoma patients. Eur Urol 74(6):792–800. https://doi.org/10.1016/j.eururo.2018.08.001

Dethlefsen L, Relman DA (2011) Incomplete recovery and individualized responses of the human distal gut microbiota to repeated antibiotic perturbation. Proc Natl Acad Sci USA 108(Suppl 1):4554–4561. https://doi.org/10.1073/pnas.1000087107

Devkota S et al (2012) Dietary-fat-induced taurocholic acid promotes pathobiont expansion and colitis in Il10−/− mice. Nature 487(7405):104–108. https://doi.org/10.1038/nature11225

Dominguez-Bello MG, Costello EK, Contreras M, Magris M, Hidalgo G, Fierer N, Knight R (2010) Delivery mode shapes the acquisition and structure of the initial microbiota across multiple body habitats in newborns. Proc Natl Acad Sci USA 107(26):11971–11975. https://doi.org/10.1073/pnas.1002601107

Duan Y, Young R, Schnabl B (2022) Bacteriophages and their potential for treatment of gastrointestinal diseases. Nat Rev Gastroenterol Hepatol. 19(2):135–144. https://doi.org/10.1038/s41575-021-00536-z. Epub 2021 Nov 15. PMID: 34782783; PMCID: PMC8966578

Duller S, Moissl-Eichinger C (2024) Archaea in the human microbiome and potential effects on human infectious disease. Emerg Infect Dis 30(8):1505–1513. https://doi.org/10.3201/eid3008.240181. PMID: 39043386; PMCID: PMC11286065

Eaden JA, Abrams KR, Mayberry JF (2001) The risk of colorectal cancer in ulcerative colitis: a meta-analysis. Gut 48(4):526–535. https://doi.org/10.1136/gut.48.4.526

Edge SB, Compton CC (2010) The American joint committee on cancer: the 7th edition of the AJCC cancer staging manual and the future of TNM. Ann Surg Oncol 17(6):1471–1474. https://doi.org/10.1245/s10434-010-0985-4

El-Serag HB (2012) Epidemiology of viral hepatitis and hepatocellular carcinoma. Gastroenterology 142(6):1264–73.e1. https://doi.org/10.1053/j.gastro.2011.12.061

Fan X, Alekseyenko AV, Wu J et al (2018) Human oral microbiome and prospective risk for pancreatic cancer: a population-based nested case-control study. Gut 67(1):120–127. https://doi.org/10.1136/gutjnl-2016-312580

Ford AC, Yuan Y, Moayyedi P (2020) Helicobacter pylori eradication therapy to prevent gastric cancer: systematic review and meta-analysis. Gut 69(12):2113–2121. https://doi.org/10.1136/gutjnl-2020-320839. Epub 2020 Mar 23

Foster JA, Rinaman L, Cryan JF (2017) Stress & the gut-brain axis: regulation by the microbiome. Neurobiol Stress 7:124–136. https://doi.org/10.1016/j.ynstr.2017.03.001

Francino MP (2015) Antibiotics and the human gut microbiome: dysbioses and accumulation of resistances. Front Microbiol 6:1543. https://doi.org/10.3389/fmicb.2015.01543

García-Rivero JL (2020) The microbiome and asthma. Arch Bronconeumol (Engl Ed) 56(1):1–2. https://doi.org/10.1016/j.arbres.2019.01.016. English, Spanish, Epub 2019 Feb 23

Garrett WS (2015) Cancer and the microbiota. Science 348(6230):80–86. https://doi.org/10.1126/science.aaa4972

Geller LT, Barzily-Rokni M, Danino T et al (2017) Potential role of intratumor bacteria in mediating tumor resistance to the chemotherapeutic drug gemcitabine. Science 357(6356):1156–1160. https://doi.org/10.1126/science.aah5043

Gibson GR, Hutkins R, Sanders ME et al (2017) Expert consensus document: the international scientific association for probiotics and prebiotics (ISAPP) consensus statement on the definition and scope of prebiotics. Nat Rev Gastroenterol Hepatol 14(8):491–502. https://doi.org/10.1038/nrgastro.2017.75

Goedert JJ, Jones G, Hua X et al (2015) Gut microbiota and the risk of breast cancer: a pilot study. J Natl Cancer Inst 107(8):djv147. https://doi.org/10.1093/jnci/djv147

Gopalakrishnan V, Spencer CN, Nezi L, Reuben A, Andrews MC, Karpinets TV et al (2018) Gut microbiome modulates response to anti–PD-1 immunotherapy in melanoma patients. Science 359(6371):97–103. https://doi.org/10.1126/science.aan4236

Grice EA, Segre JA (2011) The human microbiome: our second genome. Annu Rev Genomics Hum Genet 12:231–252. https://doi.org/10.1146/annurev-genom-090410-132430

Grivennikov SI, Greten FR, Karin M (2010) Immunity, inflammation, and cancer. Cell 140(6):883–899. https://doi.org/10.1016/j.cell.2010.01.025

Hanahan D, Weinberg RA (2011) Hallmarks of cancer: the next generation. Cell 144(5):646–674. https://doi.org/10.1016/j.cell.2011.02.013

Hou K, Wu ZX, Chen XY, Wang JQ, Zhang D, Xiao C, Zhu D, Koya JB, Wei L, Li J, Chen ZS (2022) Microbiota in health and diseases. Signal Transduct Target Ther 7(1):135. https://doi.org/10.1038/s41392-022-00974-4. PMID: 35461318; PMCID: PMC9034083

Human Microbiome Project Consortium (2012) Structure, function and diversity of the healthy human microbiome. Nature 2012486(7402):207–214. https://doi.org/10.1038/nature11234. PMID: 22699609; PMCID: PMC3564958

Irfan M, Delgado RZR, Frias-Lopez J (2020) The oral microbiome and cancer. Front Immunol 11:591088. https://doi.org/10.3389/fimmu.2020.591088. PMID: 33193429; PMCID: PMC7645040

Iwasaki A, Medzhitov R (2015) Control of adaptive immunity by the innate immune system. Nature 517(7534):392–400. https://doi.org/10.1038/nature14139

Johnson CH, Dejea CM, Edler D, Hoang LT, Santidrian AF, Fedorova I et al (2015) Metabolism links bacterial biofilms and colon carcinogenesis. Cell Metab 21(6):891–897. https://doi.org/10.1016/j.cmet.2015.04.011

Jordan VC (2003) Tamoxifen: a most unlikely pioneering medicine. Nat Rev Drug Discov 2(3):205–213. https://doi.org/10.1038/nrd1031

Kamada N, Chen GY, Inohara N, Núñez G (2013) Control of pathogens and pathobionts by the gut microbiota. Nat Rev Immunol 13(5):321–335

Kelly CR, Khoruts A, Staley C, Sadowsky MJ, Abd M, Alani M, Bakow B, Curran P, McKenney J, Tisch A, Reinert SE, Machan JT, Brandt LJ (2016) Effect of fecal microbiota transplantation on recurrence in multiply recurrent clostridium difficile infection: a randomized trial. Ann Intern Med. 165(9):609–616. https://doi.org/10.7326/M16-0271. Epub 2016 . PMID: 27547925; PMCID: PMC5909820

Koh A, De Vadder F, Kovatcheva-Datchary P, Bäckhed F (2016) From dietary fiber to host physiology: short-chain fatty acids as central bacterial metabolites. Cell 165(6):1332–1345. https://doi.org/10.1016/j.cell.2016.05.041

Kostic AD, Gevers D, Pedamallu CS, Michaud M, Duke F, Earl AM et al (2012) Genomic analysis identifies association of Fusobacterium with colorectal carcinoma. Genome Res 22(2):292–298

Kostic AD, Chun E, Robertson L et al (2013) Fusobacterium nucleatum potentiates intestinal tumorigenesis and modulates the tumor-immune microenvironment. Cell Host Microbe 14(2):207–215. https://doi.org/10.1016/j.chom.2013.07.007

Kusters JG, van Vliet AH, Kuipers EJ (2006) Pathogenesis of Helicobacter pylori infection. Clin Microbiol Rev 19(3):449–490
Lahouty M, Fadaee M, Aghaei R, Alizadeh F, Jafari A, Sharifi Y (2025) Gut microbiome and colorectal cancer: from pathogenesis to treatment. Pathol Res Pract 271:156034. https://doi.org/10.1016/j.prp.2025.156034. Epub 2025 May 21
LeBlanc JG, Milani C, de Giori GS, Sesma F, van Sinderen D, Ventura M (2013) Bacteria as vitamin suppliers to their host: a gut microbiota perspective. Curr Opin Biotechnol 24(2): 160–168. https://doi.org/10.1016/j.copbio.2012.08.005
Ley RE, Turnbaugh PJ, Klein S, Gordon JI (2006) Microbial ecology: human gut microbes associated with obesity. Nature. 2006444(7122):1022–1023. https://doi.org/10.1038/4441022a
Longley DB, Johnston PG (2005) Molecular mechanisms of drug resistance. J Pathol 205(2): 275–292. https://doi.org/10.1002/path.1706
Longoria O, Beije N, de Bono JS (2024) PARP inhibitors for prostate cancer. Semin Oncol 51(1–2): 25–35. https://doi.org/10.1053/j.seminoncol.2023.09.003. Epub 2023 Sep 25
Louis P, Hold GL, Flint HJ (2014) The gut microbiota, bacterial metabolites and colorectal cancer. Nat Rev Microbiol 12:661–672. https://doi.org/10.1038/nrmicro3344
Marchesi JR, Ravel J (2015) The vocabulary of microbiome research: a proposal. Microbiome 3(1): 31. https://doi.org/10.1186/s40168-015-0094-5. PMID: 26229597; PMCID: PMC4520061
Marchesi JR, Adams DH, Fava F et al (2016) The gut microbiome and host health: a new clinical frontier. Gut 65(2):295–303
Marco ML, Heeney D, Binda S et al (2017) Health benefits of fermented foods: microbiota and beyond. Curr Opin Biotechnol 44:94–102. https://doi.org/10.1016/j.copbio.2016.11.010
Matson V, Fessler J, Bao R, Chongsuwat T, Zha Y, Alegre ML et al (2018) The commensal microbiome is associated with anti–PD-1 efficacy in metastatic melanoma patients. Science 359(6371):104–108. https://doi.org/10.1126/science.aao3290
Maude SL, Frey N, Shaw PA et al (2014) Chimeric antigen receptor T cells for sustained remissions in leukemia. N Engl J Med 371(16):1507–1517. https://doi.org/10.1056/NEJMoa1407222
Mazmanian SK, Liu CH, Tzianabos AO, Kasper DL (2005) An immunomodulatory molecule of symbiotic bacteria directs maturation of the host immune system. Cell 122(1):107–118. https://doi.org/10.1016/j.cell.2005.05.007
Ou S, Wang H, Tao Y, Luo K, Ye J, Ran S, Guan Z, Wang Y, Hu H, Huang R (2022) Fusobacterium nucleatum and colorectal cancer: from phenomenon to mechanism. Front Cell Infect Microbiol 12:1020583. https://doi.org/10.3389/fcimb.2022.1020583. PMID: 36523635; PMCID: PMC9745098
Plottel CS, Blaser MJ (2011) Microbiome and malignancy. Cell Host Microbe 10(4):324–335. https://doi.org/10.1016/j.chom.2011.10.003
Plummer M, Franceschi S, Vignat J, Forman D, de Martel C (2015) Global burden of gastric cancer attributable to Helicobacter pylori. Int J Cancer 136(2):487–490. https://doi.org/10.1002/ijc.28999
Postow MA, Callahan MK, Wolchok JD (2015) Immune checkpoint blockade in cancer therapy. J Clin Oncol 33(17):1974–1982. https://doi.org/10.1200/JCO.2014.59.4358
Pushalkar S, Hundeyin M, Daley D, Zambirinis CP, Kurz E, Mishra A et al (2018) The pancreatic cancer microbiome promotes oncogenesis by induction of innate and adaptive immune suppression. Cancer Discov 8(4):403–416. https://doi.org/10.1158/2159-8290.CD-17-1134
Qin J et al (2010) A human gut microbial gene catalogue established by metagenomic sequencing. Nature 464(7285):59–65. https://doi.org/10.1038/nature08821
Rafter J (2003) Probiotics and colon cancer. Best Pract Res Clin Gastroenterol 17(5):849–859. https://doi.org/10.1016/S1521-6918(03)00078-7
Ren Z, Jiang J, Xie H, Li A, Lu H, Xu S et al (2017) Gut microbial profile analysis by MiSeq sequencing of pancreatic carcinoma patients in China. Oncotarget 8(56):95176–95191
Reyes VE (2023) Helicobacter pylori and Its Role in Gastric Cancer. Microorganisms. 11(5):1312. https://doi.org/10.3390/microorganisms11051312. PMID: 37317287; PMCID: PMC10220541

Ridlon JM, Kang DJ, Hylemon PB, Bajaj JS (2014) Bile acids and the gut microbiome. Curr Opin Gastroenterol 30(3):332–338. https://doi.org/10.1097/MOG.0000000000000057. PMID: 24625896; PMCID: PMC4215539

Round JL, Mazmanian SK (2009) The gut microbiota shapes intestinal immune responses. Nat Rev Immunol 9(5):313–323. https://doi.org/10.1038/nri2515

Routy B, Le Chatelier E, Derosa L, Duong CPM, Alou MT, Daillère R et al (2018) Gut microbiome influences efficacy of PD-1–based immunotherapy against epithelial tumors. Science 359(6371):91–97. https://doi.org/10.1126/science.aan3706

Rubinstein MR, Wang X, Liu W et al (2013) Fusobacterium nucleatum promotes colorectal carcinogenesis by modulating E-cadherin/β-catenin signaling via its FadA adhesin. Cell Host Microbe 14(2):195–206. https://doi.org/10.1016/j.chom.2013.07.012

Sartor RB (2008) Microbial influences in inflammatory bowel diseases. Gastroenterology 134(3): 577–594. https://doi.org/10.1053/j.gastro.2007.11.059

Sawyers C (2004) Targeted cancer therapy. Nature 432(7015):294–297. https://doi.org/10.1038/nature03095

Schwabe RF, Jobin C (2013a) The microbiome and cancer. Nat Rev Cancer 13(11):800–812. https://doi.org/10.1038/nrc3610

Schwabe RF, Jobin C (2013b) The microbiome and cancer. Nat Rev Cancer 13(11):800–812

Sekirov I, Russell SL, Antunes LC, Finlay BB (2010) Gut microbiota in health and disease. Physiol Rev 90(3):859–904. https://doi.org/10.1152/physrev.00045.2009

Smits LP et al (2013) Therapeutic potential of fecal microbiota transplantation. Gastroenterology 145(5):946–953. https://doi.org/10.1053/j.gastro.2013.08.058

Sokol H et al (2008) Faecalibacterium prausnitzii is an anti-inflammatory commensal bacterium identified by gut microbiota analysis of Crohn disease patients. Proc Natl Acad Sci USA 105(43):16731–16736. https://doi.org/10.1073/pnas.0804812105

Sung H, Ferlay J, Siegel RL et al (2021) Global cancer statistics 2020: GLOBOCAN estimates of incidence and mortality worldwide for 36 cancers in 185 countries. CA Cancer J Clin 71(3): 209–249. https://doi.org/10.3322/caac.21660

Thapa B, De Sarkar N, Giri S, Sharma K, Kim M, Kilari D (2024) Integrating PARP inhibitors in mCRPC therapy: current strategies and emerging trends. Cancer Manag Res 16:1267–1283. https://doi.org/10.2147/CMAR.S411023. PMID: 39308935; PMCID: PMC11416116

Tilg H, Kaser A (2011) Gut microbiome, obesity, and metabolic dysfunction. J Clin Invest 121(6): 2126–2132. https://doi.org/10.1172/JCI58109

Turnbaugh PJ, Ley RE, Mahowald MA et al (2006) An obesity-associated gut microbiome with increased capacity for energy harvest. Nature 444(7122):1027–1031. https://doi.org/10.1038/nature05414

Turnbaugh PJ et al (2007) The human microbiome project. Nature 449(7164):804–810. https://doi.org/10.1038/nature06244

Vetizou M, Pitt JM, Daillère R et al (2015) Anticancer immunotherapy by CTLA-4 blockade relies on the gut microbiota. Science 350(6264):1079–1084. https://doi.org/10.1126/science.aad1329

Viaud S, Saccheri F, Mignot G et al (2013) The intestinal microbiota modulates the anticancer immune effects of cyclophosphamide. Science 342(6161):971–976. https://doi.org/10.1126/science.1240537

Wu S, Rhee KJ, Albesiano E et al (2009) A human colonic commensal promotes colon tumorigenesis via activation of T helper type 17 T cell responses. Nat Med 15(9):1016–1022. https://doi.org/10.1038/nm.2015

Wu M, Chen X, Lu Q, Yao X (2024) Fecal microbiota transplantation for the treatment of chronic inflammatory skin diseases. Heliyon 10(18):e37432. https://doi.org/10.1016/j.heliyon.2024.e37432

Yan X, Zheng C, Tang Z, Feng Y, Wang B (2025) Multidrug resistance of helicobacter pylori and its impact on the diagnosis and treatment of gastrointestinal diseases and countermeasures. Sheng Wu Gong Cheng Xue Bao 41(4):1240–1251. Chinese. https://doi.org/10.13345/j.cjb.240851

Yoshimoto S, Loo TM, Atarashi K et al (2013) Obesity-induced gut microbial metabolite promotes liver cancer through senescence secretome. Nature 499(7456):97–101. https://doi.org/10.1038/nature12347

Yu LX, Schwabe RF (2017) The gut microbiome and liver cancer: mechanisms and clinical translation. Nat Rev Gastroenterol Hepatol 14(9):527–539

Yu J, Feng Q, Wong SH et al (2017) Metagenomic analysis of faecal microbiome as a tool towards targeted non-invasive biomarkers for colorectal cancer. Gut 66(1):70–78. https://doi.org/10.1136/gutjnl-2015-309800

Zalatan JG, Petrini L, Geiger R (2024) Engineering bacteria for cancer immunotherapy. Curr Opin Biotechnol 85:103061. https://doi.org/10.1016/j.copbio.2023.103061

Zeevi D et al (2015) Personalized nutrition by prediction of glycemic responses. Cell 163(5): 1079–1094. https://doi.org/10.1016/j.cell.2015.11.001

Zeller G, Tap J, Voigt AY et al (2014) Potential of fecal microbiota for early-stage detection of colorectal cancer. Mol Syst Biol 10(11):766. https://doi.org/10.15252/msb.20145645

Zhan T, Rindtorff N, Betge J, Ebert MP, Boutros M (2019) CRISPR/Cas9 for cancer research and therapy. Semin Cancer Biol 55:106–119. https://doi.org/10.1016/j.semcancer.2018.04.001

Zhang F, Chen C, Hu J, Su R, Zhang J, Han Z, Chen H, Li Y (2019) Molecular mechanism of helicobacter pylori-induced autophagy in gastric cancer. Oncol Lett 18(6):6221–6227. https://doi.org/10.3892/ol.2019.10976

Zhang S, Li C, Liu J, Geng F, Shi X, Li Q, Lu Z, Pan Y (2020) Fusobacterium nucleatum promotes epithelial-mesenchymal transiton through regulation of the lncRNA MIR4435-2HG/miR-296-5p/Akt2/SNAI1 signaling pathway. FEBS J 287(18):4032–4047. https://doi.org/10.1111/febs.15233

Zitvogel L, Daillère R, Roberti MP et al (2015a) Anticancer effects of the microbiome and its products. Nat Rev Microbiol 13(6):457–468. https://doi.org/10.1038/nrmicro3440

Zitvogel L, Galluzzi L, Viaud S, Vétizou M, Daillère R, Merad M et al (2015b) Cancer and the gut microbiota: an unexpected link. Immunity 42(6):1003–1013

Zitvogel L et al (2016) Microbiome and anticancer immunosurveillance. Cell 165(2):276–287. https://doi.org/10.1016/j.cell.2016.03.001

Microbiota and Cancer Therapy

4

Abstract

The human microbiota is a central regulator of cancer development and response to treatment. Once believed to be incidental the complex interaction between immune system and host associated microbial communities is now recognized as a critical component of oncogenesis and treatment modulation. This chapter explores how microbial dysbiosis causes immunological dysfunction, metabolic abnormalities, and chronic inflammation which in turn contribute to the development of cancer. It focus mainly on microbiota associated cancers such as colorectal, gastric and hepato-cellular cancers. It also discusses how the microbiota influences immunological responses, drug metabolism, and systemic inflammation all of which have an effect on the efficacy and toxicity of cancer treatments like immunotherapy, chemotherapy and radiation. But despite significant progress, there are still significant gaps in our mechanistic understanding of host microbiota interactions in cancer therapy. Current studies are frequently limited by interindividual variability, lack of longitudinal data, and inconsistent microbiota analysis methodologies. Moreover, microbiota profiling in oncology is not sufficiently clinically integrated and the effects of microbial modulation strategies remain poorly understood. This chapter presents new information from preclinical and clinical studies that link specific microbial taxa and metabolites to the outcomes of treatment responses in a novel manner. It also highlights novel therapeutic strategies such as dietary modifications, targeted probiotics, and fecal microbiota transplantation as potential adjuncts to conventional cancer treatment. Moreover, a promising strategy to enhance therapeutic efficacy and lessen side effects is sensible microbiota modulation based on the patient's microbial signature. In order to fully utilize the therapeutic potential of the human microbiome in cancer care, the chapter ends by highlighting the necessity of multidisciplinary research projects combining oncology, microbiology, and computational biology.

R. C. Sobti et al., *Microbiota Unveiled: Science, Health, and Future Therapies*,
https://doi.org/10.1007/978-981-95-9031-5_4

Keywords

Cancer · Dysbiosis · Fecal microbiota transplantation · Gut microbiota · Immunotherapy and microbiome

4.1 Introduction

The vast and diverse population of microorganisms that inhabit the human body is collectively referred to as the microbiota. These microorganisms specifically include: bacteria, viruses, fungi, and archaea. They are most abundant and metabolically active in the gastrointestinal tract (Lynch and Pedersen 2016). The microbiota is far from being a passive bystander and is essential for maintaining host health because of its roles in immune regulation, digestion, nutrient absorption, and pathogen defense (Sommer and Bäckhed 2013). Recent years have seen a growing body of evidence demonstrating the influence of the microbiota on the onset of cancer and the outcomes of treatment. Dysbiosis, i.e., imbalance in the composition or function of the microbiota, has been associated with carcinogenesis through mechanisms like the production of genotoxic metabolites, modulation of the host immune system, and chronic inflammation (Schwabe and Jobin 2013). Certain microbial species such as *Helicobacter pylori* have been definitively linked to gastric cancer, while gut dysbiosis has been linked to colorectal and hepatocellular carcinomas (Garrett 2015). In addition to its role in the initiation and progression of cancer, the microbiota serves a pivotal role in a significant role in the toxicity and efficacy of cancer treatments. According to recent studies, a patient's microbiota composition may have an impact on how they react to radiation, chemotherapy, and especially immunotherapy. For instance, it has been demonstrated that certain gut microbes elevate the effectiveness of immune checkpoint inhibitors, and using antibiotics during treatment has been linked to a number of negative side effects (Fig. 4.1) (Gopalakrishnan et al. 2018). These findings emphasize how crucial it is to treat cancer by considering the microbiome as a dynamic and critically important component. Understanding the reciprocal relationships between microbiota and cancer treatments can effectively facilitate enhancing patient outcomes through the use of microbial modulation techniques such as diet, probiotics, prebiotics, and fecal microbiota transplantation. As we enter the era of personalized medicine, integrating microbiota profiling into cancer management practices may lead to new therapeutic opportunities and a reexamination of conventional treatment paradigms (Fig. 4.1).

4.1.1 Human Microbiota Overview

Trillions of microorganisms, including bacteria, viruses, fungi, and archaea, make up the human microbiota. These microbes reside on and within the human body, taking up residence in areas such as the skin, genitourinary tract, respiratory tract, oral cavity, and most frequently the gastrointestinal tract (Sender et al. 2016). The gut

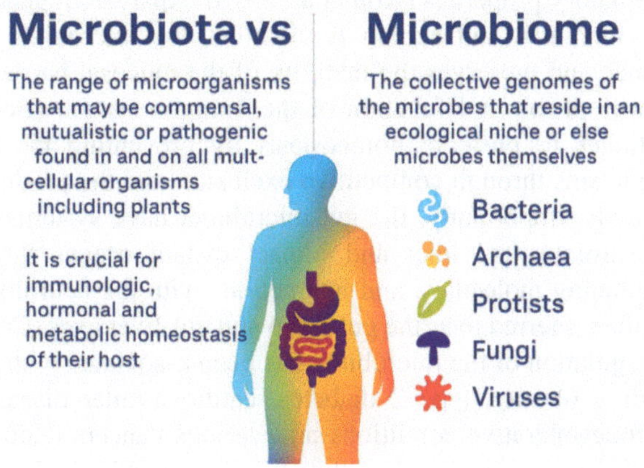

Fig. 4.1 Microbiota vs. microbiome: this figure illustrates the range of microorganisms (bacteria, archaea, protists, fungi, and viruses) that are present in and on multicellular organisms and are crucial for immunologic, hormonal, and metabolic homeostasis. The term microbiome refers to the collective genomes of microorganisms within an ecological niche or to the microbes themselves

microbiota is a metabolically active and extremely diverse ecosystem that is particularly significant to human physiology. It influences immune system development, vitamin synthesis, digestion as well as pathogen defense (Sommer and Bäckhed 2013). According to Dominguez-Bello et al. (2019), colonization begins at birth and is impacted by a number of variables such as the baby's feeding patterns, the way the baby was delivered, antibiotic exposure, diet, geography, and lifestyle. A healthy microbiota maintains host homeostasis whereas dysbiosis, i.e., loss of microbial diversity or a change in microbial composition has been linked to a variety of diseases including cancer, metabolic syndrome, neurological disorders, and inflammatory bowel disease (IBD) (Lynch and Pedersen 2016). Advances in metagenomics and next-generation sequencing have revolutionized the study of microbiomes by enabling high-resolution analysis of microbial communities and revealing complex host-microbe relationships that are increasingly recognized as critical to health and disease (Qin et al. 2010). As our knowledge of the human microbiota grows, it is now acknowledged as a substantial contributor to systemic health and a possible target for therapeutic intervention.

4.1.2 The Importance of the Microbiota to Human Health

The human microbiota is vital to maintaining overall health because it supports a wide variety of physiological, metabolic, and immunological functions. One of its primary roles is fermentation and breaking down of complex dietary polysaccharides that the human host would not otherwise be able to digest and leads to the formation of short-chain fatty acids (SCFAs) such as butyrate, acetate, and propionate. In addition to giving colonocytes energy, these SCFAs also have immunomodulatory and anti-inflammatory properties (Koh et al. 2016). Apart from generating essential nutrients like vitamin K and certain B vitamins, the microbiota also influences immune tolerance and preserves the integrity of the mucosal barrier both of which support the development and function of the immune system (Belkaid and Hand 2014). It facilitates to preserve homeostasis by preventing the colonization of pathogenic organisms through competitive exclusion and the production of antimicrobial compounds. Importantly, the gut microbiota have systemic effects that go beyond the gastrointestinal tract and impact distant organs through microbial metabolites, signaling molecules, and interactions with the neurological and endocrine systems often referred to as the gut-brain and gut-liver axes (Cryan et al. 2019). Dysbiosis/dysregulation of the microbiota has been associated with a wide range of diseases including obesity, type 2 diabetes, cardiovascular disease, autoimmune disorders, neurodegenerative conditions and various cancers (Zuo and Ng 2018). The discovery that the microbiota is a "functioning organ" has altered the paradigm in medicine. As a result, microbiome-based diagnostics and therapies have been developed with the goal of enhancing human health and reestablishing microbial equilibrium (Fig. 4.2).

Figure 4.2 illustrates central functions of the microbiota as depicted by the central human silhouette with the gut highlighted: 1. Defense against pathogens through competitive exclusion and antimicrobial production 2. Systemic effects via the gut-brain and gut-liver axes 3. Consequences of dysbiosis including links to obesity, diabetes, cardiovascular disease, autoimmune disorders, neurodegeneration, and cancers 4. Synthesis of essential nutrients like vitamin K and B vitamins

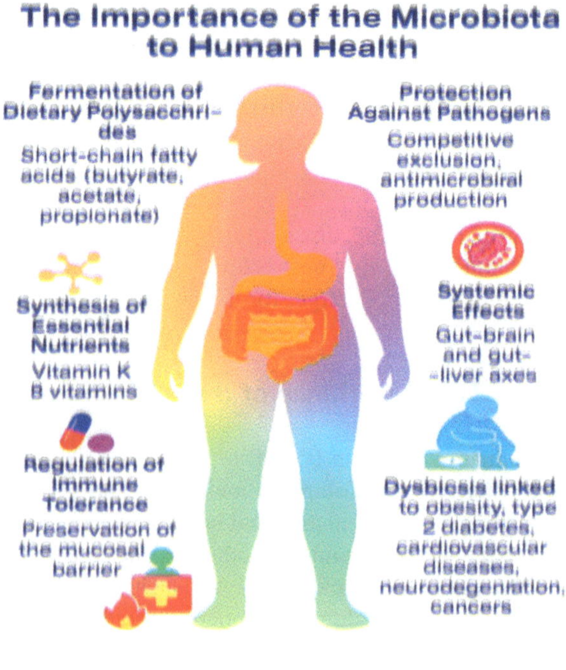

Fig. 4.2 The importance of the microbiota to human health

5. Fermentation of dietary polysaccharides into short-chain fatty acids (acetate, butyrate, and propionate) that provide energy and anti-inflammatory effects

4.1.3 Connection Between Cancer and Microbiota

Recent studies have shown a strong correlation between the human microbiota and the development, progression, and response to treatment of cancer. Microbial communities particularly in the gut can effectively influence carcinogenesis through a number of mechanisms including immune system modulation, the production of genotoxic metabolites, chronic inflammation, and disruption of epithelial barrier function (Garrett 2015). Some pathogenic bacteria such as *Fusobacterium nucleatum* have been associated with colorectal cancer because of their proinflammatory and immune-suppressive characteristics, while *Helicobacter pylori* has long been known to cause gastric cancer (Kostic et al. 2013). Microbial metabolites such as hydrogen sulfide and secondary bile acids can effectively contribute to oncogenesis by damaging DNA or fostering an environment that supports the growth of tumors (Louis et al. 2014). Conversely, beneficial microorganisms and their byproducts like short-chain fatty acids (butyrate) may have antitumorigenic effects by promoting cancer cell death and regulating host immunity (Schwabe and Jobin 2013). Moreover, recent studies suggest that the microbiota may influence the efficacy and toxicity of cancer therapies, particularly immunotherapy, in which particular microbial profiles have been associated with improved responses to immune checkpoint inhibitors (Gopalakrishnan et al. 2018). The growing recognition of these links in cancer prevention, diagnosis, and treatment has led to the development of microbiome modulation methods such as probiotics, prebiotics, and fecal microbiota transplantation (Fig. 4.3).

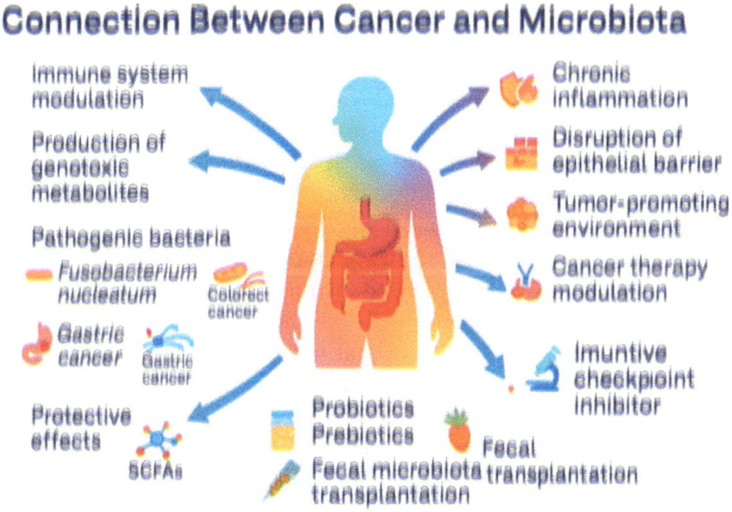

Fig. 4.3 Connection between cancer and microbiota

Figure 4.3 illustrates cancer and microbiota relationship where the arrows encircling the central human silhouette with the gut highlighted indicate the following mechanisms: 1. Immune system modulation, genotoxic metabolites, chronic inflammation, and disruption of the epithelial barrier 2. Pathogenic bacteria like *Helicobacter pylori* (gastric cancer) and *Fusobacterium nucleatum* (colorectal cancer) 3. Harmful metabolites (hydrogen sulfide and secondary bile acids) that promote tumor environments 4. Protective effects of beneficial microbes and SCFAs 5. Influence on cancer therapies especially immune checkpoint inhibitors. Probiotics, prebiotics, and fecal microbiota transplantation are among the therapeutic interventions highlighted at the bottom

4.2 The Humans Microbiota

It is diverse and dynamic ecosystem of microorganisms that live in and on the skin, gastrointestinal tract, respiratory tract, genitourinary system, and oral cavity. The majority of these organisms are bacteria, but there are also viruses, fungi, and archaea. The most metabolically active and densely populated of these is the gut microbiota which is vital to homeostasis and host physiology. Microbial cells which are approximately as many as or even more numerous than human cells are thought to make up between 10^{13} and 10^{14} of the human body. It is estimated that there are over 1000 distinct species and three million genes in the gut microbiota alone (Qin et al. 2010; Sender et al. 2016). Instead of being passive bystanders, these microbial communities actively participate in critical functions such as the synthesis of short-chain fatty acids, the synthesis of essential vitamins, the digestion of complex polysaccharides, and the development and regulation of the immune system (Sommer and Bäckhed 2013; Belkaid and Hand 2014). The highly individualized composition of the microbiota is influenced by a number of factors including age, geography, antibiotic use, genetics, and most importantly diet. Advances in metagenomics and high-throughput sequencing have enabled detailed taxonomic and functional profiling which has altered our understanding of the roles of these microbial ecosystems in health and disease. Importantly, dysbiosis has increasingly been associated with a number of diseases such as cancer, neurological disorders, metabolic syndromes, autoimmune diseases, and gastrointestinal disorders. Therefore, the human microbiota is now understood to be a dynamic and significant factor in health and disease, rather than just a passive community of microorganisms. To develop diagnostics, prophylactics, and therapeutic interventions that target the microbiota, it is necessary to comprehend the mechanisms through which dysbiosis influences pathophysiology (Fig. 4.4).

Figure 4.4 illustrates 1. Gastrointestinal tract microbiota: which is a home to Firmicutes, Bacteroidetes, Actinobacteria, and Proteobacteria. They aid in the synthesis of vitamins, production of SCFA, and immune system regulation; their dysbiosis is linked to a higher risk of cancer 2. Skin microbiota (*Staphylococcus, Corynebacterium*, and *Propionibacterium*); influenced by skin type, hygiene, climate, and cosmetics, their imbalance is linked to chronic inflammation 3. Oral cavity

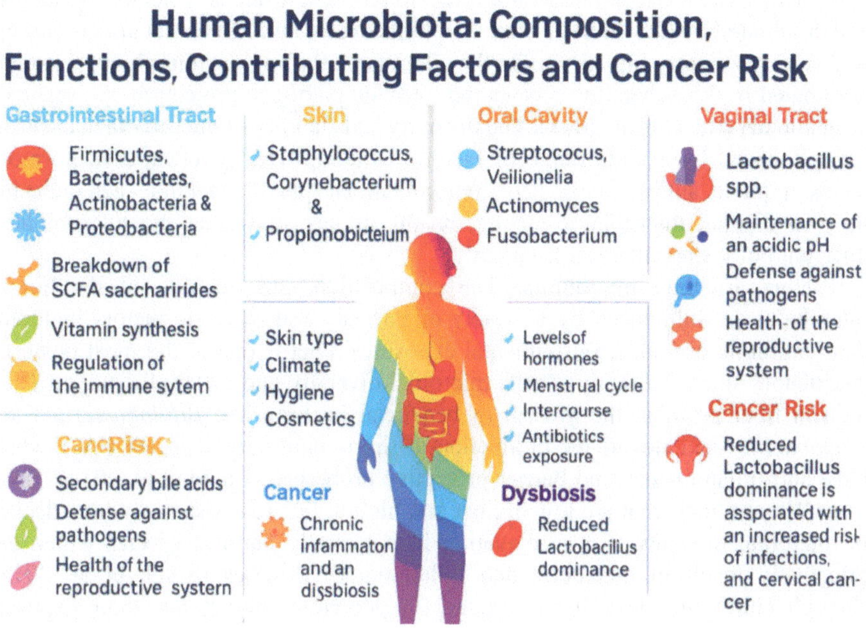

Fig. 4.4 Human microbiota-composition, function, contributing factors, and cancer risks

microbiota (*Streptococcus, Veillonella, Actinomyces*, and *Fusobacterium*); influenced by the menstrual cycle, hormones, sexual activity, and antibiotics. Their reduced dominance increases the risk of infection and cervical cancer 4. Vaginal tract microbiota: (*Lactobacillus*) species maintain an acidic pH and reproductive health.

4.2.1 Composition and Diversity

The human microbiota composition varies greatly within and between individuals. Each anatomical niche has a unique environment that affects the makeup and potential of the microbial communities that live there. The gut contains the most diverse and densely populated microbiota, with bacteria primarily belonging to the phyla Firmicutes, Bacteroidetes, Actinobacteria, and Proteobacteria. Our immune regulations, nutritional metabolism as well as defense mechanism, are all dependent on it (Qin et al. 2010; Human Microbiome Project Consortium 2012). However, the skin microbiota is more affected by environmental exposure and is dominated by species of Staphylococcus, Corynebacterium, and Propionibacterium which vary according to the type of skin (dry, moist, or oily) and body region (Grice and Segre 2011). The oral microbiota, which supports both systemic as well as oral health, is one of the most complex microbial ecosystems with over 700 species, primarily from the Streptococcus, Veillonella, Actinomyces, and Fusobacterium genera (Dewhirst et al. 2010). The nasal and respiratory tract microbiota which is typically less diverse and dominated by Corynebacterium, Staphylococcus, and Moraxella affects susceptibility to respiratory infections. On the other hand, the vaginal microbiota is usually dominated by Lactobacillus species that provide pathogen protection and maintain an acidic pH (Ravel et al. 2011). The diversity and stability of microbiota in each site are influenced by a wide range of factors including age, genetics, diet, hormone levels, hygiene, antibiotic use, and environment. Despite the fact that each niche has its own distinct microbial signature, recent studies show that the microbiota at different body sites interacts frequently.

Factors affecting microbiota: The composition and function of the human microbiota are influenced by a variety of intrinsic and extrinsic factors including diet, antibiotic use and environmental exposures. Diet is one of the most powerful modulators since it directly affects microbial diversity and metabolic output. Diets high in fiber promote the growth of beneficial bacteria like *Bifidobacterium* and *Lactobacillus* and elevate the production of short-chain fatty acids (SCFAs), which have anti-inflammatory and barrier-protective properties (David et al. 2014). However, Western diets that are low in fiber and high in fat are associated with a decline in microbial diversity and an elevation in potentially harmful species which can effectively result in metabolic and inflammatory diseases (Singh et al. 2017). Though antibiotics have been essential in preventing infections, their exposure especially in early life can effectively alter the path of microbial colonization and elevate a person's susceptibility to chronic diseases. Environmental elements such as pollution, location, cleanliness, and animal exposure also affect the microbiota composition. For example, the microbiome of people living in rural or agrarian societies is generally more diverse than that of people living in urban areas. Increased microbial exposure and dietary differences are most likely to blame for this (Yatsunenko et al. 2012). Microbiota variability is also influenced by other factors, such as breastfeeding, stress, physical activity, and the delivery method (caesarean vs. vaginal). Understanding how these factors impact microbial communities is crucial for the development of microbiome-targeted interventions intended to maintain or restore health (Fig. 4.5; Table 4.1).

4.2.2 Role in Immune Modulation and Inflammation

The development, training, and regulation of the host immune system depend on the human microbiota, which maintains a delicate balance between tolerance and immune activation. Commensal microbes aid in the early development of immune cells such as mature gut-associated lymphoid tissue and regulatory T cells (Tregs). Immune tolerance and the prevention of autoimmunity depend on these processes (Belkaid and Hand 2014). The microbiota affects both innate and adaptive immunity through the production of microbial-associated molecular patterns that interact with host pattern recognition receptors, including Toll-like receptors (TLRs) and NOD-like receptors. These interactions regulate the production of cytokines, antimicrobial peptides, and other immune mediators to preserve a healthy immune

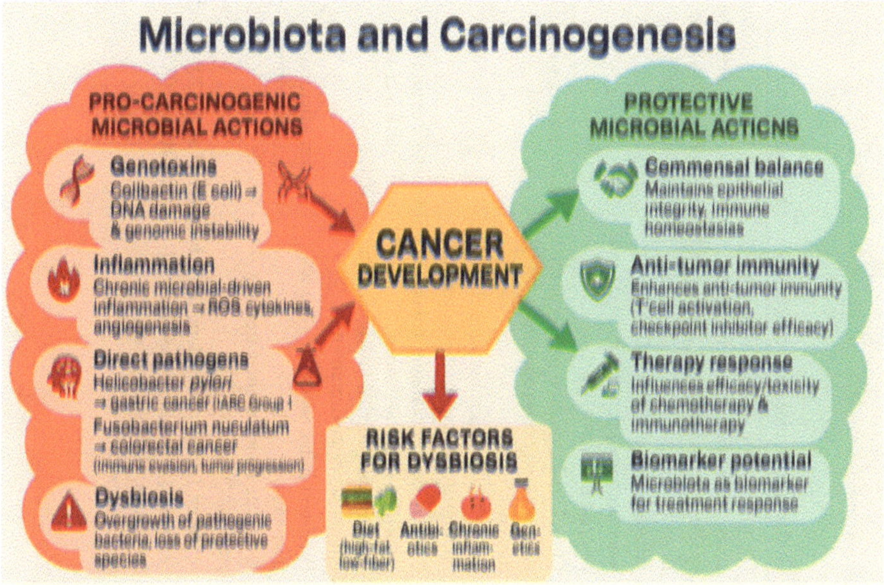

Fig. 4.5 Microbiota and carcinogenesis

response to pathogens while allowing beneficial microbes (Thaiss et al. 2016). Furthermore, microbial metabolites like butyrate and short-chain fatty acids have anti-inflammatory qualities by promoting Treg differentiation and inhibiting histone deacetylases (HDACs), which changes the expression of genes associated with inflammation (Furusawa et al. 2013). Numerous illnesses, including cancer, obesity, type 2 diabetes, and inflammatory bowel disease (IBD), are characterized by persistent low-grade inflammation. These immunoregulatory systems may be impacted by dysbiosis or modifications in the composition of microorganisms (Honda and Littman 2016). Because the microbiota is essential for immune homeostasis and inflammatory regulation, in addition to being a passive resident, it may be a promising therapeutic target for immune-mediated and inflammatory diseases.

4.3 Microbiota and Carcinogenesis

An increasing body of evidence suggests that the human microbiota serves a pivotal role in a critical role in carcinogenesis influencing the onset, metastasis, and response to treatment of cancer through a number of biological pathways. Certain microbial species can effectively directly promote tumorigenesis by producing genotoxins such as colibactin (produced by certain strains of *Escherichia coli*), which damage host cell DNA and cause genomic instability (Arthur et al. 2012). Because it can effectively lead to increased oxidative stress, cellular proliferation, and angiogenesis all of which encourage malignant transformation, chronic microbial-driven

Table 4.1 Human microbiota: composition, functions, contributing factors, and cancer risk

Body site	Dominant microbial taxa	Core functions	Key influencing factors	Dysbiosis outcomes and cancer link	Key references
Gastrointestinal tract	*Firmicutes, Bacteroidetes, Actinobacteria,* and *Proteobacteria*	Breakdown of polysaccharides SCFA synthesis vitamin synthesis Regulation of the immune system	Diet (fiber vs. fat) Antibiotic use Genetics Age Delivery method Geography	Overproduction of secondary bile acids, ROS, and pro-carcinogenic species (*F. nucleatum, B. fragilis* and *E. coli*) Inflammation and DNA damage, associated with pancreatic, liver, and colorectal cancers	Qin et al. (2010), Louis et al. (2014), Schwabe and Jobin (2013), and Xiao et al. (2025)
Skin	*Staphylococcus, Corynebacterium,* and *Propionibacterium*	Defense of barriers Exclusion of pathogens Antimicrobial peptide activation	Skin type Climate Hygiene Cosmetics	Chronic inflammation and weakened skin immunity can result from dysbiosis Possible connection between chronic inflammatory states and skin cancers	Grice and Segre (2011)
Oral cavity	*Streptococcus, Veillonella, Actinomyces,* and *Fusobacterium*	Systemic and oral immunity Decomposition of food ingredients	Saliva, diet, oral hygiene, dental health and smoking	*F. nucleatum* enrichment linked to colorectal and oral cancers encourages epithelial invasion and an inflammatory environment	Dewhirst et al. (2010) and Tjalsma et al. (2012)
Respiratory tract	*Corynebacterium, Staphylococcus,* and *Moraxella*	Defense against pathogens Modulation of mucosal immunity	Infections, smoking, pollution, and antibiotics	Persistent inflammation may result from chronic dysbiosis Associated with an increased risk of respiratory cancer	Human Microbiome Project Consortium (2012)
Vaginal tract	*Lactobacillus* spp.	Maintenance of an acidic pH Defense against pathogens Health of the reproductive system	Levels of hormones Menstrual cycle Intercourse Antibiotics exposure and hygiene	Reduced *Lactobacillus* dominance is associated with an increased risk of infections, HPV persistence and cervical cancer.	Ravel et al. (2011) and Xiao et al. (2025)

inflammation is also critically important in the tumor microenvironment (Garrett 2015). Because they cause chronic gastritis and compromise the integrity of the epithelium *Helicobacter pylori* and other bacteria, they have been classified by the International Agency for Research on Cancer (IARC) as group 1 carcinogens which are directly linked to gastric cancer (Wroblewski et al. 2010). Species like *Fusobacterium nucleatum* have been found to be more common in tumor tissues in colorectal cancer due to their interactions with immune and epithelial cells, where they facilitate tumor progression and immune evasion (Kostic et al. 2013). Furthermore, microbial metabolites such as hydrogen sulfide and secondary bile acids can effectively cause cancer by changing gene expression and damaging DNA. Conversely, commensal microbes also contribute to protection by increasing antitumor immunity and producing metabolites like butyrate that lower inflammation. The composition of the microbiota may also affect the efficacy and toxicity of cancer treatments, particularly immunotherapies and chemotherapies making it a potential biomarker and therapeutic target. Notwithstanding these findings, much remains to be discovered regarding the causal relationships, individual susceptibility, and safe and effective microbiota manipulation for the prevention and treatment of cancer.

Figure 4.5 illustrates the dual role of the human microbiota in the development of cancer. Procarcinogenic microbial actions on the left include the production of genotoxin (colibactin from *E. coli*), chronic inflammation caused by reactive oxygen species, direct pathogenic effects (*Fusobacterium nucleatum* in colorectal cancer and *Helicobacter pylori* in gastric cancer), and dysbiosis, i.e., overabundance of harmful microbes and loss of protective commensal. It is also highlighted that by preserving commensal balance, boosting anti-tumor immunity, enhancing therapy response (chemotherapy and immunotherapy) and by acting as biomarkers for treatment outcomes, protective microbial actions neutralize these effects. Imbalance between these conflicting forces is the focal point of the central hub which highlights the development of cancer. Diet, antibiotic use, chronic inflammation, and genetics are among the modifiable and nonmodifiable risk factors for dysbiosis that are shown at the bottom.

4.3.1 Microbial Dysbiosis and Cancer Risk

Microbial dysbiosis which has been found to be a significant risk factor for the development of cancer is an imbalance in the diversity, composition or function of the microbial communities. In a healthy host, the microbiota maintains a symbiotic relationship that supports immunological response, metabolic balance, and epithelial integrity. However, if this balance is disturbed for instance by food, antibiotics exposure, infections, or inflammation, it can effectively lead to an overabundance of pathogenic bacteria, a decrease in beneficial species, and a protumorigenic microenvironment (Schwabe and Jobin 2013). Through a variety of interconnected mechanisms, dysbiosis contributes to carcinogenesis by altering host immunity, generating toxins, and metabolites that harm DNA and encouraging chronic inflammation. For instance, dysbiosis in colorectal cancer-CRC is indicated by elevated levels of *Escherichia coli*, *Bacteroides fragilis*, and *Fusobacterium nucleatum*, all of which have been linked to proinflammatory and procarcinogenic properties (Tjalsma et al. 2012). These microbes can effectively impair the integrity of the epithelial barrier, elevate inflammatory cytokines, and interfere with the cell cycle. Furthermore, dysbiosis can effectively alter bile acid metabolism and generate reactive oxygen species (ROS), both of which can effectively cause mutagenesis and the emergence of cancer (Louis et al. 2014). Importantly, dysbiosis affects the entire body; correlations have been discovered between it and extraintestinal cancers such as breast, liver, and pancreatic cancer, suggesting that microbial imbalance affects the entire body (Xiao et al. 2025). Understanding the unique dysbiosis patterns linked to different cancer types could facilitate with risk assessment, microbiome-based diagnosis, and tailored prevention.

4.3.2 Mechanisms Connecting Carcinogenesis and Microbial Dysbiosis

The development of cancer is significantly influenced by microbial dysbiosis, an imbalance in the microbiota's composition and function, through a number of interconnected mechanisms (Fig. 4.1) including; 1. Direct damage to DNA caused by bacterial genotoxins such as cytolethal distending toxin (CDT) and colibactin 2. Chronic inflammation: Immune signaling pathways are activated by proinflammatory microbes such as *Helicobacter pylori* and *Fusobacterium nucleatum* which promote tumor growth and the production of microbial metabolites some of which (deoxycholic acid) promote cancer and others (butyrate) that may have protective effects but are frequently reduced in dysbiosis. When combined, these mechanisms cause genomic instability, upset cellular homeostasis, and foster an environment conducive to tumor growth (Grivennikov et al. 2010; Nougayrède et al. 2006; Yoshimoto et al. 2013).

Chronic inflammation: Chronic inflammation is a central mechanism linking microbial dysbiosis and cancer. Mucosal homeostasis and immune system balance are typically maintained by the microbiota. However, proinflammatory bacteria such as *Fusobacterium nucleatum*, Enterotoxigenic Bacteroides fragilis (ETBF), and *Helicobacter pylori* are enriched in a dysbiotic state leading to persistent inflammatory signaling. These bacteria activate innate immune receptors, such as Toll-like receptors (TLRs), which in turn cause an upregulation of the inflammatory cytokines TNF-α, IL-6, IL-17, and IL-1β. These cytokines promote immunological evasion, cell division, and angiogenesis (Grivennikov et al. 2010). For example, *F. nucleatum* promotes tumorigenesis by binding to epithelial E-cadherin with its FadA adhesin, activating the β-catenin and NF-κB pathways and increasing pro-inflammatory gene expression (Rubinstein et al. 2013). Similarly, CagA and VacA toxins are activated in chronic gastritis brought on by an *H. pylori* infection in the stomach causing damage to the integrity of the epithelium and promoting the development of gastric cancer (Wroblewski et al. 2010). This continuous immune activation results in a

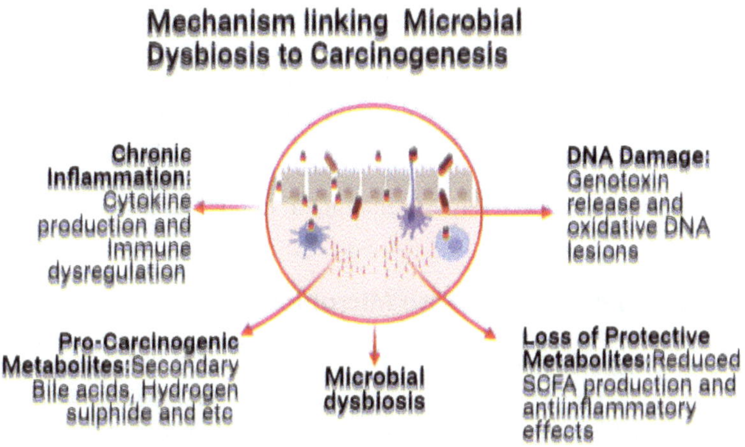

Fig. 4.6 Mechanism linking microbial dysbiosis to carcinogenesis

tumor-supportive microenvironment with oxidative stress and reduced immune surveillance (Fig. 4.6).

DNA damage: dysbiosis can effectively directly harm DNA through oxidative stress and bacterial genotoxins. In some strains of *Escherichia coli* that carry the polyketide synthase (pks) genomic island, the genotoxin colibactin can effectively result in mutations, chromosomal abnormalities, and DNA double-strand breaks (Nougayrède et al. 2006). These DNA lesions promote genomic instability, a hallmark of cancer if treatment is not received. *Helicobacter hepaticus* and *Campylobacter jejuni* produce cytolethal distending toxin (CDT), which interferes with the host cell cycle and DNA repair mechanisms, thereby accelerating the development of cancer. Moreover, inflammatory responses brought on by dysbiosis generate reactive oxygen and nitrogen species (ROS/RNS), which cause oxidative DNA damage and epigenetic changes (Garrett 2015). Genotoxic stress and inflammation work together to produce a mutagenic environment that supports the growth and metastasis of cancer (Fig. 4.6).

Production of microbial metabolite: The microbiota-host cancer axis is mediated in part by microbial metabolites. Beneficial gut microbes like *Faecalibacterium prausnitzii* and *Roseburia* species ferment dietary fibers to produce short-chain fatty acids (SCFAs), particularly butyrate, propionate, and acetate. By blocking histone deacetylases (HDACs), butyrate in particular has anti-inflammatory, and antiproliferative properties that encourage tumor cells to undergo apoptosis and cell-cycle arrest (Donohoe et al. 2012). But, because the number of bacteria that produce SCFA is often decreased in dysbiosis, these protective advantages are lessened.

However, the development of cancer is associated with secondary bile acids such as deoxycholic acid (DCA) and lithocholic acid (LCA), which are produced when microorganisms convert primary bile acids. DCA specifically promotes DNA damage, epithelial barrier dysfunction, and inflammatory signaling through activation of NF-κB and STAT3. According to Yoshimoto et al. (2013), hepatocellular carcinoma and the senescence associated secretary phenotype were exacerbated in obese mice due to elevated DCA levels. In addition, certain gut bacteria generate ammonia, hydrogen sulfide, and nitrosamines, which can effectively damage DNA or change signaling pathways, therefore encouraging the growth of tumors (Figure 4.6).

Figure 4.6 illustrates microbial imbalance or dysbiosis mediated chronic inflammation, DNA damage, procarcinogenic metabolites production and depletion of protective metabolites collectively contributing to tumor initiation and progression.

4.3.3 Cancers Associated with Microbiota

Recent developments in the study of microbiome have shed light on the significant role that the gut and organ-specific microbiota play in carcinogenesis. Liver, stomach, and colorectal cancers are among the most researched cancers in this area, and they all show intricate relationships between microbial communities and the host's immune, metabolic, and epithelial environments.

The microbiota and liver cancer: Dysbiosis of the gut microbiota has been closely associated with hepatocellular carcinoma (HCC), the most prevalent type of liver cancer. The portal vein, which carries endotoxins, immune mediators, and microbial metabolites directly from the gut, provides the liver with about 70% of its blood supply. Chronic inflammation and fibrosis are critically important predisposing factors to HCC and dysbiosis frequently leads to an increased translocation of lipopolysaccharide; a component of Gram negative bacteria which activates hepatic Toll-like receptor 4 (TLR4) pathways (Seki and Schnabl 2012). In addition, a decline in bacteria that produce short-chain fatty acids, like *Faecalibacterium prausnitzii* and *Roseburia* damages the intestinal barrier and encourages leaky gut syndrome. By permitting bacterial products to enter hepatic tissues, this condition worsens liver inflammation and promotes the growth of tumors (Yoshimoto et al. 2013). Furthermore, because of their capacity to alter the function of hepatic natural killer T (NKT) cells and their capacity to damage DNA, secondary bile acids generated by specific Clostridium species have been linked to the development of liver cancer (Fig. 4.7) (Ma et al. 2018).

Figure 4.7 illustrates the development of hepatocellular carcinoma due to dysbiosis of the gut microbiota through a series of interconnected processes along the gut-liver axis. The portal vein which transports endotoxins, microbial metabolites, and immune mediators from the gut provides the liver with almost 70% of its blood supply. Substantial highlights are: 1. Increased lipopolysaccharide translocation by Gram negative bacteria sets off hepatic TLR4 signaling leading to fibrosis and chronic inflammation which are two significant risk factors for hepatocellular carcinoma 2. Decreased abundance of SCFAs-producing bacteria (*Faecalibacterium prausnitzii* and *Roseburia*) compromising integrity of the intestinal barrier encouraging leaky gut and making it easier for bacterial products to enter hepatic tissues 3. Secondary bile acids produced by *Clostridium* species damage DNA and disrupt the activity of hepatic natural killer T cells which further encourages the growth of tumors 4. These metabolic and microbial changes work

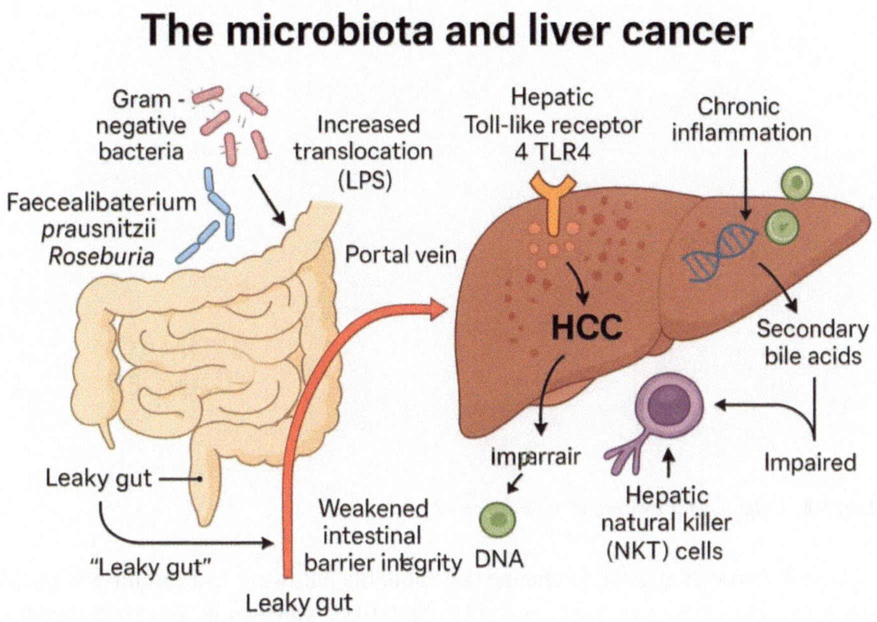

Fig. 4.7 The microbiota and liver cancer

together to produce a proinflammatory and procarcinogenic microenvironment that promotes the development of hepatocellular carcinoma.

Helicobacter pylori and stomach cancer: One of the oldest and best-established cancers associated with microbial etiology is stomach or gastric cancer which is mainly caused by *Helicobacter pylori* infection. Because of its strong causal link to noncardia gastric adenocarcinoma the World Health Organization has designated *H. pylori* as a Group I carcinogen. Persistent inflammation, glandular atrophy, and ultimately intestinal metaplasia, i.e., a precursor to carcinoma are the results of *H. pylori*'s chronic colonization of the stomach mucosa (Peek and Crabtree 2006). Cytotoxin-associated gene A (CagA) and vacuolating cytotoxin A (VacA) two of H. pylori's virulence factors, are essential for compromising the integrity of epithelial cells, altering cell signaling and fostering genomic instability. Furthermore, by generating a pro-inflammatory or nitrosating environment, changed gastric microbiota composition often brought on by antibiotic treatment or *H. pylori* eradication can effectively also aid in the development of cancer (Fig. 4.8) (Ferreira et al. 2018).

Figure 4.8 illustrates "Non-cardia Gastric Adenocarcinoma" caused by *Helicobacter pylori* infection and is classified as a Group I carcinogen by World Health Organization. The figure also highlights "intestinal metaplasia" a precursor to carcinoma caused by persistent inflammation, glandular atrophy, and chronic colonization of the stomach mucosa. The virulence factors (vacuolating cytotoxin A and

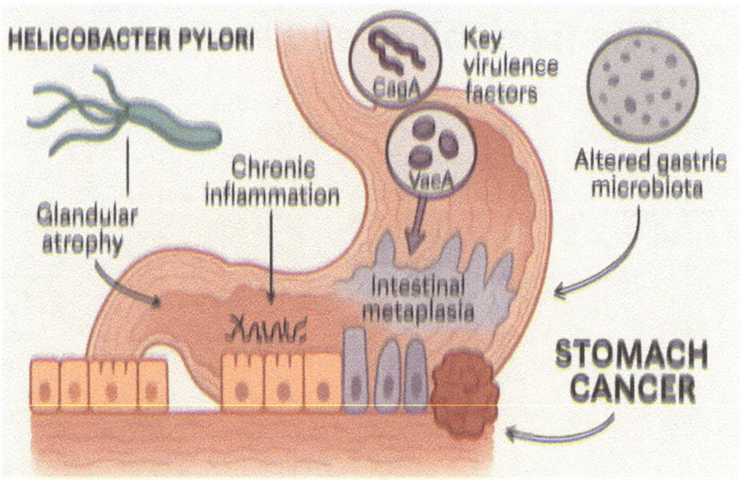

Fig. 4.8 *Helicobacter pylori* and stomach cancer

cytotoxin-associated gene A) change cell signaling pathways, compromise epithelial integrity, and cause genomic instability. Moreover, changes in the composition of the gastric microbiota, which are commonly caused by antibiotic exposure or the removal of *H. pylori* may result in proinflammatory or nitrosating conditions that further promote carcinogenesis

Gut microbiota and colorectal cancer: When it comes to the gut microbiota, colorectal cancer is arguably the most researched cancer. Proinflammatory and genotoxic bacteria like *Fusobacterium nucleatum*, *Escherichia coli*, and *Bacteroides fragilis* are more prevalent in CRC patients with dysbiosis (Kostic et al. 2013). Through the FadA adhesion molecule, *F. nucleatum* has been demonstrated to attach to and infiltrate colonic epithelial cells, triggering β-catenin signaling pathways and accelerating tumor growth (Rubinstein et al. 2013). Similarly, strains of *E. coli* that produce the genotoxin colibactin break DNA double strands in epithelial cells, which start the mutagenesis process (Cuevas-Ramos et al. 2010). According to Toprak et al. (2006), enterotoxigenic *B. fragilis* (ETBF) secretes a metalloprotease toxin (BFT) that promotes carcinogenesis by inducing immune activation, barrier disruption, and epithelial cell proliferation. Crucially, microbial metabolites that are increased in dysbiotic conditions such as hydrogen sulfide, ammonia, and secondary bile acids can effectively have cytotoxic and cancer-causing effects. On the other hand, CRC-associated microbiota usually has lower levels of healthy SCFAs like butyrate which are recognized for their anti-inflammatory and antitumorigenic qualities (Fig. 4.9) (Louis et al. 2014).

Figure 4.9 illustrates close association between "Gut Microbiota Dysbiosis" and "Colorectal Cancer" characterized by a higher frequency of genotoxic and proinflammatory bacteria (*Bacteroides fragilis*, *Escherichia coli*, and *Fusobacterium nucleatum*). *F. nucleatum* attaches itself to colonic epithelial cells via FadA adhesion molecule initiating β-catenin signaling pathways and also

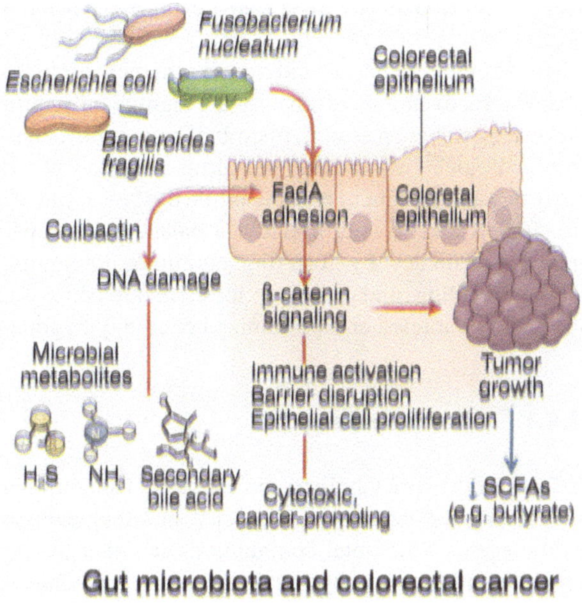

Fig. 4.9 Gut microbiota and colorectal cancer

encouraging cell proliferation ultimately leading to the development of tumors. Certain strains of *Escherichia coli* produce genotoxin (colibactin) which breaks DNA double strands and initiates mutagenesis. Enterotoxigenic *B. fragilis* secretes the BFT metallo-protease toxin which promotes epithelial proliferation, activates the immune system, and dissolves epithelial barriers. Moreover, dysbiosis is associated with increased levels of microbial metabolites that are cytotoxic and carcinogenic (hydrogen sulfide, ammonia, and secondary bile acids). However, beneficial short-chain fatty acids (butyrate) which have been demonstrated to have anti-inflammatory and antitumorigenic qualities are typically found in lower concentrations in the microbiota associated with colorectal cancer.

Immuno-modulation in cancer driven by microbiota: The microbiota influences systemic and tumor-specific immunity in addition to local epithelial interactions in all three types of cancer. By modifying T cell responses or attracting myeloid-derived suppressor cells (MDSCs), for example, some bacterial species can effectively prevent anti-tumor immunity. Conversely, advantageous microorganisms can effectively enhance immune checkpoint inhibitor responses, indicating a possible path for microbiota-targeted adjuvant therapy in oncology (Gopalakrishnan et al. 2018). So this can effectively be concluded that there is growing evidence of the complex and reciprocal relationship between microbiota and carcinogenesis. The pathophysiology of liver and colorectal cancer is largely influenced by dysbiosis and related metabolic and immunological abnormalities even though pathogens such as *H. pylori* are direct microbial oncogenic triggers. A better comprehension of these processes could facilitate the creation of risk assessment instruments, microbiome-targeted cancer treatments, and even diagnostics based on microbiota.

4.4 Microbiota and Cancer Treatment Modulation

Chemotherapy, radiation and immunotherapy are just a few of the cancer treatments whose effectiveness and toxicity are significantly influenced by the gut microbiota. According to new research, microbial communities have an impact on inflammation, immunological activation, and drug metabolism which in turn affects patient reactions and treatment results. While dysbiosis can effectively jeopardize therapeutic success and worsen toxicity, a balanced and diverse microbiota may enhance treatment efficacy by fostering antitumor immunity, and lowering side effects. Comprehending and utilizing these relationships creates new opportunities for microbiota-targeted and customized oncology treatments (Table 4.2; Fig. 4.10).

4.4.1 Chemotherapy

Effects on drug metabolism and efficacy: The gut microbiota plays a crucial role in influencing the metabolism, toxicity, and therapeutic effectiveness of chemotherapeutic agents. Microbial communities can alter the pharmacokinetics of anticancer drugs by producing enzymes that activate, inactivate, or degrade these compounds. For example, gut bacteria can metabolize drugs such as 5-fluorouracil and cyclophosphamide, thereby affecting their bioavailability and clinical outcomes (Viaud et al. 2013).

In addition to drug metabolism, the microbiota also modulates the host immune response, which is essential for optimal chemotherapy efficacy. A balanced and diverse microbiota can enhance antitumor immunity by improving the immune system's ability to recognize and eliminate cancer cells, thereby increasing treatment effectiveness (Gopalakrishnan et al. 2018).

Modulation of Side Effects: The gut microbiota also plays a key role in regulating the adverse effects of chemotherapy. Dysbiosis can exacerbate gastrointestinal toxicity, leading to symptoms such as diarrhea, nausea, and vomiting. This is primarily due to increased intestinal permeability and inflammation associated with microbial imbalance (Routy et al., 2018b). In contrast, a healthy microbiota helps maintain intestinal homeostasis, reduces inflammation, and may alleviate the severity of chemotherapy-induced side effects, thereby improving patient tolerance to treatment.

4.4.2 Radiotherapy

Similar to chemotherapy, radiotherapy may affect the microbiota directly or indirectly. The toxicity and effectiveness of radiation therapy may be impacted by dysbiosis and altered gut microbiota brought on by ionizing radiation exposure. According to recent research, the gut microbiota can effectively influence the tumor microenvironment, including immune response modulation, which is crucial for the effectiveness of radiation therapy. In addition, by boosting the immune system, some

Table 4.2 Microbiota and cancer treatment modulation

Type of therapy	Influence of microbiota	Mechanisms at play	Important microbes and results	Clinical consequences	References
Chemotherapy	Changes in the toxicity and effectiveness of drugs	Drugs are activated or inactivated by microbial enzymes. Changes in inflammation and mucosal immunity	Gemcitabine is rendered inactive by *Gammaproteobacteria* through cytidine deaminase *Barnesiella intestinihominis* and *Enterococcus hirae* increase the effectiveness of cyclophosphamide	Drug selection can be improved by microbiota screening; pre/probiotics may lessen GI side effects and mucositis	Geller et al. (2017) and Viaud et al. (2013)
Radiation therapy	Impact on side effects and radiation sensitivity	Microbiota influence immune responses after radiation; microbial metabolites, such as SCFAs, shield the epithelial barrier	Radiation induced colitis is lessened by *Faecalibacterium prausnitzii* and butyrate producers Dysbiosis is associated with poor healing and increased toxicity	Restoring the microbiota may lessen GI symptoms brought on by radiation and speed healing	Gerassy-Vainberg et al. (2018)
Checkpoint inhibitors in immunotherapy	ICI response enhancement or suppression	T cells are cross-primed by bacterial antigens SCFAs derived from microorganisms influence the production of IFN-γ and regulatory T cells	*Akkermansia muciniphila* increases the effectiveness of anti-PD-1 *Faecalibacterium* and *Bifidobacterium longum* are linked to improved melanoma response	Probiotics and FMT may be used to increase immunotherapy effectiveness in non-responders	Routy et al. (2018a) and Gopalakrishnan et al. (2018)
Focused treatment	Possible interactions between drugs and microbiota	Kinase expression or activity may be influenced by microbiota Tumor microenvironment modification through immune-microbial interactions	The impact of microbiota on the metabolism of tyrosine kinase inhibitors (TKIs) is being investigated (Pinato et al. 2021)	In the future, precision oncology may use personalized microbiota profiling to guide targeted therapy	
Strategies for microbiota modulation	Enhancement of treatment results and toxicity control	Restoring eubiosis; educating the immune system; and enhancing mucosal integrity	FMT enhanced melanoma patients' ICI response Probiotics (*Bifidobacterium* and *Lactobacillus*) have been demonstrated to lessen GI side effects from chemotherapy and radiation	In cancer treatment, FMT, diet, pre/probiotics, and antibiotic management may be used as adjuncts	Baruch et al. (2021) and Mego et al. (2015)

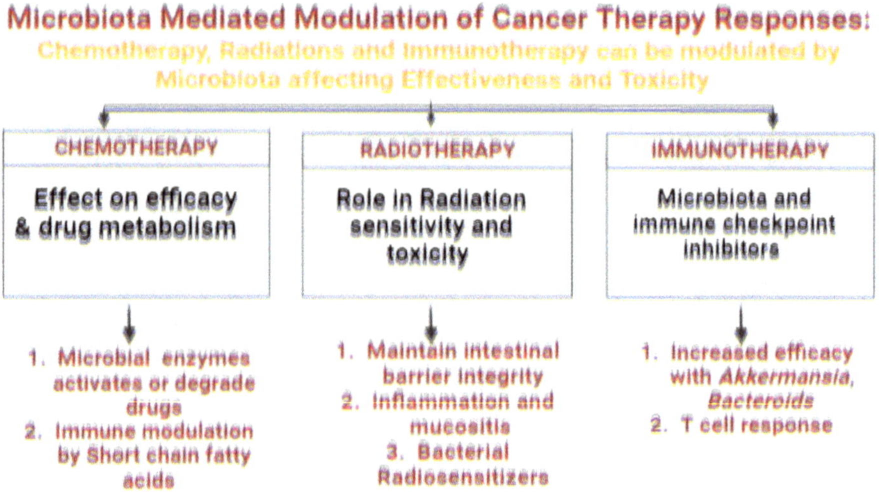

Fig. 4.10 Microbiota-mediated modulation of cancer therapies: this figure illustrates chemotherapy, radiotherapy and immunotherapy can effectively be modulated by microbiota affecting effectiveness and toxicity

bacterial species may make radiation more effective, while others may exacerbate its side effects, like intestinal inflammation and mucositis. The role of microbiota in radiation sensitivity and toxicity: the composition of the gut microbiota has a substantial impact on the toxicity that results from radiation treatment and the sensitivity of cancer cells to it. It has been demonstrated that some bacterial populations either sensitize or protect tumors from radiation-induced damage through their interactions with the immune system. Gut bacteria such as Firmicutes and Bacteroides fragilis, for instance, have been shown to influence radiation responses in a manner that heightens tumor sensitivity (Gopalakrishnan et al. 2018).

However, by increasing inflammation and immunological suppression, a disrupted microbiome can effectively exacerbate radiation-induced toxicity. Furthermore, minimizing the potential side effects of radiation therapy depends on the intestinal barrier and the microbiota's ability to maintain immunological homeostasis.

4.4.3 Immunotherapy

Immunotherapy has revolutionized the treatment of cancer by utilizing the body's immune system to identify and destroy tumor cells. The effectiveness of immunotherapy, particularly immune checkpoint inhibitors (ICIs) like CTLA-4 and anti-PD-1/PD-L1, has been shown to be impacted by the gut microbiota. Research indicates that the microbiota serves a pivotal role in a crucial role in controlling immune responses, which impacts the effectiveness of these therapies.

Microbiota and immune checkpoint inhibitors: The microbiota influences the efficacy of immune checkpoint inhibitors by altering the immune system. Immunocheckpoint inhibitors such as anti-PD-1/PD-L1 and anti-CTLA-4 facilitate the immune system recognize and fight cancer cells by blocking inhibitory signals on immune cells. However, the composition of the gut microbiota has been shown to have a significant impact on the effectiveness of these drugs. Some bacterial species, such as *Bacteroides*, *Akkermansia muciniphila* and *Faecalibacterium prausnitzii* have been associated with enhanced anti-tumor immune responses and better outcomes for patients undergoing immunotherapy (Routy et al. 2018a). These bacteria are thought to strengthen the immune system by promoting T-cell activation and proliferation, which are necessary for immune checkpoint inhibitors to work. However, a dysbiotic or unbalanced microbiome may result in less-than-ideal treatment outcomes due to weakened immune responses.

4.4.4 Studies Linking Microbiome Diversity and Treatment Response

The diversity of the microbiome has been directly linked to immunotherapy efficacy. Several studies have shown that a more diverse microbiome can effectively enhance the response to immune checkpoint inhibitors, whereas a less diverse microbiome is associated with treatment resistance (Gopalakrishnan et al. 2018). Studies have shown that melanoma patients who responded well to anti-PD-1 therapy had a higher diversity of gut bacteria, suggesting that a diverse microbiota may support a robust immune response. Furthermore, studies using preclinical models for fecal microbiota transplantation have demonstrated that transferring the microbiome of immunotherapy-sensitive animals into immunotherapy-resistant ones can effectively re-sensitize recipients to treatment.

This illustrates how critically important microbiome diversity is when determining a treatment's effectiveness (Routy et al. 2018b). These findings demonstrate how modifying the microbiome may elevate immunotherapies' efficacy, leading to more customized and effective cancer treatment regimens.

4.5 Modifying the Microbiota for Better Outcomes

Recent studies suggest that altering the microbiota may enhance treatment efficacy and reduce side effects because the human microbiota significantly influences how well cancer treatments work. Several strategies to alter the microbiome have been studied, including the use of probiotics and prebiotics, fecal microbiota transplantation, dietary and lifestyle modifications, and careful antibiotic stewardship during cancer treatment (Fig. 4.11).

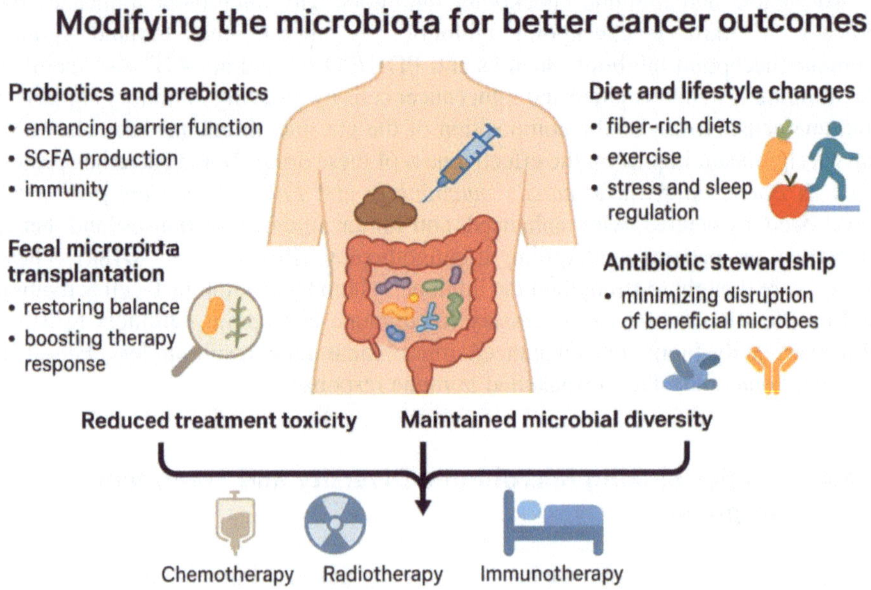

Fig. 4.11 Microbiota modification for better cancer outcomes

4.5.1 Prebiotics and Probiotics

Probiotics are live microorganisms that enhance the health of the host when taken in adequate amounts. Prebiotics are substances that support the growth and proper operation of beneficial bacteria in the gastrointestinal tract. Both prebiotics and probiotics have been studied as means of changing the gut microbiota and improving the efficacy of cancer treatments. It has been shown that probiotics such as *Lactobacillus* and *Bifidobacterium* species enhance gut barrier function, lower inflammation, and alter immune responses, all of which may enhance the effectiveness of cancer treatment. For example, by maintaining gut health and minimizing the adverse effects of chemotherapy on the gut microbiota, probiotics may reduce gastrointestinal toxicity during chemotherapy, including nausea and diarrhea. In addition, probiotics may enhance the immune response by increasing the synthesis of short-chain fatty acids which can effectively support anti-tumor immunity (Viaud et al. 2013). Prebiotics can effectively benefit beneficial bacteria by specifically promoting their growth, much like dietary fibers do. By encouraging the growth of beneficial microbial species, prebiotics may facilitate restoring the diversity and balance of the gut microbiota. According to Gopalakrishnan et al. (2018), this is crucial for the efficacy of cancer treatments, especially immunotherapy. A diet rich in prebiotics, such as fruits, vegetables, and whole grains, has been shown to enhance gut health.

4.5.2 Fecal Microbiota Transplantation

By introducing the fecal microbiota of a healthy donor, this procedure facilitates a patient's gut microbial balance. FMT has shown promise in treating gut dysbiosis caused by cancer treatments, especially chemotherapy and immunotherapy. By restoring a healthy microbiota, FMT may reduce toxicity and enhance the patient's response to treatment. According to studies, FMT can effectively successfully reverse the dysbiosis caused by chemotherapy, enhancing the gastrointestinal tract's and immune system's health (Routy et al. 2018a). FMT has also been shown to enhance the response to immunotherapy in animal models by promoting the activation of antitumor T-cells and establishing a supportive immune environment (Gopalakrishnan et al. 2018). Patients who received microbiome transplants from donors with a more diverse microbiome fared better, per early results from clinical trials examining the benefits of FMT in cancer patients undergoing treatment. The safety and effectiveness of the procedures in humans still require investigation.

4.5.3 Dietary and Lifestyle Modifications

These are essential for affecting the gut microbiota and directly affect the outcomes of cancer treatment. A balanced diet that promotes the growth of beneficial bacteria can effectively boost immunity, reduce inflammation, and enhance general health during cancer treatment. Previous studies revealed that a diet rich in fiber, polyphenols, and fermented foods can effectively elevate the number of beneficial bacteria like Lactobacillus and Bifidobacteria as well as the production of SCFAs, which have anti-inflammatory and immune-supporting properties. In addition, a diet high in fruits, vegetables, and whole grains can effectively elevate microbial diversity, which has been linked to better patient outcomes for immunotherapy (Routy et al. 2018b). However, dysbiosis can effectively be exacerbated by a diet high in processed foods and low in fiber, which can effectively impact the effectiveness of cancer treatments.

The gut microbiota can effectively also be impacted by lifestyle choices like stress reduction, physical activity, and sleep hygiene. Regular exercise has been shown to enhance immune function, elevate microbial diversity, and reduce the incidence of therapy-related side effects like fatigue and gastrointestinal distress. The microbiome may be negatively impacted by high levels of stress and sleep disturbances, which could lessen the efficacy of cancer treatments (Gopalakrishnan et al. 2018). Therefore, it might be possible to modify the microbiota and enhance the outcomes of cancer therapy by combining medical treatment with healthy lifestyle choices.

4.5.4 Antibiotic Stewardship During Cancer Therapy

Antibiotics are commonly used to treat or prevent infections during cancer treatment, particularly in patients with weakened immune systems. However, antibiotics may change the gut microbiota, which could lead to dysbiosis and compromise the effectiveness of cancer treatment. In particular, broad-spectrum antibiotics' capacity to eliminate beneficial microorganisms and encourage the growth of dangerous bacteria may worsen treatment-related toxicities and reduce therapeutic efficacy. Antibiotic stewardship, or the wise and sensible use of antibiotics, is essential to the successful treatment of cancer patients. Reducing unnecessary antibiotic use and, when possible, selecting narrow-spectrum antibiotics can effectively protect the gut microbiota. Some studies suggest that the use of antibiotics, especially PD-1 inhibitors, during immunotherapy may reduce treatment effectiveness by altering the microbiome, which in turn influences immune responses (Routy et al. 2018b). Therefore, the use of antibiotics during cancer treatment needs to be carefully monitored to ensure the reliability of that it doesn't undermine the therapeutic benefits of radiation, chemotherapy, or immunotherapy. To restore microbial balance and reduce the risk of microbiota disruption during or after antibiotic treatment, clinicians may consider alternative approaches such as FMT or probiotics. This can effectively lessen the side effects of antibiotics while improving outcomes for cancer patients.

Figure 4.11 illustrates the central tactics to alter the gut microbiota to enhance therapeutic efficacy and lessen treatment-related toxicity in cancer patients. It is highlighted that prebiotics and probiotics enhance function of gut barrier, encourage the synthesis of short-chain fatty acids, and boost immunological responses. Therefore, restoring microbial balance and increasing responsiveness to cancer treatment are two benefits of fecal microbiota transplantation. Beneficial microbes and microbial diversity are also increased by dietary and lifestyle modifications like eating high fiber diet, daily exercising and by managing stress and sleep. Antibiotic stewardship reduces interference with good bacteria. When combined, these strategies maintain microbial diversity while lowering toxicity, increasing therapy tolerance and improving the effectiveness of immunotherapy, chemotherapy and radiation therapy.

4.6 Current Research and Clinical Trials

Research on the impact of microbiota on cancer treatment outcomes is gaining traction. Numerous significant studies have shown that the microbiota can effectively influence the effectiveness of cancer treatments such as immunotherapy, chemotherapy, and radiation. As a result, additional clinical trials are being carried out to look into microbiome-based interventions to enhance therapeutic results.

Critically important research showing the impact of microbiota on treatment:

4.6.1 Gut Microbiota and Immunotherapy Response

Gopalakrishnan et al. (2018) carried out a critically important investigation that provided compelling evidence of the impact of the microbiome on the responses of melanoma patients to immune checkpoint inhibitors (ICIs). Researchers found that patients with a more diverse microbiome responded better to anti-PD-1 treatment, and that certain bacterial species, such as *Faecalibacterium prausnitzii* and *Akkermansia muciniphila,* were linked to better clinical outcomes. According to Gopalakrishnan et al. (2018), this study indicated that changing the microbiome may enhance therapeutic efficacy and emphasized the importance of microbiome composition in immunotherapy success.

4.6.2 Antibiotics and Immunotherapy Response

Routy et al. (2018a) looked into how the use of antibiotics affected how well ICIs worked. The results showed that patients who took antibiotics during immunotherapy had significantly worse responses, which were explained by alterations in their gut microbiota. This study suggests that antibiotics may need to be used more sparingly in these circumstances and emphasizes the significance of preserving a healthy microbiome in immunotherapy patients (Routy et al. 2018b).

4.6.3 Microbiota and Chemotherapy

In a groundbreaking study, Viaud et al. (2013) demonstrated that the microbiota can effectively affect how the immune system responds to chemotherapy, especially cyclophosphamide. By promoting antitumor immunity, researchers showed how a healthy microbiome could enhance the effects of chemotherapy. On the other hand, chemotherapy-induced dysbiosis, or an imbalance in the gut microbiota, resulted in a reduction in immune responses and an elevation in toxicity. This study showed how altering the microbiome could enhance the effects of chemotherapy.

4.7 Current Clinical Trials Concerning the Microbiome

Numerous ongoing clinical studies are investigating the possibility of modifying the microbiome to enhance the efficacy of cancer treatment. The interventions being investigated in these trials include probiotics, prebiotics, dietary modifications, and fecal microbiota transplantation.

4.7.1 Immunotherapy and Fecal Microbiota Transplantation

Clinical trials suggest that FMT may enhance immunotherapy response. To find out if FMT can effectively enhance melanoma patients' response to anti-PD-1 treatment, one such trial is underway (NCT03187851). Transplanting a healthy microbiome may enhance the immune system's ability to fight tumors, offering a potential way to overcome immunotherapy resistance (Gopalakrishnan et al. 2018).

4.7.2 Probiotics and Chemotherapy-Induced Toxicity

A clinical trial (NCT02402288) is examining whether probiotics can effectively reduce the gastrointestinal toxicity caused by chemotherapy. This study focuses on patients receiving chemotherapy regimens such as fluorouracil, which can effectively cause nausea, diarrhea, and mucositis. By restoring microbial balance, probiotics may reduce these side effects and enhance patient quality of life while potentially boosting chemotherapy's effectiveness.

4.7.3 Microbiome and Radiation

An ongoing trial (NCT03556426) is examining the role of the gut microbiota in colorectal cancer radiation therapy. The trial is examining whether a patient's response to radiation therapy and the severity of radiation-induced side effects are influenced by the composition of their microbiota. The researchers are also investigating the potential of probiotics to reduce side effects of radiation therapy, such as mucositis and diarrhea.

4.7.4 Personalized Microbiome Modification in Cancer Therapy

A different clinical trial (NCT03894565) is focusing on personalized microbiome interventions to enhance the outcomes of cancer immunotherapy. This study profiles the gut microbiome of cancer patients to identify microbial signatures associated with better treatment outcomes. Customizing interventions such as FMT, probiotics, and dietary modifications is the aim of the trial. Therefore, it can effectively be concluded that the connection between the microbiota and the outcomes of cancer treatments is gaining attention. Furthermore, studies have demonstrated a robust association between the microbiome and the efficacy of immunotherapy, chemotherapy, and radiation. Ongoing clinical trials are looking into new ways to change the microbiome like FMT, probiotics, prebiotics, and dietary changes, to enhance the outcomes of cancer treatment. These initiatives could soon lead to more effective and customized cancer treatments. While integrating microbiota-based interventions into cancer treatments holds great potential, a number of challenges must be addressed before these strategies can effectively be applied extensively in clinical settings. The complexity of studying the diversity of individual microbiome and the need for standardization in microbiota analysis pose significant challenges. It is also crucial to carefully navigate ethical and regulatory considerations as the field evolves.

4.8 Personalized Medicine and Microbiota Profiling

Personalized medicine has revolutionized the treatment of cancer by tailoring therapies to the unique genetic, environmental, and lifestyle traits of each patient. Even though the microbiome is a crucial component of this personalized approach, accurately identifying and treating each person's microbiota still poses a number of challenges. The primary challenge is the individual variability of the microbiome, because diet, environment, genetics and previous medical treatments are just a few of the variables that can effectively affect their microbiome. As a result, developing universal microbiota-based treatment guidelines is difficult. The need for microbiota profiling that can effectively recognize these complex relationships and predict a patient's microbiome's reaction to specific treatments is one difficulty. Personalized microbiome-based cancer treatments would require tailored interventions that can effectively be tailored to each patient's specific microbial composition, such as specific probiotics, dietary changes, or FMT. Understanding the causal relationships between the microbiome and the outcomes of cancer treatment is also essential. Further research is necessary to determine the precise mechanisms by which these microbial communities function, even though some microbiota have been associated with improved responses to immunotherapy or chemotherapy. Understanding these will be crucial to developing targeted microbiome interventions that enhance patient outcomes.

4.9 Standardization of Microbiota Analysis

Another substantial obstacle is standardizing microbiota analysis. Microbiome research currently uses a wide range of techniques, such as DNA extraction, sequencing, sample collection, and bioinformatics analysis. Because of these differences, it could be difficult to draw broad conclusions or use microbiome-based strategies in clinical settings, which could lead to inconsistent outcomes. To address this, standardized methods for microbiota analysis are desperately needed. Standardization would enhance the accuracy and consistency of microbiome profiling across different labs and clinical studies. To standardize microbiome research methods, a number of international initiatives are already underway, such as the Human Microbiome Project and the International Human Microbiome Standards Consortium. These initiatives aim to create internationally accepted guidelines for the gathering, handling, and analysis of microbiome samples in order to enable more reliable and consistent findings across investigations. However, this process is ongoing, and further effort is required to manage the intricacy of microbiome variability.

4.10 Legal and Ethical Concerns

Ethical and legal concerns become more critically important as microbiome-based therapies get closer to clinical use. Fecal microbiota transplantation raises critically important ethical issues, especially when considering donor selection, pathogen screening, and the long-term effects of microbiota transfer from one individual to another. Ensuring the safety and quality control of FMT is crucial because poorly tested or uncontrolled microbiota may introduce dangerous bacteria or viruses into the recipient's system and cause unintended consequences. Regulatory approval for microbiome-based treatments is another critically important concern. The microbiome is complicated, so changing it could have unexpected consequences. Currently, the U.S. Food and Drug Administration and the European Medicines Agency are creating frameworks for microbiome-based product approval. The lack of clear guidelines for clinical efficacy, safety, and quality control continues to impede widespread use. Regulatory agencies will need to set clear guidelines for assessing the safety and effectiveness of microbiome-based therapies in clinical settings. Privacy and ownership of microbiome data have also been questioned. The gathering, storing, and use of extremely private genetic data for microbiome profiling raise significant concerns. Establishing privacy protections and open informed consent processes is necessary to ensure the reliability of that people's microbiome data is handled ethically.

4.11 Considering the Future

The future of microbiome-based cancer treatment is bright, despite the challenges. A more personalized, microbiome-targeted approach is becoming popular, wherein therapies are tailored to the particular microbial composition of each patient. The following are significant future developments: Improved microbiome diagnostics: It is expected that techniques such as metagenomic sequencing and microbiome biomarkers will become more accurate and accessible, enabling more accurate, tailored treatments and better patient microbiome profiling. As our knowledge of the microbiome's role in cancer treatment grows, the use of probiotics, prebiotics and FMT as common supplements to chemotherapy, immunotherapy, and radiation therapy may enhance therapeutic efficacy and reduce side effects. Research on the interactions between specific microbes and cancer medications may lead to the development of novel therapeutic agents that are specifically intended to enhance or change the role of the microbiome in cancer treatment. The use of microbiome-based techniques in oncology is a novel new approach to cancer treatment. To achieve this potential, however, existing barriers in the areas of regulation, standardization, and profiling must be removed. With additional study and development, microbiome-based therapies could significantly contribute to personalized cancer care, improving patient outcomes and creating safer and more efficient cancer treatments.

4.12 Conclusion

Studying the role of the microbiota in cancer treatment is an exciting and rapidly expanding field. As research progresses, it is becoming increasingly clear that the microbiome serves a pivotal role in a substantial role in the effectiveness of cancer treatments such as immunotherapy, chemotherapy, and radiation. It has been shown that the composition of the gut microbiota influences the effectiveness of immune checkpoint inhibitors, treatment-induced toxicity, and drug metabolism. In addition, therapies that alter the microbiome, such as fecal microbiota transplantation, probiotics, and prebiotics hold great potential to enhance therapeutic outcomes and reduce side effects. However, despite the growing evidence of the role of the microbiota in cancer treatment, several challenges remain. Since each person's microbiome is unique, customized techniques for microbiome profiling and treatment are needed. The development of precise regulatory frameworks and the standardization of microbiota analysis methods are essential for the safe and effective application of microbiome-based interventions in clinical practice. Ethical concerns about donor screening for FMT, the privacy of microbiome data, and the long-term effects of microbiome manipulation must also be addressed. Microbiota-informed cancer therapy is very promising because it can effectively enhance existing cancer treatments by reducing adverse side effects, improving the environment for therapeutic efficacy, and even helping to resolve treatment resistance. As ongoing clinical trials explore the relationship between the microbiome and cancer treatment, the era of personalized microbiome-based oncology is getting closer. This would enhance patient outcomes and provide more effective, individualized care by empowering medical professionals to create treatment plans that incorporate microbiome interventions. In conclusion, despite the numerous challenges that need to be addressed, the field of applying microbiome science to cancer treatment is one that demonstrates great promise. With more research and development, microbiota-informed cancer therapy may soon be used in precision oncology to provide patients with safer and more effective treatment options.

References

Arthur JC, Perez-Chanona E, Mühlbauer M et al (2012) Intestinal inflammation targets cancer-inducing activity of the microbiota. Science 338(6103):120–123. https://doi.org/10.1126/science.1224820

Baruch EN, Youngster I, Ben-Betzalel G, Ortenberg R, Lahat A, Katz L et al (2021) Fecal microbiota transplant promotes response in immunotherapy-refractory melanoma patients. Science 371(6529):602–609. https://doi.org/10.1126/science.abb5920

Belkaid Y, Hand TW (2014) Role of the microbiota in immunity and inflammation. Cell 157(1): 121–141. https://doi.org/10.1016/j.cell.2014.03.011
Cryan JF, O'Riordan KJ, Cowan CS et al (2019) The microbiota-gut-brain axis. Physiol Rev 99(4): 1877–2013. https://doi.org/10.1152/physrev.00018.2018
Cuevas-Ramos G, Petit CR, Marcq I, Boury M, Oswald E, Nougayrède JP (2010) Escherichia coli induces DNA damage in vivo and triggers genomic instability in mammalian cells. Proc Natl Acad Sci USA 107(25):11537–11542
David LA, Maurice CF, Carmody RN et al (2014) Diet rapidly and reproducibly alters the human gut microbiome. Nature 505(7484):559–563. https://doi.org/10.1038/nature12820
Dewhirst FE, Chen T, Izard J et al (2010) The human oral microbiome. J Bacteriol 192(19): 5002–5017. https://doi.org/10.1128/JB.00542-10
Dominguez-Bello MG, Godoy-Vitorino F, Knight R, Blaser MJ (2019) Role of the microbiome in human development. Gut 68(6):1108–1114. https://doi.org/10.1136/gutjnl-2018-317503
Donohoe DR, Collins LB, Wali A, Bigler R, Sun W, Bultman SJ (2012) The Warburg effect dictates the mechanism of butyrate-mediated histone acetylation and cell proliferation. Mol Cell 48(4):612–626. https://doi.org/10.1016/j.molcel.2012.08.033
Ferreira RM, Pereira-Marques J, Pinto-Ribeiro I, Ferreira RM, Pereira-Marques J, Pinto-Ribeiro I et al (2018) Gastric microbial community profiling reveals a dysbiotic cancer-associated microbiota. Gut 67(2):226–236
Furusawa Y, Obata Y, Fukuda S, et al (2013) Commensal microbe-derived butyrate induces the differentiation of colonic regulatory T cells. Nature 504:446–450. https://doi.org/10.1038/nature12721
Garrett WS (2015) Cancer and the microbiota. Science 348(6230):80–86. https://doi.org/10.1126/science.aaa4972
Geller LT, Barzily-Rokni M, Danino T, Jonas OH, Shental N, Nejman D et al (2017) Potential role of intratumor bacteria in mediating tumor resistance to the chemotherapeutic drug gemcitabine. Science 357(6356):1156–1160. https://doi.org/10.1126/science.aah5043
Gerassy-Vainberg S, Blatt A, Danin-Poleg Y, Ben-Nun A, Komarov D, Boshnak NT et al (2018) Radiation induces proinflammatory dysbiosis: transmission of inflammatory susceptibility by host cytokine induction. Gut 67(1):97–107. https://doi.org/10.1136/gutjnl-2016-312302
Gopalakrishnan V, Spencer CN, Nezi L, Reuben A, Andrews MC, Karpinets TV et al (2018) Gut microbiome modulates response to anti–PD-1 immunotherapy in melanoma patients. Science 359(6371):97–103. https://doi.org/10.1126/science.aan4236
Grice EA, Segre JA (2011) The skin microbiome. Nat Rev Microbiol 9(4):244–253. https://doi.org/10.1038/nrmicro2537
Grivennikov SI, Greten FR, Karin M (2010) Immunity, inflammation, and cancer. Cell 140(6): 883–899. https://doi.org/10.1016/j.cell.2010.01.025
Honda K, Littman DR (2016) The microbiota in adaptive immune homeostasis and disease. Nature 535:75–84. https://doi.org/10.1038/nature18848
Human Microbiome Project Consortium (2012) Structure, function and diversity of the healthy human microbiome. Nature 486(7402):207–214. https://doi.org/10.1038/nature11234
Koh A, De Vadder F, Kovatcheva-Datchary P, Bäckhed F (2016) From dietary fiber to host physiology: short-chain fatty acids as central bacterial metabolites. Cell 165(6):1332–1345. https://doi.org/10.1016/j.cell.2016.05.041
Kostic AD, Chun E, Robertson L et al (2013) Fusobacterium nucleatum potentiates intestinal tumorigenesis and modulates the tumor-immune microenvironment. Cell Host Microbe 14(2): 207–215. https://doi.org/10.1016/j.chom.2013.07.007
Louis P, Hold GL, Flint HJ (2014) The gut microbiota, bacterial metabolites and colorectal cancer. Nat Rev Microbiol 12(10):661–672. https://doi.org/10.1038/nrmicro3344
Lynch SV, Pedersen O (2016) The human intestinal microbiome in health and disease. N Engl J Med 375(24):2369–2379. https://doi.org/10.1056/NEJMra1600266
Ma C, Han M, Heinrich B, Fu Q, Zhang Q, Sandhu M et al (2018) Gut microbiome–mediated bile acid metabolism regulates liver cancer via NKT cells. Science 360(6391):eaan5931

Mego M, Holec V, Drgona L, Hainova K, Ciernikova S (2015) Probiotic bacteria in cancer patients undergoing chemotherapy and radiation therapy. Support Care Cancer 23(1):179–187. https://doi.org/10.1007/s00520-014-2334-0

Nougayrède JP, Homburg S, Taieb F, Boury M, Brzuszkiewicz E, Gottschalk G et al (2006) Escherichia coli induces DNA double-strand breaks in eukaryotic cells. Science 313(5788): 848–851. https://doi.org/10.1126/science.1127059

Peek RM, Crabtree JE (2006) Helicobacter infection and gastric neoplasia. J Pathol 208(2):233–248

Pinato DJ, Howlett S, Ottaviani D, Urus H, Patel A, Mineo T et al (2021) Association of prior antibiotic treatment with survival and response to immune checkpoint inhibitor therapy in patients with cancer. Nat Rev Gastroenterol Hepatol 18:401–418. https://doi.org/10.1038/s41575-020-00380-9

Qin J, Li R, Raes J et al (2010) A human gut microbial gene catalogue established by metagenomic sequencing. Nature 464(7285):59–65. https://doi.org/10.1038/nature08821

Ravel J, Gajer P, Abdo Z et al (2011) Vaginal microbiome of reproductive-age women. Proc Natl Acad Sci USA 108(Suppl 1):4680–4687. https://doi.org/10.1073/pnas.1002611107

Routy B, Le Chatelier E, Derosa L, Duong CPM, Alou MT, Daillère R et al (2018a) The gut microbiome influences anticancer immunotherapy responses. Science 359(6371):106–111. https://doi.org/10.1126/science.aan3708

Routy B, Le Chatelier E, Derosa L, Duong CPM, Alou MT, Daillère R et al (2018b) Gut microbiome influences efficacy of PD-1–based immunotherapy against epithelial tumors. Science 359(6371):91–97. https://doi.org/10.1126/science.aan3706

Rubinstein MR, Wang X, Liu W, Hao Y, Cai G, Han YW (2013) Fusobacterium nucleatum promotes colorectal carcinogenesis by modulating E-cadherin/β-catenin signaling via its FadA adhesin. Cell Host Microbe 14(2):195–206. https://doi.org/10.1016/j.chom.2013.07.012

Schwabe RF, Jobin C (2013) The microbiome and cancer. Nat Rev Cancer 13(11):800–812. https://doi.org/10.1038/nrc3610

Seki E, Schnabl B (2012) Role of innate immunity and the microbiota in liver fibrosis: crosstalk between the liver and gut. J Physiol 590(3):447–458

Sender R, Fuchs S, Milo R (2016) Revised estimates for the number of human and bacteria cells in the body. PLoS Biol 14(8):e1002533. https://doi.org/10.1371/journal.pbio.1002533

Singh RK, Chang HW, Yan D, Lee KM, Ucmak D, Wong K et al (2017) Influence of diet on the gut microbiome and implications for human health. J Transl Med 15:73. https://doi.org/10.1186/s12967-017-1175-y

Sommer F, Bäckhed F (2013) The gut microbiota—masters of host development and physiology. Nat Rev Microbiol 11(4):227–238. https://doi.org/10.1038/nrmicro2974

Thaiss CA, Zmora N, Levy M, et al (2016) The microbiome and innate immunity. Nature 535:65–74. https://doi.org/10.1038/nature18847

Tjalsma H, Boleij A, Marchesi JR, Dutilh BE (2012) A bacterial driver–passenger model for colorectal cancer: beyond the usual suspects. Nat Rev Microbiol 10(8):575–582. https://doi.org/10.1038/nrmicro2819

Toprak NU, Yagci A, Gulluoglu BM, Akin ML, Demirkalem P, Celenk T et al (2006) A possible role of Bacteroides fragilis enterotoxin in the aetiology of colorectal cancer. Clin Microbiol Infect 12(8):782–786

Viaud S, Saccheri F, Mignot G, Yamazaki T, Daillère R, Hannani D et al (2013) The intestinal microbiota modulates the anticancer immune effects of cyclophosphamide. Science 342(6161): 971–976. https://doi.org/10.1126/science.1240537

Wroblewski LE, Peek RM, Wilson KT (2010) Helicobacter pylori and gastric cancer: factors that modulate disease risk. Clin Microbiol Rev 23(4):713–739. https://doi.org/10.1128/CMR.00011-10

Xiao W, Chen Q, Liu C, Yu Y, Liu T, Jin Y, Ma H, Chen S, Jiang C (2025) Gut microbiota in cancer: from molecular mechanisms to precision medicine applications. iMeta 4(2):e70017. https://doi.org/10.1002/imt2.70017. PMID: 40236776; PMCID: PMC11995170

Yatsunenko T, Rey FE, Manary MJ et al (2012) Human gut microbiome viewed across age and geography. Nature 486(7402):222–227. https://doi.org/10.1038/nature11053
Yoshimoto S, Loo TM, Atarashi K, Kanda H, Sato S, Oyadomari S et al (2013) Obesity-induced gut microbial metabolite promotes liver cancer through senescence secretome. Nature 499(7456): 97–101. https://doi.org/10.1038/nature12347
Zuo T, Ng SC (2018) The gut microbiota in the pathogenesis and therapeutics of inflammatory bowel disease. Front Microbiol 9:2247. https://doi.org/10.3389/fmicb.2018.02247

Microbiota and Aging Solutions

5

Abstract

The human gut microbiota containing trillions of bacteria, archaea, viruses, and fungi is crucial for host physiology by regulating immune function, metabolism, and neurocognitive processes. With progressive aging, there are spectacular changes in microbial composition and diversity with a decline in beneficial taxa like *Bifidobacterium* and *Faecalibacterium* and an increase of proinflammatory pathobionts, resulting in dysbiosis. These changes are responsible for compromised intestinal barrier function, systemic low-level inflammation "inflammaging," and immunosenescence, thus hastening the onset of chronic disease like metabolic syndrome, type II diabetes, cardiovascular diseases, neurodegeneration, and cancer. Mechanistically, microbiota-directed metabolites like short-chain fatty acids (SCFAs) determine epigenetic regulation, mitochondrial function, nutrient sensing pathways and autophagy, all of which are central hallmarks of aging. By contrast, a diverse, SCFA-producing microbiome has been associated with healthier aging and longer longevity. Recent findings point toward the therapeutic potential of microbiota-directed interventions like dietary modulation, probiotics, prebiotics, synbiotics, postbiotics, fecal microbiota transplantation, and drug-based modulators like metformin. However, challenges are still there with respect to interindividual variation, limited longitudinal data, and translatability gaps between animal and human models. This chapter integrates current understanding of the interplay between microbiota and aging, unmasks underlying mechanisms, and assesses emerging microbiome-based therapeutic approaches toward health span enhancement and delay of age determined decline.

Keywords

Gut microbiome · Aging · Dysbiosis · Human health · Inflammaging

R. C. Sobti et al., *Microbiota Unveiled: Science, Health, and Future Therapies*,
https://doi.org/10.1007/978-981-95-9031-5_5

5.1 Introduction

The body of a human hosts a diverse and intricate population of microorganisms that together form the human microbiota, and the gut is the main region of highest microbial abundance (Gyriki et al. 2025). This symbiotic community of trillions of viruses, bacteria, fungi, and archaea starts developing at an early phase of life and keeps getting influenced by the host genetic makeup, diet, environment, and lifestyle (Yan et al. 2025). These microbes are not just passively existing, rather, they are dynamic and bidirectional in their interactions with the host that affect various physiological processes, including immune responses, metabolic processes, and even neurocognitive health (Zhang et al. 2023). The study of the relationship between microbiota and the process of ageing has been of significant interest in the biomedical sciences, especially in the wake of the increasing ageing population globally. Scientific evidence indicates that changes in the diversity and structural composition of the gut microbiota are strongly correlated with both the rate and overall direction of the aging process.

Aging is a major contributory factor to a wide range of human diseases. It is a time-dependent process which leads to a physiological decline in the function of biological systems (Zhang et al. 2023). Various disease states are associated with aging, including gastrointestinal dysfunction, metabolic derangements, gut microbiota changes, and impaired immune response (López et al. 2023). Microbial communities that inhabit the human gastrointestinal tract experience drastic compositional and functional alterations during the life span of an organism (Mariat et al. 2009). Food and lifestyle factors have been implicated to be the major contributors to changing ecology of gut microbiota ("Microbiota and Aging," 2018). The gut microbial composition, which is continuously changing during the aging process and is vital for the immunological and physiological maturation, is one of the major contributing factors behind healthy aging (Rinninella et al. 2019). Gut microbiota alterations have been extensively linked with the majority of chronic metabolic diseases such as obesity (Turnbaugh et al. 2009), malnutrition (Subramanian et al. 2014), and certain associated diseases such as cirrhosis. (Qin et al. 2014). Furthermore, the microbiome alterations have been related to intestinal disorders, such as colorectal cancer (Modeel et al. 2025; Zackular et al. 2013), inflammatory bowel disease (IBD) (Palm et al. 2013), and neurodevelopmental disorders (Hagemann et al. 2013). Throughout the human lifespan, from infancy to senescence, healthy persons undergo compositional alterations of gut microbiota, with age-linked modifications increasing and being strongly associated with age-related decline in physiological functions (N V and M 2025; Rinninella et al. 2019).

The relationship between microbiota and aging works two ways, where aging affects microbial composition and microbial dysbiosis is the cause of the age-associated decline in microbes (Ghosh et al. 2020). Evidence supports that older people possess gut microbial dysbiosis with decreased diversity and amounts of beneficial microbes, and increased presence of potentially pathogenic taxa (Ghosh et al. 2020; Thevaranjan et al. 2017). Metagenomics and systems biology also suggest that these changes in microbiota are not just the result of aging but are an active contributor to inflammation and functional decline. Longitudinal and cross-sectional research supports that those with increased microbial diversity and increased presence of health-promoting genera like *Akkermansia* and *Bifidobacterium* are more likely to age well and live longer than those with microbiota dominated by proinflammatory or opportunistic microbes, who are at risk of frailty, metabolic, and chronic disease (Gyriki et al. 2025; Ira et al. 2024).

The interaction between gut bacteria and aging is multifactorial and is influenced by genes, the environment, and lifestyle. Although the specifics are yet to be unraveled, it is clear that gut bacteria play very critical roles in key physiological domains like modulating the immune system, regulating metabolism, and communicating with the brain (Gyriki et al. 2025; O'Toole and Jeffery 2015). A series of detrimental consequences may result from changes in the gut microbiota's composition or function, whether brought on by pharmacological interventions, dietary modifications, or natural age-related physiological changes. Interestingly, this type of dysbiosis is linked to elevated intestinal permeability, persistent low-grade inflammation known as "inflammaging," and a heightened risk of age-associated pathologies (Franceschi et al. 2000; Rinninella et al. 2019).

In elderly populations, gut dysbiosis has been robustly linked with a spectrum of age-related disorders, including chronic systemic inflammation, cognitive impairment, neurodegenerative diseases, insulin resistance, cardiovascular disease, type II diabetes, and certain malignancies (Ghosh et al. 2020; Rinninella et al. 2019). For example, studies have shown that a decline in beneficial taxa such as *Bifidobacterium* and *Faecalibacterium*, coupled with an increase in proinflammatory Proteobacteria, correlates with both frailty and the progression of metabolic and neurodegenerative conditions (Badal et al. 2020; Gyriki et al. 2025). Furthermore, age-related alterations in gut microbiota have been found to contribute to compromising the integrity of intestinal barriers, thus promoting translocation of microbial constituents into systemic circulation as well as perpetuating chronic inflammation (Franceschi et al. 2000). Overall, the above findings describe the widespread role of microbial dysbiosis in the elderly and highlight the central role it plays in structuring the physiological environment of aging.

Recent scientific data from microbiome research has identified distinctive microbial signatures associated with healthier aging and longer lifespans. Long-living patients showed higher frequencies of certain bacteria, like *Bacteroides*, *Lactobacillus,* and *Bifidobacterium,* which are associated with lower cardiovascular disease, osteoporosis, and gout risks. This has been covered in detail under Table 5.1. This data is indicative of the therapeutic potential for gut microbiota modification in favor of healthy aging (Wilmanski et al. 2021). Throughout life, especially in older age, colonization by heterogeneous microbiota of species that possess anti-inflammatory metabolites like short-chain fatty acids (SCFAs), has correlated with beneficial aging outcomes (Ghosh et al. 2020). Identification of these beneficial microbial signatures is the foundation for the construction of targeted strategies, such as prebiotics, probiotics, dietary intervention, and perhaps fecal microbiota transplantation (FMT), to preserve microbial balance throughout the aging process (Rinninella et al. 2019).

Table 5.1 Table of gut microbes detected in reducing conditions related to aging

Bacterium/group	Condition reduced	Scientific name	Key findings	References
Oscillibacter	Cardiovascular disease	*Oscillibacter* spp.	Metabolizes cholesterol, lowering its levels in humans	Li et al. (2024a)
Lactobacillus genus	Cardiovascular disease and osteoporosis	*Lactobacillus plantarum, L. rhamnosus, L. fermentum*	Strains support heart, reduce inflammation, increase bone strength	Zhao et al. (2021)
Bacteroides (some spp.)	Cardiovascular and gout	*Bacteroides vulgatus, B. dorei, B. uniformis*	Some reduce atherosclerosis (CVD) and others linked to lower gout risk	Yoshida et al. (2018)
Rikenellaceae family	Gout	*Rikenellaceae* family	Protective association with lower gout risk	Tang et al. (2024)
Faecalibacterium prausnitzii	Gout	*Faecalibacterium prausnitzii*	Anti-inflammatory, linked to lower uric acid and gout improvement.	Sokol et al. (2008)
Limosilactobacillus fermentum	Gout	*Limosilactobacillus fermentum GR-3*	Supplementation reduces uric acid and inflammation.	Zhao et al. (2022)
Actinobacteria	Osteoporosis	*Actinobacteria* (e.g., *Bifidobacterium* spp.)	Higher abundance linked to better bone health.	Duffuler et al. (2024)
Burkholderiales, Bacillales	Osteoporosis	*Burkholderiales, Bacillales* order	Associated with reduced osteoporosis risk; may regulate bone metabolism.	Li et al. (2024b)

Microbiome therapeutics is a new entry into the field of GeroScience that holds the gut microbiome as a potential biomarker and a targetable predictor of aging and longevity. In the modern age, some of the approaches to targeting microbiome have their therapeutic potential highlighted as methods that can be used to change microbial composition and activity. They have demonstrated a potential to extend health span and, thereby, lifespan. The current chapter thus summarizes the molecular mechanisms using which the microbial community of the gut may affect aging, evaluates the clinical relevance of microbial shifts with aging, and examines the current and potential interventions to use the microbiome to promote healthy aging.

5.2 Correlation Between Gut Microbiota and Aging

Gut microbiota and aging association is strong and multifaceted, covering compositional, functional, and immunological aspects. The interplay between the two has been a dominant theme in recent biomedical research, reflecting an increasing recognition of the microbiota as a marker and modulator of physiological changes in aging. With advancing age, the gut microbiota experiences profound changes in composition, diversity, and function, closely related to the health status and disease vulnerability of the older population (Wu et al. 2021). These are not all population uniform but vary based on a combination of complex interactions between genetic, environmental, dietary, and lifestyle factors, leading to highly interindividual variability in older adults (Bradley and Haran 2024).

Widespread cross-sectional research has proven that aging is generally correlated with a decrease in health-promoting microbial species, including *Bifidobacterium* and *Lactobacillus*, and increase in inflammatory and potentially pathogenic bacteria, including *Proteobacteria* and *Enterobacteriaceae* family members (Ghosh et al. 2022; Gyriki et al. 2025; Wu et al. 2021). For example, studies show that elderly adults have higher abundances of *Bacteroides, Alistipes*, and *Parabacteroides*, while the beneficial species of *Ruminococcus* and *Prevotella* have often decreased. (Wu et al. 2021). Increased frailty, chronic inflammation, and the emergence of ageing disorders like neurodegeneration, metabolic syndrome, and cardiovascular disease have all been connected to this dysbiosis (Bradley and Haran 2024; Ghosh et al. 2022; Gut Microbiota and Aging n.d.).

It has been confirmed through studies that gut microbiota composition can be employed to predict biological age, which validates the association of bacteria with ageing. Genome-Wide Association Study (GWAS) of microbiota profiles revealed some microbial markers are ubiquitous at all life stages and can model and predict an individual's age (Wu et al. 2021). Of particular interest is the presence of some taxa, including *Akkermansia muciniphila* and SCFA-producing bacteria, which correlates with healthier aging and longer longevity, whereas dominance by proinflammatory taxa correlates with poorer outcomes (Gut Microbiota and Aging, n.d.). There is additional evidence from experimental approaches using animal models for a causal relationship between changes in microbiota and aging phenotypes. In landmark works, the transfer of gut microbiota from old mice to young germ-free recipients caused the induction of chronic low-grade inflammation, enhanced intestinal permeability, and compromised immune function all of which are features of aging. In particular, aged microbiota recipients had increased inflammatory cytokines like TNF-α and IL-6 and higher frequencies of Th1, Th2, and regulatory T cells in the gut and spleen (Gyriki et al. 2025). In contrast, microbiota of youthful donors provided resistance to such age-related alterations, indicating that the gut microbiota is not a passive mirror of the host's age but plays an active role in the aging process. Saeedi et al. (2025) showed a causal relationship between gut microbiota and aortic endothelial senescence, with some *Clostridium* species participating in vascular aging through production of phenylacetic acid. In contrast, transplanting young donor microbiota into aged mice has been found to rescue aging phenotypes, as testified by reduced inflammatory markers and enhanced intestinal barrier function.

These findings are corroborated by human observations, which show that the diversity and composition of the gut microbiota are strongly correlated with older adults' health. For example, centenarians and healthy older individuals tend to have more diverse and balanced microbiota, with greater abundance of beneficial taxa, than their less healthy counterparts (Bradley and Haran 2024; Gut Microbiota and Aging n.d.). Conversely, those with lower microbial diversity and higher levels of pro-inflammatory bacteria tend to be more likely to develop frailty, cognitive impairment, and chronic disease. In addition, perturbations of gut microbiota composition have been associated with the start of age-related neurodegenerative diseases such as Parkinson's and Alzheimer's. In mouse models, microbiota transplantation from patients with neurodegenerative diseases has been reported to worsen disease phenotypes, such as motor impairment and neuroinflammation (Wu et al. 2021).

The gut microbiota plays an important role in ensuring the immune response in the gut happens, develops, and is regulated. An important factor to this relationship is how the immune system reacts to the gut microbiota. The immune system's decline with ageing is known as immunosenescence, and "inflammaging" is chronic low-grade inflammation that accompanies aging (Ghosh et al. 2022; Gyriki et al. 2025; Wu et al. 2021); it is a unifying concept in microbiota and aging. Changes associated to age in gut microbiota induce this inflammation through gut barrier disruption, enabling microbes to pass into the blood, and reducing anti-inflammatory production (Thevaranjan et al. 2017). In aged mice and possibly aged humans, these microbial changes are linked to an increased permeable gut and increased inflammation (Thevaranjan et al. 2017). This forms a vicious cycle in which inflammaging further disturbs microbiota, accelerating aging and predisposing to age-related disease (Ghosh et al. 2020). This chronic inflammatory environment not only hastens immunosenescence but also leads to the development and exacerbation of various age-related diseases.

Environmental and lifestyle determinants, including dietary intake, drug therapy, and exercise, also control the interaction between microbiota and aging. High-fiber and high-polyphenol diets are related to higher SCFA-producing bacteria abundance and better metabolic and inflammatory states among older adults (Ghosh et al. 2022; Wu et al. 2021). Bradley and Haran (2024) reviewed evidence showing that the gut

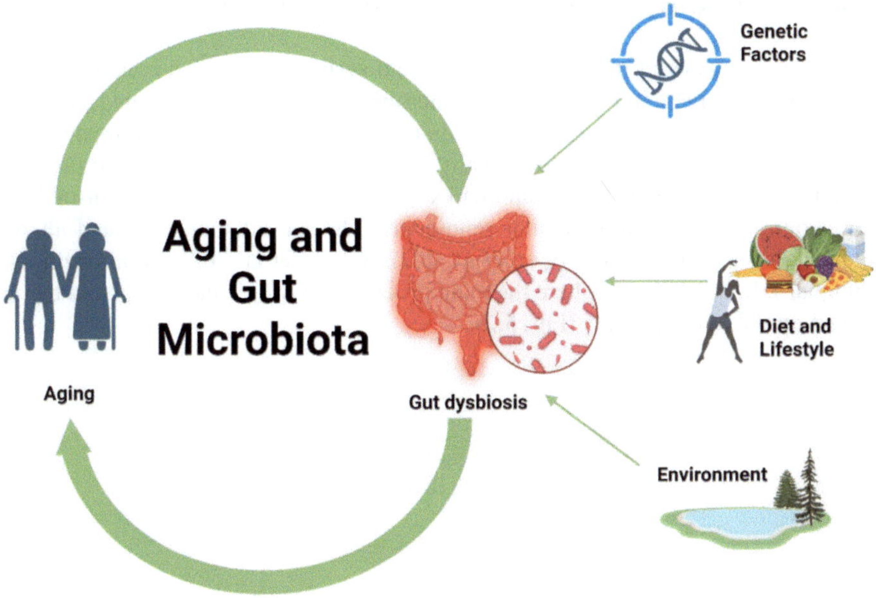

Fig. 5.1 Illustration showing the relation between gut microbiota and aging

microbiota composition of elderly individuals correlated with dietary patterns, health status, and residence type, with those consuming more plant-based diets and living independently exhibiting more favorable microbial profiles. Nevertheless, administration of antibiotics and other drugs can upset microbial ecosystems, exacerbating dysbiosis and the resultant health risks. The combined action of these external factors throughout the lifespan can account for the great variability in microbiota composition and health outcomes among elderly individuals (Bradley and Haran 2024; Wu et al. 2021).

Given the intense relationship between gut microbiota and aging, the use of the microbiota as a therapeutic target and biomarker to promote healthy aging has gained popularity. Research is underway on diet treatments, probiotics, prebiotics, and fecal microbiota transplantation for their value in restoring balance of microbes and preventing age-related decline (Ghosh et al. 2022; Gyriki et al. 2025). Although the field is in its development, these approaches are promising for enhancing the quality of life and prolonging health span in the elderly. All these have been covered under the Sect. 5.5 of this chapter (Fig. 5.1).

5.3 Mechanism Linking Microbiota to Ageing

A dynamic network of molecular, cellular, and systemic pathways is implicated in the two-way interaction between the gut microbiota and aging. Age-related dysbiosis is characterized by reduced microbial diversity with progressive enlargement of proinflammatory taxa, which functionally interact with host physiology to promote virtually all of the recognized hallmarks of aging, such as genomic instability, mitochondrial dysfunction, cellular senescence, chronic inflammation, and altered intercellular communication. (Guo et al. 2022; Gyriki et al. 2025; Kanimozhi and Sukumar 2025). We specify here the main processes by which the microbiota has an impact on the aging process (Fig. 5.2).

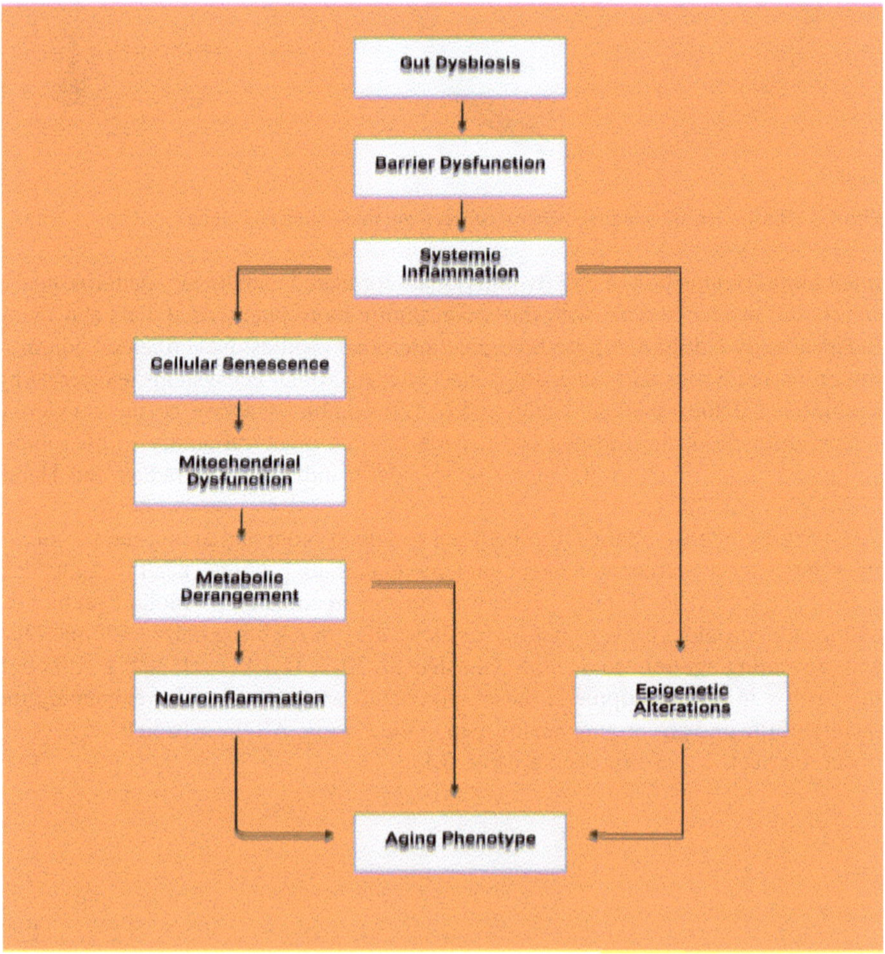

Fig. 5.2 Overview of the mechanisms linking microbiota to aging

5.3.1 Microbiota, Intestinal Barrier, and Systemic Inflammation

Old age is linked to enhanced intestinal permeability commonly referred to as, leaky gut, that lets in microbial molecules, including lipopolysaccharide (LPS), flagellins, and bacterial DNA to cross into the system circulation. This translocation initiates chronic low-grade inflammation (inflammaging) a key component of aging and a contributor to age-related diseases (Gyriki et al. 2025; Kanimozhi and Sukumar 2025; Ren et al. 2023). Said in mechanism, dysbiosis contributes to:

1. A decline in the production of short-chain fatty acids, particularly butyrate, which plays a pivotal role in preserving epithelial barrier integrity and modulating immunological responses.
2. Higher levels of pathobionts (e.g., *Proteobacteria, Desulfovibrio*), that synthesize proinflammatory metabolites and destabilize tight connections among epithelial cells (Gyriki et al. 2025; Ren et al. 2023).

This pathway is experimentally validated by the transfer of elderly microbiota into young germ-free mice, leading to alterations in gut permeability, an increase in serum IL-6 and TNF-α, and the influx of immune cells into the gut (Gyriki et al. 2025). These phenotypes can be overcome by interventions that either replace SCFA-producing bacteria or inhibit inflammatory cytokines (Gyriki et al. 2025; Ren et al. 2023) (Fig. 5.3).

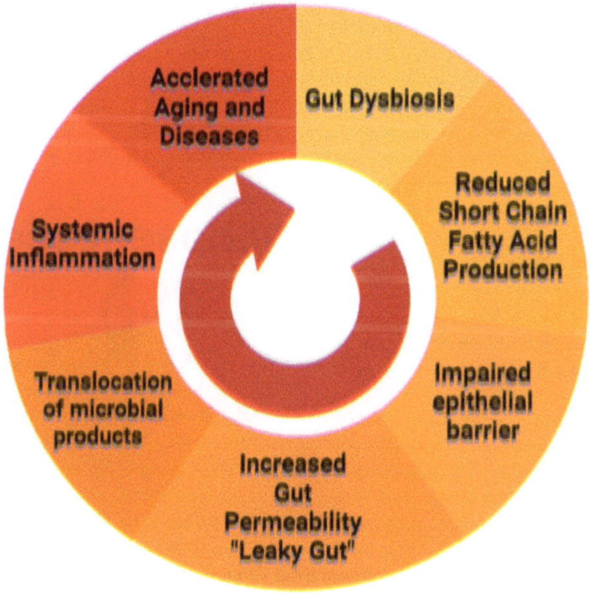

Fig. 5.3 Flowchart showing the mechanism by which gut dysbiosis leads to accelerated aging

5.3.2 Microbiota and Immunosenescence

The gut microbiota significantly influences the development and regulation of the immune system. Aging-related dysbiosis disrupts this regulation, promoting immunosenescence which is a decline in immune competence characterized by reduced pathogen defense, increased autoimmunity, and chronic inflammation (Gyriki et al. 2025; Kanimozhi and Sukumar 2025). Following are the ways to elicit it:

1. Activation of pattern recognition receptors: translocated microbial products activate Toll-like receptors (TLRs) and NOD-like receptors (NLRs) on immune cells, stimulating the NF-κB pathway and the NLRP3 inflammasome, resulting in persistent production of IL-6, TNF-α, and IL-1β cytokines (Gyriki et al. 2025; Ren et al. 2023; Xiao et al. 2025).
2. Altered T cell homeostasis: dysbiosis increases Th1, Th2, and regulatory T cell populations in the gut and spleen, as observed in both aged mice and humans (Clark and Walker 2018; Gyriki et al. 2025).
3. Impaired macrophage function: Thevaranjan et al. (2017) showed that aged microbiota impairs macrophage phagocytosis and antigen presentation, further weakening immune system (Fig. 5.4).

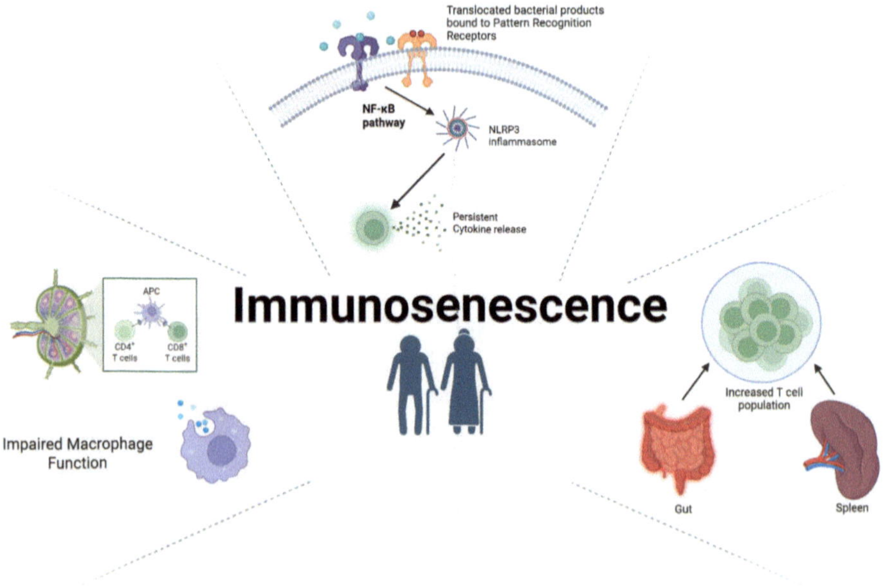

Fig. 5.4 Schematic representation showing how gut dysbiosis affects immunosenescence

5.3.3 Microbiota, Metabolism, and Nutrient Sensing

Gut microbes are central to host metabolic homeostasis, producing metabolites that influence energy balance, glucose metabolism, and lipid profiles. Aging disrupts these interactions by causing:

1. Short-chain fatty acid deficiency: butyrate, propionate, and acetate regulate glucose and lipid metabolism, enhance insulin sensitivity, and suppress inflammation via G-protein coupled receptors and histone deacetylase (HDAC) inhibition. SCFA levels in the elderly contribute to metabolic dysfunction and insulin resistance (Gyriki et al. 2025; Kanimozhi and Sukumar 2025; Ren et al. 2023).
2. Altered bile acid metabolism: dysbiosis modifies bile acid pools, impacting farnesoid X receptor and TGR5 signaling, which regulate lipid metabolism and inflammation (Gyriki et al. 2025; Kanimozhi and Sukumar 2025).
3. mTOR and insulin/IGF-1 pathways: microbial metabolites modulate the mechanistic target of rapamycin (mTOR) and insulin/IGF-1 (insulin-like growth factor-1) signaling pathways, both of which are central to aging and longevity. Caloric restriction and certain microbiota-derived metabolites suppress mTOR, promoting autophagy and lifespan extension (Shi et al. 2024) (Fig. 5.5).

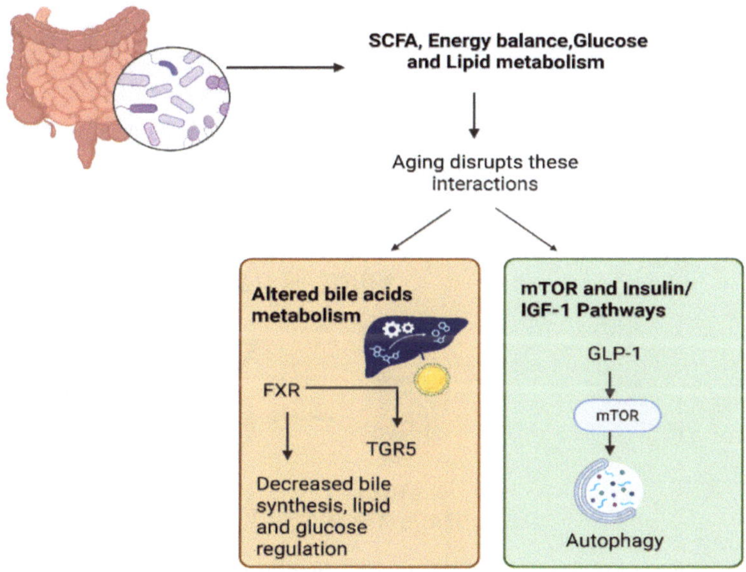

Fig. 5.5 Illustration showing the relationship between gut microbiota and metabolism, nutrient sensing

5.3.4 Microbiota and Epigenetic Regulation

The gut microbiota plays a pivotal role in the host's aging processes through the production of bioactive metabolites, i.e., short-chain fatty acids (SCFAs), including butyrate, propionate, and acetate (Gyriki et al. 2025). The metabolites are involved in epigenetic regulation by modulating gene expression without altering the inherent DNA sequence (Zhao et al. 2025). For instance, butyrate is an inhibitor of histone deacetylase, enzymes that are liable to induce chromatin condensation and transcriptional silencing. Inhibition of HDAC leads to relaxation of chromatins and induction of gene transcription for genes involved in anti-inflammatory response, cell protection, and preservation of physiological function (Jung et al. 2024). Additionally, the gut microbiota contributes to the production of nutrients like folate and methionine, which are key substrates of one-carbon metabolism. This metabolic pathway provides methyl groups required for DNA methylation, an important epigenetic modification governing gene activation and repression, thereby influencing cellular senescence and stress resistance (Gyriki et al. 2025). In conclusion, by modulating histone acetylation and DNA methylation, the gut microbiota plays a pivotal role in the regulation of inflammation, genomic stability, and healthy aging (Zhao et al. 2025) (Fig. 5.6).

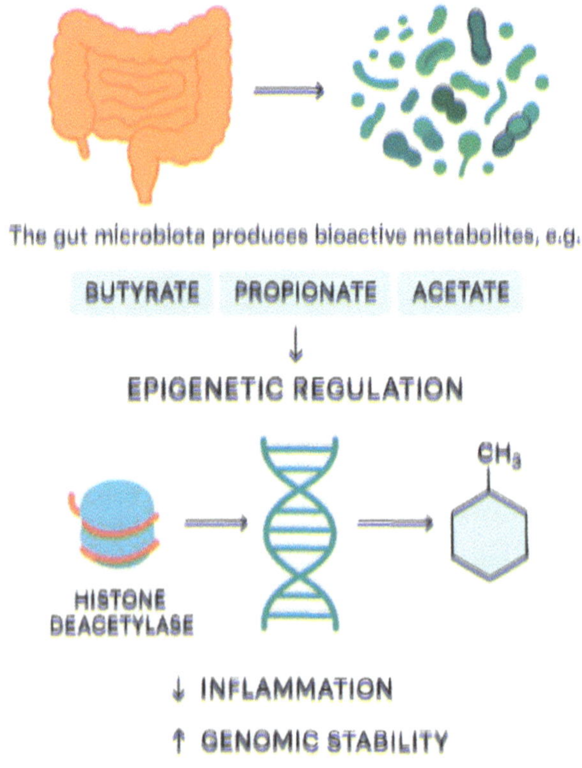

Fig. 5.6 Illustration showing how short-chain fatty acids affect epigenetic regulation

5.3.5 Microbiota, Mitochondrial Function, and Oxidative Stress

Our mitochondria's health is largely influenced by the gut microbiome. A healthy gut microbiome is a guardian of mitochondria by maintaining low levels of oxidative stress, promoting repair processes, and increasing the efficacy of energy production. When that is lost, however, mitochondrial damage and aging are increased. The following explains how this happens:

1. Regulation of oxidative stress: The gut microbiota synthesizes compounds such as short-chain fatty acids (SCFAs). These compounds regulate the activity of enzymes that combat toxic molecules known as reactive oxygen species (ROS). Excessive ROS can harm cells, and therefore, by inducing antioxidant defense, SCFAs protect mitochondria from harm (Gyriki et al. 2025; Shabani et al. 2025).
2. Facilitating mitochondrial growth and function: Certain signals from health-promoting gut microbiota induce the organism to increase mitochondrial biogenesis and make sure that they operate optimally, thereby guaranteeing that our cells possess adequate energy levels, particularly as we age (Zhao et al. 2025).
3. Dysbiosis: When the gut microbiome loses its typical balance due to aging, diet, or disease, some bacteria gain prominence. This leads to inflammation and the generation of excess ROS, which damages mitochondrial DNA. This, over time, speeds up cell aging and even contributes to the risk of age-related disease (Ghosh et al. 2022; Zhao et al. 2025).

5.3.6 Microbiota and Gut–Brain Axis

The gut microbiota is a significant contributor to interaction with the central nervous system. The gut-brain axis is bidirectional, as the gut microbiota not only sends signals to the brain but also the brain has the potential to modulate the microbiota. This is done through neuronal (nerve signal), endocrine (hormonal), and immunological (immune-related) mechanisms (Bhoomi 2024). Some metabolites such as short-chain fatty acids (SCFAs), butyrate, and propionate, tryptophan metabolites, and secondary bile acids enable this interaction by passing through the blood-brain barrier (Carabotti et al. 2015; Doenyas et al. 2025). Neurodegeneration has been linked to changes in tryptophan metabolism, which is described in detail in Sect. 5.4.1 of this chapter.

Gut to Brain Communication

Neuronal mechanisms: The brain is connected to the gut via the vagus nerve and the enteric nervous system. Gut bacteria can synthesize neuroactive compounds or communicate via these nerves and, in doing so, directly influence mood and cognition (Bhoomi 2024; Carabotti et al. 2015).

Endocrine mechanisms: The intestinal microbiome can modulate the secretion of hormones and neuropeptides, such as glucagon-like peptide 1 (GLP-1) and peptide

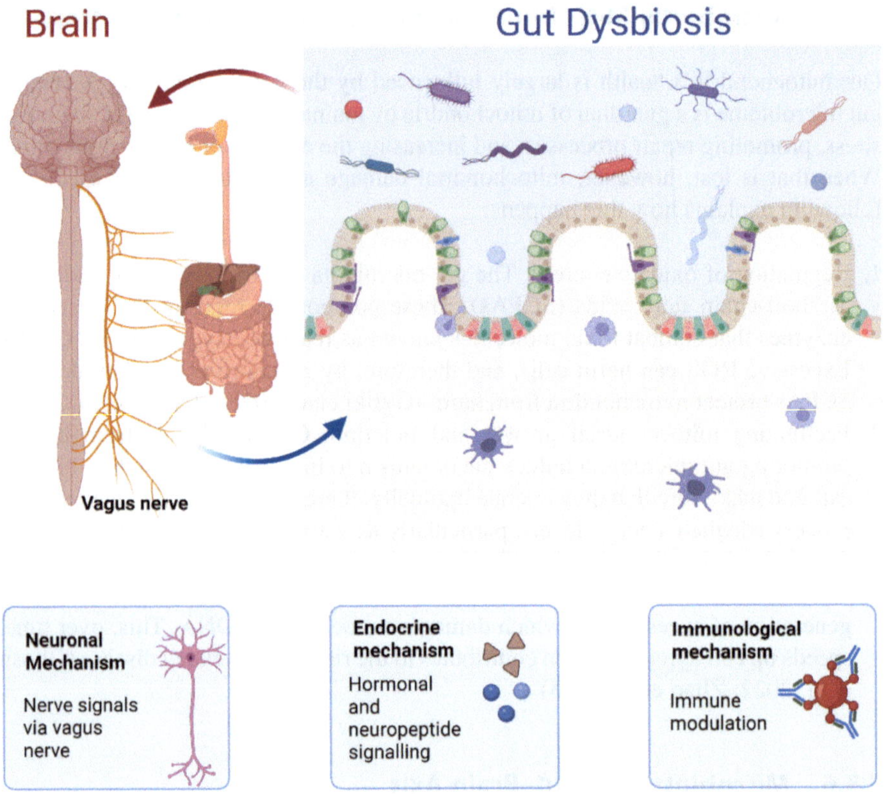

Fig. 5.7 Illustration depicting the gut-brain axis and its messengers

YY, which have access to the bloodstream and have effects on neurological functions (Loh et al. 2024).

Immunological mechanisms: The gut microbiota acts as a key controller of the immune system. Gut bacterial metabolites can have an impact in the blood, affecting peripherally and even in the brain, which is crucial in neurodegenerative disease pathology (Jiang et al. 2024; Loh et al. 2024).

When the gut dysbiosis happens, it may enhance the inflammation in the gut and in the system. This increased state of inflammation can initiate or exacerbate inflammation in the brain (neuroinflammation), an established etiology of cognitive decline and neurodegenerative disorders such as Parkinson's and Alzheimer's (Doenyas et al. 2025; Loh et al. 2024) (Fig. 5.7).

5.3.7 Microbiota and Autophagy

Autophagy is the cell's own recycling process, enabling the degradation and recycling of aged or damaged cellular components in order to preserve cellular integrity (Yu et al. 2018). During aging, this process is commonly reduced in activity, which can contribute to increased cellular damage and disease. The gut microbiota exerts its effect on autophagy primarily through two mechanisms:

1. SCFA Signaling
   Some of the intestinal microbiota ferment dietary fibers into short-chain fatty acids (SCFAs) primarily butyrate and acetate. SCFAs regulate autophagy through:
   (i) Activation of AMPK (AMP-activated protein kinase): The kinase identifies low cellular energy and initiates autophagy to balance it (Wang et al. 2022).
   (ii) Inhibition of mTOR (mechanistic target of rapamycin): mTOR is typically an inhibitor of autophagy, and upon inhibition, autophagy becomes activated (Deleyto-Seldas and Efeyan 2021).
2. Inflammasome Activation
   When autophagy goes awry, aberrant proteins and cellular debris accumulate. This has the potential to activate NF-κB, a central regulator of inflammation, which in turn activates inflammasomes which is a protein complexes that catalyze inflammation (Xiao et al. 2025). Ongoing activation of inflammasome leads to chronic inflammation and tissue damage, which promote aging and age-related disease (Fig. 5.8).

5.3.8 Pharmacological Modulation of Microbiota Related Aging Pathways

Certain drugs extend lifespan by targeting the microbiota and its metabolites. For instance, metformin alters the gut microbiome, increases SCFA production, enhances mucin secretion, and reduces gut permeability and inflammation. Acarbose and empagliflozin, on the other hand, improve glucose metabolism and vascular health by increasing beneficial SCFA-producing bacteria and reducing proinflammatory taxa (Gyriki et al. 2025).

1. Metformin
   (i) Gut microbiome modulation: Metformin modulates the gut microbiome by promoting increased beneficial bacteria, most notably *Akkermansia muciniphila* and *Bifidobacterium* species. These bacteria are important for the health of the gut lining and degrade mucin, a key component of protective mucus (Lee et al. 2021).
   (ii) Enhanced mucin secretion and barrier integrity: With metformin, the gut goblet cell number is raised, which thickens the mucus. This enhanced barrier limits the penetration of toxic compounds like, bacterial toxins, into the bloodstream, reducing inflammation and gastrointestinal disease (Lau et al. 2021).
   (iii) Increased SCFA production: Metformin increases the colonization of bacteria that ferment dietary SCFAs such as butyrate, acetate, and propionate.

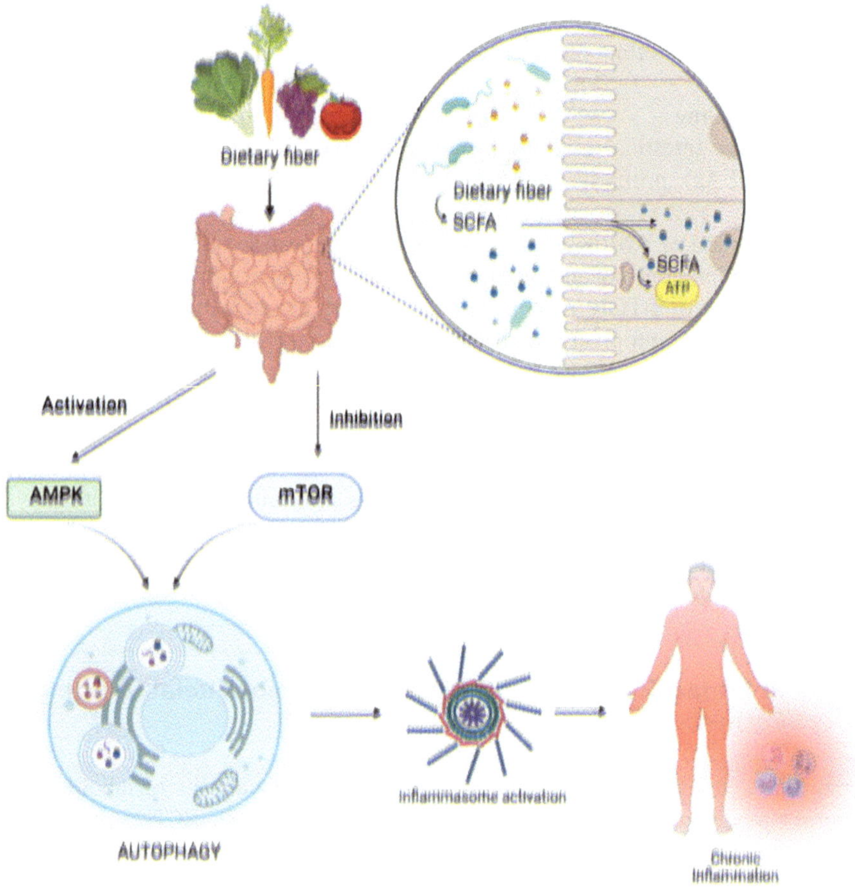

Fig. 5.8 Illustration showing the influence of gut microbiome on inflammation which in turn promotes aging

SCFAs are signaling molecules, ensure healthy metabolism, and possess anti-inflammatory effects. Moreover, they activate AMPK and suppress mTOR pathways, which regulate energy utilization and autophagy (Wang et al. 2024).

(iv) Reduced intestinal permeability and inflammation: By stimulating the growth of beneficial microbes and improving the integrity of the gut barrier, metformin reduces the permeability of inflammatory mediators, thus leading to reduced systemic inflammation (Di Vincenzo et al. 2024).

2. Acarbose
   (i) Stimulation of SCFA-producing bacteria: Acarbose retards carbohydrate digestion, allowing more amount of starch to enter the colon, to serve as substrate for carbohydrate-degrading bacteria. This supports numbers of

SCFA producers such as *Lactobacillus*, resulting in increased amounts of health-giving SCFAs in the gut (Brown et al. 2024).
   (ii) Reduction of pro-inflammatory bacteria and immune modulation: Acarbose also reduces bacteria linked with inflammation and metabolic disease. Its mechanism fosters anti-inflammatory immune cells (Treg) instead of proinflammatory cells (Th17), and it can inhibit chronic inflammation (Li et al. 2025).
   (iii) Glucose metabolism and longevity: Enhanced production of SCFAs coupled with a less inflammatory gut environment creates a synergistic effect that results in improved blood sugar control, decreased oxidative stress, and potentially enhanced lifespan (Wu et al. 2022).
3. Empagliflozin
   (i) Changing gut microbiota composition: Empagliflozin increases the diversity and abundance of SCFA-producing bacteria like *Roseburia, Eubacterium, Faecalibacterium, and Ruminococcaceae* family members. This change increases the production of healthy SCFAs (Deng et al. 2022).
   (ii) Pathogenic Bacteria Prevention: The drug also reduces the count of bacteria associated with inflammation and gut barrier disruption, including *Escherichia, Shigella, Bilophila,* and *Hungatella* (Deng et al. 2022).
   (iii) Improved metabolism and cardiovascular well-being: By modulating the microbiota to promote beneficial bacteria and suppress proinflammatory microorganisms, empagliflozin enables better glucose control, reduces inflammation, and improves vascular well-being. These changes correlate with reduced cardiovascular risk and improved metabolic profiles in diabetic individuals (Hao et al. 2023).

5.4 Microbiota-Associated Diseases in Ageing

The human gut microbiota is a dynamic consortium of bacteria, viruses, archaea, and fungi that plays a central role in the upkeep of physiological and homeostasis. Alterations in the gut microbiota composition in the course of aging, commonly termed "microbial dysbiosis," have increasingly been recognized as underpinning the etiology and exacerbation of a wide array of chronic diseases that disproportionately afflict the elderly. Reduced microbial diversity, decline in health-promoting commensals, and relative overabundance of pathobionts, which collectively give rise to systemic inflammation, metabolic imbalance, and immune dysregulation in aged organisms are characteristic features of dysbiosis alterations.

5.4.1 Neurodegenerative Disorders

One of the most compelling associations between microbiota and ageing-related diseases lies in the pathophysiology of neurodegenerative disorders, particularly Parkinson's disease (PD) and Alzheimer's disease (AD).

5.4.1.1 Parkinson's Disease

Parkinson's disease (PD) is a progressive neurodegenerative disease caused due to the degeneration of various regions of the brain because of the build-up of alpha-synuclein protein in the cell bodies of neurons, forming characteristic structures named Lewy bodies eosinophilic, round, laminated inclusions in the cytoplasm (Alfaidi et al. 2025). This buildup primarily impacts dopaminergic neurons in the midbrain's substantia nigra, and the frequency and intensity of motor symptoms are directly correlated with the extent of neuron loss (Dickson et al. 2009). Typical motor symptoms are problems initiating movement (akinesia), stiffness of muscles (rigidity), resting tremors, as well as gait and walking difficulties. Aside from these traditional motor problems, individuals with PD usually have a variety of non-motor signs and symptoms, including dementia, depression, loss of smell, and various gastrointestinal impairments. These include nausea, difficulty swallowing, salivation, constipation, and defecation difficulties. The overall clinical presentation in this disease is a combination of the direct effects of loss of dopaminergic neurons and the spread of alpha-synuclein deposition within the nervous system (Sarkar and Banerjee 2019).

The progression of Parkinson's disease has often been linked to imbalances in the gut microbiota. Research has shown that an imbalance in the gut microbiota results in progression of Parkinson's disease (Sarkar and Banerjee 2019). A study by Scheperjans et al. (2015) reported a reduced abundance of *Prevotellaceace* species relative to *Enterobacteriaceae* in the fecal microbiota of Parkinson's disease (PD) patients. *Prevotellaceace* are commensal bacteria known for their role in fermenting dietary fiber to produce mucin and short-chain fatty acids (SCFs), which play a critical function in the maintenance of gur barrier integrity (Arumugam et al. 2011). A decline in *Prevotellaceae* abundance can compromise intestinal epithelial integrity, leading to elevated gut permeability and systemic translocation of bacterial endotoxins. This endotoxemia is hypothesized to trigger or sustain pathological aggregation of α-synuclein in the enteric nervous system and may promote misfolding within the colon (Forsyth et al. 2011; Niehaus and Lange 2003). Gastrointestinal symptoms, particularly constipation, are often prodromal in PD and are thought to be mediated via vagus nerve (Svensson et al. 2015). Notably, alteration in the gut microbiota composition, including elevated levels of *Enterobacteriaceae*, have been associated with greater postural instability and motor symptoms severity in PD patients (Scheperjans et al. 2015). Elevated *Enterobacteriaceae* raised serum lipopolysaccharide levels, which are derived from their Gram-negative cell walls. Elevated absorption of LPS in PD patients elevates LPS-binding protein in blood, leading to systemic inflammation by TLR4 and NF-κB pathways. LPS and inflammatory cytokines (TNF-α, IL-1β, IL-6) disrupt

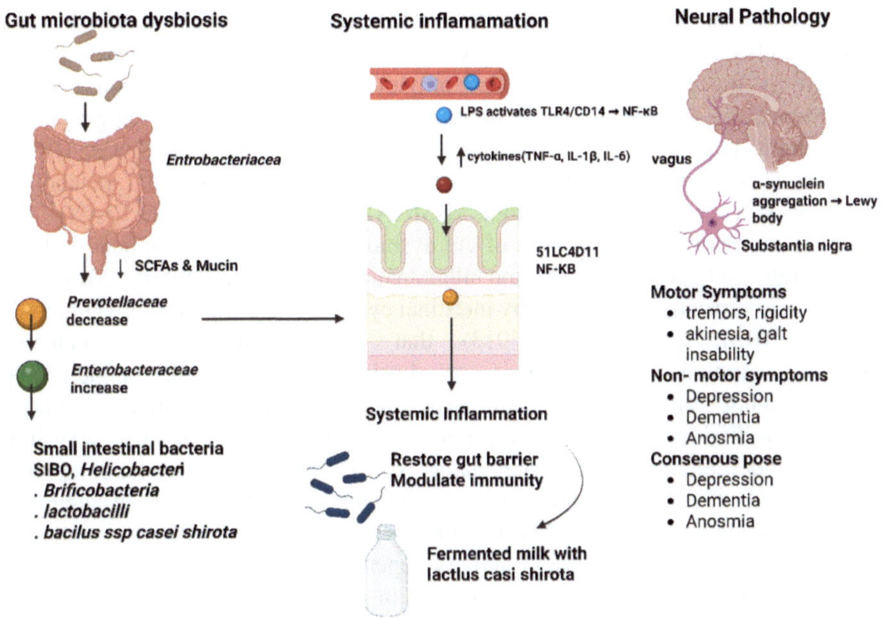

Fig. 5.9 Gut dysbiosis in Parkinson's disease decreases protective bacteria and enhances proinflammatory taxa, causing LPS-induced inflammation, barrier dysfunction, and α-synuclein aggregation. These mechanisms promote neurodegeneration and motor impairment, and probiotics potentially restore balance

the intestinal and blood-brain barriers, resulting in α-synuclein accumulation and dopaminergic neuron loss in the substantia nigra. This LPS-driven inflammation is linked to PD progression (Sarkar and Banerjee 2019). Tight junction proteins like occludins are critical in maintaining the intestinal barrier, and dysbiosis decreases occludins and increases intestinal permeability. PD patients exhibit increased proinflammatory bacteria (*Ralstonia*, *Proteobacteria*, *Enterococcaceae*) and decreased butyrate-producing, anti-inflammatory bacteria (*Blautia*, *Coprococcus*, *Roseburia*, *Faecalibacterium*). Furthermore, LPS biosynthesis genes are highly increased in the PD fecal microbiome. Motor dysfunction and Parkinson's disease (PD) have been associated with small intestinal bacterial overgrowth. *Helicobacter pylori* infection has also been suggested as a causative agent for Parkinson's disease pathophysiology. Probiotics like *Bifidobacteria*, *Lactobacilli*, and *Bacillus* spp. may reverse PD-like conditions by improving gut health and enhancing dopamine synthesis from L-tyrosine (Sarkar and Banerjee 2019). Fermented milk with *Lactobacillus casei* Shirota improves bowel movement and reduces fecal *staphylococc*i in patients with PD (Cassani et al. 2011) (Fig. 5.9).

5.4.1.2 Alzheimer's Disease (AD)

Alzheimer's disease (AD) is a progressive neurodegenerative condition marked by the gradual deterioration of cognitive abilities. It is also marked by the accumulation of misfolded amyloid-β peptides in the brain, forming senile plaques, which result from the cleavage of the amyloid-β precursor protein (AβPP). Moreover, it is also defined by the buildup of misfolded amyloid-β peptides in the brain that are by-products of amyloid-β precursor protein (AβPP) cleavage and aggregate to form senile plaques. High levels of systemic microbes like spirochetes, Herpes simplex virus type 1 (HSV-1), *Chlamydia pneumoniae* and microbial metabolites such as β-N-methylamino-L-alanine (BMAA), lipopolysaccharides (LPS), and microbial amyloids are the effects of gut dysbiosis that usually come hand in hand with heightened gut permeability and inflammation (Sarkar and Banerjee 2019). One of the toxic substances generated by intestinal cyanobacteria is β-N-methylamino-L-alanine (BMAA) (Brenner 2013) that are accountable for inducing neurodegeneration, cognitive deficit, astrogliosis, and neurofibrillary tangle formation in adult rats upon neonatal exposure. Another key microbial factor, lipopolysaccharides (LPS), derived primarily from Gram negative bacteria such as *Escherichia/Shigella* spp. is accountable for Aβ fibril formation and strongly activates NF-κB, a key regulator of neuroinflammation observed in Alzheimer's disease (AD) brains (Lukiw 2016). Toll-like receptor 4 is expressed by microglia and binds to LPS. Toll-like receptor 4 activity depends on its coreceptor CD 14, which has been demonstrated to be overexpressed in AD brain microglia. By triggering innate immunity, amyloids generated by the gut microbiota may amplify the immune response to brain Aβ42. Aging further exacerbates this by leading to heightened activation of both innate and adaptive immune responses, contributing to a persistent, low-grade inflammatory state termed "inflammaging" (Sarkar and Banerjee 2019). It was discovered that within a study there was a rise in the amount of proinflammatory *Escherichia/ shigella* spp. and fall in the amount of anti-inflammatory characterized *Eubacterium rectale* spp., in the fecal microbiome of Alzheimer's disease patients. This microbial imbalance, or dysbiosis, may contribute to enhanced peripheral inflammation and promote amyloid accumulation in the brain (Cattaneo et al. 2017). In a similar vein, *Helicobacter pylori* infection is linked to cognitive impairment, potentially through increased systemic inflammation and vascular dysfunction. Collectively, these microbial signatures highlight the critical role that bacteria play in promoting AD-linked neuroinflammation and amyloid pathology (Fig. 5.10).

5.4.1.3 Multiple Sclerosis (MS)

A chronic, immune-mediated neurological condition, multiple sclerosis (MS) is typified by neurodegeneration, axonal damage, and demyelination. Vertigo, visual loss, light headedness, discomfort, motor system weakness, impaired coordination, exhaustion, numbness, dyspepsia, depression, and cognitive decline are among the clinical presentations. The four main forms of MS are: (1) primary progressive MS (PPMS), (2) secondary progressive MS (SPMS), (3) progressive-relapsing MS (PRMS), and (4) relapsing-remitting MS (RRMS). Pathogenesis of MS is caused by combination of genetic predisposition and environmental influences. Among these, gut microbiota imbalance (dysbiosis) is increasingly recognized as a key environmental contributor to the neurological progression of MS. RRMS is marked

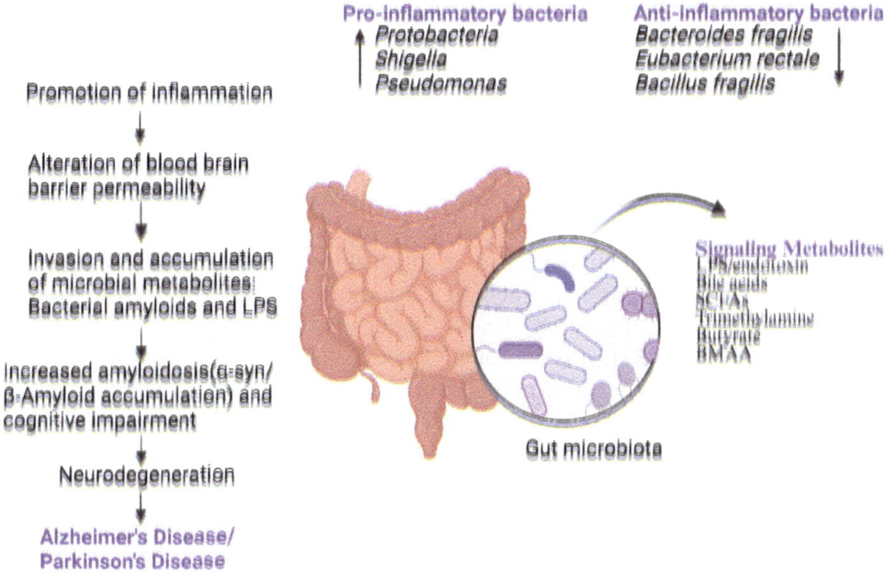

Fig. 5.10 Gut dysbiosis contributes to the onset and progression of Parkinson's and Alzheimer's diseases

by a reduction of bacteria that support anti-inflammatory immune responses, such as those mediated by regulatory T cells (Tregs), IL-10–producing CD4+ T cells, regulatory B cells, tolerogenic dendritic cells, and suppressive macrophages. On the other hand, an enrichment of bacteria that sustain pro-inflammatory activity by immune cells such as CD4+ T cells, dendritic cells, monocytes, and B cells takes place. Some gut microbes such as *Clostridium*, *Bacteroidetes*, *Prevotella*, *Parabacteroides*, and *Lactobacillus* support the generation of short-chain fatty acids (SCFAs), which serve to sustain immune cells participating in anti-inflammatory reactions (Sarkar and Banerjee 2019). Patients with RRMS have shown a reduced presence of *Parabacteroides distasonis* (Li et al. 2025) and *Prevotella copri* (Lin et al. 2024) when compared to healthy individuals.

Several probiotic interventions have been explored using the experimental autoimmune encephalomyelitis (EAE) model, a common animal model for multiple sclerosis (MS). Oral administration of *Bifidobacterium animalis* has been reported to alleviate EAE symptoms (Ezendam et al. 2008). Some *Lactobacillus* strains are found to enhance regulatory T cell subsets, expand IL-10 secretion, decrease IL-17 expression, and enhance clinical ratings in EAE models. In addition, repeated oral probiotic supplementation with *Streptococcus thermophilus*, *Bifidobacterium bifidum*, and *Lactobacillus* species in mice and rats has shown clinical improvement, Th1 and Th17 immune suppression, and augmentation of T regulatory cells (Sarkar and Banerjee 2019).

5.4.2 Diabetes Mellitus Type 2

Type 2 diabetes mellitus (T2DM) is an age-associated metabolic disorder increasingly associated with changes in gut microbiota. In db/db mice and humans, there is dysbiosis characterized by elevated pro-inflammatory and reduced butyrate-producing bacteria, e.g., *Faecalibacterium, Akkermansia*, and *Roseburia*, and high LPS biosynthesis and microbial activities associated with oxidative stress and inflammation (Gurung et al. 2020). Metabolites such as SCFAs, tryptophan derivatives, TMAO, LPS, and bile acids regulate glucose metabolism, inflammation, and gut integrity. Therapeutics such as metformin, acarbose, and empagliflozin enhance metabolic wellness and healthy aging by regulating gut microbiota—increasing SCFA production, decreasing gut permeability, and suppressing systemic inflammation. According to research conducted on *Caenorhabditis* worms, metformin may increase longevity by altering metabolism and reducing the severity and risk of a number of diabetes-related illnesses, including cancer, heart disease, and neurological problems (Gyriki et al. 2025) (Fig. 5.11).

5.4.3 Cardiovascular Disease

Atherosclerosis accounts for approximately 50% of cardiovascular mortality and is traditionally associated with hypertension, diabetes, hyperlipidemia, obesity, smoking, and aging. Cardiovascular disease (CVDs) is mostly attributable to aging not

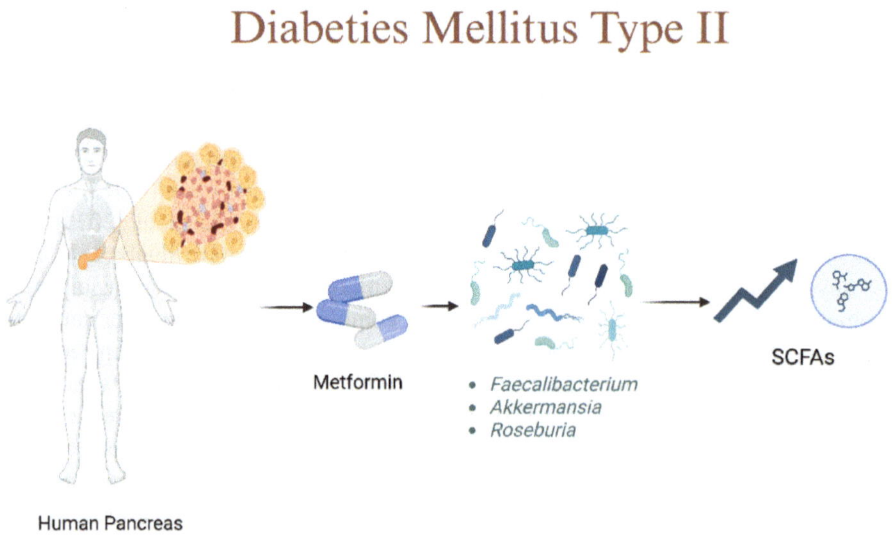

Fig. 5.11 The part played by gut microbiota in T2DM, or type 2 diabetes. Dysbiosis, which is characterized by altered levels of *Faecalibacterium, Akkermansia*, and *Roseburia*, affects gut integrity, inflammation, and glucose metabolism. Therapies such as metformin enhance metabolic health and longevity, boost the production of SCFA, and re-establish microbial balance

only due to vascular stiffening and immunosenescence, but also due to generalized remodeling of the gut microbiota with increasing age. With aging, as an individual ages, microbial diversity reduces, along with reductions in beneficial genera such as *Faecalibacterium*, *Roseburia*, and *Blautia*, possibly being replaced by potentially pathogenic taxa like *Enterobacteriaceae*, *Escherichia-Shigella*, and *Enterococcus*. These shifts in the microbiome mirror those that occur in atherosclerosis, coronary artery disease (CAD), hypertension, and heart failure (HF) and suggest that aging-related dysbiosis is the cause of increased cardiovascular risk.

Emerging data implicate gut microbiota dysbiosis as a key driver in CVD (Gyriki et al. 2025; Shen et al. 2021). In CAD, decreased *Faecalibacterium*, *Roseburia*, *Subdoligranulum*, and *E. rectale*, but increased *Escherichia-Shigella*, and *Enterococcus*, are observed routinely. Hypertension is linked with decreased microbial richness, decreased *Firmicutes/Bacteroidetes* ratio, and loss of acetate- and butyrate-producing bacteria. HF patients have decreased *Blautia*, *Collinsella*, *Faecalibacterium*, and other favorable microbes, but increased pathogenic taxa like *Campylobacter*, *Salmonella*, and *Candida* (Gyriki et al. 2025). Mechanistically, gut microbiota regulates cardiovascular health in a number of ways directly. The microbiota metabolizes dietary choline to the end product trimethylamine (TMA) and further to TMAO in the liver. Elevated TMAO is associated with elevated CVD risk, increased platelet activation, and thrombosis. Gut microbiota also causes TLR2-mediated hepatic von Willebrand factor activation, leading to thrombus formation (Jäckel et al. 2017; Reinhardt 2019). Conversely, short-chain fatty acids (SCFAs) like propionate regulate blood pressure by activating Olfr78 and GPR41 receptors and mediating vasodilation. Lipopolysaccharide (LPS), which is elevated in loss of *Bacteroides vulgatus* and *B. dorei*, causes plaque instability and atherosclerosis. Notably, supplementation of mice with *B. vulgatus* and *B. dorei* lowered circulating LPS and atherosclerosis severity (Gyriki et al. 2025).

5.4.4 Cancer

Gastric cancer, the third most frequent neoplasm and fifth leading cause of cancer death globally, is highly dependent on host age. Over time, gastric physiology significantly alters with decreased gastric acid secretion, compromised mucosal regeneration, and immune senescence, predisposing to chronic *Helicobacter pylori* colonization. Eradication of *H. pylori* has been found in 10-year follow-up studies to reduce gastric inflammation, retard atrophic gastritis and intestinal metaplasia progression, and consequently decrease risk for gastric cancer. In elderly, chronic infection with pathogenic *H. pylori* strains (CagA, VacA) aggravates endoplasmic reticulum stress, oxidative damage, and dysfunctional autophagy, thus further aggravating age-related decline in mucosal defense and facilitating carcinogenesis (Gyriki et al. 2025).

Colorectal cancer (CRC) is the second most common cause of cancer-related deaths and the third most common kind of cancer overall. It is highly associated with aging, with dramatic increases in incidence after midlife. Disruption of the gut microbiota by age is a key risk factor in this process. In older adults, microbial stability is decreased, diversity is reduced, and there is enrichment of pro-inflammatory taxa such as *Clostridium leptum* and *Clostridium coccoides* (Baidoun et al. 2021; Scanlan et al. 2008). Such community-level changes establish a procarcinogenic microenvironment in the aging colon. Particular microbiota also interacts with aged tissue to enhance malignancy. *Fusobacterium nucleatum* and *enterotoxigenic Bacteroides* fragilis activate β-catenin signaling and augment intestinal inflammation, processes more significant in aged humans where immune homeostasis is perturbed (Modeel et al. 2025; Xu et al. 2024). *Campylobacter jejuni* secretes cytolethal distending toxin that causes DNA double-strand breaks, and *E. coli*-derived colibactin induces further genotoxic stress and activates the senescence-associated secretory phenotype (SASP) in epithelial cells (Gyriki et al. 2025). During aging, this is highly detrimental because the efficiency of DNA repair diminishes and the accumulation of the SASP generates a chronic inflammatory, pro-tumorigenic environment.

Therefore, in CRC, host aging and microbial dysbiosis are synergistic: changes in the microbiota that occur with aging enhance exposure to proinflammatory and genotoxic bacterial products, whereas host aging diminishes the ability to repair damage and regulate senescence. This synergy promotes the onset and progression of the tumor in the elderly.

In pancreatic cancer (PC), gastrointestinal dysbiosis associated with aging contributes to enhanced cancer risk. In aging, microbial diversity reduces in parallel with enrichment of opportunistic pathogens. In Israeli PC patients, increased *Bacteroidetes:Firmicutes* ratios, elevated *Veillonellaceae*, *Akkermansia*, and *Odoribacter*, and decreased *Clostridiaceae*, *Lachnospiraceae*, and *Ruminococcaceae* have been noted. Likewise, Chinese cohorts have overrepresentation of pathogenic genera (*Veillonella*, *Klebsiella*, *Selenomonas*, *Prevotella*, *Enterobacter*) and loss of beneficial bacteria (*Bifidobacterium*, *Blautia*, *Anaerostipes*). Aging exacerbates these microbial imbalances by compromising gut barrier integrity, modifying bile acid metabolism, and declining host resilience, thus propagating inflammation driven by the microbiota in the pancreatic microenvironment (Gyriki et al. 2025).

Liver cancer, specifically hepatocellular carcinoma (HCC), is also influenced by aging-related microbiome alterations. With advancing age, enhanced gut permeability and modified bile acid pools lead to greater translocation of LPS and other pathogenic microbial products into the portal circulation. This persistent inflammatory stimulus induces hepatic carcinogenesis. HCC patients often have elevated *Parabacteroides*, *Clostridium*, and *Gemmiger*, and decreased *Bifidobacterium* and *Lactobacillus* (Rajapakse et al. 2023). In older adults, decreased abundance of protective commensals worsens liver inflammation and fibrosis, hastening the course from cirrhosis to carcinoma.

Combined, aging-associated alterations in the gut microbiota defined by decreasing diversity, enrichment for pro-inflammatory taxa, compromised mucosal immunity, and the buildup of microbial metabolites like LPS and secondary bile acids intersect with the signature processes of aging (DNA damage, immune senescence,

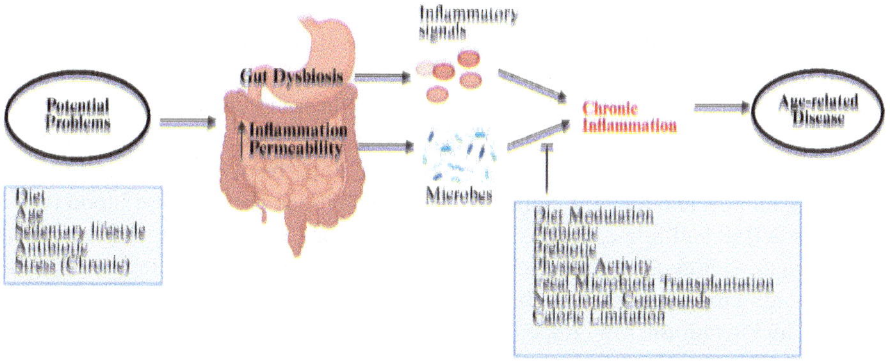

Fig. 5.12 Simplified diagram showing links between health risks, gut dysbiosis, inflammation, and age-related diseases, along with possible interventions to reduce gut-driven inflammation

SASP) to increase susceptibility to cancer. Grasping these intersections presents potential for microbiota-directed therapies in geriatric oncology, including probiotics, prebiotics, dietary modulation, and microbial metabolite–directed interventions. Such methods have the promise of postponing onset, enhancing early detection, and facilitating precision prevention of age-related cancers (Fig. 5.12).

5.5 Microbiota-Targeted Interventions to Delay Aging

With progressive aging, the human gut microbiota experiences large compositional and functional alteration that contribute to the development of chronic diseases and systemic inflammation, often referred to as "inflammaging" (Franceschi et al. 2018). Targeting the microbiota has emerged as a promising strategy to delay or mitigate age-related physiological decline. This section explores various microbiota-modulating approaches that hold therapeutic potential in promoting healthy aging.

5.5.1 Prebiotics and Probiotics

Prebiotics are selectively used by health-enhancing gut bacteria such as Lactobacillus and Bifidobacterium to stimulate their growth and activity. This benefits both patients with gastrointestinal (GI) diseases and healthy individuals. Prebiotic consumption in the elderly has been shown to reduce pro-inflammatory cytokines, increase Bifidobacterium counts, and increase anti-inflammatory cytokines, all of which reduce inflammation in the gut and facilitate immunological control. Prebiotics also elevate the production of short-chain fatty acids (SCFAs), like butyrate, which increase the absorption of minerals in the gut, such as calcium and magnesium, and suppress the growth of intestinal cancer cells (Choi et al. 2018; Boyajian et al. 2024). Recent clinical trials indicate that prebiotics modulate gut microbiota and metabolite fingerprints to improve walking speed, muscular strength, renal function, and frailty status in the elderly (Boyajian et al. 2024).

Probiotics are live bacteria that improve the health of your gut. They restore the disrupted balance of the gut microbiota due to age, disease, or antibiotics. Probiotics improved neuronal plasticity-related gene expression (e.g., BDNF), memory performance, and gut microbiota composition in aged animal models, hence establishing the gut-brain axis and aging (Choi et al. 2018). Probiotics also increase blood levels of vitamins B12 and folate, the most important for cognitive function and can retard neurological decline (Umekar et al. 2025). By strengthening the immune response and regulating the gut barrier, probiotics also reduce systemic and gut inflammation, hence reducing chronic low-grade inflammation common in aging (Ren et al. 2023).

5.5.2 Synbiotics

The synergistic effects of using probiotics and prebiotics enhance microbial activity and survival within the gut. Synbiotics boost antioxidant capacity, improve immunoglobulin production, lower inflammatory cytokines such as TNF-α and IL-6, and boost immunologic function and gastrointestinal comfort in older adults. For example, synbiotic products of inulin and *Bifidobacterium lactis* boost butyrate synthesis and gut microbiota diversity, favoring the metabolic well-being and the integrity of the mucosa (Swanson et al. 2020; Rajkumar et al. 2015; Klemashevich et al. 2014).

5.5.3 Postbiotics and Paraprobiotics

Postbiotics, offering benefits without the use of living bacteria, are bioactive compounds produced by microorganisms or derived from non-viable microbial cells. These consist of peptidoglycans, enzymes, microbial peptides, and SCFAs (Aguilar-Toalá et al. 2018). Inactivated probiotic cells, or paraprobiotics, possess a better safety profile and provide similar benefits; they are specifically well-suited to elderly or immunocompromised individuals. Postbiotics exert antibacterial, immune-modulating, and epithelial barrier-promoting activity. Blocking histone deacetylases, boosting regulatory T cell activity, and reducing systemic inflammation, butyrate, a major SCFA postbiotic, promotes gut and systemic well-being as individuals age (Segain et al. 2000; Teame et al. 2020).

Paraprobiotics and postbiotics contain stable and well-characterized molecular structures, have a longer shelf life, and pose fewer risks compared to live probiotics, making them more attractive for use as functional food ingredients and therapeutic agents (Teame et al. 2020; Monteiro et al. 2023).

5.5.4 Fecal Microbiota Transplantation

The objective of FMT is to restore microbial balance and intestinal function, which involves the transfer of gut microbiota from a healthy donor to a dysbiotic or diseased recipient. Donor fecal material, typically obtained and cryopreserved, may be given by gavage, oral capsules, or enema. In treating *Clostridium difficile* infection, FMT has been demonstrated effective in restoring microbial colonization and regular intestinal function. FMT holds tremendous potential to reverse aging-caused dysbiosis, inflammation, and dysfunction of physiological processes. Young donor-transplanted gut microbiota has been shown to exert rejuvenating effects on aged animal models of gut and brain health. It promotes gut microbial diversity, locomotor activity, and delays cognitive impairment. Astonishingly, with young donor microbiota, aged mice exhibit increased beneficial taxa such as Akkermansia, decrease in intestinal barrier injury and inflammation, and are associated with increased muscle mass and decreased frailty (Zhang et al. 2023; Sahle et al. 2024).

Immunosenescence, muscle atrophy, and cognitive decline are all influenced by changes in the gut microbiota composition with increasing age, as indicated by studies. Through modulation of healthy aging and systemic inflammation, FMT or microbiota-targeted therapies can restore a young-like microbiome. According to Novelle et al. (2025) and Zhu et al. (2024), FMT from donor healthy individuals, for instance, has been shown to restore gut barrier function, decrease systemic proinflammatory cytokines (IL-6, TNF-α), and partially reverse frailty-related characteristics in aged individuals.

Another role of intestinal microbes in aging long and healthily is also supported by experimental data from killifish models that demonstrate that transfer of microbiota from young organisms prolongs lifespan (Choi et al. 2018). Overall, in the restoration of microbial diversity, the suppression of inflammation, and enhancement of barrier and immunologic function, FMT is a promising means of gut microbial reconstitution that can delay aging-related degeneration and extend lifespan.

5.5.5 Dietary Modulation

Diet plays a major role in governing the architecture and function of the gut microbiota, and this in turn regulates healthy aging through reducing the levels of inflammation and enhancing microbial diversity.

The polyphenol, fiber, and omega-3 fatty acid-rich Mediterranean diet enhances the proliferation of beneficial gut microbes like *Bifidobacterium* and *Faecalibacterium*. Research in older people demonstrates that the diet enhances microbial richness and the production of anti-inflammatory SCFAs, reduces frailty, improves cognition, and improves metabolic health (Ghosh et al. 2020; O'Toole and Jeffery 2015).

Caloric restriction (CR): Refraining from malnutrition throughout the course of life prolongs lifespan and slows down the aging process. CR reduces systemic inflammation and enhances neuroprotection in aging models by reprogramming the gut microbiota to a more youthful profile with a higher *Bacteroidetes/Firmicutes* ratio and higher SCFA-producing bacteria (Zhang et al. 2013; Singh et al. 2024).

Fiber with vitamin D: Increased dietary fiber consumption increases microbial fermentation to produce SCFA. This supports gut barrier integrity and immunomodulation. With decreased skin synthesis and food intake, vitamin D deficiency is widespread among the elderly. It may affect calcium absorption and shift the gut microbiome to an immunosuppressive state. Fiber-dietary food supplements and vitamin D status enhance the immune system and microbiota in the elderly, advancing gut health and inflammation suppression (Choi et al. 2018; Zou et al. 2025).

5.5.6 Physical Activity and Microbiota

Physical exercise has a robust modulatory effect upon older adults' gut microbiota, both composition and functional output. Systematic reviews of cross-sectional and intervention studies show that repeated moderate exercise need not alter overall alpha/beta diversity, but enhances relative abundance of SCFA producer genera *Bifidobacterium*, *Anaerostipes*, and *Oscillospira* after programs of five weeks or longer (Min et al. 2024; Aya et al. 2023). Following a 24-week combined cardiovascular and resistance program, exercise aged 58–75 was associated with enhanced stool butyrate and in beneficial taxa including Bifidobacterium (and other yet unknown genera in *Bifidobacteriaceae*), *Oscillospira*, *Anaerostipes*, and with reductions in *Prevotella* and *Oribacterium* (Erlandson et al. 2021). Similar genus-level shifts were seen in older adults with insomnia, where higher step count was linked with higher *Bacteroides* and lower proinflammatory microbes (Magzal et al. 2022). Crucially, these changes in the microbes are accompanied by enhanced metabolic health markers, insulin sensitivity, and systemic inflammation, and hence reduced risk of age-related conditions including sarcopenia and cardiovascular disease (Dziewiecka et al. 2022). Exercise induced SCFA producer enrichment is also beneficial to the gut-brain axis, with microbiota changes linked to enhanced cognitive performance and mood in the elderly. As such, the addition of frequent, moderate intensity physical activity is a low-risk, low-cost intervention for reversing age-associated dysbiosis and to optimizing overall health (Varghese et al. 2024).

5.6 Challenges

Clinical translation and the establishment of efficacious treatments are blocked by many critical issues, despite the fact that considerable progress has been made in our understanding of how gut microbiota contribute to aging and the potential of microbiota-targeted interventions. To achieve the full potential of the microbiome to promote healthy aging, these constraints must be overcome.

5.6.1 Interindividual Variability

The large variation in microbial compositions and functional capabilities between individuals is a great challenge in microbiome studies. These differences are explained by complex host genetics, age, nutrition, environment, medication (including antibiotics), and lifestyle interactions (Falony et al. 2016). Variance in microbiota has been found in research to be due to environmental determinants and only a very small percentage is due to genetics (Rothschild et al. 2018). Frailty, comorbidity, prolonged medication, and health status worsen heterogeneity in older people (Claesson et al. 2012). Characterization of a normative "healthy" aged microbiome is further made difficult by this diversity, as it affects microbial metabolic production and immunological control. Stratification by demographic and clinical variables is needed for informative analysis (Jackson et al. 2016). A meaningful "healthy" microbiome baseline for older persons is difficult to establish, since the normal decline in microbiota with aging is non-linear and context-dependent (O'Toole and Jeffery 2015). This heterogeneity requires personalized therapies and hampers the development of microbota-modulating drugs that are efficacious across the board.

5.6.2 Longitudinal Studies and Causality

The majority of research into human microbiome aging has been cross-sectional, offering snapshots but not conclusions about causality. Therefore, it is uncertain whether microbial change is a cause or effect of aging and age-related disease. To define temporal trends of change in gut microbiota, their relationship with clinical phenotypes, and microbial predictors or contributors to the development of illness, well-controlled longitudinal cohorts are required (Valdes et al. 2018). Furthermore, a mechanistic understanding of microbiota-host interaction is still lacking despite demonstrated associations with metabolic and neurological disease (Cryan et al. 2019). To reveal the dynamics of microbe-host interactions during aging, though, long-term multiomics data integration (metagenomics, metabolomics, and transcriptomics) is essential. There are a few interventional human experimental trials that employ functional outcome measures alongside microbiological measures. Furthermore, two-way effects between microbiota and host physiology complicate showing causation and therefore the application of sophisticated experimental and computational models like gnotobiotic trials and Mendelian randomization (Zhernakova et al. 2016; Smith et al. 2013).

5.6.3 Translational Limitations from Animal Models

Mechanistic research on microbiota and aging has been made greatly easier by the use of animal models, and for convenience, rats are particularly useful. Direct extrapolation to human, however, is limited by species-specific differences in immunological aging, gut function, diet, and composition of gut microbiota. Rodent models are, while useful for the study of mechanisms such as the role of microbial metabolites in neuroimmune aging, unable to simulate the richness and diversity of the human microbiota. This is partly eased by humanized mouse models, such as transfer of human microbiota to germ-free mice. These models are still, however, unable to simulate the heterogeneous aging process of humans or interaction of human microbiome with environment. Extrapolation is necessarily limited by species-specific differences in immune system maturation, nutrition, and structure. Organoid culture or large animal models may provide additional platforms, but human clinical data are still needed (Nguyen et al. 2015).

5.6.4 Standardization of Methodologies

Reproducibility and comparability between studies remain constrained by the lack of standard methodologies in microbiome science. Standardization extends beyond the benchtop in the laboratory to bioinformatics pipelines, where taxonomic and functional microbiome profiles are significantly impacted by key analytical parameters like primer sets, reference databases, sequencing depth, and statistical approach (McLaren et al. 2019). Variation in the sampling method of the samples, DNA extraction protocol, sequencing platform, and taxonomic framework leads to varying microbial profile data, with increased difficulty in defining microbiome health or dysbiosis in older individuals (Xu et al. 2024). Although MIxS and other minimum microbiome reporting standards have been suggested, uneven uptake has been seen (Knight et al. 2018). In addition, open sharing of data and careful correction for batch effects must be done for cohort harmonization and meta-analyses. While offering enhanced taxonomic and functional resolution, new technologies like shotgun metagenomes and long-read sequencing impose added complexity. Standard operating procedures and consensus standards, therefore, need to be set and followed across the board to further progress the field.

5.6.5 Regulatory and Ethical Challenges in Interventions

Particularly if administered to vulnerable older populations, microbiome-guided interventions such as fecal microbiota transplantation (FMT) and genetically modified probiotics pose unique ethical and regulatory issues. Due to the inherent dangers of this treatment such as pathogen transmission, unforeseen ecological disruption in the host microbiota, and safety issues related to longer-term use, rigorous clinical screening of donors and clinical monitoring are required (Cammarota et al. 2017). The regulatory landscape renders translation much more difficult. The uncertain status of microbiota-derived drugs as drugs, biologics, or dietary supplements complicates approval processes, clinical uptake, and commercialization. Their successful delivery in vulnerable patient populations is significantly handicapped by uncertainty.

Current developments demonstrate complexity as well as potential. Although there is still a lack of long-term safety evidence in elderly or immunocompromised individuals, recent FDA approval of Rebyota, a standardized therapy for recurrent *Clostridium difficile* infection with FMT is a step toward regulatory clarity and underscores the necessity for regulation and safety.

5.6.6 Complexity of Microbiota-Host Interactions

The gut microbiota affects aging through complex, multifaceted processes involving epigenetic regulation, immunomodulation, metabolic control, and signaling to the brain. It is challenging to identify a single root driver or stable biomarker of microbial aging due to the extremely dynamic and nonlinear nature of these interactions, which are dependent on host genetics, environment, and time of day (O'Toole and Jeffery 2015; D'Amico et al. 2022). Emergent properties and coregulation patterns of the microbiome with the host have been identified by projecting these complex networks with systems biology approaches that integrate multiomics data (Guo et al. 2022). Machine learning and AI interpretation of host-microbiome interactions are in its infancy. Among the challenges are limitations in longitudinal datasets, heterogeneity of the data, and interpretation of predictive models (Rozera et al. 2025).

5.6.7 Lack of Personalized Approaches

Current interventions apply a one-size-fits-all strategy, which does not account for the individualistic nature of the microbiome. Without individualized profiling, interventions can have nonreproducible outcomes or even be dangerous in certain subgroups (Zeevi et al. 2015). Due to microbial interactions and heterogeneous host responses, such nonindividualized strategies risk generating less than optimal or negative outcomes. Emerging advances in deep phenotyping, metagenomics, metabolomics, and integration of host genomics, coupled with machine learning algorithms, offer promising directions toward precise individual stratification. To enhance repeatability, safety, and therapeutic efficacy, precision microbiome medicine aims to tailor interventions such as tailored probiotics, prebiotics, or dietary modifications to each individual's specific microbial and metabolic landscape (Sonnenburg and Bäckhed 2016; Hitch et al. 2022).

5.7 Conclusion

As a central controller of aging, the gut microbiota influences immunity, metabolism, cognition, and susceptibility to disease. Microbial dysbiosis, defined by decreased diversity, loss of beneficial taxa, and proliferation of pro-inflammatory microbes, is routinely associated with aging. Together, these mechanisms trigger inflammation and functional decline. A healthy, SCFA-producing microbiome enhances longevity, metabolic well-being, and cognitive resilience, based on evidence from human cohorts and model systems. Probiotics, prebiotics, synbiotics, diet modification, physical exercise, and drugs such as metformin are some of the therapies that target microbiota and have high potential to delay age-related diseases and extend lifespan. But in order to take maximum advantage of the therapeutic potential of the microbiome in the context of healthy aging, integrative, personalized, and ethically guided methods must be applied in view of interindividual heterogeneity, translational gaps, and regulatory barriers.

References

Aguilar-Toalá JE, Garcia-Varela R, Garcia HS, Mata-Haro V, González-Córdova AF, Vallejo-Cordoba B, Hernández-Mendoza A (2018) Postbiotics: an evolving term within the functional foods field. Trends Food Sci Technol 75:105–114

Alfaidi M, Barker RA, Kuan WL (2025) An update on immune-based alpha-synuclein trials in Parkinson's disease. J Neurol 272(1). https://doi.org/10.1007/s00415-024-12770-x

Arumugam M, Raes J, Pelletier E, Le Paslier D, Yamada T, Mende DR et al (2011) Enterotypes of the human gut microbiome. Nature 473(7346):174–180

Aya V, Jimenez P, Muñoz E, Ramírez JD (2023) Effects of exercise and physical activity on gut microbiota composition and function in older adults: a systematic review. BMC Geriatr 23(1): 364. https://doi.org/10.1186/s12877-023-04066-y

Badal VD, Vaccariello ED, Murray ER, Yu KE, Knight R, Jeste DV, Nguyen TT (2020) The gut microbiome, aging, and longevity: a systematic review. Nutrients 12(12):3759. https://doi.org/10.3390/nu12123759

Baidoun F, Elshiwy K, Elkeraie Y, Merjaneh Z, Khoudari G, Sarmini MT et al (2021) Colorectal cancer epidemiology: recent trends and impact on outcomes. Curr Drug Targets 22(9):998–1009

Bhoomi A (2024) The Gut-brain Axis: exploring the bidirectional communication between the Gut microbiome and the brain. J Forens Sci Res 8(1):047–057. https://doi.org/10.29328/journal.jfsr.1001064

Boyajian JL, Islam P, Abosalha A, Schaly S, Thareja R, Kassab A, Arora K, Santos M, Shum-Tim C, Prakash S (2024) Probiotics, prebiotics, synbiotics and other microbiome-based innovative therapeutics to mitigate obesity and enhance longevity via the gut-brain axis. Microbiome Res\ Report 3(3):29. https://doi.org/10.20517/mrr.2024.05

Bradley E, Haran J (2024) The human gut microbiome and aging. Gut Microbes 16(1). https://doi.org/10.1080/19490976.2024.2359677

Brenner SR (2013) Blue-green algae or cyanobacteria in the intestinal micro-flora may produce neurotoxins such as Beta-N-Methylamino-L-alanine (BMAA) which may be related to development of amyotrophic lateral sclerosis, Alzheimer's disease and Parkinson-dementia-complex in humans and equine motor neuron disease in horses. Med Hypotheses 80(1):103

Brown HA, Morris AL, Pudlo NA, Hopkins AE, Martens EC, Golob JL, Koropatkin NM (2024) Acarbose impairs gut Bacteroides growth by targeting intracellular glucosidases. MBio 15(12): e01506–e01524. https://doi.org/10.1128/mbio.01506-24

Cammarota G, Ianiro G, Tilg H, Rajilić-Stojanović M, Kump P, Satokari R, Sokol H, Arkkila P, Pintus C, Hart A, Segal J, Aloi M, Masucci L, Molinaro A, Scaldaferri F, Gasbarrini G, Lopez-Sanroman A, Link A, de Groot P, de Vos WM, Högenauer C, Malfertheiner P, Mattila E, Milosavljević T, Nieuwdorp M, Sanguinetti M, Simren M, Gasbarrini A (2017) European consensus conference on faecal microbiota transplantation in clinical practice. Gut 66(4):569–580. https://doi.org/10.1136/gutjnl-2016-313017

Carabotti M, Scirocco A, Maselli MA, Severi C (2015) The gut-brain axis: interactions between enteric microbiota, central and enteric nervous systems. Ann Gastroenterol 28(2):203–209

Cassani E, Privitera G, Pezzoli G, Pusani C, Madio C, Iorio L, Barichella M (2011) Use of probiotics for the treatment of constipation in Parkinson's disease patients. Minerva Gastroenterol Dietol 57(2):117–121

Cattaneo A, Cattane N, Galluzzi S, Provasi S, Lopizzo N, Festari C et al (2017) Association of brain amyloidosis with pro-inflammatory gut bacterial taxa and peripheral inflammation markers in cognitively impaired elderly. Neurobiol Aging 49:60–68

Choi J, Hur TY, Hong Y (2018) Influence of altered Gut microbiota composition on aging and aging-related diseases. J Lifestyle Med 8(1):1–7

Claesson MJ, Jeffery IB, Conde S, Power SE, O'Connor EM, Cusack S, Harris HM, Coakley M, Lakshminarayanan B, O'Sullivan O, Fitzgerald GF, Deane J, O'Connor M, Harnedy N, O'Connor K, O'Mahony D, van Sinderen D, Wallace M, Brennan L, Stanton C et al (2012) Gut microbiota composition correlates with diet and health in the elderly. Nature 488(7410): 178–184. https://doi.org/10.1038/nature11319

Clark RI, Walker DW (2018) Role of gut microbiota in aging-related health decline: insights from invertebrate models. Cell Mol Life Sci 75(1):93–101. https://doi.org/10.1007/s00018-017-2671-1

Cryan JF, O'Riordan KJ, Cowan CS, Sandhu KV, Bastiaanssen TF, Boehme M et al (2019) The microbiota-gut-brain axis. Physiol Rev. 99(4):1877–2013. https://doi.org/10.1152/physrev.00018.2018

D'Amico F, Decembrino N, Muratore E, Turroni S, Muggeo P, Mura R, Perruccio K, Vitale V, Zecca M, Prete A, Venturelli F, Leardini D, Brigidi P, Masetti R, Cesaro S, Zama D (2022) Oral Lactoferrin supplementation during induction chemotherapy promotes gut microbiome Eubiosis in pediatric patients with hematologic malignancies. Pharmaceutics 14(8):1705. https://doi.org/10.3390/pharmaceutics14081705

Deleyto-Seldas N, Efeyan A (2021) The mTOR–autophagy axis and the control of metabolism. Front Cell Dev Biol 9:655731. https://doi.org/10.3389/fcell.2021.655731

Deng X, Zhang C, Wang P, Wei W, Shi X, Wang P, Yang J, Wang L, Tang S, Fang Y, Liu Y, Chen Y, Zhang Y, Yuan Q, Shang J, Kan Q, Yang H, Man H, Wang D, Yuan H (2022) Cardiovascular benefits of Empagliflozin are associated with gut microbiota and plasma metabolites in type 2 diabetes. J Clin Endocrinol Metabol 107(7):1888–1896. https://doi.org/10.1210/clinem/dgac210

Di Vincenzo F, Del Gaudio A, Petito V, Lopetuso LR, Scaldaferri F (2024) Gut microbiota, intestinal permeability, and systemic inflammation: a narrative review. Intern Emerg Med 19(2):275–293. https://doi.org/10.1007/s11739-023-03374-w

Dickson DW, Fujishiro H, Orr C, DelleDonne A, Josephs KA, Frigerio R et al (2009) Neuropathology of non-motor features of Parkinson disease. Parkinsonism Relat Disord. https://doi.org/10.1016/S1353-8020(09)70769-2

Doenyas C, Clarke G, Cserjési R (2025) Gut–brain axis and neuropsychiatric health: recent advances. Sci Reports 15(1):3415., s41598-025-86858-3. https://doi.org/10.1038/s41598-025-86858-3

Duffuler P, Bhullar KS, Wu J (2024) Targeting gut microbiota in osteoporosis: impact of the microbial based functional food ingredients. Food Sci Human Wellness 13(1):1–15. https://doi.org/10.26599/FSHW.2022.9250001

Dziewiecka H, Buttar HS, Kasperska A, Ostapiuk-Karolczuk J, Domagalska M, Cichoń J, Skarpańska-Stejnborn A (2022) Physical activity induced alterations of gut microbiota in humans: a systematic review. BMC Sports Sci Med Rehabilit 14(1):122. https://doi.org/10.1186/s13102-022-00513-2

Erlandson KM, Liu J, Johnson R, Dillon S, Jankowski CM, Kroehl M, Robertson CE, Frank DN, Tuncil Y, Higgins J, Hamaker B, Wilson CC (2021) An exercise intervention alters stool microbiota and metabolites among older, sedentary adults. Therapeut Adv Infect Dis 8: 20499361211027067. https://doi.org/10.1177/20499361211027067

Ezendam J, De Klerk A, Gremmer ER, Van Loveren H (2008) Effects of Bifidobacterium animalis administered during lactation on allergic and autoimmune responses in rodents. Clin Exp Immunol 154(3):424–431

Falony G, Joossens M, Vieira-Silva S, Wang J, Darzi Y, Faust K et al (2016) Population-level analysis of gut microbiome variation. Science 352(6285):560–564

Forsyth CB, Shannon KM, Kordower JH, Voigt RM, Shaikh M, Jaglin JA et al (2011) Increased intestinal permeability correlates with sigmoid mucosa alpha-synuclein staining and endotoxin exposure markers in early Parkinson's disease. PLoS One 6(12):e28032

Franceschi C, Garagnani P, Parini P, Giuliani C, Santoro A (2018) Inflammaging: a new immune–metabolic viewpoint for age-related diseases nature reviews. Nat Rev Endocrinol 14(10):576–590. https://doi.org/10.1038/s41574-018-0059-4

Franceschí C, Bonafè M, Valensin S, Olivieri F, De Luca M, Ottaviani E, De Benedictis G (2000) Inflamm-aging: an evolutionary perspective on immunosenescence. Ann N Y Acad Sci 908(1): 244–254. https://doi.org/10.1111/j.1749-6632.2000.tb06651.x

Ghosh TS, Rampelli S, Jeffery IB, Santoro A, Neto M, Capri M, Giampieri E, Jennings A, Candela M, Turroni S, Zoetendal EG, Hermes GDA, Elodie C, Meunier N, Brugere CM, Pujos-Guillot E, Berendsen AM, De Groot LCPGM, Feskins EJM et al (2020) Mediterranean diet intervention alters the gut microbiome in older people reducing frailty and improving health status: the NU-AGE 1-year dietary intervention across five European countries. Gut 69(7): 1218–1228. https://doi.org/10.1136/gutjnl-2019-319654

Ghosh TS, Shanahan F, O'Toole PW (2022) The gut microbiome as a modulator of healthy ageing. Nature Rev Gastroenterol Hepatol 19(9):565–584. https://doi.org/10.1038/s41575-022-00605-x

Guo J, Huang X, Dou L, Yan M, Shen T, Tang W, Li J (2022) Aging and aging-related diseases: from molecular mechanisms to interventions and treatments. Signal Trans and Targeted Therapy 7(1). https://doi.org/10.1038/s41392-022-01251-0

Gurung M, Li Z, You H, Rodrigues R, Jump DB, Morgun A, Shulzhenko N (2020) Role of gut microbiota in type 2 diabetes pathophysiology. EBioMedicine 51:102590. https://doi.org/10.1016/j.ebiom.2019.11.051

Gut microbiota and aging (n.d.)

Gyriki D, Nikolaidis CG, Bezirtzoglou E, Voidarou C, Stavropoulou E, Tsigalou C (2025) The gut microbiota and aging: interactions, implications. Front Aging. https://doi.org/10.3389/fragi.2025.1452917

Hagemann N, Ackermann N, Christmann J, Brier S, Yu F, Erdmann KS (2013) The serologically defined colon cancer antigen-3 interacts with the protein tyrosine phosphatase PTPN13 and is involved in the regulation of cytokinesis. Oncogene 32(39):4602–4613. https://doi.org/10.1038/onc.2012.485

Hao H, Li Z, Qiao S, Qi Y, Xu X, Si J, Liu Y, Chang L, Shi Y, Xu B, Wei Z, Kang L (2023) Empagliflozin ameliorates atherosclerosis via regulating the intestinal flora. Atherosclerosis 371:32–40. https://doi.org/10.1016/j.atherosclerosis.2023.03.011

Hitch TCA, Hall LJ, Walsh SK, Leventhal GE, Slack E, de Wouters T, Walter J, Clavel T (2022) Microbiome-based interventions to modulate gut ecology and the immune system. Mucosal Immunol 15(6):1095–1113. https://doi.org/10.1038/s41385-022-00564-1

Ira R, Adwani J, Krishnan AO, Subramanian G, Yadav S, Shukla S, Rao S, Prakash T (2024) Understanding aging through the Lense of gut microbiome. Explorat Res Hypoth Med 9(4): 294–307. https://doi.org/10.14218/erhm.2024.00008

Jäckel S, Kiouptsi K, Lillich M, Hendrikx T, Khandagale A, Kollar B et al (2017) Gut microbiota regulate hepatic von Willebrand factor synthesis and arterial thrombus formation via toll-like receptor-2. Blood 130(4):542–553

Jackson MA, Jeffery IB, Beaumont M, Bell JT, Clark AG, Ley RE, O'Toole PW, Spector TD, Steves CJ (2016) Signatures of early frailty in the gut microbiota. Genome Med 8(1):8. https://doi.org/10.1186/s13073-016-0262-7

Jiang M, Kang L, Wang Y-L, Zhou B, Li H-Y, Yan Q, Liu Z-G (2024) Mechanisms of microbiota-gut-brain axis communication in anxiety disorders. Front Neurosci 18:1501134. https://doi.org/10.3389/fnins.2024.1501134

Jung YH, Chae CW, Han HJ (2024) The potential role of gut microbiota-derived metabolites as regulators of metabolic syndrome-associated mitochondrial and endolysosomal dysfunction in Alzheimer's disease. Exp Mol Med 56(8):1691–1702. https://doi.org/10.1038/s12276-024-01282-3

Kanimozhi NV, Sukumar M (2025) Aging through the lens of the gut microbiome: challenges and therapeutic opportunities. Arch Gerontol Geriatr Plus 2(2):100142. https://doi.org/10.1016/j.aggp.2025.100142

Klemashevich C, Wu C, Howsmon D, Alaniz RC, Lee K, Jayaraman A (2014) Rational identification of diet-derived postbiotics for improving intestinal microbiota function. Curr Opin Biotechnol 26:85–90

Knight R, Vrbanac A, Taylor BC, Aksenov A, Callewaert C, Debelius J, Gonzalez A, Kosciolek T, McCall LI, McDonald D, Melnik AV, Morton JT, Navas J, Quinn RA, Sanders JG, Swafford AD, Thompson LR, Tripathi A, Xu ZZ, Zaneveld JR et al (2018) Best practices for analysing microbiomes. Nat Rev Microbiol 16(7):410–422. https://doi.org/10.1038/s41579-018-0029-9

Lau WL, Tran T, Rhee CM, Kalantar-Zadeh K, Vaziri ND (2021) Diabetes and the gut microbiome. Semin Nephrol 41(2):104–113. https://doi.org/10.1016/j.semnephrol.2021.03.005

Lee CB, Chae SU, Jo SJ, Jerng UM, Bae SK (2021) The relationship between the gut microbiome and metformin as a key for treating type 2 diabetes mellitus. Int J Mol Sci 22(7):3566. https://doi.org/10.3390/ijms22073566

Li C, Stražar M, Mohamed AMT, Pacheco JA, Walker RL, Lebar T, Zhao S, Lockart J, Dame A, Thurimella K, Jeanfavre S, Brown EM, Ang QY, Berdy B, Sergio D, Invernizzi R, Tinoco A, Pishchany G, Vasan RS et al (2024a) Gut microbiome and metabolome profiling in Framingham heart study reveals cholesterol-metabolizing bacteria. Cell 187(8):1834–1852.e19. https://doi.org/10.1016/j.cell.2024.03.014

Li N, Wang H, Pei H, Wu Y, Li L, Ren Y, Wang S, Ma Y, Luo M, Yuan J, Li L, Qin D (2024b) Genus_Ruminococcus and order_Burkholderiales affect osteoporosis by regulating the microbiota-gut-bone axis. Front Microbiol 15:1373013. https://doi.org/10.3389/fmicb.2024.1373013

Li X, Zheng S, Xu H, Zhang Z, Han X, Wei Y, Jin H, Du X, Xu H, Li M, Zhang Z, Wang S, Sun G, Zhang D (2025) The direct and indirect inhibition of proinflammatory adipose tissue macrophages by acarbose in diet-induced obesity. Cell Reports Med 6(1):101883. https://doi.org/10.1016/j.xcrm.2024.101883

Lin Q, Dorsett Y, Mirza A, Tremlett H, Piccio L, Longbrake EE, Choileain SN, Hafler DA, Cox LM, Weiner HL, Yamamura T, Chen K, Wu Y, Zhou Y (2024) Meta-analysis identifies common gut microbiota associated with multiple sclerosis. Genome Med 16(1). https://doi.org/10.1186/s13073-024-01364-x

Loh JS, Mak WQ, Tan LKS, Ng CX, Chan HH, Yeow SH, Foo JB, Ong YS, How CW, Khaw KY (2024) Microbiota–gut–brain axis and its therapeutic applications in neurodegenerative diseases. Signal Transduct Target Ther 9(1):37. https://doi.org/10.1038/s41392-024-01743-1

López-Otín C, Blasco MA, Partridge L, Serrano M, Kroemer G (2023) Hallmarks of aging: an expanding universe. Cell 186(2):243–278. https://doi.org/10.1016/j.cell.2022.11.001

Lukiw WJ (2016) Bacteroides fragilis lipopolysaccharide and inflammatory signaling in Alzheimer's disease. Front Microbiol 7. https://doi.org/10.3389/fmicb.2016.01544

Magzal F, Shochat T, Haimov I, Tamir S, Asraf K, Tuchner-Arieli M, Even C, Agmon M (2022) Increased physical activity improves gut microbiota composition and reduces short-chain fatty acid concentrations in older adults with insomnia. Sci Reports 12(1):2265. https://doi.org/10.1038/s41598-022-05099-w

Mariat D, Firmesse O, Levenez F, Guimarães V, Sokol H, Doré J, Corthier G, Furet JP (2009) The Firmicutes/Bacteroidetes ratio of the human microbiota changes with age. BMC Microbiol 9(1):123. https://doi.org/10.1186/1471-2180-9-123

McLaren MR, Willis AD, Callahan BJ (2019) Consistent and correctable bias in metagenomic sequencing experiments. eLife 8:e46923. https://doi.org/10.7554/eLife.46923
Min L, Ablitip A, Wang R, Luciana T, Wei M, Ma X (2024) Effects of exercise on gut microbiota of adults: a systematic review and meta-analysis. Nutrients 16(7):1070. https://doi.org/10.3390/nu16071070
Modeel S, Siwach S, Dolkar P, Chaurasia M, Yadav P, Atri A, Yadav A, Negi T, Negi RK (2025) Emerging risk factors and the role of Gut microbiota in immunomodulation and therapeutic implications in colorectal cancer. Cancer Pathogen Therapy 4(01):14–30. S294971322500076X. https://doi.org/10.1016/j.cpt.2025.06.007
Monteiro SS, Schnorr CE, Pasquali MAB (2023) Paraprobiotics and Postbiotics-current state of scientific research and future trends toward the development of functional foods. Foods (Basel, Switzerland) 12(12):2394. https://doi.org/10.3390/foods12122394
N V K, M S (2025) Aging through the lens of the gut microbiome: challenges and therapeutic opportunities. Arch Gerontol Geriatr Plus 2(2):100142. https://doi.org/10.1016/j.aggp.2025.100142
Nguyen TLA, Vieira-Silva S, Liston A, Raes J (2015) How informative is the mouse for human gut microbiota research? Dis Model Mech 8(1):1–16
Niehaus I, Lange JH (2003) Endotoxin: is it an environmental factor in the cause of Parkinson's disease? Occup Environ Med 60(5):378–378
Novelle MG, Naranjo-Martínez B, López-Cánovas JL, Díaz-Ruiz A (2025) Fecal microbiota transplantation, a tool to transfer healthy longevity. Ageing Res Rev 103:102585. https://doi.org/10.1016/j.arr.2024.102585
O'Toole PW, Jeffery IB (2015) Gut microbiota and aging. Science 350(6265):1214–1215. https://doi.org/10.1126/science.aac8469
Palm F, Nagel C, Bruckmaier RM, Aurich JE, Aurich C (2013) Clinical parameters intestinal function and IGF1 concentrations in colostrum-deprived and colostrum-fed newborn pony foals. Theriogenology 80(9):1045–1051. https://doi.org/10.1016/j.theriogenology.2013.08.002
Qin N, Yang F, Li A, Prifti E, Chen Y, Shao L, Guo J, Le Chatelier E, Yao J, Wu L, Zhou J, Ni S, Liu L, Pons N, Batto JM, Kennedy SP, Leonard P, Yuan C, Ding W, Chen Y, Hu X, Zheng B, Qian G, Xu W, Ehrlich SD, Zheng S, Li L (2014) Alterations of the human gut microbiome in liver cirrhosis. Nature 513(7516):59–64. https://doi.org/10.1038/nature13568
Rajapakse J, Khatiwada S, Akon AC, Yu KL, Shen S, Zekry A (2023) Unveiling the complex relationship between gut microbiota and liver cancer: opportunities for novel therapeutic interventions. Gut Microbes 15(2):2240031
Rajkumar H, Kumar M, Das N, Kumar SN, Challa HR, Nagpal R (2015) Effect of probiotic Lactobacillus salivarius UBL S22 and prebiotic fructo-oligosaccharide on serum lipids, inflammatory markers, insulin sensitivity, and gut bacteria in healthy young volunteers: a randomized controlled single-blind pilot study. J Cardiovasc Pharmacol Ther 20(3):289–298
Reinhardt C (2019) The gut microbiota as an influencing factor of arterial thrombosis. Hamostaseologie 39(02):173–179
Ren J, Li H, Zeng G, Pang B, Wang Q, Wei J (2023) Gut microbiome-mediated mechanisms in aging-related diseases: are probiotics ready for prime time? Front Pharmacol. https://doi.org/10.3389/fphar.2023.1178596
Rinninella E, Raoul P, Cintoni M, Franceschi F, Miggiano GAD, Gasbarrini A, Mele MC (2019) What is the healthy gut microbiota composition? a changing ecosystem across age, environment, diet, and diseases. Microorganisms 7(1):14. https://doi.org/10.3390/microorganisms7010014
Rothschild D, Weissbrod O, Barkan E, Kurilshikov A, Korem T, Zeevi D, Costea PI, Godneva A, Kalka IN, Bar N, Shilo S, Lador D, Vila AV, Zmora N, Pevsner-Fischer M, Israeli D, Kosower N, Malka G, Wolf BC, Avnit-Sagi T et al (2018) Environment dominates over host genetics in shaping human gut microbiota. Nature 555(7695):210–215. https://doi.org/10.1038/nature25973

Rozera T, Pasolli E, Segata N, Ianiro G (2025) Machine learning and artificial intelligence in the multi-omics approach to gut microbiota. Gastroenterology 169(3):487–501. https://doi.org/10.1053/j.gastro.2025.02.035

Saeedi Saravi SS, Pugin B, Constancias F, Shabanian K, Spalinger M, Thomas A, Le Gludic S, Shabanian T, Karsai G, Colucci M, Menni C, Attaye I, Zhang X, Allemann MS, Lee P, Visconti A, Falchi M, Alimonti A, Ruschitzka F et al (2025) Gut microbiota-dependent increase in phenylacetic acid induces endothelial cell senescence during aging. Nature Aging 5(6): 1025–1045. https://doi.org/10.1038/s43587-025-00864-8

Sahle Z, Engidaye G, Shenkute Gebreyes D, Adenew B, Abebe TA (2024) Fecal microbiota transplantation and next-generation therapies: a review on targeting dysbiosis in metabolic disorders and beyond. SAGE Open Med 12:20503121241257486. https://doi.org/10.1177/20503121241257486

Sarkar SR, Banerjee S (2019) Gut microbiota in neurodegenerative disorders. J Neuroimmunol 328: 98–104

Scanlan PD, Shanahan F, Clune Y, Collins JK, O'Sullivan GC, O'Riordan M et al (2008) Culture-independent analysis of the gut microbiota in colorectal cancer and polyposis. Environ Microbiol 10(3):789–798

Scheperjans F, Aho V, Pereira PA, Koskinen K, Paulin L, Pekkonen E et al (2015) Gut microbiota is related to Parkinson's disease and clinical phenotype. Mov Disord 30(3):350–358

Segain JP, De La Blétiere DR, Bourreille A, Leray V, Gervois N, Rosales C et al (2000) Butyrate inhibits inflammatory responses through NFκB inhibition: implications for Crohn's disease. Gut 47(3):397–403

Shabani M, Ghoshehy A, Mottaghi AM, Chegini Z, Kerami A, Shariati A, Taati Moghadam M (2025) The relationship between gut microbiome and human diseases: mechanisms, predisposing factors and potential intervention. Front Cell Infect Microbiol 15:1516010. https://doi.org/10.3389/fcimb.2025.1516010

Shen X, Li L, Sun Z, Zang G, Zhang L, Shao C, Wang Z (2021) Gut microbiota and atherosclerosis—focusing on the plaque stability. Front Cardiovasc Med 8:668532

Shi P, Xu S, Yang Z, Wang L, Wu Y, Li Y, Zhu Z (2024) Harnessing gut microbiota for longevity: insights into mechanisms and genetic manipulation. iMetaOmics 1(2):36. https://doi.org/10.1002/imo2.36

Singh I, Anand S, Gowda DJ, Kamath A, Singh AK (2024) Caloric restriction mimetics improve gut microbiota: a promising neurotherapeutics approach for managing age-related neurodegenerative disorders. Abstract. Biogerontology 25(6):899–922. https://doi.org/10.1007/s10522-024-10128-4

Smith PM, Howitt MR, Panikov N, Michaud M, Gallini CA, Bohlooly-Y M, Glickman JN, Garrett WS (2013) The microbial metabolites, short-chain fatty acids, regulate colonic Treg cell homeostasis. Science 341(6145):569–573. https://doi.org/10.1126/science.1241165

Sokol H, Pigneur B, Watterlot L, Lakhdari O, Bermúdez-Humarán LG, Gratadoux J-J, Blugeon S, Bridonneau C, Furet J-P, Corthier G, Grangette C, Vasquez N, Pochart P, Trugnan G, Thomas G, Blottière HM, Doré J, Marteau P, Seksik P, Langella P (2008) Faecalibacterium prausnitzii is an anti-inflammatory commensal bacterium identified by gut microbiota analysis of Crohn disease patients. Proc Natl Acad Sci 105(43):16731–16736. https://doi.org/10.1073/pnas.0804812105

Sonnenburg JL, Bäckhed F (2016) Diet-microbiota interactions as moderators of human metabolism. Nature 535(7610):56–64. https://doi.org/10.1038/nature18846

Subramanian S, Huq S, Yatsunenko T, Haque R, Mahfuz M, Alam MA, Benezra A, DeStefano J, Meier MF, Muegge BD, Barratt MJ, VanArendonk LG, Zhang Q, Province MA, Petri WA Jr, Ahmed T, Gordon JI (2014) Persistent gut microbiota immaturity in malnourished Bangladeshi children. Nature 510(7505):417–421. https://doi.org/10.1038/nature13421

Svensson E, Horváth-Puhó E, Thomsen RW, Djurhuus JC, Pedersen L, Borghammer P, Sørensen HT (2015) Vagotomy and subsequent risk of Parkinson's disease. Ann Neurol 78(4):522–529

Swanson KS, Gibson GR, Hutkins R, Reimer RA, Reid G, Verbeke K et al (2020) The international scientific Association for Probiotics and Prebiotics (ISAPP) consensus statement on the definition and scope of synbiotics. Nat Rev Gastroenterol Hepatol 17(11):687–701

Tang C, Li L, Jin X, Wang J, Zou D, Hou Y, Yu X, Wang Z, Jiang H (2024) Investigating the impact of Gut microbiota on gout through Mendelian randomization. Orthop Res Rev 16:125–136. https://doi.org/10.2147/ORR.S454211

Teame T, Wang A, Xie M, Zhang Z, Yang Y, Ding Q, Gao C, Olsen RE, Ran C, Zhou Z (2020) Paraprobiotics and postbiotics of probiotic lactobacilli, their positive effects on the host and action mechanisms: a review. Front Nutr 7:570344. https://doi.org/10.3389/fnut.2020.570344

Thevaranjan N, Puchta A, Schulz C, Naidoo A, Szamosi JC, Verschoor CP, Loukov D, Schenck LP, Jury J, Foley KP, Schertzer JD, Larché MJ, Davidson DJ, Verdú EF, Surette MG, Bowdish DME (2017) Age-associated microbial dysbiosis promotes intestinal permeability, systemic inflammation, and macrophage dysfunction. Cell Host Microbe 21(4):455–466.e4. https://doi.org/10.1016/j.chom.2017.03.002

Turnbaugh PJ, Hamady M, Yatsunenko T, Cantarel BL, Duncan A, Ley RE, Sogin ML, Jones WJ, Roe BA, Affourtit JP, Egholm M, Henrissat B, Heath AC, Knight R, Gordon JI (2009) A core gut microbiome in obese and lean twins. Nature 457(7228):480–484. https://doi.org/10.1038/nature07540

Umekar M, Premchandani T, Tatode A, Qutub M, Raut N, Taksande J, Hussain UM (2025) Vitamin B12 deficiency and cognitive impairment: a comprehensive review of neurological impact. Brain Disorders 18:100220

Valdes AM, Walter J, Segal E, Spector TD (2018) Role of the gut microbiota in nutrition and health. BMJ 361:k2179. https://doi.org/10.1136/bmj.k2179

Varghese S, Rao S, Khattak A, Zamir F, Chaari A (2024) Physical exercise and the gut microbiome: a bidirectional relationship influencing health and performance. Nutrients 16(21):3663. https://doi.org/10.3390/nu16213663

Wang S, Li H, Yuan M, Fan H, Cai Z (2022) Role of AMPK in autophagy. Front Physiol 13: 1015500. https://doi.org/10.3389/fphys.2022.1015500

Wang Y, Jia X, Cong B (2024) Advances in the mechanism of metformin with wide-ranging effects on regulation of the intestinal microbiota. Front Microbiol 15:1396031. https://doi.org/10.3389/fmicb.2024.1396031

Wilmanski T, Diener C, Rappaport N, Patwardhan S, Wiedrick J, Lapidus J, Earls JC, Zimmer A, Glusman G, Robinson M, Yurkovich JT, Kado DM, Cauley JA, Zmuda J, Lane NE, Magis AT, Lovejoy JC, Hood L, Gibbons SM et al (2021) Gut microbiome pattern reflects healthy ageing and predicts survival in humans. Nat Metab 3(2):274–286. https://doi.org/10.1038/s42255-021-00348-0

Wu Y, Xu J, Rong X, Wang F, Wang H, Zhao C (2021) Gut microbiota alterations and health status in aging adults: from correlation to causation. Aging Med 4(3):206–213. https://doi.org/10.1002/agm2.12167

Wu B, Yan J, Yang J, Xia Y, Li D, Zhang F, Cao H (2022) Extension of the life span by Acarbose: is it mediated by the gut microbiota? Aging Dis 13(4):1005. https://doi.org/10.14336/AD.2022.0117

Xiao Y, Feng Y, Zhao J, Chen W, Lu W (2025) Achieving healthy aging through gut microbiota-directed dietary intervention: focusing on microbial biomarkers and host mechanisms. J Adv Res 68:179–200. https://doi.org/10.1016/j.jare.2024.03.005

Xu Z, Yeoh YK, Tun HM, Fei N, Zhang J, Morrison M, Kamm MA, Yu J, Chan FKL, Ng SC (2024) Variation in the metagenomic analysis of fecal microbiome composition calls for a standardized operating approach. Microbiol Spectrum 12(12). https://doi.org/10.1128/spectrum.01516-24

Yan Z, Guan G, Jia H, Li H, Zhuoga S, Zheng S (2025) The association between gut microbiota and accelerated aging and frailty: a Mendelian randomization study. Aging Clin Exp Res 37(1):82. https://doi.org/10.1007/s40520-025-02971-3

Yoshida N, Emoto T, Yamashita T, Watanabe H, Hayashi T, Tabata T, Hoshi N, Hatano N, Ozawa G, Sasaki N, Mizoguchi T, Amin HZ, Hirota Y, Ogawa W, Yamada T, Hirata K (2018) Bacteroides vulgatus and Bacteroides dorei reduce gut microbial lipopolysaccharide production and inhibit atherosclerosis. Circulation 138(22):2486–2498. https://doi.org/10.1161/CIRCULATIONAHA.118.033714

Yu L, Chen Y, Tooze SA (2018) Autophagy pathway: cellular and molecular mechanisms. Autophagy 14(2):207–215. https://doi.org/10.1080/15548627.2017.1378838

Zackular JP, Baxter NT, Iverson KD, Sadler WD, Petrosino JF, Chen GY, Schloss PD (2013) The gut microbiome modulates colon tumorigenesis. mBio 4(6). https://doi.org/10.1128/mBio.00692-13

Zeevi D, Korem T, Zmora N, Israeli D, Rothschild D, Weinberger A et al (2015) Personalized nutrition by prediction of glycemic responses. Cell 163(5):1079–1094

Zhang C, Li S, Yang L, Huang P, Li W, Wang S, Zhao G, Zhang M, Pang X, Yan Z, Liu Y, Zhao L (2013) Structural modulation of gut microbiota in life-long calorie-restricted mice. Nat Commun 4(1). https://doi.org/10.1038/ncomms3163

Zhang Y, Wang X, Li W, Yang Y, Wu Z, Lyu Y, Yue C (2023) Intestinal microbiota: a new perspective on delaying aging? Front Microbiol 14. https://doi.org/10.3389/fmicb.2023.1268142

Zhao X, Zhong X, Liu X, Wang X, Gao X (2021) Therapeutic and improving function of lactobacilli in the prevention and treatment of cardiovascular-related diseases: a novel perspective from gut microbiota. Front Nutrition 8:693412. https://doi.org/10.3389/fnut.2021.693412

Zhao S, Feng P, Hu X, Cao W, Liu P, Han H, Jin W, Li X (2022) Probiotic Limosilactobacillus fermentum GR-3 ameliorates human hyperuricemia via degrading and promoting excretion of uric acid. iScience 25(10):105198. https://doi.org/10.1016/j.isci.2022.105198

Zhao H, Qiu X, Wang S, Wang Y, Xie L, Xia X, Li W (2025) Multiple pathways through which the gut microbiota regulates neuronal mitochondria constitute another possible direction for depression. Front Microbiol 16:1578155. https://doi.org/10.3389/fmicb.2025.1578155

Zhernakova A, Kurilshikov A, Bonder MJ, Tigchelaar EF, Schirmer M, Vatanen T, Mujagic Z, Vila AV, Falony G, Vieira-Silva S, Wang J, Imhann F, Brandsma E, Jankipersadsing SA, Joossens M, Cenit MC, Deelen P, Swertz MA, LifeLines cohort study, Weersma RK et al (2016) Population-based metagenomics analysis reveals markers for gut microbiome composition and diversity. Science 352(6285):565–569. https://doi.org/10.1126/science.aad3369

Zhu M, Huang Y, Wang Z, Jin Z, Cao J, Zhong Q, Xiong Z (2024) Fecal microbiota transplantation attenuates frailty via gut-muscle axis in old mice. Aging Dis 16(2):1180–1198. https://doi.org/10.14336/AD.2024.0321

Zou Q, Fu B, Gou Y, Zhong G, Liu Y (2025) Vitamin D metabolites and the gut microbiota dietary index are associated with reduced all-cause mortality in US adults: a prospective cohort study. BMC Nutr 11(1). https://doi.org/10.1186/s40795-025-01211-1

Prebiotics, Probiotics, and Postbiotics in Therapy

6

Abstract

The gut microbiome is now being seen as a prime driver of human health with broad ramifications for nutrition, disease prevention, and drug development. In this chapter, the synergy of five pillars, i.e., prebiotics, probiotics, postbiotics, synbiotics and next-generation probiotics, is argued to create a new paradigm for functional nutrition and precision medicine. These bioactive metabolites interact to regulate gut ecology, broaden microbial diversity, and impact host processes, such as metabolic, neurological, and immune functions. Drawing on the latest discoveries in metagenomics, high-throughput screening, and microbial engineering, this research identifies how targeted microbiome modulation can drive gut resilience, restore microbial homeostasis, and maintain systemic health. Practical applications are discussed in parallel with future therapeutic approaches, demonstrating the potential for personalized nutrition to maximize well-being outcomes. Positioning gut health at the center of human strength, this chapter highlights the game changing potential of microbial therapeutics towards the next health revolution. As science continues to advance, these methods stand the chance to surpass conventional methods to enable precision, safe, and efficient methods for maximizing lifetime health and disease prevention.

Keywords

Prebiotic · Probiotics and postbiotics · Synbiotics and next-generation probiotics (NGPs) · Therapeutic agents · Immuno-modulation and host microbiota interactions

R. C. Sobti et al., *Microbiota Unveiled: Science, Health, and Future Therapies*,
https://doi.org/10.1007/978-981-95-9031-5_6

6.1 Introduction

6.1.1 Overview of the Human Microbiome and Its Importance for Therapy

The gut is the largest body interface with the external world. The human gut microbiome, which consists of trillions of microorganisms, is increasingly regarded as a "functional organ" that plays a central role in our health and well-being. This dynamic microbial community of yeasts, archaea, bacteria, viruses, fungi, and protozoa, that colonizes our skin and mucosal surfaces, controls an effective symbiotic relationship with the host that impacts both local and systemic bodily functions (Mar Rodríguez et al. 2015; Li et al. 2021a, b). The symbiotic relationship between human host and its bacterial inhabitants has drawn widespread research attention in recent years. These microorganisms are primarily located in the gastrointestinal (GI) tract, and along with their combined genomes, are referred to as the gut microbiome. The GI tract is not merely a digestive system but also significantly contributes to human host homeostasis, disease, and health; it also possesses immune-modulating properties and influences physiological health by the gut-brain axis. On the other hand, dysbiosis (imbalance in the makeup and function of microorganisms) has been linked to a number of diseases such as cancer, obesity, metabolic syndrome, autism spectrum disorders, neurodegenerative diseases, and inflammatory bowel disease (IBD) (Tilg et al. 2020; Kho and Lal 2018) (Fig. 6.1).

Bioactive microbial therapies including prebiotics, probiotics, postbiotics, and synbiotics have emerged into prominence in this context. These biologically friendly, nonsurgical interventions can restore microbial balance, alleviate inflammation, and play a critical role in the prophylactic and treatment of disease

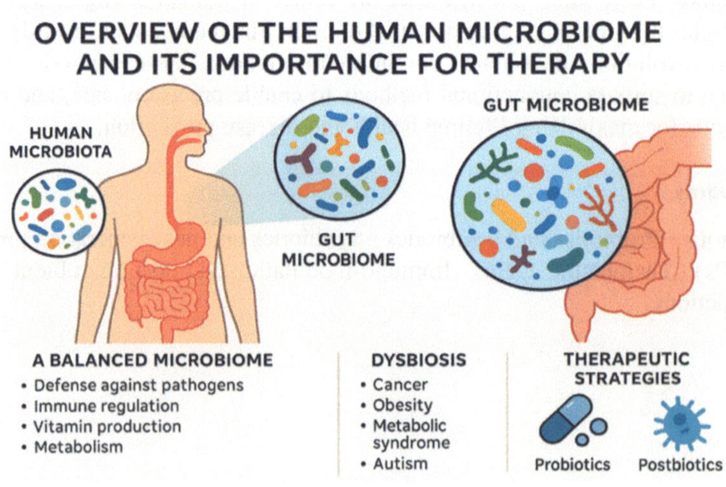

Fig. 6.1 Overview of the human microbiome and its therapeutic relevance

(Maslennikov et al. 2018). These microbial populations also produce a vast array of signaling molecules that influence host physiology. Metabolites like tryptophan derivatives and short-chain fatty acids (SCFAs) are known to regulate immune maturation, energy homeostasis, and integrity of mucosal barriers and have been implicated in disease development such as inflammatory bowel disease (IBD) (Russo et al. 2019). Therapeutic and dietary interventions involve dietary fiber content diets, fermented foods, and probiotic, prebiotic, and synbiotic supplementation in the form of functional foods or medical-grade formulations (Box 6.1).

Box 6.1: Definitions Established by Consensus Panels Such as FAO/WHO and ISAPP

- Probiotics are "*live microorganisms that, when administered in adequate amounts, confer a health benefit on the host.*"
- Prebiotics are "*substrates that are selectively utilized by host microorganisms, conferring a health benefit.*"
- Syn-biotics are "*mixtures comprising live microorganisms and substrate (s) selectively utilized by host microorganisms that confer a health benefit.*"
- Postbiotics refer to "*preparations of inanimate microorganisms and/or their components that confer a health benefit on the host.*"

However, diet is an important key factor in defining the function and structure of the microbiome. Host health and resilience are enhanced by maintaining ecological balance and microbial diversity. A healthy gut microbiota provides several health benefits which improve intestinal barrier function, and blocking gastrointestinal pathogens, gut inflammation, diarrhea, and inflammatory bowel diseases. Intervening in the gut microbiota through the application of targeted microbiological and dietary interventions promises exciting possibilities for preventive and therapeutic approaches to modern medicine as knowledge increases (Fig. 6.2).

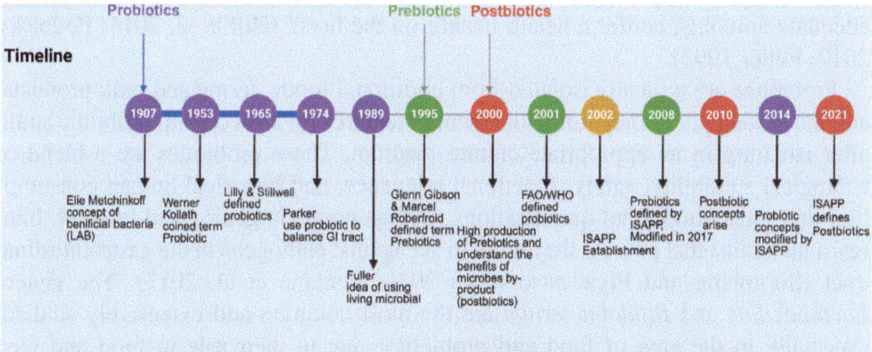

Chronological overview timeline of probiotics, prebiotics and postbiotics

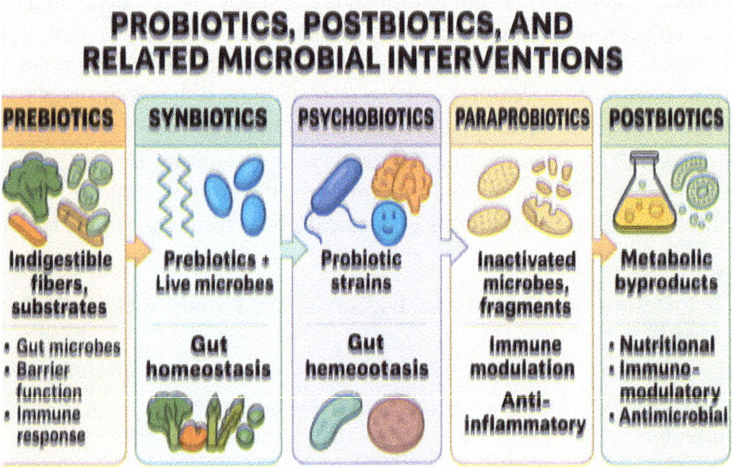

Fig. 6.2 Probiotics, postbiotics, and related microbial interventions

6.2 Probiotics

The theory that bacteria in the gut might affect health goes back as far as 1908, when Laureate Élie Metchnikoff put forth the theory that ingestion of lactic acid-producing bacteria would enhance health and increase longevity by displacing disease-promoting microbes with healthy ones (Anukam and Reid 2007). Four decades later, in 1965, Lilly and Stillwell established the term "probiotic," derived from the Greek "pro" (for) and "bios" (life), to refer to microbial-derived substances that encourage the growth of other microorganisms (Caramia and Silvi 2011). The International Scientific Association for Probiotics and Prebiotics (ISAPP), expanding upon previous Food and Agriculture Organization (FAO)/WHO definitions, defines probiotics as "live microorganisms that, when administered in adequate amounts, confer a health benefit on the host" (Hill et al. 2014; Podolsky 2012; Fuller 1992).

Probiotics are typically isolated from traditional foods, fermented milk products, and the human gut as they fulfill all essential features for a successful probiotic strain after isolating in an appropriate culture medium. These probiotics are a blend of ecological suitability, safety, functional relevance, and historical human consumption that meets important qualifications such as nonpathogenic, acid-tolerant, bile-resistant strains that possess the ability to act against pathogens in the gastrointestinal tract (Sornplang and Piyadeatsoontorn 2016; Fontana et al. 2013). The genera *Lactobacillus* and *Bifidobacterium* are the most common and extensively studied, especially in the area of food and probiotics, due to their role in food and feed development and preservation, and also to the probiotic properties are exhibited by some important strains like *Bacillus* species, *Escherichia coli*, and the yeast *Saccharomyces boulardii* (Lukjancenko et al. 2012). These microbes are commonly Gram-positive, nonpathogenic, and can withstand gastrointestinal conditions, which is vital for their efficacy. They inhibit harmful bacteria from occupying adhesion sites and competing for nutrients, and secrete antimicrobial substances such as bacteriocins, organic acids, and hydrogen peroxide; they fortify the gut barrier through the augmentation of tight junction proteins and enhancing mucosal immunity, probiotics may influence both the local and systemic immune response. Some probiotic strains may have beneficial effects on psychological health by acting on the neuroendocrine system and even producing neuroactive substances like gamma-aminobutyric acid (GABA). It has also shown to improve gastrointestinal health in many areas and frequently used as therapeutics for conditions like inflammatory bowel disease (IBD), constipation, antibiotic-associated diarrhea, and irritable bowel syndrome (IBS). In immunomodulation, probiotics can prevent respiratory and gut infections, treat allergies and metabolic health; there is also interest in their possible role in obesity, insulin sensitivity, and lipid metabolism. Their control over the gut-brain axis makes them possible adjunctive treatments for depression and anxiety.

In ancient times, probiotics were consumed by humans through fermented foods in antiquity, and this practice is still documented in human history. Food fermentation of some kind has been evident in almost every civilization, with the Greeks and Romans known to have fermented fresh fruit juice by using starter cultures to make wine and the peoples of China, Korea, and Japan have all relied significantly on fermented vegetables. Hindus have been consuming fermented milk products from prehistoric times, according to India's Vedic hymns. Numerous civilizations, including the Greeks, Egyptians, and Romans, have left numerous records indicating the fermentation of milk for cheese and butter (Kiple and Ornelas 2000; Gogineni et al. 2013). Nowadays, with the commercialization of probiotic, they are found in diverse forms such as capsules, powders, and as supplements which contain defined strains and amounts. Yakult, fermented milk, developed by Shirota in 1935 in Japan is frequently credited as the first commercially sold probiotic. The term "Yakult" originated from the Esperanto word "yahurto," meaning yogurt (Sugahara and Hirota 2021; Levin 2017). This approach maximizes safety, stability, and functionality over the traditional methods. Through fermentation, these microorganisms synthesize bioactive compounds such as SCFAs, exopolysaccharides, vitamin B-complex, and bioactive peptides, most of which contain immunomodulatory or antimicrobial activities. However, several strains are not able to withstand the acidic environment of the stomach (pH), bile salts, and enzymatic digestion after oral consumption, and the host immune mechanisms and competition with the prevailing gut microbes may also hinder in colonizing the gut. Therefore, appropriate strain selection is required, and dosages must be sufficient, products must be viable, and clinical evidence supporting health claims must be strong for effective probiotic use (Table 6.1).

Table 6.1 Probiotic sources and their functions

Probiotics source	Probiotics	Functions
Kimchi, Kefir, Yogurt, and Sauerkraut	*Lactobacillus reuteri*	Treat inflammation by reducing content of cholestasis-associated microbiota
Primarily found in supplement forms, initially identified from the feces of newborns fed with breast milk	*Bifidobacterium lactis* IDCC 4301	Suppresses cell differentiation lipid deposition
Lesaffre's baker's yeast strain collection	*Saccharomyces cerevisiae 1*-3856	Regulating intestinal microbiome and increase SCFAs
Soil and fermented foods like Kimchi and Natto	*Bacillus subtilis*	Improve the composition of intestinal microbes
Soil, vegetables, and soured milk or cheeses *C. butyricum* is also found in the human and animal gut	*Clostridium butyricum*	Alteration of gut microbiota
Originally isolated from the feces of human infants fed with breastmilk	*Bifidobacterium breve CCFM1025*	Modifies gut microbial composition and tryptophan pathways Alleviates depression and associated digestive disturbances
Isolated from the gut of a healthy adult human	*Lacticaseibacillus rhamnosus GG*	Helps prevent antibiotic-associated diarrhea children and adult populations
Extracted from a World War I soldier's feces who showed remarkable resistance to pandemic Shigella diarrhea	*Escherichia coli Nissle 1917 (EcN)*	Therapeutic approaches for inflammatory bowel disease, including ulcerative colitis
Isolated from tropical fruits like lychee and mangosteen and is naturally present in several fermented meals	*Saccharomyces boulardii*	Eradication therapies of *Helicobacter pylori* infection
Originally isolated from dairy culture collection	*Bifidobacterium lactis BB-12*	Immune modulation by Stimulating IgA production

6.2.1 Probiotics in Therapy

6.2.1.1 Mechanisms of Action

By altering the gut environment, modifying immune responses and offering defense against infections probiotics promote host health through a number of complex mechanisms (Fig. 6.3; Table 6.2). The main ways that probiotics works are as follows:

1. Pathogen competitive exclusion: in which advantageous bacteria outcompete harmful organisms for nutrients and binding sites on the intestinal epithelium. It is one of the main ways that probiotics shield the host. This makes it difficult for pathogens to colonize or spread. *Salmonella, Escherichia coli,* and *Clostridium* difficile cannot adhere to mucosal surfaces because probiotics like *Lactobacillus*

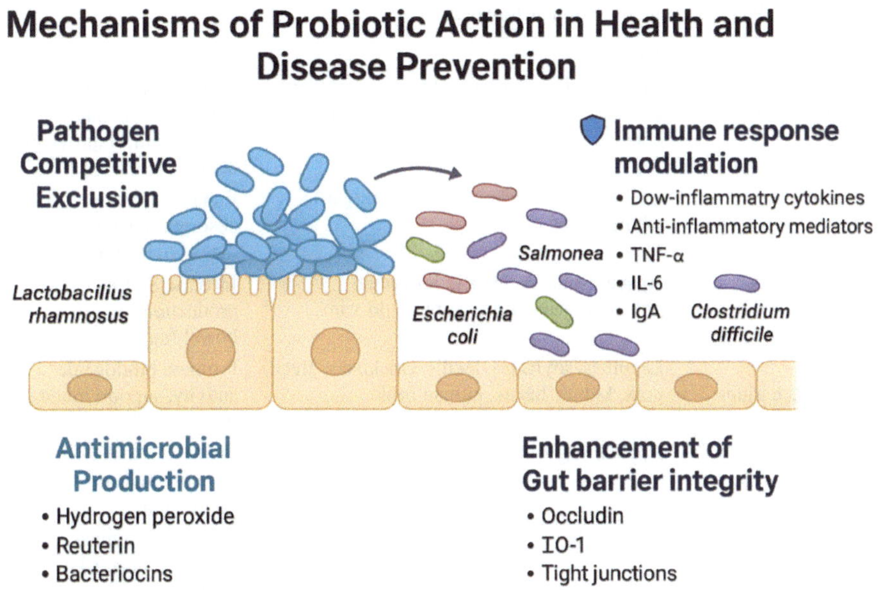

Fig. 6.3 Mechanisms of action of probiotics in host health

rhamnosus and *Bifidobacterium longum* adhere to them and form biofilms that serve as a physical barrier (Lebeer et al. 2010). Furthermore, by generating lactic acid and other organic acids some probiotic strains can effectively change the pH in the area, preventing the growth of pathogens that are sensitive to pH. In the intestinal and vaginal mucosa, where microbial balance depends on preserving a slightly acidic environment, this mechanism is particularly crucial (Porcari et al. 2025).

2. Immune response modification: the immunomodulatory effects of probiotics are well established. They affect both innate and adaptive immunity through their interactions with epithelial cells and antigen-presenting cells including dendritic cells and macrophages. They can effectively suppress proinflammatory mediators like TNF-α, IL-6 and IL-1β while increasing the synthesis of anti-inflammatory cytokines like transforming growth factor-β (TGF-β) and IL-10 (Kwon et al. 2010). In addition, probiotics can effectively enhance regulatory T cell (Treg) maturation and activation which fosters immune tolerance particularly in autoimmune and allergic diseases (O'Mahony et al. 2005). There are also some strains that elevate the production of pattern recognition receptors (PRRs) including IgA, natural killer (NK) cells, and Toll-like receptors (TLRs) which enables the immune system to react to threats more efficiently without causing undue inflammation (Bron et al. 2017).
3. Improvement of gut barrier integrity: toxins and dangerous bacteria cannot pass through the intestinal epithelial barrier. Probiotics strengthen the gut barrier in a number of ways including increased expression of tight junction proteins like

Table 6.2 Prebiotic sources and their functions

Pre-biotics	Sources	Functions
Fructo-oligosaccharides (FOS)	Onion, leek, asparagus, chicory, Jerusalem artichoke, garlic, oat, wheat, barley, rye, honey	Stimulate growth of Bifidobacteria and Lactobacilli, improve gut health
Fructans	Onion, banana, wheat, artichokes, garlic, whole foods	Enhance mineral absorption, support digestive health
Inulin	Agave, banana, chicory, Jerusalem artichoke, dandelion, onion, wild yam	Increase SCFA production, improve bowel function
Galacto-oligosaccharides (GOS)	Legumes, soybeans, lentils, chickpeas, green peas, kidney beans, human milk	Promote bifidogenic activity, support infant gut flora
Pectins	Apples, sugar beet pulp	Increase viscosity in gut, promote SCFA production
β-Glucans	Oats, barley	Modulate immune system, reduce LDL cholesterol
Iso-malto oligosaccharides	Miso, soy sauce, sake, honey	Support beneficial microbes, improve digestion
Xylo-oligosaccharides (XOS)	Bamboo shoots, fruits, vegetables, milk, honey	Enhance gut flora, exhibit anti-inflammatory activity
Arabinoxylan oligosaccharides (AXOS)	Cereals	Improve colonic fermentation, reduce intestinal inflammation

occludin, claudins, and zonula occludens-1 (ZO-1). For example, *Lactobacillus plantarum* and *Bifidobacterium infantis* have been shown to preserve tight junction organization and epithelial integrity in models of gut damage brought on by inflammation and stress (Anderson et al. 2010). A more intact epithelial lining leads to better nutrient absorption, less systemic inflammation, and decreased endotoxemia.

4. Production of antimicrobial compounds: probiotic microorganisms produce a variety of antimicrobial compounds that directly inhibit harmful bacteria. These include: acids like lactic acid and acetic acids that decrease pH, hydrogen peroxide that degrades microorganisms' cell walls, small antimicrobial peptides with a variety of functions called bacteriocins. Reuterin, a potent antimicrobial compound that inhibits *Salmonella*, *Listeria,* and *Clostridium* species, is produced by *Lactobacillus reuteri*. These chemicals function by weakening the bacterial membrane, disrupting the cell wall's synthesis or stopping the pathogens production of proteins and DNA (Negash and Tsehai 2020). Probiotics can effectively prevent infections without making people resistant to antibiotics

because they have endogenous antimicrobial activity. Because of these complex mechanisms which include immune signaling, ecological competition, epithelial repair, and biochemical inhibition probiotics are believed to be beneficial in both therapeutic and preventive settings. The fact that these effects differ depending on the strain emphasizes how critically important it is to carefully choose probiotic strains before utilizing them in clinical settings.

Figure 6.3 illustrates various ways through which probiotics promote health: (1) Competitive exclusion of pathogens by means of acid production that inhibits pathogen growth, competition for nutrients and colonization of epithelial surfaces. (2) Immune modulation that stimulates regulatory T cells and anti-inflammatory responses while suppressing proinflammatory cytokines (3) Enhancement of gut barrier integrity by upregulating the tight junction protein which decreases permeability and inflammation and also strengthens the gut barrier integrity (4) Antimicrobial production like lactic acid, hydrogen peroxide, bacteriocins, and reuterin that directly inhibit harmful microorganisms.

6.2.1.2 Probiotics Therapeutic Applications

Probiotics are becoming more widely acknowledged as helpful dietary supplements for the treatment of a number of illnesses. They have the ability to prevent, support, and even treat gastrointestinal, metabolic, neurological, allergic, and infectious diseases due to their strain-specific and disease-targeted activities (Fig. 6.4).

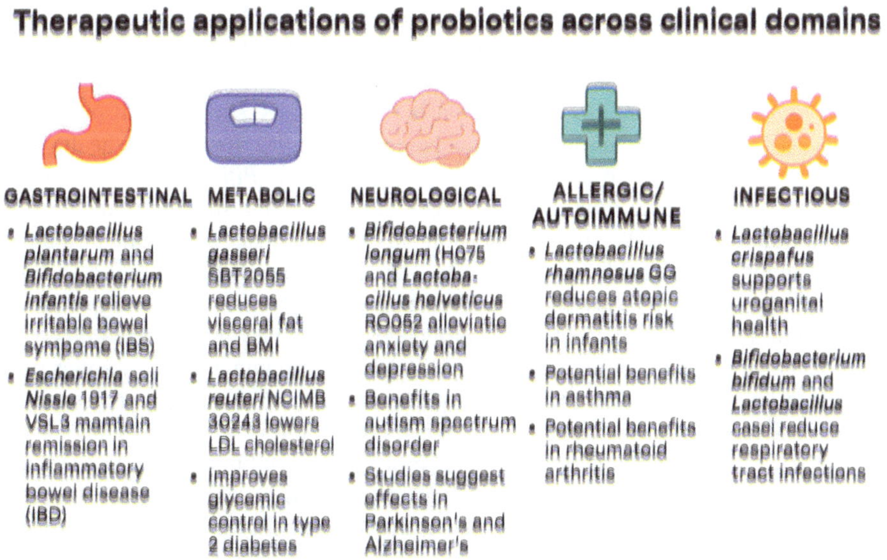

Fig. 6.4 Therapeutic applications of probiotics

1. Intestinal conditions: The majority of probiotic research has focused on gastrointestinal health. They aid in the restoration of microbial balance, the maintenance of mucosal integrity, and the control of the local immune response. Numerous studies have demonstrated that probiotic strains such as *Lactobacillus plantarum* and *Bifidobacterium infantis* can effectively facilitate patients with irritable bowel syndrome (IBS) with their abdominal pain, irregular bowel habits, and bloating (O'Mahony et al. 2005; Lyra et al. 2010). In patients with inflammatory bowel disease or IBD, probiotics like *Escherichia coli* Nissle 1917 and the multistrain formulation VSL#3 have shown promise in maintaining remission and reducing inflammation in Crohn's disease and ulcerative colitis (Miele et al. 2009; Jadhav et al. 2023). Diarrhea: both infectious as well as antibiotic-associated diarrhea can effectively be treated with probiotics. Two of the most researched strains *Saccharomyces boulardii* and *Lactobacillus rhamnosus* GG show a decrease in the severity and duration of symptoms (Szajewska and Kolodziej 2015). Constipation: it has been demonstrated that strains of *Bifidobacterium* lactis elevate the frequency and consistency of stools (O'Callaghan and Van Sinderen 2016).
2. Metabolic disorders: Probiotics are a helpful tool for managing metabolic syndrome because they can effectively change body weight, lipid profiles, and glucose metabolism. For ex, (a) obesity: probiotics may regulate energy harvest, reduce fat accumulation, and control adipokine levels. Clinical trials showed that *Lactobacillus gasseri* SBT2055 reduced visceral fat and Body Mass Index (BMI) (Kadooka et al. 2010). Probiotics aid in better glycemic control for individuals with type 2 diabetes mellitus (T2DM) by enhancing insulin sensitivity and reducing systemic inflammation. A meta-analysis found that probiotic supplementation significantly reduced fasting blood glucose and HbA1c in patients with type 2 diabetes (Peng et al. 2024) (b). Fat metabolism: it has been found that certain strains, such as *Lactobacillus reuteri* NCIMB 30242, lower total and LDL cholesterol by means of bile salt deconjugation and SCFA synthesis (Jones et al. 2012).
3. Nervous system conditions: The gut-brain axis has shown that probiotics (Psychobiotics) have intriguing new potential benefits for mental health and cognitive function such as (a) depression and anxiety: *Bifidobacterium longum* R0175 and *Lactobacillus helveticus* R0052 have been shown to reduce the symptoms of anxiety and depression in both human and animal studies (Messaoudi et al. 2011). (b) Cognitive function and autism: probiotic therapy has shown promise in lowering gastrointestinal symptoms and altering behavior in children with autism spectrum disorders despite the fact that the research is still in its early phases (Tomova et al. 2015). (c) Neuroinflammation: probiotics may reduce levels of proinflammatory cytokines and oxidative stress markers which are connected to neurodegenerative diseases like Parkinson's and Alzheimer's.
4. Allergic and autoimmune diseases: Probiotics are helpful in preventing or reducing allergic and autoimmune reactions because they can effectively affect immune tolerance such as (a) eczema and atopic dermatitis: prenatal and postnatal administration of *Lactobacillus rhamnosus* GG considerably decreased the incidence of atopic dermatitis in high-risk infants (Kalliomäki et al. 2001).

(b) Asthma: although human data are conflicting certain probiotic strains have been demonstrated to modulate IgE levels and lessen airway inflammation in murine models (Song et al. 2024). Large-scale clinical trials are still required but preliminary research indicates that probiotics may be able to modify gut microbiota and systemic inflammation in RA (Vaghef-Mehrabany et al. 2014).

5. Infections: By increasing mucosal immunity and generating antimicrobial compounds probiotics can effectively prevent and treat a variety of infections like. (1) Urogenital health: probiotics like *Lactobacillus crispatus* preserve vaginal eubiosis and guard against recurrent Urinary Tract Infection (UTIs) (Reid 2008). (2) Respiratory infections: Oral administration of *Bifidobacterium bifidum* and *Lactobacillus casei* has been linked to a decreased incidence and duration of respiratory tract infections and common colds (Zhao et al. 2022). (3) Oral health: *Streptococcus salivarius* K12-containing probiotic lozenges inhibit bacteria that cause gingivitis and dental caries and lessen halitosis (Burton et al. 2006). Probiotics are now acknowledged as disease-modifying agents in a number of clinical domains in addition to general health. Because of their strain-specific and condition-dependent effects, therapeutic efficacy depends on careful selection, dosage, and delivery. The potential for incorporating probiotics into standard medical practice seems to be growing as the body of evidence keeps expanding.

Figure 6.4 illustrates therapeutic applications of probiotics. Probiotics have effects in a variety of clinical domains that are strain specific and condition targeted. In gastrointestinal disorders, strains like *Bifidobacterium lactis* enhance constipation while *Lactobacillus plantarum*, *Bifidobacterium infantis* and *E. coli* Nissle 1917 reduce infectious or antibiotic-associated diarrhea, support IBD remission, and alleviate IBS symptoms. Probiotics such as *Lactobacillus gasseri* and *Lactobacillus reuteri* influence glycemic control, lipid metabolism, and body weight in individuals with type 2 diabetes and obesity. Psychobiotics (*B. longum* and *L. helveticus*) have been shown to enhance neurodegeneration and autism while lowering anxiety and depression. Probiotics may also modulate autoimmunity and affect immune tolerance, reducing the risk of asthma, eczema, and atopic dermatitis. Moreover, they strengthen the host's defenses against infections, for example, *Streptococcus salivarius* K12 promotes oral health, *Bifidobacterium bifidum* lowers respiratory infections, and *Lactobacillus crispatus* supports urogenital health. Probiotics as a whole are acknowledged as possible disease-modifying substances with growing therapeutic applications.

6.2.1.3 Delivery Systems and Formulations

Probiotics' therapeutic efficacy depends on their efficient delivery to the site of action which is usually the gastrointestinal tract. Probiotic formulation and delivery techniques need to be carefully considered because of their sensitivity to environmental factors like temperature, oxygen, gastric acid, and bile salts. The most widely used forms in clinical and commercial settings are discussed in this section along with developments in encapsulation technologies that enhance stability, targeted release, and viability (Fig. 6.5).

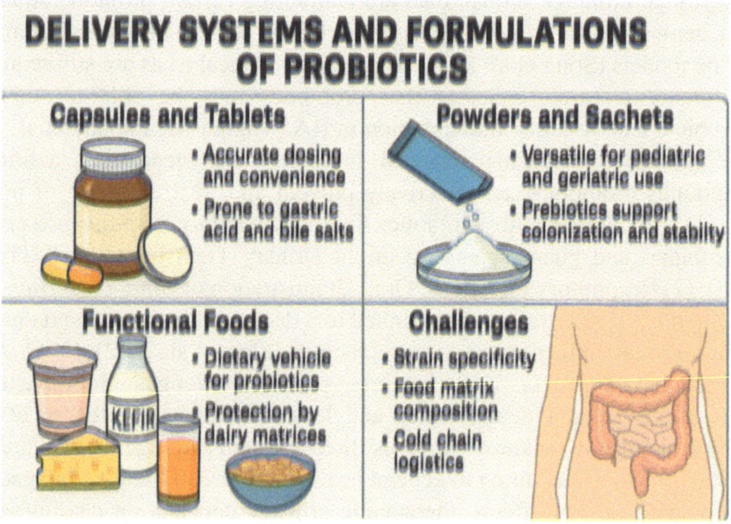

Fig. 6.5 Delivery systems and formulations of probiotics

1. Conventional formulations, functional foods, powders, capsules, and tablets: because of their ease of use, accuracy in dosing and convenience probiotics are frequently taken as tablets or capsules. Usually made with excipients that guard against pH and moisture, they contain bacteria that have been freeze dried. However, exposure to bile salts and stomach acid can effectively still result in viability loss (Saarela et al. 2000).
2. Powders and sachets: powdered formulations can effectively be combined with liquids or foods and are very versatile for use in both pediatric and geriatric settings. These are frequently lyophilized (freeze dried) probiotics that may contain prebiotics (fructo-oligosaccharides and inulin) to enhance gut colonization and shelf stability (Champagne et al. 2018).
3. Functional foods: a convenient and user-friendly way to add probiotics to daily diets is through functional foods such as probiotic-enriched yogurt, kefir, cheese, ice creams, fruit juices and cereals. In order to keep probiotic strains safe during digestion and storage the food matrix is essential. Particularly, dairy-based matrices offer superior protection against stomach acid (Tripathi and Giri 2014). However, survival in functional foods depends on the matrix and strain and cold chain logistics are frequently needed. As a result, functional foods might not be appropriate in every socioeconomic or geographic setting.

Figure 6.5 illustrates probiotic formulation techniques and delivery system. Probiotics are made to be more active and survive longer while being stored and moving through the digestive system. Tablets and capsules are examples of traditional systems that provide accurate dosage but may lose their effectiveness when exposed to stomach acid and bile salts. Both pediatric and geriatric settings can effectively benefit from the use of powders and sachets which are often lyophilized and combined with prebiotics. Functional foods (yogurt, kefir, cheese, juices, and cereals) offer a dietary path for daily probiotic intake while dairy matrices offer additional protection. These delivery methods strike a balance between strain-specific survival, stability, and convenience.

6.2.1.4 Technologies for Targeted Delivery Through Encapsulation

Encapsulation enhances the ability of probiotic cells to withstand processing, storage, and passage through the gastrointestinal tract by encasing them in a protective matrix or coating. This is especially helpful for guaranteeing that the product is delivered to the intestine in a viable and active state (Fig. 6.6).

1. The process of microencapsulation: At the microscale (1–1000 μm), probiotics are encapsulated in lipid based or polymeric materials. Common ingredients include pectin, gelatin, chitosan, and alginate. These substances create semipermeable membranes that shield the microorganisms from oxygen, bile, and stomach acid while permitting the free passage of nutrients (Cook et al. 2012). According to studies, the survival rates of *Bifidobacterium bifidum* and *Lactobacillus acidophilus* in capsule form are noticeably higher than those of nonencapsulated forms (Gbassi et al. 2009).
2. Delivery using nanoparticles: A cutting-edge technique for precisely delivering probiotics is nanotechnology. Probiotics can effectively be added to biodegradable nanoparticles (lipid nanoparticles or Poly(lactic-co-glycolic acid)-PLGA) via nanoencapsulation which provides targeted release, improved bioavailability, and extended shelf life. This method may be used to treat diseases like inflammatory bowel disease and colon cancer even though it is still in the experimental stage.
3. Coating of the enteric system: Enteric-coated capsules allow for targeted colonic delivery because they are made to resist stomach acid and break down in the more alkaline environment of the intestine. Methacrylic acid polymers (MAP), cellulose acetate phthalate (CAP), and hydroxypropyl methylcellulose phthalate (HPMCP) are examples of coating agents (Kailasapathy 2002).
4. Spray drying and freeze drying: These are standard methods for preserving probiotics. Despite providing superior survival, freeze drying (lyophilization) is

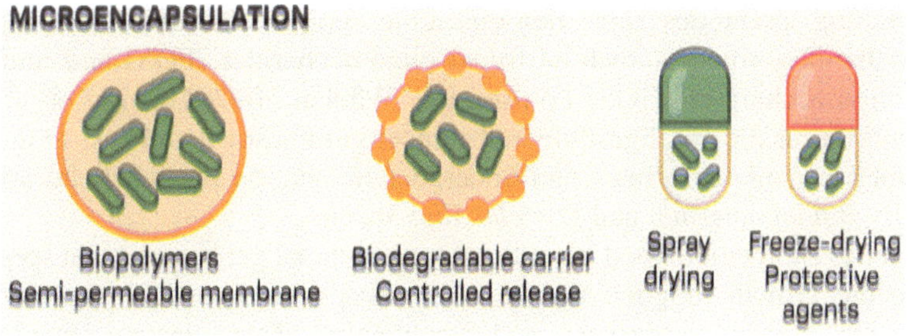

Fig. 6.6 Encapsulation technologies for targeted probiotic delivery

costly and energy intensive. In contrast, spray drying is less expensive but may cause thermal stress to cells. During drying, protective agents like maltodextrin, trehalose, and skim milk powder are frequently used (Afzaal et al. 2020). The formulation strategy has a significant impact on the efficacy of probiotic therapy. Although the market is dominated by traditional forms like capsules and functional foods, current encapsulation technologies are making it more and more possible to customize delivery systems for particular clinical outcomes and health targets. Personalized delivery methods that optimize therapeutic effect while reducing microbial loss are probably going to be central to the future of probiotic therapy.

Figure 6.6 illustrates encapsulation technologies for targeted probiotic delivery: by protecting cells from adverse conditions, encapsulation increases probiotic survival during storage and gastrointestinal transit. Techniques include enteric-coated capsules that release probiotics in the intestine and resist gastric acid, lipid or PLGA-based nanoparticle systems that enhance bioavailability and enable targeted release and microencapsulation with alginate, chitosan or gelatin to create protective membranes. Stabilization methods like freeze-drying and spray-drying are also frequently used and viability is increased by protective agents like maltodextrin and trehalose. More accurate, strain-specific therapeutic applications are made possible by these technologies.

6.3 Prebiotics

The term prebiotics was originally coined in 1995 by Marcel Roberfroid and Glenn Gibson, who defined them as nonviable food ingredients that provide a health benefit to the host by modifying the microbiota. This initial definition highlighted the functional nexus between diet and the gut microbiota, which has been the mainstay of contemporary nutrition science. In 2017, International Scientific Association for Probiotics and Prebiotics (ISAPP) termed prebiotics as a substrate that host microorganisms selectively use to provide health benefits. In addition to carbohydrates, this broader understanding includes noncarbohydrate compounds like polyphenols, polyunsaturated fatty acids, and particular bioactive peptides that can be transformed by gut microbes into beneficial metabolites. Most classic examples of prebiotics are non-digestible polysaccharides, but especially oligosaccharides, fructans such as fructo oligosaccharides (FOS) and inulin, and galactooligosaccharides (GOS) (Wilson and Whelan 2017; Guarino et al. 2020). These substances are not digestible in the upper gut intestinal and pass to the colon, where they become substrates and undergo fermentation by beneficial microbes, especially *Bifidobacterium* and *Lactobacillus* species.

Prebiotics are inedible food ingredients for beneficial gut bacteria that specifically enhance the growth of good microbes to reshape the intestinal microbiota and increase its diversity, robustness, and functionality. This selective stimulating of good bacteria results in a healthier microbial environment and exerts widespread beneficial effects on the body. Prebiotic are having extensive health benefits, such as induce microbial diversity and metabolic activity within the gut that results in numerous physiological benefits. These are improved bowel and digestion function like better bowel motility and improved digestion, pathogen inhibition by reduced growth of pathogenic microbes and immunity strengthening by increased mucosal immunity further has positive neurophysiological effects which connects with better sleep quality and act on the gut-brain axis affecting mood, stress, and cognition (Nasiri et al. 2025; Varanoske et al. 2022).

Probiotic like short-chain fatty acids (SCFAs) has a number of mechanisms by which most of benefits occur is the generation of SCFAs such as acetate, propionate, and butyrate by gut microbiota. These SCFAs are critical mediators of host-microbe interactions, providing anti-inflammatory effects, immunomodulatory functions, regulation of energy and subdue infectious bacteria and vitalize the flourishment of beneficial bacteria, butyrate act as a primary potential source for colonic epithelial cells, maintaining integrity of the intestinal barrier and conferring potent anti-inflammatory effects.

Outside of the gut, prebiotics make a substantial contribution to cardio metabolic well-being through enhanced insulin sensitivity, better glucose metabolism, and controlled lipid profiles, additionally plays a part in immune modulation via activation of gut-associated lymphoid tissue (GALT) and induction of cytokine and chemokine production also enhanced absorption of minerals such as calcium and magnesium, aiding bone density, potentially decreasing systemic toxin burden in chronic kidney disease and promoting skin hydration and exfoliation. In addition, prebiotics are currently being researched as treatment options for metabolic disorders like nonalcoholic fatty liver disease (NAFLD), type I and type II diabetes, and obesity (Carpi et al. 2022; Loman et al. 2018). The multiverse of prebiotic activity including microbiota modulation, immune modulation, and metabolic adjustment makes them an integral part of preventive nutrition and a valuable adjuvant clinical therapy. By demonstrating how the diet can optimize host-microbiome relationships and enhance health, they serve as a platform for interaction between nutrition, microbiology, and medicine. Their ability to specifically target desired microbial populations and trigger the release of advantageous metabolites is evidence of their potency.

6.4 Postbiotics

The International Scientific Association for Probiotics and Prebiotics (ISAPP) in 2021 officially defined Postbiotics as the product of dead microbes and their fractions with demonstrated or claimed host benefits. This was a groundbreaking shift in microbiome science, moving away from the exclusive use of live microbes in therapeutic approaches. Postbiotics is a broad category of microbial-derived compounds, like cell wall apparatuses, bioactive proteins, peptides, and some other metabolites, non-viable microbial cells, SCFAs, polyamines, and bacteriocins. Examples of which include Para probiotics which are referred to as “ghost probiotics," being inactive bacterial cells that are beneficial for health, and fermented infant formulas, being fermented with probiotic bacteria but often do not contain live cells at the time of consumption. One of the primary advantages of postbiotics is that they are more stable. In contrast to live probiotics, they are far less vulnerable to processing conditions, temperature, and pH variation, gastric acidity. This stability is advantageous for inclusion in food, medicines, and animal health products. Their longer shelf-life and lower risk of microbial translocation or infection also render them a safer alternative, especially for the vulnerable groups.

Postbiotics exert their effects through several mechanisms, usually comparable to those of probiotics and prebiotics. They assist in altering the gut microbiota composition, inhibiting pathogenic bacteria, stimulating beneficial microbial growth, supporting gut barrier function, facilitating nutrient uptake, and they also have anti-inflammatory, antioxidant, antiproliferative, and immunomodulatory properties. Postbiotics like SCFAs, tryptophan metabolites, and bacteriocins are crucial for immune modulation, antimicrobial qualities, and cell signaling and also important in physiological regulation. There are two types of postbiotic health benefits: direct effects and indirect effects. Inducing digestive enzymes and enhancing mucosal immunity are two examples of the direct effects that postbiotic substances have on the gut tissues of the host while indirect effect consists of stimulation of beneficial bacteria and inhibition of pathogenic bacteria, both ways complement each other to maintain a healthy gut ecosystem.

In addition, postbiotics have significant potential for use in animal livestock and veterinary medicine. They offer the potential to optimize feed efficiency, gut health, and minimize the utilization of antibiotics in beef and poultry operations. As a nonviable particle, they circumvent the challenge associated with keeping the microbes viable, but provide functional bioactivity that improves nutrient absorption, physiological function, immune defense mechanism by the host. Whereas postbiotics possess many advantages, their application is contingent upon a number of factors, including host-specific responses, bioactive moiety standardization, and production technologies used. They are relatively more stable than live microorganisms and can be produced in large quantities. However, regulatory clearance and consumer acceptance remain contingent upon strict quality assurance, precise molecular characterization, and strong clinical validation.

In summary, postbiotics represent a next-generation microbiome modulating tool. Their stability, safety, and functional flexibility combine to make them extremely appealing for a wide range of therapeutic and nutritional purposes. With continued research on their mechanisms, clinical effectiveness, and large-scale manufacture, postbiotics are set to be at the forefront of microbiome-based health solutions in the future (Table 6.3).

6.4.1 Postbiotic Composition

Postbiotic preparations contain a wealth of physiologically active substances that through a variety of mechanisms, support disease prevention or health maintenance.

Table 6.3 Postbiotic sources and their functions

Source/species	Postbiotics	Functions
Lactobacillus acidophilus	Peptidoglycan	Treatment of diarrhea Anti-inflammatory effect
Bifidobacterium BB12	Heat inactivated cells	Interference with the production of Streptococcus mutans biofilms
Bacillus pumilus SE5	Intermediate	Suppress growth of pathogenic bacteria
Clostridium butyricum sp	SCFAs	Modulate intestinal microbiota composition
Pediococcus acidilactici NCDC 252	Exopolysaccharides	Antioxidant and anticancer activity on human colon cancer cell line (HCT116)
Saccharomyces boulardii	Cell-free supernatants	Anti-inflammatory effect Antioxidant activity
Lactococcus sp.	Enzymes	Reducing intestinal inflammatory
Lactiplantibacillus plantarum	Bacteriocins	Antibacterial Antiobesity Helps manage diabetes
Bifidobacterium bifidum	Vitamins	Antioxidant Supports healthy endothelial function Helps lower blood sugar
Akkermansia muciniphila	Extracellular vesicles	Antiobesity Prevent inflammation
Lacticaseibacillus rhamnosus	Teichoic acids	Exhibits immunomodulatory activity along with antiobesity and anti-inflammatory effects

Critically important elements consist of (1) Short-chain fatty acids: the gut microbiota ferments dietary fibers to produce butyrate, propionate, and acetate. SCFAs are essential in: (a) changing the pH and microbiota balance of the gut, (b) providing colonocytes with energy especially butyrate and controlling immunological and inflammatory responses (Koh et al. 2016; Fukuda et al. 2011). (2) Bacteriocins and antimicrobial peptides: these tiny proteins provide protective effects by preventing the growth of harmful bacteria and fungi. For instance, *Lactobacillus reuteri* produces reuterin which has potent antibacterial qualities (Gänzle 2019). (3) Enzymes: especially for those who are lactose intolerant enzymes such as β-galactosidase (lactase) enhance digestion. Other enzymes facilitate the body absorb nutrients or break down toxins. (4) Cell wall components: immunoactive structures called lipoteichoic acids (LTAs), peptidoglycans, and teichoic acids interact with pattern recognition receptors (PRRs) like TLR2 and NOD2 to modify both innate and adaptive immune responses (Molina et al. 2019). (5) Organic acids: by lowering intestinal pH, lactic acid, formic acid, and succinic acid, they inhibit the growth of pathogenic microorganisms and promote the colonization of advantageous strains. (6) Neurotransmitters and vitamins: postbiotics support neurological health by containing vitamin B group compounds and even neuroactive molecules like GABA and serotonin precursors (Aguilar-Toalá et al.

2018). (7) Bacterial extracellular vesicles (EVs) have the ability to modify host immunity, signaling, and gene expression by delivering useful molecules to host cells such as proteins, lipids, and RNAs (Villageliu and Samuelson 2022).

6.4.2 Mechanisms of Action

Postbiotics interact with the host through intricate molecular and cellular processes to produce a variety of bioactivities. They work without colonization or viability in contrast to probiotics. Signaling molecules, microbial cell components, and bioactive metabolites are the main mediators of their actions.

1. Inhibition of inflammation: postbiotics control inflammation by increasing anti-inflammatory mediators like IL-10 and decreasing proinflammatory cytokines like TNF-α, IL-6, and IL-1β. Lactobacillus and Bifidobacterium cell wall fragments such as peptidoglycans and lipoteichoic acids can effectively mediate these effects by interacting with host cell Toll-like receptors (Tsilingiri and Rescigno 2013). For instance, it has been demonstrated that heat-killed Lactobacillus casei reduces inflammation in colitis models by preventing NF-κB activation.
2. Properties of antioxidants: enzymatic antioxidants (catalase and superoxide dismutase), SCFAs, and phenolic compounds that scavenge free radicals and lessen oxidative stress are found in many postbiotic preparations. According to Aguilar-Toalá et al. (2018), these characteristics facilitate to protect cells and lower the risk of developing chronic illnesses like cancer and cardiovascular diseases. For example, glutathione peroxidase and other cellular antioxidant defenses are activated by butyrate (Wang et al. 2012).
3. Alteration of immune responses in the host: both innate and adaptive immunity are impacted by postbiotics. Using propionate and butyrate to stimulate regulatory T cells (Tregs), improving mucosal IgA secretion and controlling the activity of macrophages and dendritic cells. These outcomes strengthen pathogen defense, lower autoimmunity, and enhance immune surveillance. It has been demonstrated that EVs derived from Lactobacillus stimulate Treg induction and anti-inflammatory macrophage polarization (Molina et al. 2019).
4. Receptor-mediated signaling pathways: postbiotic elements engage in interactions with host receptors including TLRs, i.e., initiating cascades of immune signaling, GPR41 and GPR43: SCFA receptors that control metabolism and the immune system, microbial tryptophan metabolites activate the aryl hydrocarbon receptor (AhR) which modifies mucosal homeostasis. In the gut and surrounding tissues, these interactions support immunological tolerance, homeostasis, and epithelial integrity (Koh et al. 2016).

Figure 6.7 illustrates the use of postbiotics in therapy. Postbiotics are nonviable microbial cells, their structural components or their metabolic byproducts that promote host health. Cell-free supernatants, inactivated bacteria (heat-killed

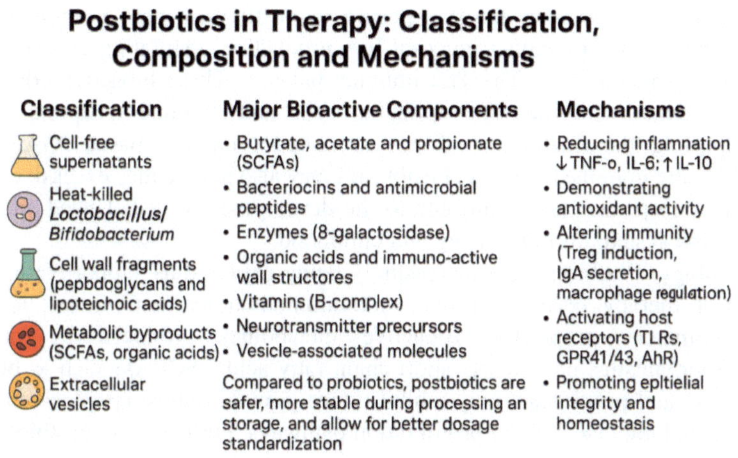

Fig. 6.7 Postbiotics in therapy: classification, composition, and mechanisms

Lactobacillus/Bifidobacterium), cell wall fragments (peptidoglycans and lipoteichoic acids), metabolic byproducts (SCFAs and organic acids) and extracellular vesicles are among them. Substantial bioactive components include butyrate, acetate and propionate (SCFAs), bacteriocins, antimicrobial peptides, enzymes (β-galactosidase), organic acids, immuno-active wall structures, vitamins (B-complex), neurotransmitter precursors and vesicle-associated molecules. Compared to probiotics, postbiotics are safer, more stable during processing and storage and allow for better dosage standardization. By reducing inflammation (↓ TNF-α, IL-6; ↑ IL-10), demonstrating antioxidant activity, altering immunity (Treg induction, IgA secretion, macrophage regulation), and activating host receptors (TLRs, GPR41/43, AhR), they promote epithelial integrity and homeostasis.

6.4.3 Postbiotic's Therapeutic Uses

(1) Health of children: Postbiotics may contribute significantly to neonatal and pediatric health by modulating gut microbiota, reducing intestinal inflammation, strengthening epithelial barrier integrity, and enhancing host immune defenses. Several studies have suggested that postbiotics can lower the incidence and severity of necrotizing enterocolitis (NEC), a life-threatening gastrointestinal disorder commonly observed in preterm infants. Heat-killed strains of *Bifidobacterium bifidum* and *Lactobacillus acidophilus* are considered safer alternatives to live probiotics in vulnerable neonates because they minimize the possibility of bacterial translocation, systemic infection, and probiotic-associated sepsis while still retaining beneficial immunomodulatory effects. Additionally, postbiotics may improve nutrient absorption and support overall intestinal development during early life.

(2) Immune development: Early exposure to postbiotics may promote immune maturation by stimulating mucosal IgA production, regulating cytokine secretion, and maintaining Th1/Th2 immune balance. These bioactive compounds can interact with intestinal epithelial cells and immune receptors, thereby supporting the maturation of gut-associated lymphoid tissue (GALT) and enhancing immune tolerance. Postbiotics may also help reduce excessive inflammatory responses and contribute to the development of a balanced and stable immune system during infancy and childhood.

(3) Oncology: Postbiotics are increasingly being explored as supportive agents in cancer therapy because of their immunomodulatory, antioxidant, and anti-inflammatory properties. Bioactive metabolites produced by probiotic microorganisms, particularly short-chain fatty acids (SCFAs) such as butyrate, acetate, and propionate, may inhibit tumor growth, induce apoptosis in cancer cells, and regulate cellular proliferation pathways. Furthermore, postbiotics may enhance gut microbiota balance and improve the host immune response during chemotherapy or radiotherapy, thereby reducing treatment-associated inflammation and gastrointestinal complications. Although current findings are promising, additional clinical studies are required to establish their long-term therapeutic efficacy and safety in oncology. (Canani et al. 2011; Jobin 2018).

(4) Health of the heart: Postbiotics are helpful in cardiovascular conditions because they have anti-inflammatory, lipid-lowering and endothelial protective properties. By controlling hepatic lipid metabolism and encouraging bile acid excretion, butyrate and propionate reduce serum cholesterol. *Lactobacillus plantarum* cell-free extracts elevate nitric oxide availability and decrease TNF-α to reduce atherosclerotic lesions.

(5) Skin and oral health: Postbiotics promote the healing of the oral and cutaneous mucosa.

(6) Benefits for dermatology and wound healing: *Lactobacillus rhamnosus* lysates have been shown to decrease eczema symptoms, elevate keratinocyte proliferation, and decrease inflammation (Lee et al. 2025).

(7) Oral health: According to Haukioja (2010), postbiotic mouthwashes that contain bacteriocins or acidic metabolites inhibit *Streptococcus mutans* and lessen dental plaque and halitosis. A new class of biotherapeutics called postbiotics provides multifactorial bioactivity, stability, and safety without the dangers of live bacteria. Strong clinical potential is indicated by their anti-inflammatory, immunomodulatory, and antioxidant properties as well as their varied uses in cardiology, dermatology, oncology, and pediatrics. To confirm these results and direct regulatory approvals, standardized formulations and additional human trials are required. In order to choose safe, efficient, and context-appropriate interventions, it is becoming more and more crucial to compare probiotics or live microorganisms and postbiotics/nonviable microbes or their metabolites as microbial therapeutics research progresses. Although they both seek to enhance the host health, their features like safety, stability, regulatory status, and clinical suitability all differ greatly (Fig. 6.8).

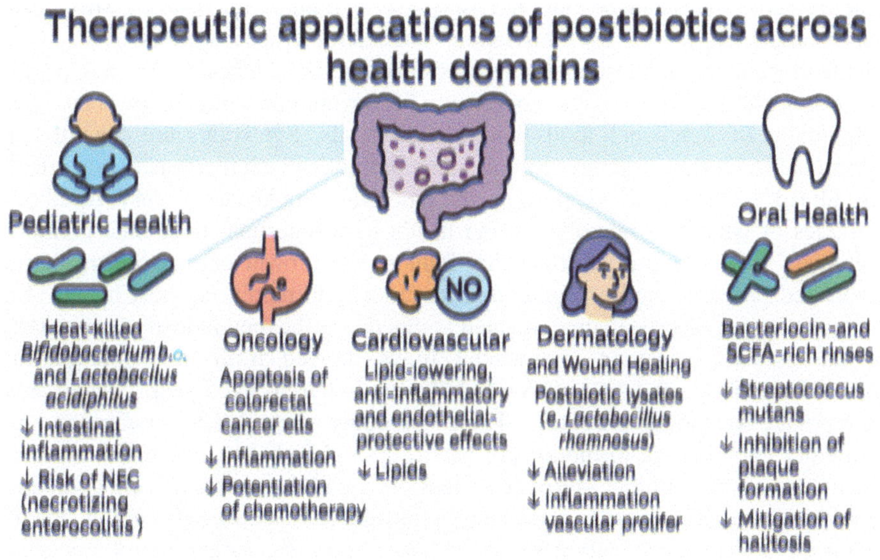

Fig. 6.8 Postbiotics and their therapeutic applications

Figure 6.8 illustrates postbiotics which are the nonviable components or bioactive metabolites of probiotics. They provide a number of health benefits without the dangers of live microorganisms. They balance immune development, enhance mucosal IgA responses, and lessen necrotizing enterocolitis (NEC) in pediatric care. Short-chain fatty acids (butyrate and propionate) have anticancer properties by inducing apoptosis in colorectal cancer cells and enhancing immunotherapy responses. The cardiovascular benefits include reduced cholesterol, better lipid metabolism, and vascular protection. Postbiotic lysates reduce halitosis and inhibit cariogenic bacteria like *Streptococcus mutans* in oral health; they also lessen inflammation and promote tissue repair in wound healing and dermatology. These findings all support the use of postbiotics as stable, safe, and effective treatments in the areas of pediatrics, cardiology, dermatology, oncology, and oral medicine.

6.5 Synbiotics

Synbiotics represent a synergistic strategy for microbiota modulation by combining probiotics and prebiotics in a manner that enhances the survival, colonization, and metabolic activity of beneficial microorganisms in the gastrointestinal tract. This combined approach improves the stability and functional efficacy of probiotic strains while simultaneously promoting a favorable gut microbial balance. Synbiotics have gained considerable attention in the development of functional foods and nutraceuticals because of their potential to improve digestive health, strengthen immune responses, enhance nutrient absorption, and reduce the risk of gastrointestinal disorders. By selectively stimulating beneficial microbes, synbiotics may contribute to maintaining intestinal homeostasis and overall host health. As more people become health conscious in the modern era, the development of synbiotics has been a highlight of functional foods and nutraceuticals. Synbiotics are regarded as nutritional foods that combine prebiotics and probiotics such that synergistic effects are already achievable, supporting the survival, implantation, and metabolic activities of beneficial microbes in the host's gastrointestinal tract. The prebiotic ingredient is a selective substrate that induces the growth or metabolism of the associated probiotic strain, augmenting its efficacy in modifying host health. The synbiotic strategy has been implemented in practice in the food industry, particularly in Europe, where certain commercial foods derived from milk like Aktifit, Switzerland; Proghurt, Austria; Vifit, UK and Belgium and Fysiq, Netherlands, incorporate synbiotic products. These products demonstrate prebiotics and probiotics combination to induce gut health through enhanced microbial viability and colonization. Synbiotics provide a superior overall therapeutic benefit compared to probiotics or prebiotics. By providing a nutritional substrate (e.g., inulin, fructo-oligosaccharides) specifically designed to foster health-enhancing microbes such as *Lactobacillus* and *Bifidobacterium*, synbiotics are capable of stimulating the numbers of these genera and restoring and maintaining intestinal microbial balance. Also, synbiotics are associated with detoxification of harmful metabolites such as nitrosamines, deactivation of carcinogens, and prevention of gastrointestinal diseases like diarrhea and constipation. Mechanistically, synbiotics augment ecological resistance of probiotic microorganisms.

The prebiotic substrates enhance the survival of probiotics against environmental stresses like gastric pH, bile salts, oxygen stress, and temperature variations, the normal barriers to microbial survival within the GI tract. This enhanced survival not only prolongs the shelf life of probiotic survival within the gut but also amplifies their functional effect on host physiology. Though synbiotics research continues, preliminary studies recognize potential therapeutic applications, especially in extensive inflammatory diseases like inflammatory bowel disease (IBD). Though limited clinical trials have reviewed synbiotics effectiveness in the treatment of IBD, the theoretical basis is a suitable foundation for research to be ongoing. The synergistic action of immune modification, barrier fortification, and microbiota repopulation positions synbiotics as the next level of gut health therapy. It is essential that the term "synbiotic" be strictly reserved for instances where the prebiotic functions to enhance specifically the related probiotic strain, such that true synergism is possible. In that manner, synbiotics are not a random mixture of contents because they represent a specific, strategic intervention in microbiome modulation. Simply put, synbiotics can be "a supplement containing both a prebiotic and probiotic that interact to enhance beneficial microflora of the human gut." Through such synergistic action, synbiotics allow for the restoration of a harmonious intestinal ecosystem and exert more therapeutic scope than single microbial or substrate treatments (Fig. 6.9).

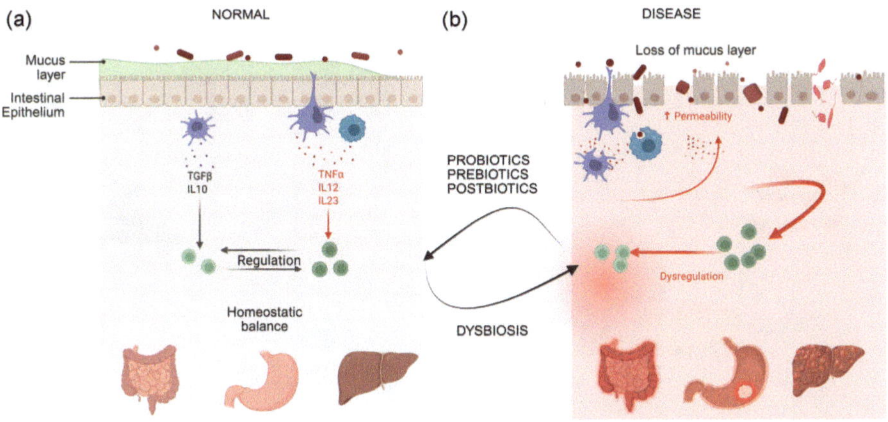

Fig. 6.9 Impact of gut microbiota balance and dysbiosis on intestinal health. (**a**) In a healthy state, the intestinal epithelium is protected by an intact mucus layer and a balanced gut microbiota, maintaining homeostasis through immune regulation. Anti-inflammatory cytokines such as TGF-β and IL-10 promote regulatory immune responses, while proinflammatory cytokines such as TNF-α, IL-12, IL-23 are kept under control, preserving barrier integrity and organ health. (**b**) In disease states, disruption of the mucus layer and dysbiosis increase gut permeability, triggering uncontrolled immune activation and elevated pro-inflammatory cytokine production. This dysregulation contributes to chronic inflammation, which can affect the intestine, stomach, and liver. Probiotics, prebiotics, and postbiotics can help rebound microbiota ecosystems, vitalize functional integrity, and modulate immunological pathways, reducing the progression of disease

6.6 Beyond Traditional Probiotics

Conventional probiotics, mainly in the form of lactic acid fermenting bacteria like *Lactobacillus* spp. and *Bifidobacterium* spp., are widely known for their health advantages in general wellness. These microbes have been ingested for decades as fermented foods and supplements. Nonetheless, most of these strains were chosen empirically either randomly or through cultural experience and not by strict scientific screening (Stasiak-Różańska et al. 2021). Regardless of their well-documented biological safety and potential to promote health, their overall disease prevention or treatment impact is frequently statistically insignificant (Ng et al. 2023). In addition, Lactic acid bacteria (LAB) obtained from food sources tend to be sensitive to environmental stresses. They cannot tolerate the adverse conditions of the stomach environment of human beings, oxidative stress upon storage in the fridge, and long-term acidic pH stress post fermentation. Moreover, traditional probiotics are not disease-specific (Sarao and Arora 2017; Al-Fakhrany and Elekhnawy 2024). Most of the commercially licensed probiotics thus far fall under the category of LAB group, specifically *Lactobacillus* spp, which have lactic acid produced that yields an unfavorable condition for pathogenic bacteria. The bacteria also yield useful secondary metabolites including exopolysaccharides, bacteriocins, and enzymes (Salas-Jara et al. 2016). Conventional probiotics have a limited effect on the intestinal microbiota of human, with minimal ability to fundamentally change its structure or

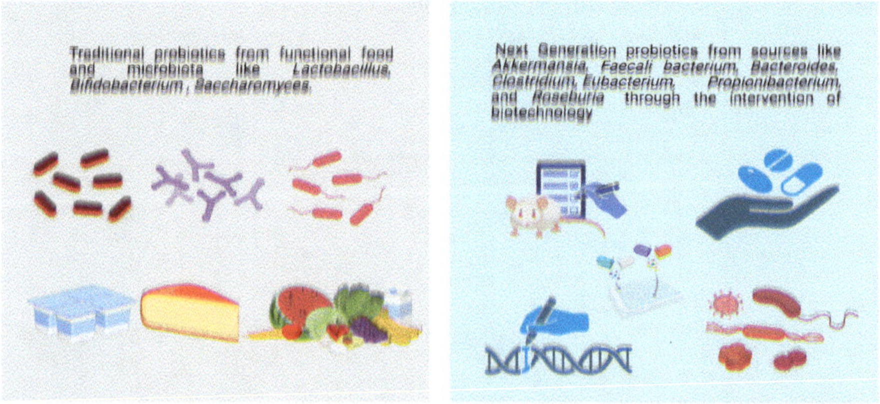

Fig. 6.10 Diagrammatic illustration of different sources of traditional and next-generation probiotics

Table 6.4 Difference between conventional and next-generation probiotics

Conventional probiotics	Next-generation probiotics
Derived from food, fermented products, or human gut flora	Derived from commensal microbes or novel sources without prior use in humans
They play roll in pathogen inhibition, gut flora balance, and vitamin production	NGP are strain-specific; they modulate immune response, gut-brain axis, and metabolic pathways
Maintenance gut health digestion, immunity	Work as disease-specific applications
Generally Recognized as Safe (GRAS), or QPS in Europe	Requires detailed safety and efficacy testing, regulated as live biotherapeutic products (LBPs)
Usually administered through foods, supplements	pharmaceutical-grade products with precise dose, delivery form capsules, encapsulated forms
Use culture-based and traditional taxonomy	Use advanced approaches like next-generation sequencing (NGS)

function. This directs us toward the growing need for suitably selected bacterial strains, improved delivery systems, and the identification of new, disease-specific microbes from non-traditional food and dairy products (Neef and Sanz 2013).

Probiotic product development and research in recent years shifted considerably away from the classic preoccupation with *Lactobacillus* strains towards the use of other gut-sorted bacteria possessing superior capabilities. This has generated next-generation probiotics (NGPs), which are microbial strains identified using advanced scientific methods and engineered to elicit targeted health outcomes (Singh and Natraj 2021). High-throughput sequencing technologies such as metagenomics and metatranscriptomics is progressing rapidly in gut microbiome science, is now enabling researchers to study microbial populations in healthy and pathological patients. Such technologies have expanded the therapeutic scope of probiotics by isolating commensal gut bacteria with significant functions in health promotion (Zhang et al. 2022; Arora and Baldi 2015; Xie et al. 2013). The NGPs consist of mostly gut-derived strains that possess qualities like enhanced adhesion to intestinal epithelial cells, insensitive to bile's acidic pH, and enhanced stability that make their efficacy better to modulate the overall host immunity (Torp et al. 2022). These traits, coupled with their disease-targeting profile, place NGPs as very promising innovation over conventional probiotic (Fig. 6.10; Table 6.4).

6.7 Next-Generation Probiotics

Next-generation probiotics are a game-changer in the area of microbiome-based medicine, beyond the conventional, sometimes empirically or randomly selected, probiotics. Unlike the traditional, who were originally picked for overall digestive health due to their occurrence in fermented foods, NGPs are specifically identified using advanced scientific techniques such as genomics and metagenomics. This enables researchers to identify specific beneficial bacteria that can specifically control the body's interactions with its microbes (Suez et al. 2020; Qin et al. 2010; Singh and Natraj 2021). NGPs are live microbial cultures identified by comprehensive scrutiny of the microbiome of the body. When taken in proper dosages, they are specifically aimed at delivering quantifiable benefits to health. Though they are biological products, they differ from vaccines, and they mainly play preventative or therapeutic functions (Al-Fakhrany and Elekhnawy 2024). One major difference is in their origin and regulation. Traditional probiotics have centuries of history of safe food use, whereas the majority of the NGP strains are novel additions to the human microbiota without prior existing safety information when used as foods. They are selected based on their specific functions, rather than on their traditional use (Bottacini et al. 2017). This represents a significant shift from broad, species-level probiotics to very specific, strain-level biotherapeutics. Thus, conventional probiotics are usually regulated as foods or dietary supplements, whereas NGPs are live biotherapeutic products (LBPs) and regulated by rigorous pharmaceutical regulatory systems (Shyam et al. 2021). An LBP, as described by the U.S. Food and Drug Administration (FDA) in 2012, is a living organism containing biological product, and is not a vaccine, for the treatment, cure, or prophylactic of human disease. Hence, the development of NGPs requires rigorous clinical testing, such as standardization of dose, safety profiling, efficacy testing, and long-term postmarket surveillance. This regulatory mechanism is instituted to ensure that NGPs are held to the same standards of safety and effectiveness as all other therapeutic agents (Cordaillat-Simmons et al. 2020).

Some of the NGP candidates are uncovered through microbiome research which are not commonly linked to traditional probiotics. This includes *Akkermansia muciniphila, Bacteroides thetaiotaomicron, Faecalibacterium prausnitzii, Roseburia intestinalis, Christensenella minuta, Eubacterium species, Prevotella copri, Bacteroides fragilis,* and *Parabacteroides goldsteinii* (Phùng and Karbowiak 2024; Abouelela and Helmy 2024).

Certain *Bacteroides* species, for example, have exhibited potent anti-inflammatory properties, showing they hold promise for therapy (Hiippala et al. 2020). Also in the same way, certain *Clostridium* species are already well established to be significant to digestive health, such as enhancing the intestinal barrier and generating healthy short-chain fatty acids (SCFAs) (Guo et al. 2020). Although of vast promise, the growth and engineering of NGPs are highly technically demanding. Most candidate strains are highly oxygen-sensitive, making it difficult to grow, stabilize, and store them during production. Withstanding their viability during production, storage, and shipping continues to be a major barrier to broader clinical use. In addition to the effects on metabolism, certain NGPs have also been found useful in clearing antimicrobial-resistant pathogens from the human gut and guarding against multidrug-resistant infections. These are thought to be products of direct microbiological interactions and immunomodulation, which make them potential candidates as alternates or as adjunct therapeutics to traditional antimicrobials (Chang et al. 2019).

Since the industry is progressing at such a hectic pace, regulatory processes need to keep up to achieve this harmony between innovation and safety. The appropriate development and application of NGPs in clinical level depend on reliable risk-benefit evaluations and globally agreed standards (Khan and Malik 2019). Their potential for transforming the treatment of chronic disease, infectious pathogens, and microbiome dysfunction has opened a quantum leap of advancement in precision microbial medicine (Fig. 6.11).

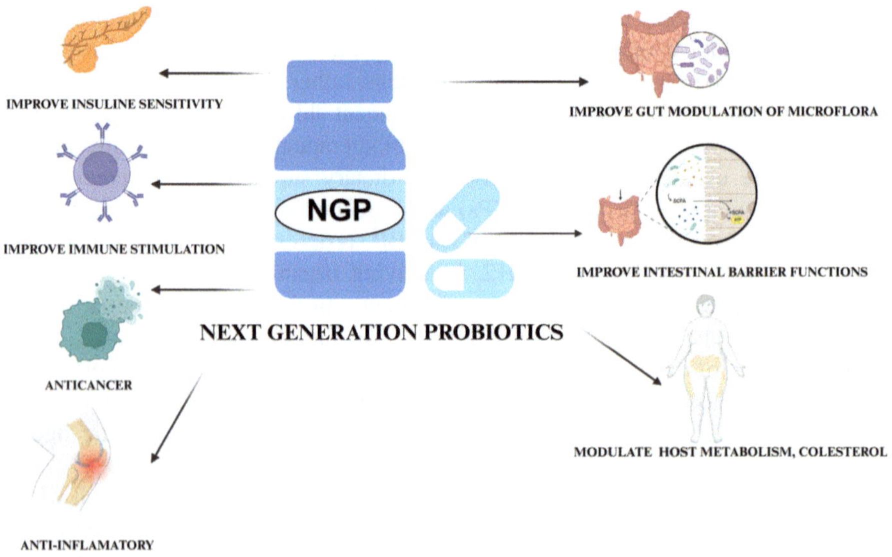

Fig. 6.11 Diagrammatic representation of next-generation probiotic's function

6.7.1 Development of Next-Generation Probiotics

Next-generation sequencing advances have revolutionized the identification and characterization of next-generation probiotics (NGPs), whereby scientists are able to identify and even culture bacterial strains of the human microbiome that were hitherto uncultivable or unrecognized. Unlike conventional probiotics, which were traditionally isolated from fermented milk or fecal matter, NGPs are usually derived directly from the gut microbiome and are designed as host-specific, targeted therapeutic treatments. Several bacterial candidates such as *Eubacterium, Akkermansia, Bacteroides, Clostridium, Faecali bacterium, Propionibacterium*, and *Roseburia* were characterized as potential NGP strains due to their roles in immune control, gut barrier function, and restoration of microbial homeostasis in pathological situations (Jan et al. 2024; Abouelela and Helmy 2024). Next-generation probiotics (NGPs) selection begins with isolating strains from target habitats, facilitated by next-generation sequencing (NGS) technologies that allow screening for candidates. Advances in culturomics, employing high-throughput culture systems in anaerobic environments, have also enabled the successful cultivation of previously "unculturable" microbes. The candidate strains identified then undergo rigorous taxonomic identification, typically by NGS followed by comprehensive functional validation to confirm that they are safe and possess therapeutic potential. Various strategies, e.g., coisolation and selective culture, make the process easier. In particular, NGP strain selection is made on the basis of several key parameters such as the ability to withstand food processing and storage, survive in gastrointestinal tract, have quantifiable health impacts, and have strain-specific resistance to environmental stress (Tripathi and Giri 2014). For next-generation probiotics (NGPs) to hold translational value, candidate strains must meet several minimum qualifications. These are that they must withstand storage and processing, be stress-resistant in the environment, display detectable health-benefiting functions, and convey strain-specific safety (Tripathi and Giri 2014). NGPs are also under strict regulatory evaluation before entering the clinical phase, similar to the FDA model for live biotherapeutic products (LBPs) (Shyam et al. 2021; O'Toole et al. 2017). NGPs are a paradigm in probiotics that are living microbial therapeutics providing precision-based, long-term sustainable solutions to the treatment of long-term conditions such as autoimmune diseases, metabolic syndrome, gastrointestinal disease, and microbial dysbiosis (Zhai et al. 2019). There is mounting popularity for the new biotherapeutics fueled by mounting understanding of host–microbiota interactions and the potential for precision microbiome interventions to transform modern-day healthcare.

Fecal samples are collected from healthy human donors and proceeds through next-generation sequencing and bioinformatics towards the selection of candidate strains. Isolated bacteria are grown and selected bacteria are in vitro functional screened and in vivo tested in animal models finally the Candidate strains are advanced towards human clinical trials, leading to the manufacture of NGP-based therapeutic products (Fig. 6.12; Table 6.5).

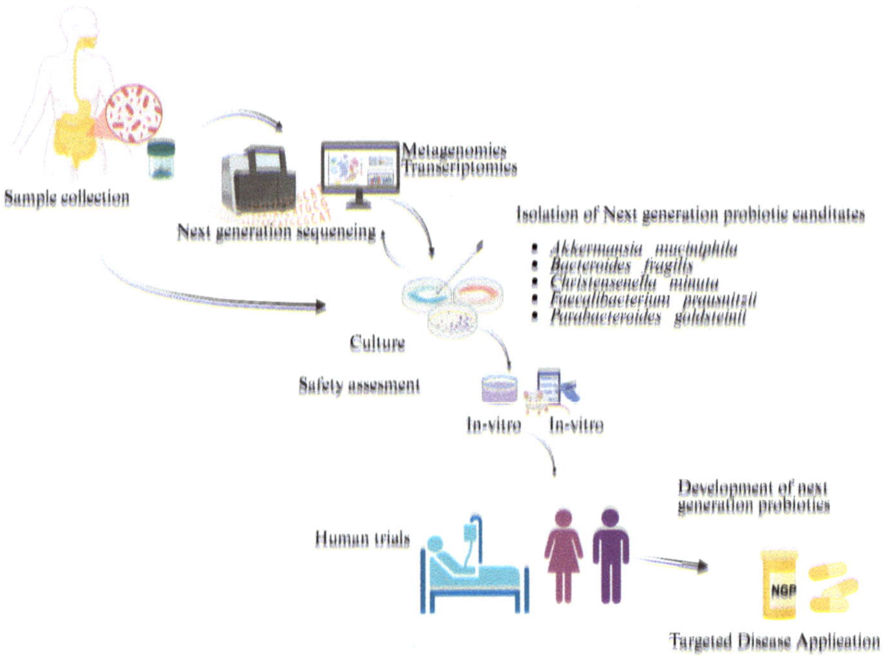

Fig. 6.12 Diagrammatic illustration of the pipeline for the discovery and development of next-generation probiotics (NGPs)

Table 6.5 Some of the important candidates for NGP mechanism and their functions

Bacteria strain	Shapes	Key features	Functions and mechanism	Treatment
Bifidobacterium spp.	Rod-shaped, branched	Gram-positive, anaerobic	Activates DCs and CD8+Tcells produces butyrate, can transport anti-cancer genes, modulates gut immunity	Cancer therapy adjunct (anti-PD-L1), colitis, CRC
Prevotella copri	Rod-shaped	Gram-negative, anaerobic	Produces succinate (improves glucose homeostasis), produces BCAAs (may cause insulin resistance)	Potential prediabetes management
Akkermansia muciniphila	Oval or coccus-like	Gram-negative, anaerobic	Improves gut barrier, influences cancer immunotherapy	Obesity, T2D, metabolic syndrome, adjunct in cancer ICB
Bacteroides fragilis	Rod-shape	Gram-negative, anaerobic	PSA induces CD4+ FoxP3 Tregs via dendritic cells, reduces systemic inflammation	Cancer, neuroinflammation, immune modulation

(continued)

Table 6.5 (continued)

Bacteria strain	Shapes	Key features	Functions and mechanism	Treatment
Christensenella minuta	Rod-shaped	Gram-positive, anaerobic	Alters host microbiota, reduces body weight gain	Obesity and metabolic disorders
Faecalibacterium prausnitzii	Rod-shaped	Gram-positive, anaerobic	Butyrate production, modulates immune response, enhances cancer immunotherapy efficacy	IBD, CRC, adjunct in cancer immunotherapy
Parabacteroides goldsteinii	Rod-shaped	Gram-negative, anaerobic	Enhances Treg and IL-10 production, reduces gut permeability decreases inflammation	Obesity, metabolic syndrome, liver inflammation
Eubacterium limosum	Rod-shaped	Gram-positive, anaerobic	Produces SCFAs, reduces DSS-induced colitis	Colitis, potential role in HSCT recovery
Enterococcus hirae	Spherical (coccus)	Gram-positive, facultative anaerobe	Translocate to lymph nodes, enhances memory T cells, modulates immune system	Adjunct to PD-1/ PD-L1 ICB; caution due to pathogenicity
Enterococcus faecium	Spherical (coccus)	Gram-positive, facultative anaerobe	Increases T cell activation, modulates gut microbiota	Adjunct to ICB, caution due to nosocomial infection risk
Collinsella aerofaciens	Rod-shaped	Gram-positive, anaerobic	Modulates immune response, mechanism under further study	Adjunct to ICB (needs more validation)
Burkholderia cepacia	Rod-shaped	Gram-negative, aerobic/ facultative	Stimulates Th1 responses, synergistic with *B. fragilis* in enhancing anticancer immunity	Adjunct in cancer ICB, caution due to opportunistic infection

6.7.2 Challenges

NGPs identification is the most difficult step since it is entirely based on assumptions strategy and largely relies on correlation studies which compares the relative abundance levels of microorganisms between diseased and healthy individuals. After the identified of desired microorganism, it should be validated for safety, several technological aspects and functions by *in vivo*, *in vitro* models and proper biocomputational approaches. The unusual growth conditions have to be considered in order to obtain myriad yield of live biotherapeutic products (LBP) and formulation has to make the culture viable and bioactive until they are consumed. The product so developed has to be tested for safety, effectiveness, and efficiency and confirmation of dose by clinical trials and has to clear the regulatory approval process. Finally, the specialists have already introduced the term NGP, which is still in its infancy but important and needs large-scale research for the exploitation of NGP as preventive medicine (Stanton et al. 2003; Fuller 2012; Yao et al. 2020; O'Toole et al. 2017). Some key points have been given below (Fig. 6.13; Tables 6.6 and 6.7; Box 6.2):

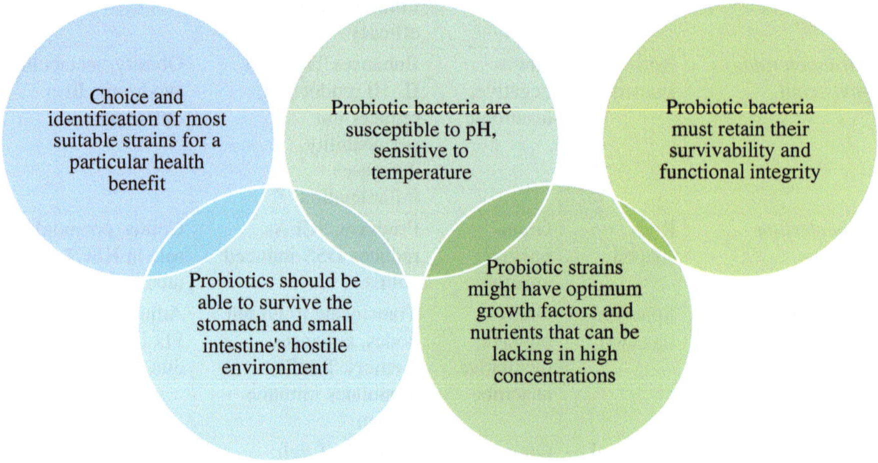

Fig. 6.13 Challenges in characterization of NGP

Table 6.6 Next-generation probiotics in the treatment of human disease

Disease	Benefits of NGPs in the treatment of human disease
Metabolic diseases 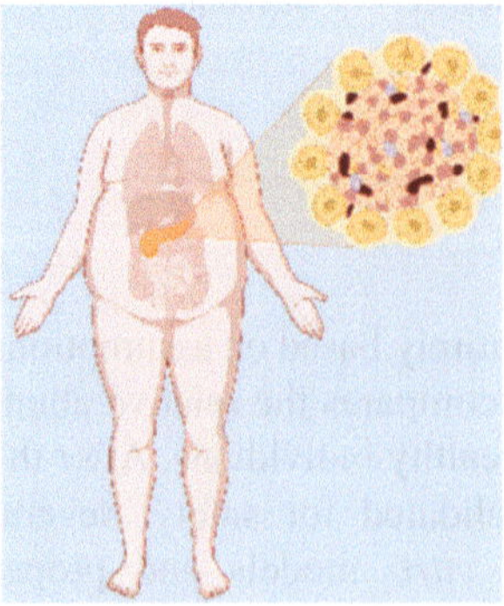	(a) Improvement of insulin sensitivity, lipid metabolism, liver function, and reduction of hepatic steatosis (b) Modulation of gut microbiota (c) Reduction of body weight and fat accumulation (d) Enhancement of intestinal barrier function (e) Anti-inflammatory effects (f) Regulation of bile acid and sterol metabolism (g) Antioxidant effects and metabolic gene regulation

(continued)

Table 6.6 (continued)

Disease	Benefits of NGPs in the treatment of human disease
Cardiac disease 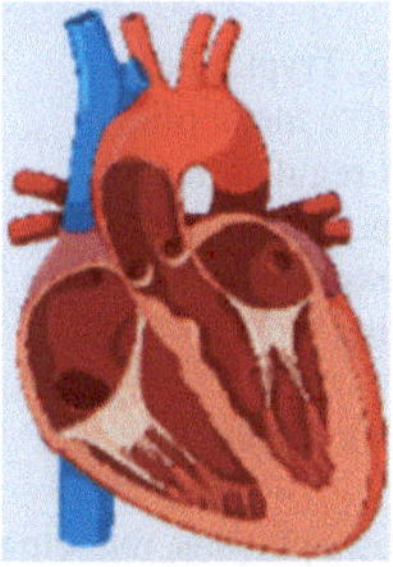	(a) Improvement of insulin sensitivity (b) Conversion of dietary compounds into cardioprotective metabolites (c) Reversal of obesity-related complications (d) Reduction of systemic inflammation (e) Reversal of obesity-related complications which are CVD risk factors
Cancer 	(a) Induction of apoptosis in cancer cells (b) Antiproliferative effects, cancer cell growth inhibition (c) Modulation of inflammation and cytokine expression (d) Selective exclusion of harmful and tumor-promoting microorganisms (e) Enhancement of response to cancer immunotherapy (f) Suppression of tumor microenvironment and inhibition of cancer cell migration
Gastrointestinal diseases 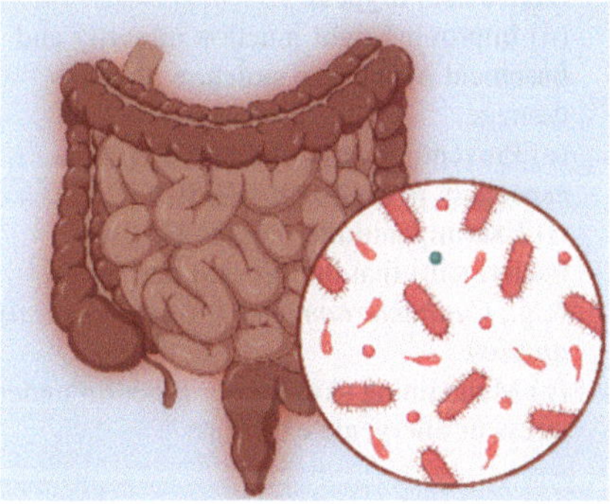	(a) Restoring gut microbiota balance (b) Competing with pathogenic bacteria for nutrients (c) Reducing growth and activity of pathogenic bacteria (d) Reduce gut inflammation by stimulating the immune system (e) Modulating gut motility, improving symptoms of constipation and diarrhea (f) Improving gut permeability and intestinal barrier (g) Reducing gut hypersensitivity caused by stress or inflammation (h) Improving digestion of lactose intolerance

(continued)

Table 6.6 (continued)

Disease	Benefits of NGPs in the treatment of human disease
Neurophysiological diseases 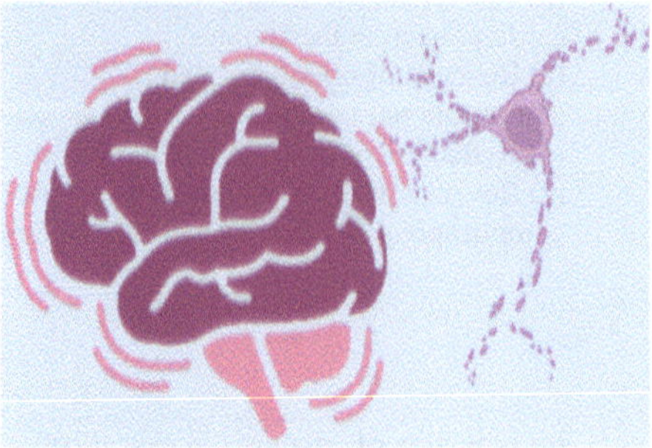	(a) Modulate gut–brain axis signaling, including: Vagus nerve signaling, Tryptophan metabolism, Neuroactive metabolite production, HPA axis regulation, GABAergic system modulation (b) Regulate protein expression and brain signaling by Influencing BDNF–TrkB signaling pathway (c) Reduce Alzheimer's disease symptoms by: Decreasing amyloid-beta plaques, reducing neuroinflammation, enhancing antioxidant capacity (d) Support neurodevelopmental disorders: autism spectrum disorder (ASD) (e) Provide neuroprotective effects: inhibit beta-amyloid fibril formation (linked to Alzheimer's), exhibit antioxidant properties, reduce oxidative stress
Skin diseases 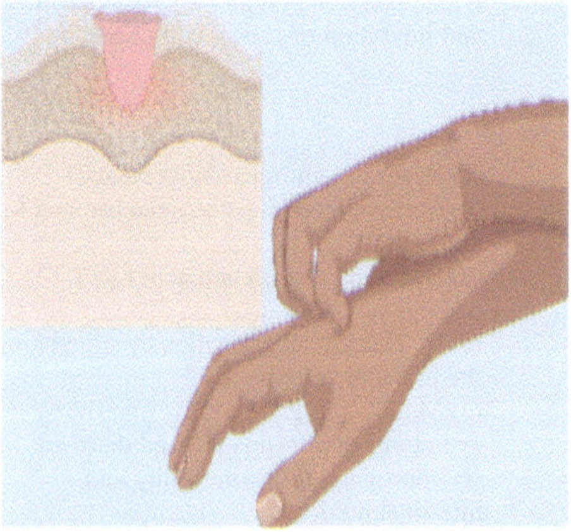	(a) Reducing skin inflammation (e.g., in acne, atopic dermatitis, rosacea) (b) Enhancing wound healing through: increased wound contraction, accelerated epithelialization, and increased collagen content (c) Inhibiting trans epidermal water loss to reduce skin dryness (d) Improving tight junction integrity and basement membrane structure after UV damage (e) Preventing early childhood eczema, especially in high-risk infants (f) Exerting antimicrobial effects via bacteriocins that inhibit skin pathogens (e.g., *Cutibacterium acnes*, *Staphylococcus aureus*) (g) Maintaining skin barrier function under stress or allergenic exposure

Table 6.7 Regulatory authority for the safety of probiotics in different countries

Country/ region	Probiotic category	Regulatory authority
USA	Dietary supplements, live biotherapeutic agents, drugs, medical food	FDA, DSHEA
Japan	Functional food	Food for Specified Health Use (FOSHU)
India	Functional food	Food Safety and Standards Authority of India (FSSAI)
China	Functional food	State Food and Drug Administration (SFDA)
Malaysia	Functional food	Food Safety and Quality Division (FSQD), NPCB, Drug Control Authority
Europe	Biotherapeutic agent	European Food Safety Authority (EFSA)
Germany, Belgium	Biotherapeutic agent	European Food Safety Authority (EFSA)
Italy	Dietary food	National Regulatory Authorities
Canada	Natural health products	Canadian Food and Drugs Act (Natural Health Products Regulations)
Denmark, Sweden, Finland	Food supplements	National Food Safety Authorities
UK	Health claims related to probiotics	Joint Health Claims Initiative (JHCI)
Brazil	Regulatory approval for probiotics	National Health Surveillance Agency
Denmark, Sweden, Finland	Food supplements	National Food Safety Authorities

Box 6.2: Some of the Clinical Trials Conducted on the Benefit of NGPs on Mice and Human for the Treatment of Several Diseases

- *Akkermansia muciniphila* can break down intestinal mucin components, causing a competitive inhibitory effect on other pathogens that break down mucin also function as a marker for an optimal metabolic profile in a host (Macchione et al. 2019)
- *Bacteroides fragilis*, also from human faces, conferred protection against antibiotic-associated diarrhea in rats by augmenting intestinal defenses (Zhang et al. 2018)

(continued)

Box 6.2 (continued)

- *Bacteroides species* of human feces were reported for the first time as cholesterol-reducing bacteria. Likewise, *Bacteroides fragilis*, sourced from a healthy breast-fed infant, was proven to increase phagocytosis, polarize macrophages M1, and ameliorate immune imbalance-related condition, inflammatory disease, and mental disorders (Gérard et al. 2007; Deng et al. 2016)
- *Clostridium butyricum* in ulcerative colitis patients, where clinical trial data showed it to be effective against pouchitis prevention and modifying the gut microbiota profile. In another a certain bacterial strain suppressed *Vibrio parahaemolyticus* infection in mice (Yasueda et al. 2016)
- *Lactococcus lactis* could potentially be involved in preventing Parkinsonism by suppressing ferroptosis and repairing dysbiosis and oxidative stress in humans (Yue et al. 2022)
- Vaginal commensal bacterium *Lactobacillus plantarum 5BL* was shown to induce apoptosis, with strong anticancer activity against numerous cancer cell lines, such as cervical (HeLa), colon (HT-29), gastric (AGS), and breast (MCF-7) (Nami et al. 2014)
- Mixed *Bifidobacterium strains*, mirroring human skin and intestinal flora, proved a powerful anticancer activity against colon adenocarcinoma cell lines without any toxic effect on normal rat epithelial cells. In murine models, the blend suppressed the levels of colon cancer biomarkers like epidermal growth factor receptors (EGFR) and cyclooxygenase-2 (COX-2), causing disease stabilization, maintenance of colon tissue integrity, and suppression of cancer development and metastasis. Their increased antiproliferative activity is thought to be due to the quorum sensing activity of these blended bacteria (Parisa et al. 2020; Sapkota et al. 2025)

6.8 Gastrointestinal Disease

Gastrointestinal (GI) diseases can be defined as a broad range of disorders that weaken the usual functioning of the digestive tract including esophagus, stomach, intestines, and peripheral organs such as the gall bladder, pancreas, and liver. These conditions may arise due to genetic factors, infections, dietary imbalances, immune dysfunction, or modulation in the intestinal microbiota (Nagao-Kitamoto et al. 2016; Bhattacharjee 2022). Common GI diseases include peptic ulcer disease (PUD), inflammatory bowel disease (IBD), gastroesophageal reflux disease (GERD), and, irritable bowel syndrome (IBS). These can significantly affect nutrient absorption and mucosal integrity of gut wall. It has been proposed that many GI disorders are closely linked to imbalances in the gut microbiota which is leading to chronic inflammation, immune dysregulation, and epithelial barrier dysfunction. As a result, microbiota-targeted therapeutic strategies such as administration of probiotics and prebiotics are gaining prominence over conventional treatment strategies (Bhattacharjee 2022; Nagao-Kitamoto et al. 2016; Ullah et al. 2025). There has been a substantial elevation in the use of prebiotics and probiotics in the treatment of gut diseases caused by disruption of gut microbiota. Probiotics refers to the use of living microbes to re-establish the gut microbiota and increase gut barrier integrity, decrease inflammation and modulation of immune responses. On the other hand, prebiotics refer to nondigestible fibers which support and stimulate the growth of beneficial microbes in gut which results in increased production of short-chain fatty acids (SCFAs) which improves gut barrier integrity, enhances immune activity, and overall gut health (Abbasi et al. 2022).

Overview of Treatment Modalities

I. The following therapeutic inventions are usually followed for the treatment of gastrointestinal and systemic diseases:
II. Pharmacological treatments: it includes the use of medicinal compounds such as antibiotics, inhibit proton pump and others with similar properties. For example, in case of inflammatory bowel disease (IBD), medications include 5-aminosalicyclic acids, immune-modulators and corticosteroids (Li et al. 2021a, b).
III. Nutritional treatments: this strategy aims to improve diseased conditions by enhancing food, such as by adding fibers. This dietary intervention stimulates growth of beneficial bacteria hence increasing production of SCFAs (Parada Venegas et al. 2019).
IV. Microbiota-targeted treatments: they are of following types:
V. Probiotics: they consist of live microbes which are having positive effects for gut health. These beneficial microbes help in improving gut barrier integrity and enhance the production of SCFAs which in turn improve disease symptoms (Abbasi et al. 2022).
VI. Prebiotics: they are composed of compounds that stimulate growth of beneficial bacteria hence improving health of gut by re-establishing microbiota (Parada Venegas et al. 2019).
VII. Postbiotics: they are composed of inactivate microbes or their components which are comparatively stable and confers health benefits such as reducing inflammation and pathogenic microbes (Aguilar-Toalá et al. 2020).
VIII. Fecal microbiota transplantation (FMT): it is a therapeutic intervention which includes transferring microbes from healthy donor to recipient resulting in re-establishment of healthy microbiota (Liu et al. 2023) (Fig. 6.14).

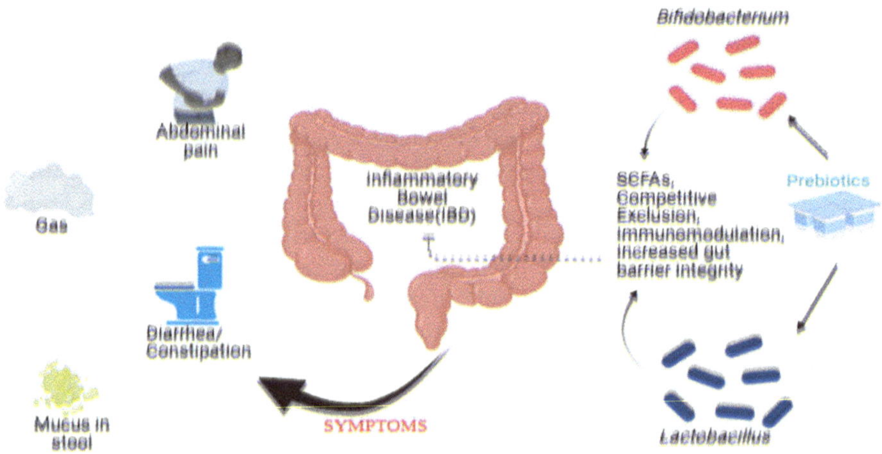

Fig. 6.14 Diagram illustrating the symptoms association with inflammatory bowel disease (IBD) such as abdominal pain, diarrhea or constipation, gas, and mucus in the stool and microbial interactions along with the beneficial role of prebiotics which help in promoting the growth of beneficial bacteria, that have therapeutic effects through the SCFAs, competitive exclusion of pathogens, immune modulation, and gut barrier enhancement

6.8.1 Role of Prebiotics, Probiotics and Postbiotics in Treating Gastrointestinal Disease

6.8.1.1 Inflammatory Bowel Disease (IBD)

Inflammatory bowel disease (IBD) is a chronic and complicated condition that causes inflammation in the digestive tract. The two distinct types are Crohn's disease, which can cause digestive system and ulcerative colitis, which specifically targets the colon and rectum. IBD occurs when the immune system overreacts, leading to ongoing inflammation and damage to intestinal tissues. This may happen because the body becomes less tolerant of the normal microbes living in the gut, resulting in less variety of gut bacteria, especially those that thrive without oxygen (Martyniak et al. 2021; De Souza and Fiocchi 2016). In IBD patients, the composition of intestinal microbes is typically disrupted harmless species such as *Lactobacillus* and *Bifidobacterium* being reduced, but potentially pathogenic species such as *Bacteroides*, *Escherichia*, and *Enterococci* being increased (Zhang et al. 2023). Microbe imbalance has serious consequences, particularly with regard to its influence on short-chain fatty acid (SCFAs) production. SCFAs are synthesized when gut bacteria ferment carbohydrate fiber, and they play a key role in maintaining intestinal health. They provide a primary source of energy to colonocytes, suppress inflammation by inhibiting NF-κB signaling and other proinflammatory signaling, and drive the differentiation of regulatory T cells, critical for controlling intestinal inflammation (Parada Venegas et al. 2019). It also drives the differentiation of regulatory T cells, which are important in controlling intestinal inflammation. Butyrate and acetate are some of the SCFAs (Recharla et al. 2023). Probiotics help restore and maintain a healthy gut microbiota by increasing beneficial bacteria like *Bifidobacterium* and *Lactobacillus* and inhibiting pathogens. This involves competitive exclusion where they compete for adhesion sites and nutrients with IBD-causing pathogens (Martyniak et al. 2021). They can modulate immune responses in the gut by interacting with immune cells and affecting cytokine levels. They can act by increasing the production of anti-inflammatory cytokines (IL-10, TGF-β) and reducing the levels of proinflammatory cytokines (TNF-α, IL-6, IL-8, IL-17, MCP-1) (Cristofori et al. 2021). This modulation can influence antigen-presenting cells (dendritic cells), T cells (including regulatory T cells or Tregs), and B cell maturation (Zhang et al. 2023). Their metabolites can improve gut barrier integrity by promoting mucus secretion and regulating tight junction proteins. Examples include *Bifidobacterium longum* 536, VSL#3 (a consortium of 8 probiotic strains), *Escherichia coli* Nissle 1917, etc. (Abraamyan et al. 2024).

Prebiotics withstand digestion and absorption in the upper GI tract, enabling them to pass into the colon where they are fermented by the gut microbes. Prebiotics enhance the growth and functions of beneficial gut bacteria like *Lactobacillus, Bifidobacterium, Clostridium, Lactococcus*, and *Bacteroides* while suppressing the growth of harmful microbes (Li et al. 2023). Their fermentation by gut microbes generates short-chain fatty acids (SCFAs) such as essentially butyrate, acetate, and propionate. Butyrate serves as a key energy source for colon cells, supports the repair of the intestinal lining and enhances mucus secretion (Parada Venegas et al. 2019). Prebiotics also modulate immune function by stimulating anti-inflammatory cytokines like IL-10 and decreasing pro-inflammatory cytokines such as TNF-α and IL-6. Prebiotics include things like fructooligosaccharides, lactulose, oligofructose, ispaghula husk, and germinated barley foodstuff (Liu et al. 2023)

6.8.1.2 Irritable Bowel Syndrome (IBS)

Irritable Bowel Syndrome (IBS) is a prevalent gastrointestinal disorder characterized by repeated cycles of discomfort in the abdominal area often accompanied by unusual change in stool form or frequency. Bloating and flatulence are among the most common and distressing symptoms experienced by patients with IBS (Ford et al. 2023). One notable microbial change in these patients is the sharp reduction of *Bifidobacterium longum* 35624, a probiotic strain extensively studied for its ability to relieve IBS symptoms. Such imbalances also affect the production of short-chain fatty acids (SCFAs), particularly butyrate and acetate, which are critical for upholding the integrity of the gut barrier. As a result, dysbiosis leads to broader metabolic alterations, reflected in shifts in bile acids, SCFAs, and tryptophan metabolites (Leylabadlo et al. 2022). The microbiome is involved in the production of metabolites that influence immune system development and mucosal integrity. Abnormal immune activation and imbalances in specific cytokines such as an altered IL-10/IL-12 ratio have been observed in IBS patients (Verna and Lucak 2010).

Probiotics when administered in adequate amounts provide beneficial effects via several mechanisms including modulating gut microbiota composition, competing with pathogens, enhancing gut barrier function and alteration of the immune system. It also eases the painful symptoms of IBS (Martyniak et al. 2021). Examples include

Bifidobacterium infantis 35624, *Lactobacillus plantarum* DSM 9843, *Lactobacillus reuteri* ATCC 55730, *Bacillus coagulans* MTCC 5856, etc. While probiotics are generally considered safe, there are concerns like of bacteremia due to the transfer of bacteria from the intestinal lumen to the blood. Research on using prebiotics specifically for IBS is limited and has shown clashing or inconclusive results. Some studies suggest that prebiotics could potentially worsen symptoms of IBS. For instance, the addition of fermentation substrates from prebiotics could increase intestinal gas production which is a common IBS symptom (Abraamyan et al. 2024).

6.8.1.3 Gastroesophageal Reflux Disease (GERD)

Gastroesophageal reflux disease is a prevalent digestive disorder which results from the retrograde flow of gastric contents into the esophagus, oropharynx, or respiratory tract. It is a chronic and relapsing condition where the backflow of stomach contents causes symptoms or complications that interfere with daily life. Treatment typically includes dietary and lifestyle modifications along with medications such as proton pump inhibitors (PPIs), which reduce stomach acid production. However, long-term PPI use has been associated with a notable reduction in gut microbiome, an increase in pathogenic bacterial populations, and a heightened risk of small intestinal bacterial overgrowth (SIBO) (Cheng and Ouwehand 2020). Probiotics are being viewed as a promising way for controlling the symptoms of GERD and preventing its relapse. Probiotics act through multiple mechanisms, including direct interaction with gut microbiota, metabolic activities from their enzymes, enhancement of the gut barrier and stimulation of gastric emptying by targeting stomach mucosal receptors (Zang et al. 2020). Notable examples include *Bifidobacterium bifidum* YIT 10347, which adheres to gastric cells and increases the production of mucin; *Lactobacillus johnsonii* No. 1088, which decreases the production of gastric acid; and *Lactobacillus gasseri* LG21, which raises the levels of pepsinogen, which may facilitate digestion and shorten the time it takes for the stomach to transit (Herdiana 2023; Plaza-Díaz et al. 2017).

6.8.1.4 Peptic Ulcer Disease (PUD)

Peptic ulcer disease (PUD) happens when the protective lining of the stomach or small intestine becomes injured, typically due to the reverse flow of stomach acid. Infection with the *Helicobacter pylori* (*H. pylori*) bacterium is one of its primary causes, as it not only leads to ulcer development but also elevate the risk of gastric cancer in the long run. Other than direct damage to gastric mucosa, *H. pylori* infection can lead to derangement of balance in gut microbiota, and thus make an individual prone to illnesses such as small intestinal bacterial overgrowth (SIBO). Microbial diversity loss undermines the body's protective mechanisms against colonization by pathogenic microorganisms, making it susceptible to infection. Dysbiosis also decreases the production of the protective short-chain fatty acids (SCFAs) that play an important role in preserving mucosal barrier integrity and healing. The altered microbial environment also leads to enhanced immune activation, facilitating chronic inflammation that blocks healing of the gastric epithelium (Gong et al. 2018). Probiotic supplementation replenishes microbial balance and enhances efficacy and tolerability of routine therapy. Clinical trials have proven that co-administration of probiotics enhances *H. pylori* eradication rates by approximately 13% and reduces the incidence of treatment side effects, such as antibiotic-associated diarrhea, up to 41%. For example, the randomized controlled trial using *Lactobacillus rhamnosus* GG demonstrated higher eradication and lower gastrointestinal disturbances than monotherapy with antibiotics. Similarly, *Lactobacillus reuteri* DSM 17938 has been found to significantly suppress *H. pylori* colonization through the production of reuterin, a broad-spectrum antimicrobial compound. Another strain, *Saccharomyces boulardii*, a probiotic yeast, has shown the ability to increase eradication rates while also preventing diarrhea and gut dysbiosis during triple therapy. Mechanistically, these probiotics exert their benefits through multiple pathways producing antibacterial compounds (like bacteriocins), inhibiting bacterial adhesion to the gastric mucosa, regulating inflammatory responses, and enhancing mucosal barrier function. For instance, *Lactobacillus johnsonii* NCC 533 and *Lactobacillus acidophilus* LB have demonstrated protective effects in mouse models by reducing gastric inflammation and H. pylori colonization. Beyond their antimicrobial role, probiotics like *Bifidobacterium bifidum* and *Bifidobacterium longum* contribute to gut health by modulating immune responses, promoting tight junction integrity, and restoring overall microbiota diversity (Liang et al. 2022; Ojima et al. 2020) (Fig. 6.15).

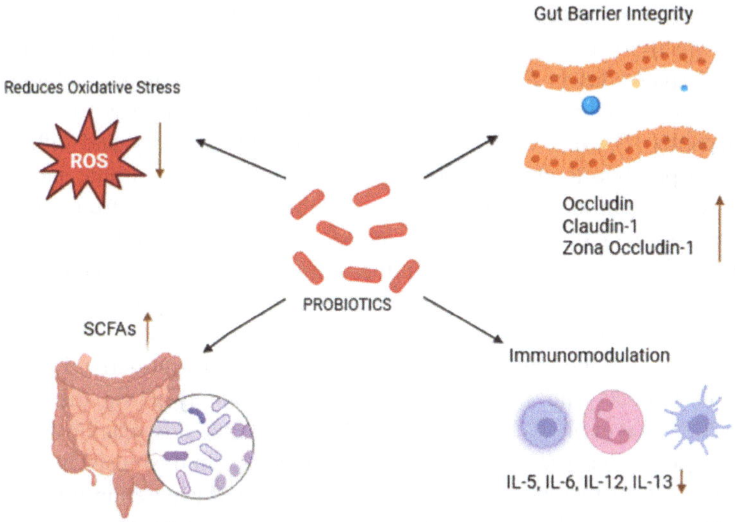

Fig. 6.15 Diagrammatic illustrating diverse mechanisms of probiotics conferring health benefits, they enhance gut barrier integrity by upregulating occludin, claudin-1, and zonula occludens-1 tight junction proteins and thereby reduce intestinal permeability. Probiotics also induce the production of short-chain fatty acids (SCFAs), which are involved in intestinal well-being and metabolic homeostasis. In addition to that, they also modulate immune responses by lowering pro-inflammatory cytokines such as IL-5, IL-6, IL-12, and IL-13. Additionally, probiotics counteract oxidative stress by lowering reactive oxygen species (ROS) levels, which defends against cellular and tissue damage. All these processes collectively support digestive health and general well-being

6.9 Systemic Illness

6.9.1 The Role of Prebiotics, Probiotics, and Postbiotics in Treating Systemic Illnesses

Systemic disease is defined by pathogen infections that involve more than one organ system. While such illnesses are not limited to the gut, physiology of the gut and composition of enteric microbiota are important in their initiation and exacerbation. A few examples include liver disease such as nonalcoholic fatty liver disease, cardiovascular disease, autoimmune disease, neuroinflammatory disease, and metabolic syndromes such as obesity and type 2 diabetes mellitus. Increasing evidence suggests that the central gut microbiota participation in the etiopathogenesis of the diseases, by means of microbial metabolite biosynthesis, immune system modulation, and communication along gut–organ axes, i.e., the gut–liver axis (Campaniello et al. 2023).

6.9.1.1 Liver Diseases

Liver disease encompasses a wide array of disorders, ranging from reversible accumulation of fat (steatosis) to severe, irreparable damage such as fibrosis, cirrhosis, and liver cancer. The three most common are alcoholic liver disease (ALD), nonalcoholic fatty liver disease (NAFLD), and cirrhosis and hepatic encephalopathy as complications. NAFLD is increasingly prevalent worldwide, mainly due to physical inactivity and unhealthy dietary habits. It is characterized by lipid accumulation in hepatocytes in the absence of significant alcohol intake. Although commonly asymptomatic in the early stages, NAFLD can progress to nonalcoholic steatohepatitis (NASH), characterized by hepatocellular injury and inflammation, and further progress to cirrhosis. Similarly, ALD arises from chronic and excessive alcohol consumption, following a comparable progression from steatosis to steatohepatitis and fibrosis. In advanced stages, it is frequently complicated by coagulopathy and encephalopathy, further compromising liver function. Cirrhosis involves widespread scarring of liver tissue which eventually leads to loss of liver function. The liver and gut are anatomically and functionally connected via the gut-liver axis, a communication system through which gut-derived metabolites, microbial products and immune signals travel to the liver via the portal vein (Wang 2025; Yu et al. 2025). Under healthy conditions, this relationship supports metabolic homeostasis and immune balance. However, in dysbiosis, the gut barrier function is altered thereby promoting inflammation and generating toxic microbial metabolites causing liver diseases. A compromised intestinal barrier allows pathogen-associated molecular patterns (PAMPs) such as lipopolysaccharide (LPS), bacterial DNA and peptidoglycan to enter the portal circulation and reach the liver. These PAMPs activate Toll-like receptors (TLRs) on hepatic cells particularly TLR4 thereby triggering pro-inflammatory signaling pathways that contribute to hepatocyte injury and fibrogenesis (Gao et al. 2019).

Probiotic supplementation can counteract the process of barrier disintegration by enhancing populations of beneficial bacteria such as *Lactobacillus* and

Bifidobacterium which support mucosal integrity, inhibit pathogenic overgrowth and suppress endotoxin production. They help restore gut barrier function by promoting tight junction protein expression and enhancing mucus secretion, thereby reducing "leaky gut" and the translocation of harmful microbial products. Certain bacteria strains, such as *Lactobacillus plantarum*, *Lactobacillus rhamnosus GG*, *Lactobacillus acidophilus*, and *Bifidobacterium longum* are known for their ability to hepatic oxidative stress reduction, modulate immune responses and decrease systemic inflammation which is a characteristic of various hepatic disorders. In ALD, probiotics have been shown to reduce endotoxemia by reestablishing intestinal flora and preventing the proliferation of Gram-negative bacteria. In NAFLD, they influence lipid metabolism, reduce hepatic steatosis and improve insulin sensitivity through their anti-inflammatory and antioxidant effects. Yeast-based probiotics like *Saccharomyces boulardii* have demonstrated similar benefits especially in reducing gut inflammation and enhancing mucosal repair (Pan et al. 2024; Wang 2025; Yu et al. 2025). Specific prebiotics such as inulin, galactooligosaccharides (GOS) and fructooligosaccharides (FOS) have shown the capacity to increase the number of *Bifidobacterium* and *Lactobacillus* species. This enhances the release of short-chain fatty acids (SCFAs) like acetate, propionate and butyrate, which contribute to decreasing inflammation by enhancing intestinal barrier integrity, reducing oxidative stress, and regulating lipid metabolism. Natural sources of prebiotics include garlic, onions, chicory root, bananas and whole grains while purified forms are widely used in dietary, nutritional supplements and functional foods. As a non-invasive, diet-based approach, prebiotics offer a safe and supportive strategy for liver disease management particularly when integrated with microbiota-focused interventions such as probiotics (Pan et al. 2024).

6.9.1.2 Metabolic Diseases

Metabolic disorders including obesity and type II diabetes mellitus have risen as a major public health challenge often interlinked with one another and affecting millions worldwide. Obesity acts as a common precursor to T2DM and is also strongly linked with liver diseases like nonalcoholic fatty liver disease (NAFLD). As these disorders progress, they disrupt normal metabolic homeostasis and are often accompanied by systemic complications including cardiovascular, hepatic, and renal dysfunction severely damaging organs resulting in death. Besides genetic and lifestyle causes, a significant causative factor is dysbiosis of microbiota. Obese individuals often show an increased ratio of Firmicutes to Bacteroidetes resulting in greater energy harvest from food and excessive fat deposition. This microbial imbalance also leads to loss of intestinal barrier integrity allowing bacterial components to enter the blood. The resulting systemic inflammation promotes insulin resistance and elevates metabolic dysfunction. Individuals with T2DM have decreased levels of beneficial microbes such as *Akkermansia*, *Bifidobacterium*, and *Faecalibacterium* and an increase in proinflammatory species like *Ruminococcus* and *Fusobacterium*. These changes impair glucose metabolism which reduce insulin sensitivity causing metabolic disorders (Li et al. 2021a, b). Probiotic interventions include specific strains of *Lactobacillus* and *Bifidobacterium* which have been shown to improve lipid profiles, decrease in fasting blood glucose levels and elevates the production of beneficial metabolites like short-chain fatty acids (SCFAs). For example, *Lactobacillus gasseri* has been associated with reductions in body fat and BMI in overweight individuals while *Lactobacillus reuteri* and *Bifidobacterium lactis* have demonstrated potential in regulating insulin sensitivity and glucose levels. These microbes exert their effects by promoting SCFA production, suppressing LPS translocation and modulating hormones involved in satiety and glucose regulation (Lee et al. 2024). They help strengthen the gut mucosal barrier by promoting specific tight junction proteins such as ZO-1 and occludin thereby inhibiting the systemic spread of microbial toxins (Wang et al. 2024). Prebiotics play a significant role in the management of metabolic disturbances such as obesity, type II diabetes mellitus, and accompanying dysfunctions via the prebiotic-selective stimulation of growth of beneficial bacteria such as *Bifidobacterium, Lactobacillus*, and *Akkermansia muciniphila*. In obesity, supplementation with compounds such as oligofructose-enriched inulin has been found to reduce body weight, BMI, and visceral and subcutaneous adiposity. These influences are largely attributed to improved appetite control and greater satiety, resulting in reduced total energy intake. Beyond inulin, other prebiotics such as unsaturated alginate oligosaccharides and polysaccharides derived from acorn or sago have demonstrated the ability to improve obesity-associated metabolic parameters and protect against insulin resistance in high-fat diet models. Chicory root inulin has been corelated with improved blood glucose control, liver function, calcium homeostasis, and cardiovascular markers in diabetic individuals. Prebiotics also help alleviate metabolic endotoxemia by reducing circulating levels of LPS and inflammatory markers like plasminogen activator inhibitor-1 (PAI-1) contributing to improved immune and metabolic profiles (Li et al. 2020; Ning et al. 2017) (Fig. 6.16).

Prebiotics, probiotics, and postbiotics act synergistically to promote intestinal health and immune system. Prebiotics induce the growth of beneficial microbes, whereas probiotics exert direct action on gut microbiota equilibrium. Postbiotics, which are the result of microbial metabolism, also play a role in immune modulation. These agents induce SCFA production as well as mucus secretion, enhancing the integrity of the intestinal barrier. They also augment the secretion of antimicrobial peptides such as α-defensins and IgA, fortifying mucosal defense. In interactions with dendritic cells, the bioactive cause the differentiation of naive T cells into T helper 1 (Th1), T helper 2 (Th2), or regulatory T cell subsets. Immune modulation through this effect generates enhanced cell-mediated immunity (enhanced IL-12, IFN-γ, IL-2) and modulated humoral responses (controlled IL-13, IL-4, IL-5, IL-6) to achieve well-balanced immune function and reduced inflammation.

6.9.1.3 Autoimmune Diseases

Autoimmune diseases like celiac disease and systemic lupus erythematosus (SLE) occur when the body's immune system attacks its own tissues mistaking them as foreign and damages them. SLE is a complex disease that affects many organs. Celiac disease is a condition where consuming gluten triggers an immune attack on

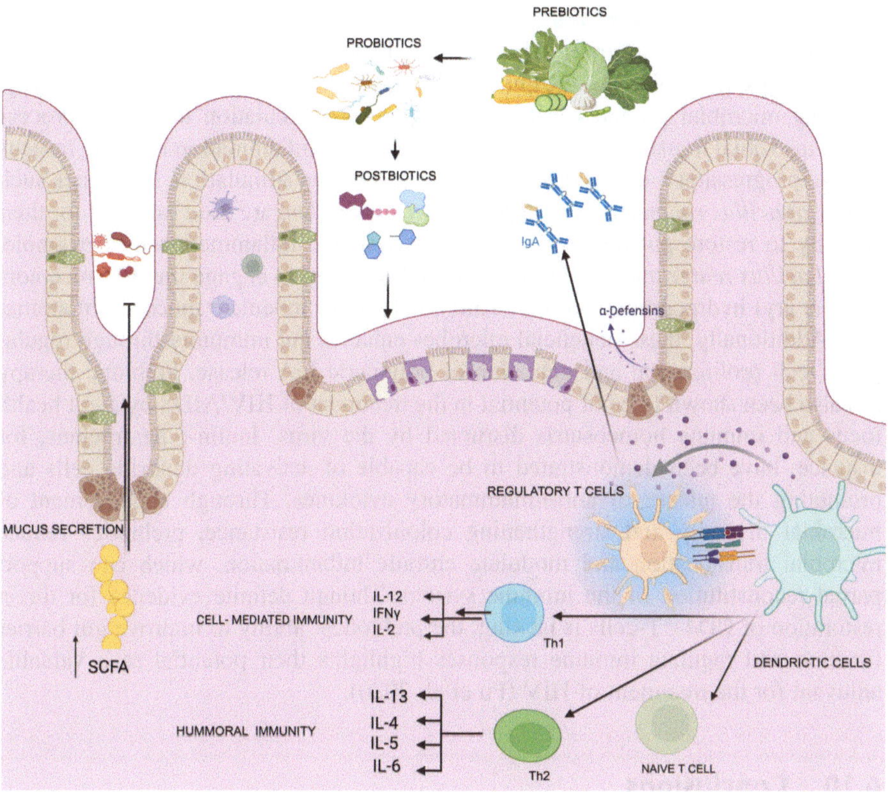

Fig. 6.16 Immunomodulatory effects of probiotics, prebiotics, and postbiotics in gut health

the small intestine. This damage can be worsened by an altered gut microbiota conditions like decrease in the number of good bacteria such as *Bifidobacterium* and enhanced levels of certain bacteria such as *Ruminococcus gnavus* (Habiballah et al. 2025). Probiotic strains like *Bifidobacterium* and *Lactobacillus* are observed to reduce inflammation, balance immune responses, and improve symptoms. These beneficial effects are linked to improvements in gut barrier function and reductions in harmful immune activity. Probiotics are increasingly being recognized as powerful therapeutic agents in the treatment of autoimmune diseases, primarily based on their ability to reconstitute the balance of gut microbes and modulate immune responses. Experimental studies have demonstrated that a few strains such as *Bifidobacterium bifidum*, *Lactobacillus casei*, and *Lactobacillus reuteri* exert desirable effects by suppressing inflammation and enhancing immune tolerance. In systemic lupus erythematosus (SLE), probiotics have been shown to improve the proliferation of regulatory T cells (Tregs) decrease the production of autoantibodies, and inflammation responses. Similarly, in celiac disease, probiotics may assist to improve gut barrier function and immune homeostasis, at least in part, by the production of protective metabolites such as short-chain fatty acids (SCFAs), which regulate inflammation and maintain mucosal integrity (Liu et al. 2018).

6.9.1.4 Infectious Diseases

HIV infection results in progressive immunodepletion by infecting and destroying CD4+ T cells, particularly of the gut. This damage compromises gut barrier integrity, allowing microbial products to translocate into the circulation through a process called microbial translocation. The resultant systemic inflammation not only hastens disease progression but also provokes chronic immune stimulation. Probiotics such as *Lactobacillus rhamnosus* and *Bifidobacterium bifidum* are used assessed for their capacity to restore gut integrity and reduce systemic inflammation. For example, *Lactobacillus reuteri* metabolites have been discovered to engage immune receptors like the aryl hydrocarbon receptor (AhR), helping to modulate mucosal inflammation. Additionally, these beneficial microbes enhance gut immunity through regulation T-cell proliferation and induction of polymeric IgA release. Prebiotic therapy has also been shown to be of potential in the treatment of HIV/AIDS by a gut health focus and immune homeostasis disrupted by the virus. Inulin-type fructans, for instance, have been demonstrated to be capable of activating dendritic cells and promoting the release of anti-inflammatory cytokines. Through enhancement of microbial diversity and strengthening colonization resistance, prebiotics reduce microbial translocation and modulate chronic inflammation, which can support partial reconstitution of the immune system. Though definite evidence for direct restoration of CD4+ T-cells is lacking, the prebiotics' ability to improve gut barrier function and regulate immune responses highlights their potential as a valuable adjuvant for the treatment of HIV (Fu et al. 2020).

6.10 Conclusions

Based on the complementary functions of probiotics, prebiotics, synbiotics, postbiotics, and next-generation probiotics (NGPs), the fields of nutrition, gut health, and microbiome treatment are undergoing a revolutionary shift. Each of these plays a distinct but complementary role in the overall health of the host. They achieve this by regulating gut microbiota, direct action on the balance and composition of the helpful bacteria within our gastrointestinal tract, boosting immune function, enhancing metabolic and neurological health, facilitating more effective energy processing and brain well-being.

Live viable health-promoting microbes called probiotics and prebiotics, the substrates that supply them with nutrients, have been employed for thousands of years to repopulate the gut with a healthy population of microbes. When combined strategically, they form synbiotics, which act in concert with several synergistic actions that have a very powerful influence on gut survival, colonization, and metabolic function of the beneficial bacteria. A more recent and just as important category of compounds is postbiotics. They are nonviable microbial cells and the metabolites therefrom, offering a safe, stable, and functionally active source of inducing various desirable effects, including anti-inflammatory effects, helping to reduce inflammation in the body, immunomodulatory effects, controlling and stabilizing the immune system, metabolic effects, being beneficial to the manner in which the body processes nutrients, even without living organisms.

One of the noteworthy advances in this field is the introduction of next-generation probiotics (NGPs). NGPs are rationally selected gut-tuned microbes identified by cutting-edge techniques like high-throughput sequencing, metagenomics, and functional screening. Unlike traditional food-based probiotics, which are primarily consumed for general health benefits, next-generation probiotics (NGPs) are designed for specified health ailments and are more efficient to have increased, adhesion improved adhesion on the gut wall, acid/bile tolerance, greater resistance during the hostile digestive tract, and targeted activities against particular health issues. This aimed specificity has opened doors to therapeutic application in many diseases, both chronic and acute, psychiatric and neurological illness, and metabolic syndromes. Together, these microbial products offer a multifaced solution to maintaining and restoring health. They not only supplement beneficial organisms but also reshape the gut community and guide systemic function. This is a powerful integration of nutrition, microbiology, and medicine. Yet to reach their full therapeutic promise, further clinical evidence, safety trials, regulatory harmonization, and technological enhancement are needed. As scientific understanding continues to reveal the profound interconnections between the microbiome and human health, this bioactive stand poised to play an increasingly crucial role in personalized medicine, preventive dietetics, and microbiome-guided interventions.

In summary, probiotics and their modern counterparts, synbiotics, postbiotics, and next-probiotics, are not standalone solutions. Instead, they are interconnected pillars of an integrative microbiome-based model of health. Their judicious addition to diets, therapies, and public health programs has enormous potential for disease control, wellness promotion, and shaping the future of functional nutrition.

References

Abbasi S, Aghdaei HA, Zali MR, Kenarkoohi A (2022) Probiotics and prebiotics in gastrointestinal disorders: mechanisms, efficacy, and safety. Front Nutr 9:925468

Abouelela ME, Helmy YA (2024) Next-generation probiotics as novel therapeutics for improving human health: current trends and future perspectives. Microorganisms 12(3):430

Abraamyan F, Gharibyan H, Hovhannisyan L, Gabrielyan A (2024) A comprehensive analysis of efficacy and safety of probiotics in gastrointestinal diseases. J Gastroenterol Hepatol Res. Advance online publication

Afzaal M, Saeed F, Hussain S, Mohamed AA, Alamri MS, Ahmad A et al (2020) Survival and storage stability of encapsulated probiotic under simulated digestion conditions and on dried apple snacks. Food Sci Nutr 8(10):5392–5401

Aguilar-Toalá JE, Garcia-Varela R, Garcia HS, Mata-Haro V, González-Córdova AF, Vallejo-Cordoba B, Hernández-Mendoza A (2018) Postbiotics: an evolving term within the functional foods field. Trends Food Sci Technol 75:105–114

Aguilar-Toalá JE, Garcia-Varela R, Garcia HS, Mata-Haro V, González-Córdova AF, Vallejo-Cordoba B, Hernández-Mendoza A (2020) Postbiotics: an evolving term within the functional foods field. Trends Food Sci Technol 92:26–39

Al-Fakhrany OM, Elekhnawy E (2024) Next-generation probiotics: the upcoming biotherapeutics. Mol Biol Rep 51(1):505

Anderson RC, Cookson AL, McNabb WC, Park Z, McCann MJ, Kelly WJ, Roy NC (2010) Lactobacillus plantarum MB452 enhances the function of the intestinal barrier by increasing the expression levels of genes involved in tight junction formation. BMC Microbiol 10(1):316

Anukam KC, Reid G (2007) Probiotics: 100 years (1907–2007) after Elie Metchnikoff's observation. Commun Curr Res Educ Topics Trends Appl Microbiol 1:466–474

Arora M, Baldi A (2015) Regulatory categories of probiotics across the globe: a review representing existing and recommended categorization. Indian J Med Microbiol 33:S2–S10

Bhattacharjee G (2022) Gut microbiota in gastrointestinal diseases. Curr Opin Gastroenterol 38(2): 109–115

Bottacini F, Van Sinderen D, Ventura M (2017) Omics of bifidobacteria: research and insights into their health-promoting activities. Biochem J 474(24):4137–4152

Bron PA, Kleerebezem M, Brummer RJ, Cani PD, Mercenier A, MacDonald TT et al (2017) Can probiotics modulate human disease by impacting intestinal barrier function? Br J Nutr 117(1): 93–107

Burton JP, Chilcott CN, Moore CJ, Speiser G, Tagg JR (2006) A preliminary study of the effect of probiotic Streptococcus salivarius K12 on oral malodour parameters. J Appl Microbiol 100(4): 754–764

Campaniello D, Bevilacqua A, Speranza B, Racioppo A, Sinigaglia M, Corbo MR (2023) A narrative review on the use of probiotics in several diseases. evidence and perspectives. Front Nutr 10:Article 1209238

Canani RB, Di Costanzo M, Leone L, Bedogni G, Brambilla P, Cianfarani S et al (2011) Epigenetic mechanisms elicited by nutrition in early life. Nutr Res Rev 24(2):198–205

Caramia G, Silvi S (2011) Probiotics: from the ancient wisdom to the actual therapeutical and nutraceutical perspective. In: Probiotic bacteria and enteric infections: cytoprotection by probiotic bacteria. Springer, Dordrecht, pp 3–37

Carpi RZ, Barbalho SM, Sloan KP, Laurindo LF, Gonzaga HF, Grippa PC et al (2022) The effects of probiotics, prebiotics and synbiotics in non-alcoholic fat liver disease (NAFLD) and non-alcoholic steatohepatitis (NASH): a systematic review. Int J Mol Sci 23(15):8805

Champagne CP, da Cruz AG, Daga M (2018) Strategies to improve the functionality of probiotics in supplements and foods. Curr Opin Food Sci 22:160–166

Chang CJ, Lin TL, Tsai YL, Wu TR, Lai WF, Lu CC, Lai HC (2019) Next generation probiotics in disease amelioration. J Food Drug Anal 27(3):615–622

Cheng J, Ouwehand AC (2020) Gastroesophageal reflux disease and probiotics: systematic review on the efficacy of probiotics in alleviating symptoms. J Clin Gastroenterol 54(8):655–664. [PMC Article]

Cook MT, Tzortzis G, Charalampopoulos D, Khutoryanskiy VV (2012) Microencapsulation of probiotics for gastrointestinal delivery. J Control Release 162(1):56–67

Cordaillat-Simmons M, Rouanet A, Pot B (2020) Live biotherapeutic products: the importance of a defined regulatory framework. Exp Mol Med 52(9):1397–1406

Cristofori F, Dargenio VN, Dargenio C, Miniello VL, Barone M, Francavilla R (2021) Anti-inflammatory and immunomodulatory effects of probiotics in gastrointestinal disorders. Front Immunol 12:578386

De Souza HS, Fiocchi C (2016) Immunopathogenesis of IBD: current state of the art. Nat Rev Gastroenterol Hepatol 13(1):13–27

Deng H, Li Z, Tan Y, Guo Z, Liu Y, Wang Y et al (2016) A novel strain of Bacteroides fragilis enhances phagocytosis and polarises M1 macrophages. Sci Rep 6(1):29401

Fontana L, Bermudez-Brito M, Plaza-Diaz J, Munoz-Quezada S, Gil A (2013) Sources, isolation, characterisation and evaluation of probiotics. Br J Nutr 109(S2):S35–S50

Ford AC, Wright-Hughes A, Alderson SL, Ow PL, Ridd MJ, Foy R et al (2023) Amitriptyline at low-dose and titrated for irritable bowel syndrome as second-line treatment in primary care (ATLANTIS): a randomised, double-blind, placebo-controlled, phase 3 trial. Lancet 402 (10414):1773–1785

Fu Y-S, Chu Q-S, Ashuro AA, Di D-S, Zhang Q, Liu X-M, Fan Y-G (2020) The effect of probiotics, prebiotics, and synbiotics on CD4 counts in HIV-infected patients: a systematic review and meta-analysis. BioMed Res Int, Article 7947342

Fukuda S, Toh H, Hase K, Oshima K, Nakanishi Y, Yoshimura K et al (2011) Bifidobacteria can protect from enteropathogenic infection through production of acetate. Nature 469(7331): 543–547

Fuller R (1992) History and development of probiotics. In: Probiotics: the scientific basis. Springer Netherlands, Dordrecht, pp 1–8

Fuller R (2012) Probiotics: the scientific basis. Springer, Berlin/Heidelberg

Gänzle M (2019) Fermented foods. In: Food microbiology: fundamentals and frontiers, pp 855–900

Gao B, Ahmad MF, Nagy LE, Tsukamoto H, Zakhari S (2019) Inflammatory pathways in alcoholic steatohepatitis. Gut Liver 13(4):403–415

Gbassi GK, Vandamme T, Ennahar S, Marchioni E (2009) Microencapsulation of Lactobacillus plantarum spp in an alginate matrix coated with whey proteins. Int J Food Microbiol 129(1): 103–105

Gérard P, Lepercq P, Leclerc M, Gavini F, Raibaud P, Juste C (2007) Bacteroides sp. strain D8, the first cholesterol-reducing bacterium isolated from human feces. Appl Environ Microbiol 73(18): 5742–5749

Gogineni VK, Morrow LE, Gregory PJ, Malesker MA (2013) Probiotics: history and evolution. J Anc Dis Prev Rem 1(2):1–7

Gong D, Shi Y, Chen D (2018) Peptic ulcer disease and Helicobacter pylori infection: microbial dysbiosis and pathogenesis. J Gastrointest Disord 5(3):111–120

Guarino MPL, Altomare A, Emerenziani S, Di Rosa C, Ribolsi M, Balestrieri P et al (2020) Mechanisms of action of prebiotics and their effects on gastro-intestinal disorders in adults. Nutrients 12(4):1037

Guo P, Zhang K, Ma X, He P (2020) Clostridium species as probiotics: potentials and challenges. J Anim Sci Biotechnol 11(1):24

Habiballah DN, Li F, Jiang L (2025) Immune metabolic restoration in systemic lupus erythematosus: role of probiotics L. casei and L. reuteri. Front Immunol 16:1602235

Haukioja A (2010) Probiotics and oral health. Eur J Dentistry 4(3):348–355

Herdiana Y (2023) Functional food in relation to Gastroesophageal Reflux Disease (GERD). Nutrients 15(16):Article 3583

Hiippala K, Jouhten H, Ronkainen A, Hartikainen A, Kainulainen V, Jalanka J, Satokari R (2018) The potential of gut commensals in reinforcing intestinal barrier function and alleviating inflammation. Nutrients 10(8):988

Hill C, Guarner F, Reid G, Gibson GR, Merenstein DJ, Pot B et al (2014) Expert consensus document. The international scientific association for probiotics and prebiotics consensus statement on the scope and appropriate use of the term probiotic. Nat Rev Gastroenterol Hepatol 11(8):506–514

Jadhav A, Jagtap S, Vyavahare S, Sharbidre A, Kunchiraman B (2023) Reviewing the potential of probiotics, prebiotics and synbiotics: advancements in treatment of ulcerative colitis. Front Cell Infect Microbiol 13:1268041

Jan T, Negi R, Sharma B, Kumar S, Singh S, Rai AK et al (2024) Next generation probiotics for human health: An emerging perspective. Heliyon

Jobin C (2018) Precision medicine using microbiota. Science 359(6371):32–34

Jones ML, Martoni CJ, Parent M, Prakash S (2012) Cholesterol-lowering efficacy of a microencapsulated bile salt hydrolase-active Lactobacillus reuteri NCIMB 30242 yoghurt formulation in hypercholesterolaemic adults. Br J Nutr 107(10):1505–1513

Kadooka Y, Sato M, Imaizumi K, Ogawa A, Ikuyama K, Akai Y et al (2010) Regulation of abdominal adiposity by probiotics (Lactobacillus gasseri SBT2055) in adults with obese tendencies in a randomized controlled trial. Eur J Clin Nutr 64(6):636–643

Kailasapathy K (2002) Microencapsulation of probiotic bacteria: technology and potential applications. Curr Issues Intest Microbiol 3(2):39–48

Kalliomäki M, Salminen S, Arvilommi H, Kero P, Koskinen P, Isolauri E (2001) Probiotics in primary prevention of atopic disease: a randomised placebo-controlled trial. Lancet 357(9262): 1076–1079

Khan ST, Malik A (2019) Next-generation probiotics their molecular taxonomy and health benefits. In: Health and safety aspects of food processing technologies. Springer, Cham, pp 471–500

Kiple KF, Ornelas KC (eds) (2000) The Cambridge world history of food, vol 2. Cambridge University Press

Kho ZY, Lal SK (2018) The human gut microbiome—a potential controller of wellness and disease. Front Microbiol 9:356589

Koh A, De Vadder F, Kovatcheva-Datchary P, Bäckhed F (2016) From dietary fiber to host physiology: short-chain fatty acids as key bacterial metabolites. Cell 165(6):1332–1345

Kwon HK, Lee CG, So JS, Chae CS, Hwang JS, Sahoo A et al (2010) Generation of regulatory dendritic cells and CD4+ Foxp3+ T cells by probiotics administration suppresses immune disorders. Proc Natl Acad Sci 107(5):2159–2164

Lebeer S, Vanderleyden J, De Keersmaecker SC (2010) Host interactions of probiotic bacterial surface molecules: comparison with commensals and pathogens. Nat Rev Microbiol 8(3): 171–184

Lee SH, Lee KY, Kim SH (2024) Bifidobacterium lactis IDCC 4301 (B. lactis Fit™) effects on obesity and metabolic health parameters. Mol Nutr Food Res 67(3):e2200789

Lee C, Pei L, Park H, Kim H, Huh CS (2025) Skin protection effects of Lactobacillus paragasseri HN910 lysate and the role of alanine. Probiot Antimicrob Proteins:1–18

Levin R (2017) Probiotics: the first 10 000 years. In: Probiotic dairy products, pp 17–36

Leylabadlo HE, Heravi FS, Soltani E, Abbasi A, Kafil HS, Parsaei M et al (2022) The role of gut microbiota in the treatment of irritable bowel syndrome. Rev Res Med Microbiol 33(1):e89–e104

Li S, Wang L, Liu B, He N (2020) Unsaturated alginate oligosaccharides attenuated obesity-related metabolic abnormalities by modulating gut microbiota in high-fat-diet mice. Food Funct 11(5): 4443–4454

Li P, Zheng Y, Chen X, Zhou D (2021a) Pharmacotherapy for inflammatory bowel disease: advances and perspectives. Front Pharmacol 12:670677

Li H-Y, Zhou D-D, Gan R-Y, Huang S-Y, Zhao C-N, Shang A, Xu X-Y, Li H-B (2021b) Effects and mechanisms of probiotics, prebiotics, synbiotics, and postbiotics on metabolic diseases targeting gut microbiota: a narrative review. Nutrients 13(9):Article 3211

Li Y, Li W, Zeng Z, Han Y, Chen Q, Dong X et al (2023) Lasso peptide MccY alleviates non-typhoidal salmonellae-induced mouse gut inflammation via regulation of intestinal barrier function and gut microbiota. Microbiol Spectrum 11(6):e01784–e01723

Liang B, Huang J, Liu S, Huang H (2022) Current and future perspectives for Helicobacter pylori eradication therapies. Front Cell Infect Microbiol 12:1042070

Liu Y, Alookaran JJ, Rhoads JM (2018) Probiotics in autoimmune and inflammatory disorders. Nutrients 10(10):Article 1537

Liu J, Ding W, Yuan Y (2023) Microbiota-targeted interventions for gastrointestinal disorders: current advances and future perspectives. Front Cell Infect Microbiol 13:1174523

Loman BR, Hernandez-Saavedra D, An R, Rector RS (2018) Prebiotic and probiotic treatment of nonalcoholic fatty liver disease: a systematic review and meta-analysis. Nutr Rev 76(11): 822–839

Lukjancenko O, Ussery DW, Wassenaar TM (2012) Comparative genomics of Bifidobacterium, Lactobacillus and related probiotic genera. Microb Ecol 63(3):651–673

Lyra A, Krogius-Kurikka L, Nikkilä J, Malinen E, Kajander K, Kurikka K et al (2010) Effect of a multispecies probiotic supplement on quantity of irritable bowel syndrome-related intestinal microbial phylotypes. BMC Gastroenterol 10(1):110

Macchione IG, Lopetuso LR, Ianiro G, Napoli M, Gibiino G, Rizzatti G et al (2019) Akkermansia muciniphila: key player in metabolic and gastrointestinal disorders. Eur Rev Med Pharmacol Sci 23(18)
Mar Rodríguez M, Pérez D, Javier Chaves F, Esteve E, Marin-Garcia P, Xifra G et al (2015) Obesity changes the human gut mycobiome. Sci Rep 5(1):14600
Martyniak M, Ligeza J, Walkowiak J (2021) Exploring the gut microbiome's role in inflammatory bowel disease. Int J Mol Sci 22(7):3714
Maslennikov R, Pavlov C, Ivashkin V (2018) Small intestinal bacterial overgrowth in cirrhosis: systematic review and meta-analysis. Hepatol Int 12(6):567–576
Messaoudi M, Lalonde R, Violle N, Javelot H, Desor D, Nejdi A et al (2011) Assessment of psychotropic-like properties of a probiotic formulation (Lactobacillus helveticus R0052 and Bifidobacterium longum R0175) in rats and human subjects. Br J Nutr 105(5):755–764
Miele E, Pascarella F, Giannetti E, Quaglietta L, Baldassano RN, Staiano A (2009) Effect of a probiotic preparation (VSL# 3) on induction and maintenance of remission in children with ulcerative colitis. Off J Am College Gastroenteroll ACG 104(2):437–443
Molina-Tijeras JA, Gálvez J, Rodríguez-Cabezas ME (2019) The immunomodulatory properties of extracellular vesicles derived from probiotics: a novel approach for the management of gastrointestinal diseases. Nutrients 11(5):1038
Nagao-Kitamoto H, Kitamoto S, Kuffa P, Kamada N (2016) Pathogenic role of the gut microbiota in gastrointestinal diseases. Intest Res 14(2):127–138
Nami Y, Abdullah N, Haghshenas B, Radiah D, Rosli R, Khosroushahi AY (2014) Assessment of probiotic potential and anticancer activity of newly isolated vaginal bacterium Lactobacillus plantarum 5BL. Microbiol Immunol 58(9):492–502
Nasiri H, Valedkarimi Z, Akbari M (2025) Gut–brain–immune axis implications for mental health and disease. Mol Cell Biochem:1–23
Neef A, Sanz Y (2013) Future for probiotic science in functional food and dietary supplement development. Curr Opin Clin Nutr Metab Care 16(6):679–687
Negash AW, Tsehai BA (2020) Current applications of bacteriocin. Int J Microbiol 2020(1): 4374891
Ng QX, Lim YL, Yaow CYL, Ng WK, Thumboo J, Liew TM (2023) Effect of probiotic supplementation on gut microbiota pubs.acs.org/JAFC review in patients with major depressive disorders: a systematic review. Nutrients 15(6):1351
Ning C, Zhang S, Wang S (2017) Chicory inulin ameliorates type 2 diabetes mellitus and glucolipid metabolism by modulating insulin receptor substrate and MAPK pathways. Mol Nutr Food Res 61(10)
O'Callaghan A, Van Sinderen D (2016) Bifidobacteria and their role as members of the human gut microbiota. Front Microbiol 7:925
O'Mahony L, McCarthy J, Kelly P, Hurley G, Luo F, Chen K et al (2005) Lactobacillus and bifidobacterium in irritable bowel syndrome: symptom responses and relationship to cytokine profiles. Gastroenterology 128(3):541–551
O'Toole PW, Marchesi JR, Hill C (2017) Next-generation probiotics: the spectrum from probiotics to live biotherapeutics. Nat Microbiol 2(5):1–6
Ojima MN, Nakanishi Y, Baba Y (2020) Bifidobacterium bifidum suppresses gut inflammation by modulating host immune response. Front Microbiol 11:1349
Pan Y, Zhang Y, Li X, Chen H, Wang J, Liu Q (2024) Efficacy of probiotics, prebiotics, and synbiotics on liver injury and metabolism in non-alcoholic fatty liver disease: a meta-analysis. BMC Gastroenterol 24:356
Parada Venegas D, De la Fuente MK, Landskron G, González MJ, Quera R, Dijkstra G, Harmsen HJM, Faber KN, Hermoso MA (2019) Short chain fatty acids (SCFAs) mediated gut epithelial and immune regulation and its relevance for inflammatory bowel diseases. Front Immunol 10: 277

Parisa A, Roya G, Mahdi R, Shabnam R, Maryam E, Malihe T (2020) Anti-cancer effects of Bifidobacterium species in colon cancer cells and a mouse model of carcinogenesis. PLoS One 15(5):e0232930

Peng X, Xian H, Ge N, Hou L, Tang T, Xie D et al (2024) Effect of probiotics on glycemic control and lipid profiles in patients with type 2 diabetes mellitus: a randomized, double blind, controlled trial. Front Endocrinol 15:1440286

Phùng TTT, Karbowiak T (2024) Biomacromolecules-based encapsulation technologies for the protection of oxygen-sensitive next-generation probiotics. Food Biomacromol 1(2):88–106

Plaza-Díaz J, Ruiz-Ojeda FJ, Vilchez-Padial LM, Gil A (2017) Evidence of the anti-inflammatory effects of probiotics and synbiotics in intestinal chronic diseases. Nutrients 9(5):Article 555

Podolsky SH (2012) Metchnikoff and the microbiome. Lancet 380(9856):1810–1811

Porcari S, Mullish BH, Asnicar F, Ng SC, Zhao L, Hansen R et al (2025) International consensus statement on microbiome testing in clinical practice. Lancet Gastroenterol Hepatol 10(2): 154–167

Qin J, Li R, Raes J, Arumugam M, Burgdorf KS, Manichanh C et al (2010) A human gut microbial gene catalogue established by metagenomic sequencing. Nature 464(7285):59–65

Recharla N, Chen W, Qin Y (2023) Gut microbial metabolite butyrate and its therapeutic role in inflammatory bowel disease. Front Microbiol 14:1472776

Reid G (2008) Probiotic Lactobacilli for urogenital health in women. J Clin Gastroenterol 42: S234–S236

Russo E, Giudici F, Fiorindi C, Ficari F, Scaringi S, Amedei A (2019) Immunomodulating activity and therapeutic effects of short chain fatty acids and tryptophan post-biotics in inflammatory bowel disease. Front Immunol 10:2754

Saarela M, Mogensen G, Fonden R, Mättö J, Mattila-Sandholm T (2000) Probiotic bacteria: safety, functional and technological properties. J Biotechnol 84(3):197–215

Salas-Jara MJ, Ilabaca A, Vega M, García A (2016) Biofilm forming Lactobacillus: new challenges for the development of probiotics. Microorganisms 4(3):35

Sapkota H, Dasgupta S, Roy B, Pathan EK (2025) Human commensal bacteria: next-generation pro-and post-biotics for anticancer therapy. Front Bioscience-Elite 17(1):26809

Sarao LK, Arora M (2017) Probiotics, prebiotics, and microencapsulation: a review. Crit Rev Food Sci Nutr 57(2):344–371

Shyam KU, Krishnan R, Jeena K, Vijaysunderdeva G, Prasad KP (2021) Next-generation probiotics-future therapeutics for sustainable aquaculture. Aquaculture 25(4):23–26

Singh TP, Natraj BH (2021) Next-generation probiotics: a promising approach towards designing personalized medicine. Crit Rev Microbiol 47(4):479–498

Song XL, Liang J, Lin SZ, Xie YW, Ke CH, Ao D et al (2024) Gut-lung axis and asthma: a historical review on mechanism and future perspective. Clin Transl Allergy 14(5):e12356

Sornplang P, Piyadeatsoontorn S (2016) Probiotic isolates from unconventional sources: a review. J Anim Sci Technol *58*(1):26

Stanton C, Desmond C, Coakley M, Collins JK, Fitzgerald G, Ross RP (2003) Challenges facing development of probiotic containing functional foods. In: Handbook of fermented functional foods. CRC Press, Boca Raton, pp 27–58

Stasiak-Różańska L, Berthold-Pluta A, Pluta AS, Dasiewicz K, Garbowska M (2021) Effect of simulated gastrointestinal tract conditions on survivability of probiotic bacteria present in commercial preparations. Int J Environ Res Public Health 18(3):1108

Suez J, Zmora N, Elinav E (2020) Probiotics in the next-generation sequencing era. Gut Microbes 11(1):77–93

Sugahara H, Hirota T (2021) Probiotic beverages in Japan (some history and current developments). In: Probiotic beverages. Academic, pp 35–48

Szajewska H, Kołodziej M (2015) Systematic review with meta-analysis: Saccharomyces boulardii in the prevention of antibiotic-associated diarrhoea. Aliment Pharmacol Ther 42(7):793–801

Tilg H, Zmora N, Adolph TE, Elinav E (2020) The intestinal microbiota fuelling metabolic inflammation. Nat Rev Immunol 20(1):40–54

Tomova A, Husarova V, Lakatosova S, Bakos J, Vlkova B, Babinska K, Ostatnikova D (2015) Gastrointestinal microbiota in children with autism in Slovakia. Physiol Behav 138:179–187

Torp AM, Bahl MI, Boisen A, Licht TR (2022) Optimizing oral delivery of next generation probiotics. Trends Food Sci Technol 119:101–109

Tripathi MK, Giri SK (2014) Probiotic functional foods: survival of probiotics during processing and storage. J Funct Foods 9:225–241

Tsilingiri K, Rescigno M (2013) Postbiotics: what else? Benefic Microbes 4(1):101–107

Ullah F, Arbab M, Chang X, Bibi F, Muhammad A, Rehman A, Suleman M, Ullah A, Hassan F, Tian Y, Li Q (2025) Gut microbiota therapy in gastrointestinal diseases. Front Cell Dev Biol 13: 1514636

Vaghef-Mehrabany E, Alipour B, Homayouni-Rad A, Sharif SK, Asghari-Jafarabadi M, Zavvari S (2014) Probiotic supplementation improves inflammatory status in patients with rheumatoid arthritis. Nutrition 30(4):430–435

Varanoske AN, McClung HL, Sepowitz JJ, Halagarda CJ, Farina EK, Berryman CE et al (2022) Stress and the gut-brain axis: cognitive performance, mood state, and biomarkers of blood-brain barrier and intestinal permeability following severe physical and psychological stress. Brain Behav Immun 101:383–393

Verna EC, Lucak S (2010) Use of probiotics in gastrointestinal disorders: what to recommend? Ther Adv Gastroenterol 3(5):307–319. https://doi.org/10.1177/1756283X10373814

Villageliu DN, Samuelson DR (2022) The role of bacterial membrane vesicles in human health and disease. Front Microbiol 13:828704

Wang J (2025) Gut-liver axis in liver disease: from basic science to clinical applications and therapeutics. Mol Med Rep 28(1):Article 13375

Wang HB, Wang PY, Wang X, Wan YL, Liu YC (2012) Butyrate enhances intestinal epithelial barrier function via up-regulation of tight junction protein Claudin-1 transcription. Dig Dis Sci 57(12):3126–3135

Wang K, Xu T, Wang J, Zhang X (2024) Enhancing gut microbiota and microbial function with inulin supplementation to improve obesity-related metabolic abnormalities. Nat Commun 15: 1598

Wilson B, Whelan K (2017) Prebiotic inulin-type fructans and galacto-oligosaccharides: definition, specificity, function, and application in gastrointestinal disorders. J Gastroenterol Hepatol 32: 64–68

Xie G, Zhang S, Zheng X, Jia W (2013) Metabolomics approaches for characterizing metabolic interactions between host and its commensal microbes. Electrophoresis 34(19):2787–2798

Yao M, Xie J, Du H, McClements DJ, Xiao H, Li L (2020) Progress in microencapsulation of probiotics: a review. Compr Rev Food Sci Food Saf 19(2):857–874

Yasueda A, Mizushima T, Nezu R, Sumi R, Tanaka M, Nishimura J et al (2016) The effect of Clostridium butyricum MIYAIRI on the prevention of pouchitis and alteration of the microbiota profile in patients with ulcerative colitis. Surg Today 46:939–949

Yu J-X, Wu J, Chen X, Zang S-G, Li X-B, Wu L-P, Xuan S-H (2025) Gut microbiota in liver diseases: initiation, development and therapy. Front Med 12:1615839

Yue M, Wei J, Chen W, Hong D, Chen T, Fang X (2022) Neurotrophic role of the next-generation probiotic strain L. lactis MG1363-pMG36e-GLP-1 on Parkinson's disease via inhibiting ferroptosis. Nutrients 14(22):4886

Zang M, Sun X, Li M, Zhang B (2020) Microbiota-targeted therapies in digestive diseases: mechanisms of probiotics and their application in GERD treatment. Front Microbiol 11:1132

Zhai Q, Feng S, Arjan N, Chen W (2019) A next generation probiotic, Akkermansia muciniphila. Crit Rev Food Sci Nutr 59(19):3227–3236

Zhang W, Zhu B, Xu J, Liu Y, Qiu E, Li Z et al (2018) Bacteroides fragilis protects against antibiotic-associated diarrhea in rats by modulating intestinal defenses. Front Immunol 9:1040
Zhang H, Duan Y, Cai F, Cao D, Wang L, Qiao Z et al (2022) Next-generation probiotics: microflora intervention to human diseases. Biomed Res Int 2022(1):5633403
Zhang D, Yang Y, Lu W (2023) Regulatory role of short-chain fatty acids in inflammatory bowel disease: mechanisms and therapeutic potential. Cell Commun Signal 20(1):64
Zhao Y, Dong BR, Hao Q (2022) Probiotics for preventing acute upper respiratory tract infections. Cochrane Database Syst Rev 8

Role of Soil and Plant Microbiome in Sustainable Agriculture

7

Abstract

Microbiome linked to plants and soils are essential in determining the resilience, productivity and general health of agricultural ecosystems. With a focus on their contributions to sustainable agriculture, this chapter examines the intricate relationships that exist between plants and their microbial partners in the rhizosphere, phyllosphere, and endosphere. We start by providing an overview of the functional diversity and composition of soil and plant microbiota, emphasizing their contributions to disease prevention, nutrient cycling, and plant growth. There is a thorough discussion of mechanisms like microbial mediated immunity, mycorrhizal symbiosis, phosphate solubilization, and biological nitrogen fixation which demonstrate how these organic processes have the capacity to enhance crop productivity and ecological balance while lowering reliance on chemical inputs. The chapter also looks at how the diversity and function of microbiomes are affected by contemporary farming methods like tillage, monocropping, pesticide utilize and synthetic fertilization. The potential for creating targeted and environmentally friendly microbial interventions is examined in relation to developments in microbiome engineering such as; synthetic microbial consortia, precision agriculture instruments and omics-based technologies (metagenomics, transcriptomics and metabolomics). Successful field uses of microbial inoculants to elevate crop resilience and soil fertility are demonstrated by real world case studies. There are still large research gaps in spite of these developments. The ecological dynamics of microbial communities in field settings are still poorly understood. The commercial microbial products performance varies and formulation and delivery methods are not standardized. Furthermore, there are obstacles to widespread adoption because the regulatory environment surrounding microbiome based agricultural inputs is still developing. In order to create resilient, site specific microbial solutions future perspectives include combining microbiome science with artificial intelligence, predictive modeling and climate

R. C. Sobti et al., *Microbiota Unveiled: Science, Health, and Future Therapies*,
https://doi.org/10.1007/978-981-95-9031-5_7

smart agriculture. To close the gap between research and application long term field trials, microbiome preservation techniques and farmer participation must also be prioritized. All things considered, this chapter emphasizes how soil and plant microbiome has the capacity to be transformed to create a sustainable and biologically integrated agricultural future.

Keywords

Microbiome · Rhizosphere · Phyllosphere · Endosphere · Microbial symbiosis

7.1 Introduction

Microbiome play highly integral role in enhancing the soil health, it encourages plant growth and lessens the need for chemical fertilizers and pesticides so the microbiome is essential to sustainable agriculture. Together, the soil and plant microbiome enhance nutrient availability, enhance nutrient cycling, and strengthen plant resistance to environmental stressors. While soil microorganism's aid in nitrogen fixation, organic matter breakdown and pathogen suppression, plant-associated microbiome especially in the rhizosphere enhance nutrient uptake, encourage plant growth, and provide disease resistance. Sustainable agriculture has the capacity to elevate crop yields, enhance soil fertility, and lessen the environmental impact of conventional farming methods by utilizing the potential of these microbial communities.

7.1.1 Overview of Sustainable Agriculture

Sustainable agriculture refers to farming practices that meet current food and fiber needs without jeopardizing the ability of future generations to meet their own. It emphasizes integrating economically viable, environmentally and socially responsible strategies that maintain ecosystem health and elevate productivity. Crop rotation, conservation tillage, organic farming, and agroecology are some of the methods used in sustainable agriculture to lessen the adverse environmental effects of farming operations, including soil erosion, water pollution, and biodiversity loss (Gomiero et al. 2011). Because it increases soil fertility, water retention, and strengthens resilience to climate extreme all of which support the long-term viability and productivity of agricultural systems maintaining sustainable agriculture. Through the promotion of biological processes that naturally enhance plant growth and disease resistance such as those mediated by soil and plant microbiome, sustainable agriculture also aims to reduce dependency on synthetic fertilizers and chemicals (Altieri 2002).

7.2 Microbiome Significance in Agriculture

Microbiome enhance soil health, elevate crop productivity, and help suppress disease, so their communities (bacteria, fungi, viruses, and archaea) are essential to sustainable agriculture. Plant growth, nutrient uptake, and pathogen resistance are all directly impacted by interactions between plants and the microbiome that is associated with them especially the rhizosphere and endosphere microbiome. The soil's microorganisms aid in the cycling of nutrients (phosphorus solubilization and nitrogen fixation), enhance soil structure, and boost soil stability and water retention. Plant microbiome associated with plant roots, stems, and leaves aid in plant growth by producing hormones that support growth, disease resistance, and increased tolerance to environmental stressors (Berendsen et al. 2012). By helping crops adjust to changing environmental conditions, reducing the need for chemical inputs, and mitigating the effects of climate change, plant microbiome symbioses have the capacity to elevate the resilience of agricultural systems (Mendes et al. 2013). Because they enhance ecosystem services and lessen the need for artificial fertilizers and pesticides, microbiome is therefore essential to reaching the objectives of sustainable agriculture.

7.3 The Differences Between the Microbiome of Plants and Soil

The term microbiome refers to the entire population of microorganisms and their genetic material that live in a specific environment. For instance, the term human microbiome refers to the microbial communities that live on and within the human body. In agricultural systems, the term microbiome also refers to the microorganisms that reside in soil, on plants and within plant tissues. Importantly, soil and plant microbiome interact dynamically rather than existing as separate entities. The soil microbiome is the diverse community of microorganisms that live in the soil environment. These microbiomes interact with the soil matrix to affect the soil's fertility, structure and nutrient cycling. The breakdown of pollutants, fixing nitrogen in the atmosphere, solubilizing phosphorus, and decomposing organic matter all depend on soil microorganisms to maintain soil fertility and health (Berg et al. 2014). The soil microbiome has the capacity to also influence plant growth by influencing nutrient availability and employing biological control mechanisms to protect plants from disease (Lugtenberg and Kamilova 2009). The term plant microbiome refers to a collection of microorganisms that inhabit various plant tissues such as the rhizosphere (area surrounding roots), phyllosphere (surface of leaves), and endosphere (internal plant tissues). The plant microbiome affects growth and development and is essential to plant health through nutrient uptake, disease resistance, and stress tolerance (Vandenkoornhuyse et al. 2015).

Microorganisms interact with plant roots in the rhizosphere microbiome, an essential component of the plant microbiome that enhances nutrient uptake, particularly for essential nutrients like potassium, phosphorus, and nitrogen. Furthermore, rhizosphere microbes have the capacity to produce compounds like phytohormones that support plant growth and protect plants from pathogens by either outcompeting harmful microbes or producing antibiotics. Therefore, it has the capacity to be claimed that while the soil microbiome regulates soil health and nutrient availability, the plant microbiome directly affects plant growth and stress tolerance. The rhizosphere microbiome, the main interface between soil and plant microbiomes, mediates integral interactions that promote plant health and agricultural sustainability. Understanding the complex interactions between this microbiome has the capacity to enhance farming practices and aid in the development of sustainable agriculture solutions.

7.4 The Soil Microbiome

7.4.1 Composition

The term soil microbiome refers to the diverse group of microorganisms (bacteria, fungi, viruses, and archaea) that inhabit the soil environment. These microorganisms maintain soil health, promote plant growth, and support ecosystem processes. The soil microbiome is incredibly diverse due to the wide variations in microbial populations across soil types, climates, and agricultural practices. Soil microbial communities are influenced by a variety of factors, such as soil pH, moisture content, organic matter content, and plant root presence. The rhizosphere often contains more diverse and abundant soil microbial communities than bulk soil (Fig. 7.1) (Berendsen et al. 2012; Fierer et al. 2012).

1. Bacteria: Which comprise a wide variety of species engaged in processes like pathogen suppression, organic matter breakdown, and nutrient cycling, are the most common and diverse group in the soil microbiome. Common soil bacteria belong to the genera *Pseudomonas, Bacillus, Rhizobium*, and *Actinobacteria* (Mendes et al. 2013).
2. Fungi: Soil fungi, like those in the genera *Aspergillus* and *Trichoderma*, play a substantial role in the decomposition of organic matter and the formation of soil aggregates. Furthermore, by facilitating the uptake of nutrients and water, mycorrhizal fungi are essential to symbiotic relationships with plants.
3. Archaea: Although less common than bacteria, archaea are essential to soil ecosystems, particularly for nitrogen cycling and methane oxidation. For example, the genus *Methanobacterium* aids in the critical process of methane reduction in wetland soils.
4. Viruses: *Bacteriophages*, the main type of soil virus, infect bacteria and have the power to regulate bacterial populations, which has an impact on microbial diversity and community structure. Through horizontal gene transfer, these interactions might also support microbial evolution.

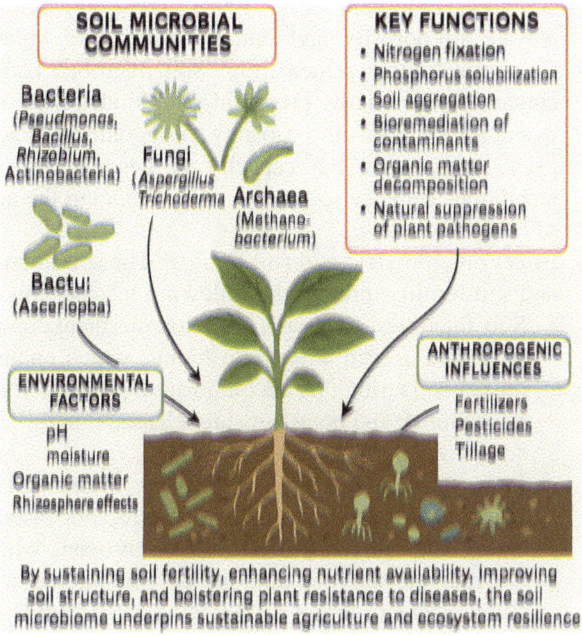

Fig. 7.1 Soil microbiome: composition and functions

Chemical fertilizers, pesticides, and tillage methods are examples of anthropogenic factors that have an impact on soil microbiome. According to Lautenberg and Kamilova (2009), these actions may disrupt microbial communities and lower biodiversity.

7.4.2 Functions of the Soil Microbiome

The soil microbiome primarily supports soil health and plant productivity in the following ways:

1. Nutrient cycling: (a) Nitrogen fixation: one nutrient that plants require for growth is nitrogen. In order to convert atmospheric nitrogen into a form that plants have the capacity to utilize, legumes develop symbiotic relationships with nitrogen-fixing bacteria like *Rhizobium* species. This process reduces the need for synthetic nitrogen fertilizers (b). Phosphorus solubilization: two kinds of soil bacteria and fungi *Bacillus* and *Penicillium* generate acids that break down phosphorus minerals, thereby making more phosphorus available to plants. This process is crucial in soils with low phosphorus availability (Richardson et al. 2009).
2. Aggregation and soil structure: Microbial activity significantly affects soil structure. Microorganisms, particularly fungi, produce extracellular polysaccharides that bind soil particles together, promoting soil aggregation and improving soil

structure (Balesdent et al. 1998). These aggregates enhance water retention, elevate soil porosity, and reduce erosion all of which promote plant growth.
3. Contamination bioremediation: Soil microbes facilitate the bioremediation of contaminated soils by promoting the biodegradation of organic pollutants. For example, fungi has the capacity to degrade pesticides and heavy metals, while pseudomonas has the capacity to degrade petroleum hydrocarbons. These techniques enhance soil health while reducing the adverse environmental effects of human activity.
4. Effect on crop yield and soil fertility: The soil microbiome influences crop yield and soil fertility through the following mechanisms:
   - Soil microbes are responsible for the breakdown of organic matter such as plant and animal waste into humus which enhances soil structure and adds nutrients. During this decomposition process, vital nutrients such as nitrogen, phosphorus, and potassium are released into the soil for plant uptake (Bardgett and van der Putten 2014).
   - Increasing soil nutrient availability: When soil microorganisms are active, plants have greater access to nutrients. For instance, nitrogen-fixing bacteria enrich the soil with bioavailable nitrogen, whereas phosphate-solubilizing bacteria elevate the amount of phosphorus available in the soil. These techniques encourage more ecologically friendly farming practices by reducing dependency on chemical fertilizers.
   - Effect on plant disease suppression: Soil microbes have the capacity to prevent plant diseases by generating antibiotics, competing for resources or bolstering plant defenses. For example, bacteria in the genus *Bacillus* produce compounds that stop plant pathogens from growing, while fungal species such as *Trichoderma* elevate plant resistance to fungal pathogens (Lugtenberg and Kamilova 2009). This natural disease suppression encourages more ecologically friendly farming methods by reducing the need for chemical pesticides. Therefore, it has the capacity to be concluded that maintaining soil health, enhancing nutrient availability, encouraging the decomposition of organic matter, and strengthening plant resistance to disease and stress all depend on the soil microbiome. Utilizing the potential of the soil microbiome is essential to achieving sustainable agricultural practices that boost crop yields, promote environmental health and utilize fewer synthetic chemicals (Fig. 7.1)

Figure 7.1 illustrates soil microbiome composition (bacteria, fungi, viruses, and archaea) and their role in maintaining soil health and plant productivity. Among them, fungi (*Trichoderma* and mycorrhizae) break down organic matter and enhance soil aggregation while bacteria (*Rhizobium* & *Bacillus*) control nutrient cycling and pathogen suppression. Bacteriophages control microbial populations while archaea aid in the cycling of nitrogen and the reduction of methane. These communities work together to enhance soil fertility, nutrient availability, bioremediation, and natural disease resistance, laying the groundwork for sustainable agriculture.

7.5 Plant Microbiome

7.5.1 Structure and Significance

The term plant microbiome refers to the diverse community of microorganisms that inhabit the rhizosphere, roots, stems, leaves, flowers, and fruits. These microbes form dynamic relationships with their host that has a significant impact on the plant's growth, development, and health. The highest microbial density is found in the rhizosphere, but plants also contain microbial communities in their internal tissues such as endophytes which are microorganisms that reside inside plant tissues and epiphytes which are microorganisms that reside on the outside of the plant (Berendsen et al. 2012). (1) Bacteria: The most prevalent kind of microorganism linked to plants is bacteria. They promote growth, prevent disease, and cycle nutrients by colonizing plant surfaces and rhizospheres. According to Mendes et al. (2013), some examples are *Bacillus* (pathogen inhibitors), *Rhizobium* (nitrogen fixers), and *Pseudomonas* (growth promoters). (2) Fungi: Mycorrhizal fungi, in particular, form symbiotic connections with plant roots, enhancing the plants' resistance to infections and facilitating the uptake of nutrients especially phosphorus. Other fungi that suppress disease include Trichoderma. (3) Viruses: in the plant microbiome, viruses are less well understood. Plant-associated viruses have the potential to alter plant physiology and stress responses by influencing microbial dynamics and plant health. (4) Archaea and other microbes: although they are less prevalent, archaea are frequently found in plant rhizospheres and play a role in nutrient cycling processes such as nitrogen fixation. Furthermore, although their precise functions are still unclear, protists and other microbes have the capacity to interact with plants.

7.5.2 The Plant Microbiome Roles

Numerous biological processes essential to plant growth and survival are supported by plant microbiome which offers plants several advantages (Tables 7.1 and 7.3):

1. Plant growth promotion: (a). Nitrogen fixation: Some bacteria, like *Rhizobium* species, develop symbiotic relationships with legumes in which they fix atmospheric nitrogen into forms that plants have the capacity to utilize. This enhances soil health and lessens the need for artificial fertilizers (b). Phytohormone production: Auxins, cytokinins, and gibberellins are among the phytohormones produced by plant-associated microbes like Pseudomonas and Bacillus that influence plant growth, root development, and general plant vigor (Lugtenberg and Kamilova 2009). These microbes enhance the uptake of water and nutrients by promoting the growth of roots and shoots.
2. Pathogen defense and disease resistance: (a) Antimicrobial metabolites: a range of beneficial microorganisms produce antimicrobial compounds that stop plant pathogens from proliferating. For example, Bacillus and Pseudomonas produce

Table 7.1 Plant-microbe interaction mechanisms

Mechanism/ biological agents	Description	Key microbes involved	Functions/benefits	References
Quorum sensing	Microbial behavior is coordinated through the use of signaling molecules. Controls the nutrient cycle, pathogen resistance, and biofilm formation	Numerous bacteria linked to plants	Regulates the microbial population, improves communication, and controls biofilm	Mendes et al. (2013)
Biofilm formation	In order to create protective layers, microbial communities adhere to plant surfaces	*Bacillus* and *Pseudomonas*	Suppress infections, encourage plant development, and boost resistance to stress	Mendes et al. (2013)
Rhizosphere microbiome	Microbially active soil zone that is influenced by roots and produced by root exudates	Various types of soil fungi and bacteria	Increases microbial diversity, nutrient uptake, and promotes plant health	Philippot et al. (2013)
Plant Growth Promoting Rhizobacteria (PGPR)	Beneficial bacteria that help absorb nutrients and produce hormones in the rhizosphere	*Azospirillum*, *Pseudomonas*, and *Bacillus*	Disease prevention, phosphate solubilization and nitrogen fixation	Glick (2012)
Mycorrhizal fungi	Fungi that use hyphal networks to establish symbiotic relationships with plant roots	Arbuscular Mycorrhizal Fungi (AMF)	Increase nutrient and water uptake and strengthen resistance to drought and salinity	Smith and Read (2008)
Microbial cross talk	To draw in and mold microbial communities, plants emit chemical cues	Numerous microbiological species	Increases the diversity and functions of endophytes and modifies the root microbiome	Mendes et al. (2013)
Root exudate Mediated Microbial recruitment	Certain microorganisms are drawn to organic acids, sugars, and flavonoids	Beneficial fungi and bacteria that fix nitrogen	Inhibits pathogens and promotes good bacteria	Bais et al. (2006)
Biological control: natural agents	Pathogens are suppressed by beneficial microbes through metabolite production or competition	*Bacillus*, *Pseudomonas*, and *Trichoderma*	Mycoparasitism, antagonistic interactions, and systemic resistance induction (ISR)	Haas and Défago (2005)

(continued)

Table 7.1 (continued)

Mechanism/ biological agents	Description	Key microbes involved	Functions/benefits	References
Biocontrol strategies	Immuno-induction, parasitism, competition, and antibiosis	Biocontrol fungi and PGPR	Reduced disease and enhanced plant resistance	Combined sources

antibiotics that prevent harmful bacteria, viruses, and fungi from growing in soil or on plant surfaces (Raupach and Kloepper 1998); (b) Induced systemic resistance: the microbiota of a plant has the capacity to strengthen its defenses against infection. By releasing elicitors, beneficial microbes strengthen plant defense mechanisms and elevate resistance to bacterial, viral and fungal diseases (Pieterse et al. 2014).

3. Impact on plant growth and development: Through a variety of processes such as symbiosis, defense activation and direct influence on plant growth the plant microbiome significantly affects plant productivity and health. Rhizobia, mycorrhizal fungi, and other beneficial microbes in symbiosis: Rhizobia: these nitrogen fixing bacteria grow as nodules on the roots of legumes and transform atmospheric nitrogen into ammonium, which the plant has the capacity to utilize to grow. According to previous studies, this symbiosis greatly increases soil fertility and lessens the need for chemical nitrogen fertilizers. Species from the genera *Rhizophagus* and *Glomus* are examples of mycorrhizal fungi, which coexist symbiotically with plant roots. These fungi elevate the surface area available for absorbing water and nutrients especially in nutrient-poor soils by spreading their root systems.

7.5.3 Plant Microbe Interaction Mechanisms

1. Quorum sensing: It is a mechanism used by many microorganisms including those related to plants to coordinate their behavior. It involves the release and response of signaling molecules by microbial cells. Quorum sensing has the ability to control microbial biofilm formation, pathogen resistance, and nutrient cycling in the context of plant microbiomes (Fig. 7.2; Table 7.1).
2. Formation of biofilm: Communities of microorganisms known as biofilms cling to the surfaces of plants including roots and leaves to form protective coatings that have the capacity to aid in halting the colonization of pathogens. By outcompeting dangerous microorganisms and generating antimicrobial compounds, biofilm-forming bacteria such as *Pseudomonas* and *Bacillus* are known to suppress pathogens and encourage plant growth (Mendes et al. 2013). The dynamic relationship between the soil microbiome and plant roots affects

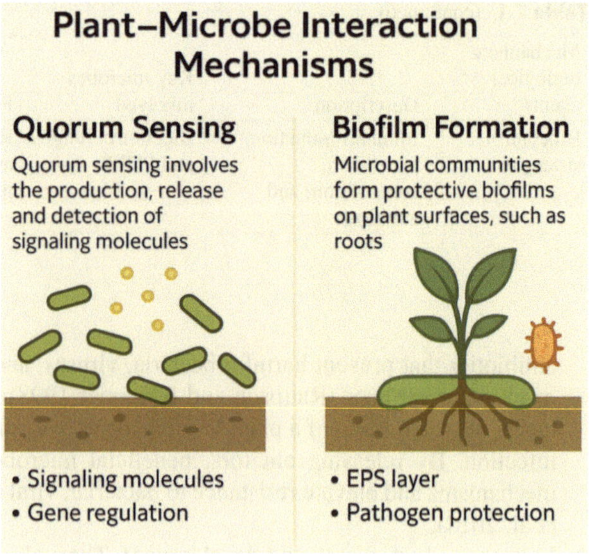

Fig. 7.2 Plant-microbe interaction mechanisms

plant nutrition, disease resistance, and overall productivity. Fostering beneficial microbial communities makes plants more resilient to biotic and abiotic stressors, leading to healthier and more productive crops (Fig. 7.2; Table 7.1).

Figure 7.2 illustrates the pivotal mechanisms of plant-microbe interactions. On the left-hand side (1) quorum sensing: it is the process of producing, releasing, and detecting signaling molecules that control microbial gene expression and allow community behavior to be coordinated while on the right-hand side; (2) Biofilm formation illustrates microbial communities forming protective biofilms on plant surfaces like roots which are composed of an extracellular polymeric substance layer that enhances pathogen protection and supports plant health.

7.5.4 The Soil Plant System Microbial Interactions

1. Rhizosphere microbiome: The rhizosphere is a small area of soil that is directly influenced by the microbial activity and secretions from roots. This area is an integral interface where plant roots and soil microorganisms interact dynamically to form a hotspot of biochemical activity and microbial diversity that has a substantial impact on plant health, nutrient uptake, and soil fertility. The rhizosphere is home to dense microbial populations which are often 10–100 times higher than bulk soil due to the presence of mucilage, sloughed off root cells, and root exudates (organic compounds secreted by roots) that serve as nutrients for microbes (Philippot et al. 2013).
2. Rhizobacteria and fungi: Rhizobacteria particularly plant growth-promoting rhizobacteria (PGPR) such as *Bacillus, Azospirillum,* and *Pseudomonas* create

phytohormones (indole-3-acetic acid) aid in the absorption of nutrients (phosphorus solubilization, nitrogen fixation), and inhibit pathogens by generating antibiotics (Glick 2012). By establishing extensive hyphal networks, mycorrhizal fungi such as arbuscular mycorrhizal fungi (AMF) enhance plant resistance to salinity and drought, boost phosphorus and micronutrient uptake, and foster disease resistance (Smith and Read 2008).

7.5.5 Interactions Between the Microbiome of Plants and Soil

The microbial populations in plants and soil are not isolated; rather, they form a networked system where cross-talk mechanisms help with microbial recruitment, communication, and functional integration (Table 7.1).

1. Microbial cross talk: Plant roots selectively draw soil microbes that colonize the rhizosphere by means of chemical cues and nutrient gradients. This cross-talk impacts both the root microbiome and the composition and function of endophytic communities (Mendes et al. 2013).
2. Microbial diversity and root exudates: Plants release a variety of compounds such as sugars, amino acids, organic acids, and flavonoids, which act as chemical cues for specific microbial communities. These root exudates affect the metabolic activity, diversity, and composition of the rhizosphere microbiome (Bais et al. 2006). For example, certain exudates draw in helpful microorganisms like bacteria that fix nitrogen, while others inhibit infections or serve as symbiotic signals.
3. Biological control of plant pathogens: The soil plant microbiome system plays a foundational role in a crucial role in lowering plant diseases by employing natural microorganisms to either strengthen the plant immune response or stop the growth of pathogens (Fig. 7.3; Tables 7.1 and 7.2).
4. Natural microbial agents: Microbial antagonists compete with pathogens for nutrients and space, produce antimicrobial metabolites, or induce systemic resistance in plants. Key biocontrol microbes include: (1) Trichoderma spp.: fungi that inhibit pathogenic fungi by secreting enzymes, producing secondary metabolites and myco-parasitism. (2) Bacillus species: produce lipopeptides, siderophores, and enzymes that prevent the growth of pathogens and their biofilms (Haas and Défago 2005). (3) Pseudomonas species: well known for generating antibiotics, phenazines, and hydrogen cyanide which are useful in treating a range of plant illnesses.
5. Strategies for biocontrol: (1) Competition: In the rhizosphere, beneficial microbes outcompete pathogens for ecological niches. (2) Antibiosis: The production of antibiotics and antifungal substances by PGPR or fungi. (3) Parasitism: Pathogens are directly attacked by microbes like Trichoderma. (4) Induction of host resistance: To prevent pathogen invasion, plant immune systems are activated.

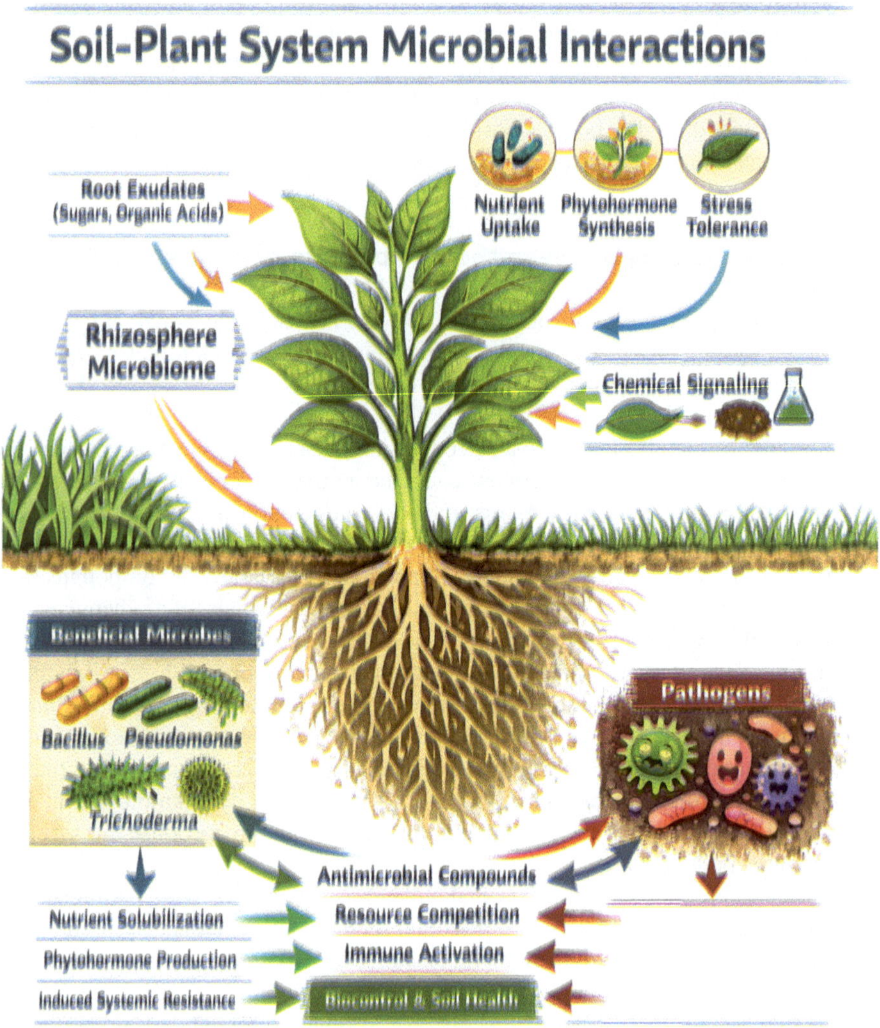

Fig. 7.3 Soil-plant system microbial interactions

Figure 7.3 illustrates how the rhizosphere microbiome which is nourished by root exudates promotes nutrient uptake, phytohormone synthesis, and stress tolerance to support plant growth. Microbial diversity and functional integration are impacted by chemical signaling between plants and soil. Beneficial microorganisms such as *Bacillus, Pseudomonas*, and *Trichoderma* compete with pathogens, produce antimicrobials, and trigger plant immune responses to provide biological control improving soil health and crop resilience.

Table 7.2 Typical microbial inoculants and their uses in agriculture

Microbial group	Exemplary genera/species	Principal purpose	Utilization in farming	References
Nitrogen-fixing bacteria	*Azotobacter*, *Azospirillum* *Rhizobium*	Fixation of atmospheric nitrogen by biological means	Increases the amount of nitrogen in soil and decreases the use of synthetic fertilizers	Vessey (2003) and Bashan and de-Bashan (2010)
Phosphate-solubilizing bacteria	*Burkholderia*, *Pseudomonas* *Bacillus*	Insoluble phosphates' solubilization in soil	Increases availability of phosphorus for plants	Sharma et al. (2013)
Potassium-solubilizing microbes	*Frateuria aurantia*, *Bacillus mucilaginosus*	Releases potassium from mica and other minerals	Increases crop resilience and potassium nutrition	Mahanty et al. (2017)
Mycorrhizal fungi	*Rhizophagus*, *Funneliformis*, and *Glomus*	Increases the absorption of water and nutrients, particularly phosphorus	Establishes symbiotic connections with plant roots	Smith and Read (2008)
Rhizobacteria that promote plant growth	*Streptomyces*, *Bacillus*, and *Pseudomonas*	Increases stress tolerance and produces phytohormones (auxins and cytokinins)	Increases resistance to abiotic stress and promotes plant growth	Glick (2012) and Lugtenberg and Kamilova (2009)
Biocontrol agents	*Trichoderma* spp., *Bacillus subtilis*, *Pseudomonas fluorescens*	Suppresses plant pathogens via competition, antibiosis and ISR	Controls root and foliar diseases naturally	Mahanty et al. (2017)
Organic matter degraders	*Bacillus*, *Actinobacteria*, and *Cellulomonas*	Breakdown of cellulose, lignin, and organic residues	Increases nutrient cycling and humus formation	Mahanty et al. (2017)
Sulfur-oxidizing bacteria	*Thiobacillus*	Transforms elemental sulfur into sulfate that plants can use	Fixes soils that lack sulfur	Mahanty et al. (2017)
Drought-resistant microbes	*Paenibacillus polymyxa* and *Bacillus Megaterium*	Increases root growth under water stress and produces exo-polysaccharides	Increases the resilience of crops in arid and semi-arid areas	Mahanty et al. (2017)

7.6 Techniques Based on the Microbiome for Sustainable Agriculture

The application of microbiome-based technologies offers promising prospects for agricultural sustainability by increasing crop productivity, decreasing chemical inputs, and improving ecosystem resilience. By utilizing advantageous microbial communities, these strategies promote soil fertility, plant health, and environmental sustainability (Fig. 7.4; Tables 7.2 and 7.3). (1) Inoculants and biofertilizers: microbial inoculants, sometimes referred to as biofertilizers, are blends of beneficial microorganisms that are sprayed on plants or soil to boost nutrient availability, encourage plant growth, and fortify disease resistance. (2) Manufacturing and application: common inoculant species include Arbuscular Mycorrhizal Fungi (AMF), phosphate-solubilizing bacteria (PSB), Rhizobium, *Azospirillum*, and *Azotobacter* (nitrogen fixation). These organisms form symbiotic relationships with crops by providing hormones that promote growth and facilitate crop nutrient absorption (Bashan et al. 2014; Vessey 2003). These products are increasingly being used as greener alternatives to synthetic fertilizers because they reduce greenhouse gas emissions and enhance soil health over time. (3) Crop enhancement

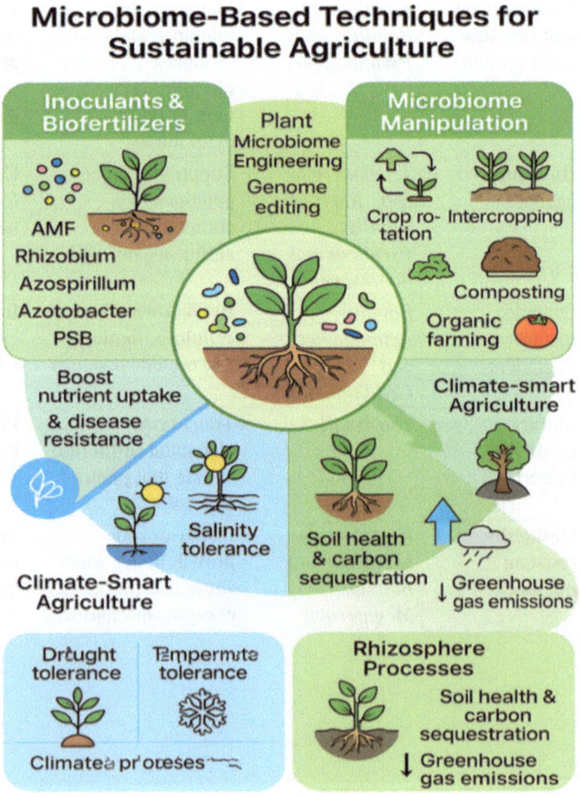

Fig. 7.4 Microbiome-based techniques for sustainable agriculture

Table 7.3 Difficulties in using microbiome in sustainable agriculture

Category of challenge	Important concerns	Implications	References
Variability and complexity	Crop species, soil type, and climate all affect microbiome diversity; field results are unpredictable	Microbial inoculants' inconsistent performance Standardization challenges	Fierer and Jackson (2006), Bender et al. (2016), and Compant et al. (2019)
Safety and regulatory concerns	Diverse international laws Potential for GM or non-native microbes to disturb the environment	Public opposition to GM microbes; approval and commercialization delays	Kaminsky et al. (2019)
Implementation and scaling	Translation issues from the lab to the field Insufficient technical infrastructure and training for farmers	Farmers poor adoption Large-scale applications have low efficiency	Vejan et al. (2016) and Schlaeppi and Bulgarelli (2015)

through microbiome manipulation: microbiome manipulation refers to methods of improving or changing the natural microbial communities associated with crops in order to enhance agricultural outcomes. Controlling the microbiota in the soil: crop rotation and intercropping encourage microbial diversity and reduce pathogen accumulation. Composting increases soil organic matter and fosters the growth of beneficial microorganisms by introducing decomposer microbes (Lazcano et al. 2021). Organic farming practices that preserve soil structure and microbial habitats like reduced tillage and the utilization of cover crops elevate microbial activity and resilience (van Bruggen et al. 2015) (4) Plant microbiome engineering: new advances in genome editing and synthetic biology allow the introduction of specific beneficial traits in microbes or the formation of microbial consortia in order to optimize plant-microbe interactions (Mueller and Sachs 2015). To elevate abiotic stress resistance or nitrogen fixation, modified bacteria have the capacity to be injected into seeds or roots. (5) Sustainability and climate change: because microbiome elevate resource efficiency and aid in plant adaptation to environmental stressors, they are essential to climate-smart agriculture. According to Ngumbi and Kloepper (2016), certain microbes help plants develop drought tolerance by promoting root growth, blocking stomata to prevent water loss and generating stress-relieving substances like exo-polysaccharides and osmo-protectants. It is well known that strains of *Bacillus* and *Pseudomonas* become more resistant to temperature and salinity fluctuations. (6) Soil health and carbon sequestration: microbial activity in the rhizosphere aids in the formation of organic carbon in the soil and the sequestration of carbon, which lowers atmospheric CO_2 levels (Banerjee et al. 2019). Healthy microbiomes prevent land degradation and boost soil resilience to climate-induced stresses like erosion, desertification, and nutrient depletion.

Figure 7.4 illustrates (1) Methods for increasing crop productivity, lowering chemical inputs, and strengthening ecosystem resilience by utilizing beneficial microbial communities. (2) Plant microbiome engineering through genome editing and synthetic consortia to enhance nitrogen fixation and stress tolerance. (3) Microbiome manipulation through crop rotation, intercropping, composting, and organic farming to promote microbial diversity. (4) Use of inoculants and biofertilizers (AMF, *Rhizobium, Azospirillum, Azotobacter* & PSB) to elevate nutrient uptake and disease resistance. (5) Microbial processes in the rhizosphere contributing to soil health and carbon sequestration, lowering greenhouse gas emissions and lessening the effects of climate change. (6) Microbial applications also help climate-smart agriculture by enhancing resistance to drought, salinity, and temperature.

7.7 Difficulties in Using Microbiome in Agriculture

7.7.1 Variability and Complexity

The agricultural microbiome is incredibly dynamic and complex and is influenced by a wide range of biotic and abiotic factors including soil type, crop species, climate, agricultural practices, and seasonal variability. Over time and space, these factors may drastically change the makeup and role of microbial communities (Fierer and Jackson 2006; van der Heijden et al. 2008). Because of variations in pH, moisture content or competition from native microbes' beneficial microorganisms like the nitrogen fixing Rhizobium may thrive in one type of soil but struggle to colonize another. Furthermore, erratic interactions between native microbiota and plant genotypes and microbial inoculants have the capacity to lead to inconsistent performance which often yields different laboratory and field results (Bender et al. 2016; Compant et al. 2019). These disparities pose a significant obstacle to the widespread adoption and confidence in microbiome-based technologies in agriculture (Fig. 7.5; Table 7.3).

7.7.2 Safety and Legal Issues

The commercialization and utilization of microbial products such as genetically modified microbes, biopesticides and biofertilizers are restricted by stringent regulatory frameworks that vary between countries and regions. Strict efficacy, safety, and environmental risk assessments are usually required by regulatory bodies which has the capacity to hinder innovation and pose difficulties for researchers and startups. Furthermore, using nonnative or genetically modified (GM) microorganisms raises biosafety concerns. The potential for ecological disruption, horizontal gene transfer, or outcompeting of naturally occurring beneficial microbes could have long term effects on the environment (Kaminsky et al. 2019). Ethical and public perception concerns complicate the adoption and management of genetically modified microbial tools in agriculture (Fig. 7.5; Table 7.3).

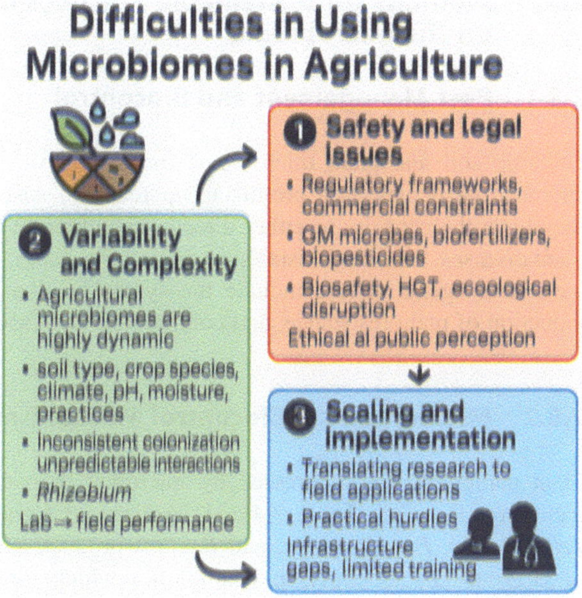

Fig. 7.5 Challenges in applying microbiome to agriculture

7.7.3 Scaling and Implementation

Converting microbiome research into field-ready, scalable applications remains a formidable challenge. The variety and unpredictability of real agricultural environments cannot be replicated by microbial technologies because they are typically developed in controlled environments (Schlaeppi and Bulgarelli 2015). Closing this lab to field gap requires the development of dependable formulations, shelf-stable inoculants, and delivery systems that are inexpensive and simple for farmers to utilize. Another substantial barrier is a lack of infrastructure, knowledge, and training among stakeholders, particularly smallholder farmers in developing countries (Vejan et al. 2016). Effective implementation requires interdisciplinary collaboration among microbiologists, agronomists, legislators, and local communities to develop context-specific solutions that enhance regional farming practices (Fig. 7.5; Table 7.3).

Figure 7.5 illustrates the main obstacles to microbiome-based technologies. It illustrates that the lab and field results are inconsistent due to differences in soil, climate, crop species, and native microbial interactions. Due to ethical and ecological concerns, the commercialization of genetically modified microbes, biofertilizers, and biopesticides is restricted by safety and regulatory barriers. The development of stable formulations, reasonably priced delivery systems, and sufficient farmer training especially in areas with limited resources continue to be obstacles to scaling and implementation.

7.8 Methods for Sustainable Practices Based on the Microbiome

7.8.1 Pest Management and Biocontrol

Utilizing soil and plant microbiome for biocontrol, which reduces the need for chemical pesticides, is a promising approach to sustainable agriculture. Because they have the capacity to either directly or indirectly control pest populations and plant diseases, microorganisms such as bacteria, fungi, and viruses are referred to as biocontrol agents. These include the synthesis of antimicrobial compounds, the induction of plant resistance, and competition for resources.

7.8.2 Microorganism Biocontrol Mechanisms

Plant pathogens are suppressed by microbial biocontrol agents in several ways including (Fig. 7.6; Table 7.4): (1) Antibiosis: Certain microorganisms such as *Trichoderma, Pseudomonas*, and *Bacillus* produce antimicrobial compounds such as enzymes, bacteriocins, and antibiotics which directly prevent the growth of harmful pathogens. For instance, by generating lipopeptides that are poisonous to a variety of bacteria and fungi, *Bacillus subtilis* efficiently manages illnesses like *Fusarium* wilt and *Rhizoctonia* root rot (Vejan et al. 2016). (2) Competition for resources: Pathogens and beneficial microorganisms fight for nutrients and space on plant surfaces or in the rhizosphere. For example, the genus *Pseudomonas* out-competes soil borne pathogens by more efficiently colonizing plant roots and

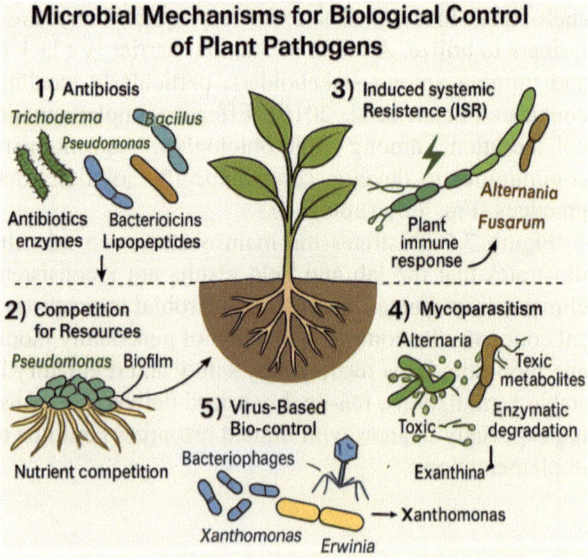

Fig. 7.6 Microbial biocontrol mechanisms

Table 7.4 Microorganism biocontrol mechanisms in plant health management

Mechanism	Description	Example	References
Antibiosis	Production of antimicrobial substances that stop the growth of pathogens, such as bacteriocins, enzymes, and antibiotics	Lipopeptides produced by *Bacillus subtilis* regulate Rhizoctonia root rot and *Fusarium* wilt	Vejan et al. (2016)
Competition for resources	Pathogens are outcompeted by beneficial microbes for nutrients and space on plant surfaces or in the rhizosphere	*Pseudomonas* species form biofilms to stop infections by colonizing roots more efficiently than soilborne pathogens	Van Dijk et al. (2008)
Induced systemic resistance	Beneficial microorganisms activate plant defense mechanisms, increasing resistance to infections	*Pythium* and *Fusarium* susceptibility is decreased by *Pseudomonas fluorescens* and PGPR	–
Mycoparasitism	Through direct contact and enzymatic lysis, fungi parasitize and eliminate other fungal pathogens	*Trichoderma harzianum* produces toxic metabolites and secretes enzymes to target *Alternaria* and *Fusarium*	Vejan et al. (2016)
Virus-based biocontrol	Utilizing *bacteriophages* to target particular crop bacterial pathogens	Phages are used to treat bacterial illnesses brought on by *Erwinia* and *Xanthomonas*	Vandenkoornhuyse et al. (2015)

forming biofilms that stop pathogen invasion. (3) Induced systemic resistance (ISR): Certain advantageous microorganisms have the capacity to trigger defenses in plants, improving their resistance to disease. *Pseudomonas fluorescens* and *Rhizobacterium* that promote plant growth (PGPR) reduce a plant's vulnerability to diseases like *Fusarium* and *Pythium*. These microbes lessen the need for chemical pesticides by enhancing the plant's resistance to biotic and abiotic stresses through the development of systemic resistance. (4) Mycoparasitism: Fungi like Trichoderma spp. parasitize and kill other fungal pathogens by direct contact, enzymatic breakdown of pathogen cell walls and the production of toxic metabolites. For example, Trichoderma harzianum is commonly used as a biocontrol agent against soilborne fungi such as species of Alternaria and Fusarium (Vejan et al. 2016). (5) Virus-based biocontrol: Phages, also known as *Bacteriophages*, are viruses that target bacterial pathogens specifically. Are a useful tool for managing bacterial plant diseases caused by *Xanthomonas* and *Erwinia* species in agriculture. Because of its environmental safety and specificity, this biocontrol strategy is becoming more and more well liked (Vandenkoornhuyse et al. 2015).

Figure 7.6 illustrates primary ways through which beneficial microorganism's shield plants from diseases. Antibiosis is the process by which bacteria (*Bacillus, Pseudomonas,* and *Trichoderma*) produce antimicrobial substances (lipopeptides, antibiotics, and enzymes). When beneficial microbes colonize roots and occupy niches more successfully than pathogens competition for resources arises. These microbes frequently form biofilms that prevent infection. Plant growth-promoting rhizobacteria (*Pseudomonas fluorescens*) cause induced systemic resistance which primes plant defenses against fungi (*Pythium* and *Fusarium*). Through enzymatic cell wall degradation and toxic metabolites fungi (*Trichoderma harzianum*) utilize myco-parasitism to directly target pathogens. Lastly, bacteriophages that target bacterial pathogens (*Xanthomonas* and *Erwinia*) are used in virus-based biocontrol. These processes when used in combinations enhance plant immunity, lessen reliance on pesticides, and support sustainable agriculture.

7.8.3 Microbiome-Based Biocontrol Methods over Traditional Chemical Pesticides

(1) Environmental safety: Compared to chemical pesticides, biocontrol agents are often more environmentally friendly. They do not harm humans, animals, or beneficial nontarget organisms such as organic pest predators and pollinators. (2) Sustainability: Biocontrol reduces the adverse environmental effects of agriculture. Microbial agents provide an effective substitute for chemical pesticides by diversifying pest management methods and reducing dependency on chemicals. (3) Decreased risk of resistance: Natural microbial control agents lessen the chance that pathogens will become resistant. For instance, it is challenging for pathogens to concurrently adapt to several biocontrol mechanisms due to the variety of microbial interactions in soil and on plant surfaces. (4) Complementary to organic farming: Microbial biocontrol is especially beneficial in organic farming systems because they employ fewer chemical pesticides. Using biocontrol agents based on microbiome has the capacity to significantly elevate crop protection while adhering to organic farming practices (Fig. 7.7).

Figure 7.7 illustrates two aspects of microbiome-based biocontrol in agriculture. On the plus side, microbial agents provide safe alternatives to chemical pesticides that are safe for pollinators, animals, humans, and other beneficial organisms. They reduce chemical inputs to promote sustainability, limit the development of pathogen resistance through a variety of mechanisms and integrate seamlessly with organic farming systems. On the down side, field variables like soil variability, moisture content, and temperature usually reduce efficacy when compared to lab results. Stability, shelf life, and farmer-friendly formulations are issues in mass production and regulatory frameworks demand extensive safety and efficacy testing. Furthermore, consumer perception and acceptance slow adoption. Future developments in metagenomics, microbial consortia design and CRISPR-based genome editing may help overcome these limitations and elevate the scalability of microbiome-driven pest management.

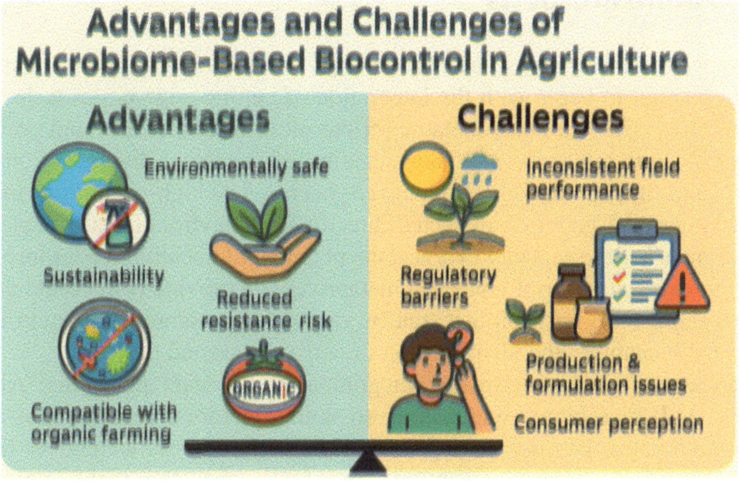

Fig. 7.7 Advantages and challenges of microbiome-based biocontrol

7.9 Challenges in Implementing Microbial Biocontrol

Despite their potential, microbial biocontrol agents face several challenges in large-scale agricultural applications: (1) Field conditions: Although laboratory and greenhouse tests often show good results, field conditions such as variations in soil composition, temperature, and moisture has the capacity to affect the efficacy of biocontrol agents. Environmental factors that affect this microbe performances frequently limit their effectiveness in specific locations or climates. (2) Regulatory barriers: Before using biocontrol agents in agriculture, many countries need regulatory approval which could delay their commercial availability. Regulatory agencies usually require extensive safety and efficacy testing which has the capacity to be expensive and time consuming. (3) Mass production and formulation: Making and creating microbial biocontrol agents for commercial use has the capacity to be challenging. It is necessary to optimize aspects like shelf life, microbial stability, and ease of application prior to broad adoption. (4) Consumer perception and acceptance: Despite the fact that biocontrol is regarded as a safer alternative to chemical pesticides; consumer acceptance of biocontrol products varies. Initiatives for public awareness and education are crucial to accelerating the adoption of these sustainable farming methods. The future of microbiome-based biocontrol depends on the application of cutting-edge technologies such as metagenomics, synthetic biology, and gene editing. By sequencing the microbiome of plants and soils, scientists have the capacity to identify novel microbial strains with potent biocontrol potential and tailor microbial consortia for specific agricultural applications. Furthermore, CRISPR technology may facilitate the development of microbial strains that are more effective at controlling pests and infections in a range of environmental conditions.

7.10 Future Prospects

(1) Advances in the study of the microbiome: A revolutionary era in microbiome research is about to begin with the integration of cutting-edge omics technologies. Tools such as metagenomics, metatranscriptomics, metabolomics, and proteomics have significantly advanced our understanding of microbial diversity, community structure, and function in soil and plant ecosystems. The high-resolution characterization of microbial genomes extracted directly from environmental samples using metagenomics has the capacity to reveal the potential of uncultured microorganisms. Furthermore, metatranscriptomics and metaproteomic provide insight into the functional dynamics of microbiomes under various agronomic and environmental conditions by enabling the identification of active genes and proteins.

Future trends show a growing integration of big data platforms, AI-driven analytics, and microbiome engineering to tailor microbial consortia for specific crops and climates. By maximizing microbial function, these techniques have the capacity to aid in disease prevention, stress tolerance, and yield improvement (Jansson and Hofmockel 2018). Next-generation sequencing and portable sequencers have enabled data-driven precision agriculture and real-time microbiome monitoring. (2) Integrating other sustainable farming techniques with the microbiome: The future of microbiome-based agriculture depends on its integration with broader agroecological ideas. Techniques such as conservation tillage, cover crops, intercropping, and organic amendments not only enhance soil structure and organic matter content, but they also promote the development of beneficial microbial communities. For example, no-till farming reduces soil disturbance and preserves microbial habitats, while composting adds organic substrates and beneficial microbes that encourage microbial growth. Microbiome-based interventions are used in conjunction with synergistic methods such as crop rotation and integrated pest management to create resilient and productive agro-ecosystems. Microbial inoculants and organic fertilizers have proven to be more effective at promoting plant growth and disease resistance than synthetic inputs alone (Timmusk et al. 2017). This convergence advances the goals of climate-smart agriculture by improving soil health, reducing greenhouse gas emissions, and fostering long-term sustainability. (3) Strategies for teamwork: To fully utilize microbiomes in sustainable agriculture, interdisciplinary collaboration is required. Microbiologists, agronomists, soil scientists, plant pathologists, bioinformaticians, and environmental scientists will need to work together to develop trustworthy and field-relevant microbial solutions (Sessitsch and Mitter 2015). Furthermore, communication with farmers, lawmakers, and agricultural extension services is crucial for knowledge transfer and real-world implementation. With the aid of capacity-building programs like training sessions, knowledge-sharing websites, and participatory research, farmers have the capacity to adopt microbiome-based practices. The establishment of microbial consortia that are locally adapted to a range of agroecosystems will be made possible by funding public-private partnerships and open-access microbial libraries, which will encourage innovation.

7.11 Conclusion

The growing body of evidence highlights the critical role the microbiome plays in promoting sustainable agriculture. The crucial functions that soil and plant microbiome play in nutrient cycling, disease prevention, and boosting plant resilience are the biological foundations of fruitful agro-ecosystems. The intricate relationships that these microbial communities—which comprise bacteria, fungi, viruses, archaea, and protozoa—have with their environment and host plants have a significant effect on soil fertility and plant health. Examples of microbiome-based methods that offer sustainable alternatives to synthetic chemicals and fertilizers include microbial inoculants, biofertilizers, and biocontrol agents. Because of their ability to fix atmospheric nitrogen, solubilize phosphates, produce phytohormones, and fight off pathogens, they have the potential to address significant agricultural issues. Furthermore, farming systems have the capacity to become more resilient by enhancing soil structure and biodiversity through the integration of microbiomes with agroecological practices such as crop rotation, organic farming, and reduced tillage. Despite the complexity, regulatory barriers, and scalability issues associated with microbiome applications, advancements in omics technologies, bioinformatics, and systems biology are ushering in a new era in microbiome research. These developments enable precision microbiome engineering, adaptive management strategies tailored to specific agroecosystems, and real-time diagnostics. Therefore, it has the capacity to be concluded that harnessing the potential of soil and plant microbiome is both a scientific opportunity and crucial for the future of agriculture. In light of resource scarcity, climate change, and environmental degradation, sustainable food production requires an understanding of and ability to manage microbial communities. There is an urgent need for more interdisciplinary research, policy support, and farmer involvement in order to successfully incorporate microbiome innovations into agricultural practices. The future of farming lies in the microbiome and the microscopic environment beneath our feet.

References

Altieri MA (2002) Agroecology: the science of sustainable agriculture. CRC Press

Bais HP, Weir TL, Perry LG, Gilroy S, Vivanco JM (2006) The role of root exudates in rhizosphere interactions with plants and other organisms. Annu Rev Plant Biol 57:233–266

Balesdent J et al (1998) Soil organic matter turnover in a long-term field experiment. Soil Sci Soc Am J 62(4):1186–1193

Banerjee S, Walder F, Büchi L, Meyer M, Held AY, Gattinger A et al (2019) Agricultural intensification reduces microbial network complexity and the abundance of keystone taxa in roots. ISME J 13:1722–1736

Bardgett RD, van der Putten WH (2014) Belowground biodiversity and ecosystem functioning. Nature 515(7528):505–511

Bashan Y, de-Bashan LE (2010) How the plant growth-promoting bacterium Azospirillum promotes plant growth—a critical assessment. Adv Agron 108:77–136

Bashan Y, de-Bashan LE, Prabhu SR, Hernandez JP (2014) Advances in plant growth-promoting bacterial inoculant technology: formulations and practical perspectives (1998–2013). Plant Soil 378:1–33

Bender SF, Wagg C, van der Heijden MGA (2016) An underground revolution: biodiversity and soil ecological engineering for agricultural sustainability. Trends Ecol Evol 31(6):440–452

Berendsen RL et al (2012) The rhizosphere microbiome and plant growth. Trends Plant Sci 17(8): 478–486

Berg G, Grube M, Schloter M, Smalla K (2014) The plant microbiome and its importance for plant and human health. Front Microbiol 5:491. https://doi.org/10.3389/fmicb.2014.00491

Compant S, Samad A, Faist H, Sessitsch A (2019) A review on the plant microbiome: ecology, functions, and emerging trends in microbial application. J Adv Res 19:29–37

Fierer N, Jackson RB (2006) The diversity and biogeography of soil bacterial communities. Proc Natl Acad Sci USA 103(3):626–631

Fierer N, Leff JW, Adams BJ, Nielsen UN, Bates ST, Lauber CL, Owens S, Gilbert JA, Wall DH, Caporaso JG (2012) Cross-biome metagenomic analyses of soil microbial communities and their functional attributes. Proc Natl Acad Sci USA 109(52):21390–21395. https://doi.org/10.1073/pnas.1215210110

Glick BR (2012) Plant growth-promoting bacteria: mechanisms and applications. Scientifica (Cairo) 2012:963401

Gomiero T et al (2011) Environmental impact of different agricultural management practices: conventional vs. sustainable agriculture. Crit Rev Plant Sci 30(1):1–34

Haas D, Défago G (2005) Biological control of soil-borne pathogens by fluorescent pseudomonads. Nat Rev Microbiol 3(4):307–319

Jansson JK, Hofmockel KS (2018) Soil microbiomes and climate change. Nat Rev Microbiol 16(1): 22–35

Kaminsky LM, Trexler RV, Malik RJ, Hockett KL, Bell TH (2019) The inherent conflicts in developing soil microbial inoculants. Trends Biotechnol 37(2):140–151

Lazcano C, Gómez-Brandón M, Revilla P, Domínguez J (2021) Short-term effects of organic and inorganic fertilizers on soil microbial community structure and function. Biol Fertil Soils 57:69–79

Lugtenberg B, Kamilova F (2009) Plant-growth-promoting rhizobacteria. Ann Rev Microbiol 63: 541–556

Mahanty T et al (2017) Biofertilizers: potential approach for sustainable agriculture development. Environ Sci Pollut Res Int 24(4):3315–3333

Mendes R, Garbeva P, Raaijmakers JM (2013) The rhizosphere microbiome: significance of plant beneficial, plant pathogenic, and human pathogenic microorganisms. FEMS Microbiol Rev 37(5):634–663

Mueller UG, Sachs JL (2015) Engineering microbiomes to enhance plant and animal health. Trends Microbiol 23(10):606–617

Ngumbi E, Kloepper J (2016) Bacterial-mediated drought tolerance: current and future prospects. Appl Soil Ecol 105:109–125

Philippot L, Raaijmakers JM, Lemanceau P, van der Putten WH (2013) Going back to the roots: the microbial ecology of the rhizosphere. Nat Rev Microbiol 11(11):789–799

Pieterse CMJ et al (2014) Induced systemic resistance by beneficial microbes. Annu Rev Phytopathol 52:345–375

Raupach GS, Kloepper JW (1998) Mixtures of plant growth-promoting rhizobacteria for the suppression of soil-borne plant diseases. Appl Environ Microbiol 64(9):3527–3532

Richardson AE, Barea JM, McNeill AM et al (2009) Acquisition of phosphorus and nitrogen in the rhizosphere and plant growth promotion by microorganisms. Plant Soil 321:305–339. https://doi.org/10.1007/s11104-009-9895-2

Schlaeppi K, Bulgarelli D (2015) The plant microbiome at work. Mol Plant-Microbe Interact 28(3): 212–217

Sessitsch A, Mitter B (2015) 21st-century agriculture: Integration of plant microbiomes for improved crop production and food security. Microb Biotechnol 8(1):32–33
Sharma SB, Sayyed RZ, Trivedi MH, Gobi TA (2013) Phosphate solubilizing microbes: sustainable approach for managing phosphorus deficiency in agricultural soils. Springerplus 2(1):587
Smith SE, Read DJ (2008) Mycorrhizal symbiosis, 3rd edn. Academic
Timmusk S, Behers L, Muthoni J, Muraya A, Aronsson AC (2017) Perspectives and challenges of microbial application for crop improvement. Front Plant Sci 8:49
van Bruggen AHC, Semenov AM, van Diepeningen AD, de Vos OJ (2015) Relation between soil health, wave-like fluctuations in microbial populations, and soil-borne plant disease management. Appl Soil Ecol 86:147–160
van der Heijden MGA, Bardgett RD, van Straalen NM (2008) The unseen majority: soil microbes as drivers of plant diversity and productivity in terrestrial ecosystems. Ecol Lett 11(3):296–310
Vandenkoornhuyse P, Quaiser A, Duhamel M, Le Van A, Dufresne A (2015) The importance of the microbiome of the plant holobiont. New Phytol 206(4):1196–1206. https://doi.org/10.1111/nph.13312
Vejan P, Abdullah R, Khadiran T, Ismail S, Nasrulhaq BA (2016) Role of plant growth promoting rhizobacteria in agricultural sustainability-a review. Molecules 21(5):573. https://doi.org/10.3390/molecules21050573
Vessey JK (2003) Plant growth promoting rhizobacteria as biofertilizers. Plant Soil 255(2):571–586

Enhancing Food Production Through Microbiota Engineering

8

Abstract

Microbiota engineering is rapidly emerging as a revolutionary approach to enhance food production systems, offering novel solutions to persistent issues in agriculture, livestock production, and food processing. In order to enhance resilience, elevate productivity, and advance sustainability in food systems, this chapter looks at how microbial communities whether they are present in plants, soil, or the gut microbiota of livestock have the capacity to be controlled. Beneficial microorganisms such as bacteria that support plant growth and probiotics for livestock provide strategies to elevate crop yield, reduce the environmental impact of agriculture, and enhance animal health. Despite its promise many questions remain particularly about the complex, multilayered interactions between microbiota and their environment. Current studies usually lack a comprehensive, systems-based approach and focus on isolated applications or individual microbial strains rather than the larger, synergistic effects within agricultural and food ecosystems. Furthermore, the long-term ecological, economic and food safety impacts of microbiota engineering are poorly understood. Furthermore, ethical and legal issues surrounding the alteration of microbiota in food systems prevent broad adoption. Novel approaches like synthetic biology and CRISPR-based technologies are demonstrating promise as tools for precisely engineering microbial communities with desired functions. Nevertheless, there are still challenges in scaling these innovations and integrating them in different agricultural contexts. Interdisciplinary cooperation is necessary to close knowledge gaps, enhance technological applications, and establish standardized regulatory frameworks in order to fully realize the potential of microbiota engineering. Large-scale, longitudinal studies should be the main focus of future research in order to evaluate the ecological and environmental effects of engineered microbiota. The long-term viability of microbiota-based solutions will depend on research into sustainable and scalable delivery systems and how to incorporate

R. C. Sobti et al., *Microbiota Unveiled: Science, Health, and Future Therapies*,
https://doi.org/10.1007/978-981-95-9031-5_8

them into regenerative and circular agricultural practices. Future sustainable food production systems that support environmental sustainability and global food security has the capacity to benefit greatly from microbiota engineering's contribution to interdisciplinary collaboration and the closure of these research gaps.

Keywords

Microbiota engineering · Synthetic biology · Sustainable agriculture · Global food security · Regulatory framework

8.1 Introduction

Microbiota engineering is the deliberate alteration of microbial communities in a particular environment, typically to enhance specific microbial interactions for beneficial outcomes or maximize their functional output. The term encompasses a variety of techniques for controlling microbial behavior including metabolic pathway engineering, microbial community assembly, genetic modification, and environmental condition optimization. According to Bäckhed et al. (2005), these methods aim to harness the microbiota potential to achieve specific industrial, agricultural, or health results. Microbiota engineering basically seeks to understand and control the complex ecosystem of microbes that inhabit a variety of settings such as the human gut, agricultural soils, and fermented food products in order to enhance the quality and functionality of these systems (Pflughoeft and Versalovic 2012).

8.1.1 Microbial Community's Significance in Food Systems

Food production, safety, and quality are all greatly impacted by microbial communities, which are essential to food systems. Both the traditional and modern food industries utilize microbes in many stages of food processing, fermentation, and preservation. They contribute to the development of distinctive flavors, textures, and aromas in fermented foods such as kimchi, cheese, yogurt, and sauerkraut. Furthermore, microbial communities are essential for the biotransformation of food substrates, which boosts the nutritional benefits of food products especially functional foods and nutrient bioavailability. According to van der Heijden and Hartmann (2016) the soil microbiota in agricultural food production influences crop health and yield by encouraging nutrient cycling, suppressing pathogens, and increasing plant resistance to environmental stressors. Beneficial microbes also contribute to the fermentation processes that prolong food's shelf life and keep it from spoiling. The stability and health of microbial communities in food systems are crucial for ensuring safe, healthy, and sustainable food production because of these diverse roles.

8.1.2 Necessity of Microbiota Engineering

The need for microbiota engineering to address several problems in food and health systems is increasingly recognized. As the world's population grows, there is an increasing need for sustainable, nutritious food sources. Microbiota engineering presents an opportunity to enhance food security through boosting crop yields, reducing reliance on chemical pesticides and fertilizers and enabling more efficient utilization of agricultural resources. Furthermore, microbiota engineering may help create microbial systems that have the capacity to recognize and eradicate pathogens, improving food safety as concerns about food-borne illnesses elevate (Lavilla and Amarita 2025). Furthermore, our growing knowledge of the human microbiome has highlighted its significance in immunity, digestion, and metabolic regulation. Microbiota engineering in the human gut has the potential to treat or prevent a number of diseases such as diabetes, obesity, and gastrointestinal disorders through dietary modifications or probiotic treatments. Similarly, microbiota engineering in animal agriculture may result in more sustainable meat production systems by enhancing livestock productivity, health, and disease resistance. Another reason for the importance of microbiota engineering is its potential to solve environmental issues such as reducing greenhouse gas emissions from agricultural activities and improving waste management by allowing microorganisms to break down organic matter. By creating microbial communities for specific roles in the soil, food processing systems, or human bodies' microbiota engineering holds promise for enhancing productivity, sustainability, and health in a number of domains (Fig. 8.1; Table 8.1).

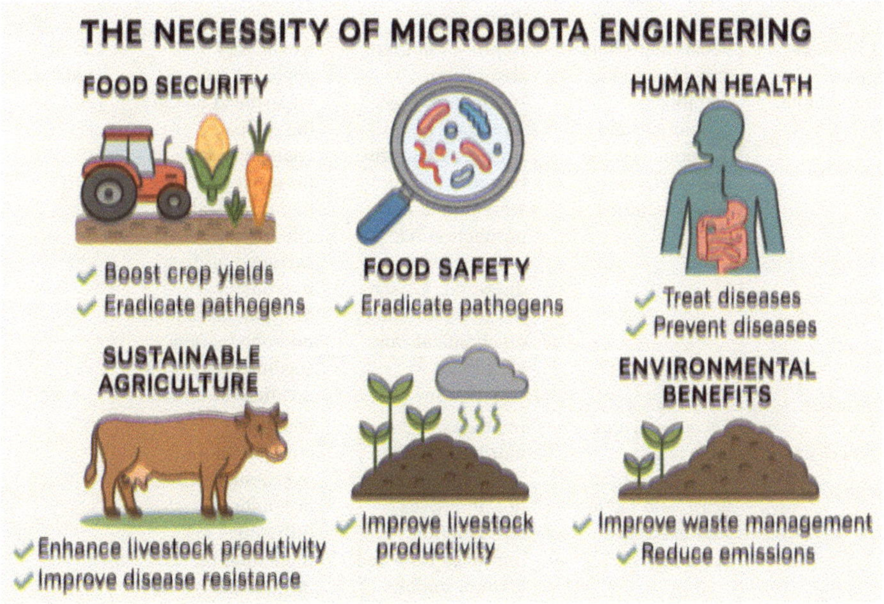

Fig. 8.1 The necessity of microbiota engineering

Table 8.1 Microbiota engineering in food production systems

Application area	Key microbiota or technologies	Specific benefits	Challenges or limitations	References
Plant microbiota	*Rhizobacteria*, Mycorrhizal Fungi, and plant growth-promoting bacteria	Improves soil microbial balance, drought resistance, biotic and abiotic stress resistance, root growth, and nutrient uptake, enhances plant health, and lessens reliance on synthetic fertilizers	Soil-related variations in microbial efficacy; challenges in sustaining microbial populations in fluctuating conditions; and the possibility of unforeseen consequences for nontarget species	Cheng et al. (2023) and Pandey and Saharan (2025)
Soil microbiome	*Actinobacteria*, Mycorrhizal Fungi, *Bacillus* species, and nitrogen-fixing bacteria (*Rhizobium*)	Reduces the need for chemical fertilizers, improves soil nutrient cycling, boosts crop productivity, improves soil structure, and promotes long-term soil health. The microbiome diversity can boost resistance to disease	Large-scale agricultural microbiome stability issues; unclear long-term effects on ecosystem health; and the possibility of microbiome imbalances brought on by monocultures	
Livestock gut microbiota	Fecal microbiota transplantation, prebiotics, synbiotics, and probiotics (lactic acid bacteria, *Bifidobacteria*)	Improves nutrient absorption, strengthens the immune system, decreases harmful bacteria, aids in weight gain and production rates, and improves digestion. Lessens livestock's reliance on chemicals and antibiotics	Regulatory approval for widespread use in the market, challenges in preserving microbial stability, ethical issues surrounding fecal microbiota transplantation, and microbial strain-specific efficacy in a variety of animals	Cheng et al. (2023)
Food processing	*Saccharomyces cerevisiae*, *Bifidobacterium*, yeast, and lactic acid bacteria (LAB)	Increases nutritional value, prolongs food shelf life, improves fermentation, improves taste and texture, inhibits spoiling, and lowers food-borne pathogens.	Risk of contamination, overfermentation, or spoiling; difficulties in regulating the use of genetically modified microorganisms in food production;	

(continued)

Table 8.1 (continued)

Application area	Key microbiota or technologies	Specific benefits	Challenges or limitations	References
		Extensively utilized in fermented foods, dairy products, meat products, and drinks	and a lack of knowledge regarding the long-term health effects of eating food containing modified microbiota	
Food waste reduction	Microbial Consortiums, Fungi (*Aspergillus* species), and Bacteria (*Bacillus* and *Pseudomonas* species)	Helps keep food waste out of landfills and lower carbon footprints by biodegrading it into useful byproducts like compost, biofuels, and bioplastics. Complex organic matter can be effectively broken down by microbial consortia	Consumer resistance to using products made from waste, the cost-effectiveness of widespread implementation in waste management systems, and technological scalability are still issues	Dey et al. (2025)
Climate-resilient agriculture and microbiota	Plant symbionts, salt-tolerant fungi, and drought-tolerant bacteria (such as *Bacillus* and *Pseudomonas* species)	Enhances drought resistance, increases water use efficiency, and encourages plant growth in unfavorable weather conditions to support crops in stressed environments. Can aid in restoring degraded soils' productivity	Lack of field data on long term effectiveness in climate change; possibility of unforeseen ecological effects from changed soil microbial compositions; requires extensive validation	Tan et al. (2022)

Figure 8.1 illustrates main areas where microbiota engineering provides substantial advantages. It illustrates microbiota-based strategies elevate crop yields and aid in the eradication of plant diseases in the context of food security. By focusing on dangerous microorganisms, engineered microbes lower the risk of food-borne illnesses. Microbial interventions enhance disease resistance and livestock productivity in sustainable agriculture. So, by altering gut and systemic microbial interactions, microbiota engineering has the capacity to prevent and treat diseases in humans. Last but not least, better waste management and lower greenhouse gas emissions contribute to more resilient and sustainable ecosystems.

8.2 Microbiota in the Food Production Ecosystem

Many of the microorganisms in this group are critical to enhancing the sustainability, quality, and production of food. Several settings like soil, plant surfaces, and animal digestive tracts are home to these microbial communities. Plant-associated microbiota such as rhizosphere, phyllosphere, and endophytic microbes enhance plant health in agriculture by enhancing stress tolerance, inhibiting pathogens and facilitating nutrient uptake. Furthermore, the microbial communities in livestock and aquaculture systems affect digestion, immunological responses, and overall animal health. By understanding and optimizing these microbial interactions, microbiota engineering has the capacity to elevate crop yields, animal productivity, and the need for synthetic chemicals in food production. This will contribute to the development of a more sustainable farming system.

8.2.1 Plant-Associated Microbiota

Plant-associated microbiota is critical to plant health and productivity because it influences growth, nutrient availability, stress tolerance, and disease resistance. These microbial communities inhabit a variety of niches within the plant system, including the rhizosphere (root zone), phyllosphere (leaf surface), and endosphere (inside plant tissues). Understanding this microbiota is crucial to harnessing their beneficial properties to elevate crop yields and reduce the need for synthetic fertilizers and pesticides. The primary microbial communities linked to plant growth and health are the rhizosphere, phyllosphere and endophytic microbiota with a focus on plant growth-promoting rhizobacteria (Fig. 8.2; Table 8.1).

8.2.1.1 Rhizosphere, Phyllosphere, and Endophytes

The rhizosphere is the tiny region of soil that encircles and is influenced by plant roots. This zone is home to a wide variety of microorganisms including fungi, bacteria, and archaea which interact intricately with plant roots. These microbes are essential for promoting plant growth, inhibiting pathogens, and cycling nutrients. The abundance of nutrients that microorganisms receive from plant-released root exudates into the rhizosphere affects the makeup of microbial communities and fosters mutualistic relationships. Nevertheless, the phyllosphere, i.e., the surface of plant leaves, stems, and flowers is home to a unique class of microorganisms mostly bacteria and fungi. These microorganisms influence plant stress responses, nutrient uptake, and disease resistance. Because microbes in the phyllosphere often encounter more severe environmental conditions including desiccation and UV radiation; hence, specialized adaptations are required (Vorholt 2012). Endophytes are microorganisms that live inside plant tissues without posing a threat to the plant. These microbes are found in almost every plant organ including the roots, stems, and leaves. Endophytes have the capacity to promote plant health by producing bioactive compounds, promoting nutrient uptake and enhancing stress tolerance. Because

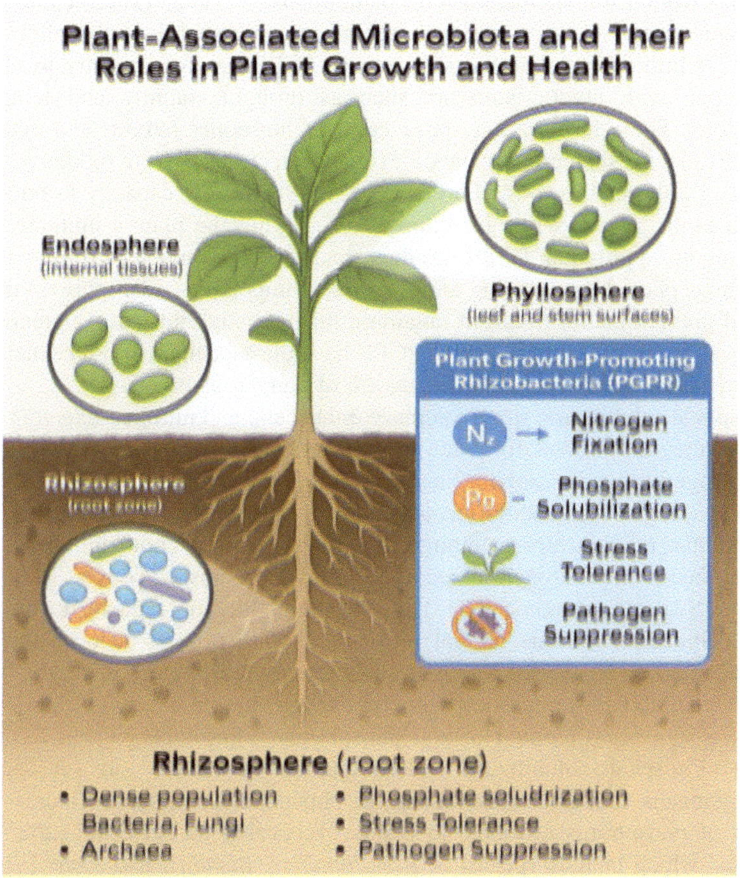

Fig. 8.2 Plant-associated microbiota

endophytic microbiota directly affects plant growth and resilience, research on plant-microbe interactions is growing in popularity.

8.2.1.2 Plant Growth-Promoting Rhizobacteria (PGPR)

This class of soil bacteria promotes plant growth through a number of methods. These bacteria colonize the rhizosphere and their presence often enhances plant yield, growth, and health. There are several ways that PGPR has the capacity to be advantageous: (1) (A). Nitrogen fixation: some PGPR species, like *Azotobacter* and *Rhizobium*, has the capacity to fix nitrogen from the atmosphere and transform it into a form that plants have the capacity to utilize to grow (Glick 2012). This process is especially integral in nutrient poor soils where nitrogen is a limiting factor for plant growth. (B). Phosphate solubilization: despite being a necessary nutrient for plants, a significant amount of soil phosphorus is insoluble and unavailable to them. (2) *Bacillus* and *Pseudomonas* species are examples of PGPR that has the capacity to degrade phosphate from insoluble materials for plant uptake. Lei et al. (2025) claim that this encourages improved plant growth, especially in soils with low phosphorus availability. (3) Improvement of stress tolerance: PGPR has the capacity to also help plants cope with abiotic stressors such as drought, salinity and temperature fluctuations. These bacteria may produce plant hormones (auxins and cytokinins) that boost plant resilience or metabolites that protect plants from oxidative damage under stressful conditions. (4) Furthermore, PGPR has the capacity to boost resistance to biotic stressors like infections by fortifying the plant's immune system. (5) Pathogen suppression: PGPR has the capacity to outcompete harmful pathogens in the rhizosphere by producing antimicrobial compounds (siderophores) that bind iron and prevent pathogens from obtaining this essential nutrient (Lugtenberg and Kamilova 2009). Furthermore, some PGPR have the capacity to produce metabolites or enzymes that directly prevent the growth of dangerous bacteria.

Phosphate solubilizers, stress tolerance enhancers, and nitrogen fixers: (1) one of the most integral nutrients for plant growth is nitrogen, which many plants rely on from nitrogen-fixing microbes. Legumes and nitrogen-fixing bacteria, including those belonging to the genera *Rhizobium, Bradyrhizobium*, and *Azotobacter* coexist in symbiotic relationships, although the rhizosphere of nonleguminous plants contains free-living nitrogen fixers. These microorganisms convert atmospheric nitrogen (N2) into ammonia (NH3), which plants has the capacity to utilize and absorb (Glick 2012). This natural nitrogen fixation process reduces the use of synthetic nitrogen fertilizers, which has the capacity to be environmentally harmful. Phosphate-solubilizing microbes: while phosphate is a necessary nutrient for plant growth, a significant amount of soil phosphorus is insoluble, which limits plant uptake. Phosphate-solubilizing microorganisms (PSMs) are a class of microorganisms that solubilize inorganic phosphate compounds, increasing the amount of phosphorus available to plants. According to Lei et al. (2025), these microbes, which include species of *Pseudomonas, Bacillus,* and *Aspergillus*, produce organic acids like citric acid that break down phosphate compounds in the soil and release phosphorus ions for plant uptake. PSMs have the capacity to be used as biofertilizers to lessen the need for chemical phosphorus fertilizers and have a great deal of potential to elevate plant growth in phosphorus-deficient soils. (2) Enhancers of stress tolerance: abiotic stressors that have the capacity to drastically lower crop yields and plant growth include high temperatures, drought, and salinity. Plant-associated microbes, especially PGPR, have developed defense mechanisms to counteract these stresses. For instance, certain bacteria have the ability to produce auxins and gibberellins, two plant hormones that promote plant growth in stressful situations. Other microbes may produce osmolytes or exopolysaccharides that help plants retain water during drought stress, or they may boost the activity of antioxidant enzymes that protect plants from oxidative damage. Furthermore, PGPR has the capacity to enhance nutrient uptake and root growth, increasing a plant's resilience to environmental stressors. Thus, to enhance plant growth, nutrient availability, and stress tolerance, PGPR and other plant-associated microbiota are required. By developing beneficial relationships with plants in the rhizosphere, phyllosphere and endosphere, these microbes promote sustainable farming practices and reduce the need for chemical pesticides and fertilizers. Understanding and working with these microbial communities which include phosphate solubilizers, stress tolerance enhancers, and nitrogen fixers offers exciting opportunities to enhance food production systems and advance agricultural sustainability.

Figure 8.2 illustrates how the rhizosphere (root zone), phyllosphere (leaf and stem surfaces), and endosphere (internal tissues) are home to plant-associated microbial communities that have a direct impact on plant health and productivity. These microbes fix nitrogen (*Rhizobium, Azotobacter*), solubilize phosphate (*Pseudomonas, Bacillus, Aspergillus*), and suppress pathogens by generating siderophores and antimicrobial metabolites. They also elevate abiotic stress tolerance by producing hormones and osmolytes. By generating bioactive compounds and enhancing stress tolerance, endophytes also make a contribution. In order to lessen reliance on chemical inputs, the rhizosphere, phyllosphere, and endophytic microbiota particularly PGPR all work together to support sustainable agriculture.

8.2.2 Soil Microbiota

The fertility, health, and growth of plants depend on this community of microorganisms, which includes bacteria, fungi, archaea, and protozoa. By affecting a range of soil processes, including nutrient cycling, organic matter decomposition, and biocontrol, these microbes promote agricultural productivity and ecosystem sustainability. Because they enhance soil structure, regulate nutrient availability, and protect plants from soil-borne diseases, soil microbiotas are vital to soil ecosystems (Fig. 8.3; Table 8.1).

8.2.2.1 Contribution to Soil Fertility and Nutrient Cycling

The soil microbiota is essential to nutrient cycling because it maintains soil fertility and ensures that plants have access to the nutrients they require. Microbial communities help transform essential nutrients such as sulfur, phosphorus, and nitrogen into forms that plants have the capacity to utilize. Examples of nitrogen-fixing bacteria that convert atmospheric nitrogen into ammonia that plants have the capacity to absorb are *Rhizobium* and *Azotobacter*. Furthermore, two phosphorus-solubilizing microbes, *Bacillus* and *Pseudomonas*, release bound phosphorus from the soil for plant uptake (Lei et al. 2025). Furthermore, microorganisms enhance the soil matrix and boost fertility by assisting in the breakdown of organic matter and releasing carbon, nitrogen, and other vital elements into the soil (Fierer 2017). These processes are crucial for sustaining plant growth, especially in soils with low nutrient levels.

8.2.2.2 Decomposition of Organic Matter

The breakdown of organic matter is one of the soil microbiota's other crucial functions. Decomposers that break down plant and animal waste and add nutrients to the soil include bacteria, fungi, and actinomycetes. This process breaks down complex organic materials like proteins, cellulose, and lignin into simpler

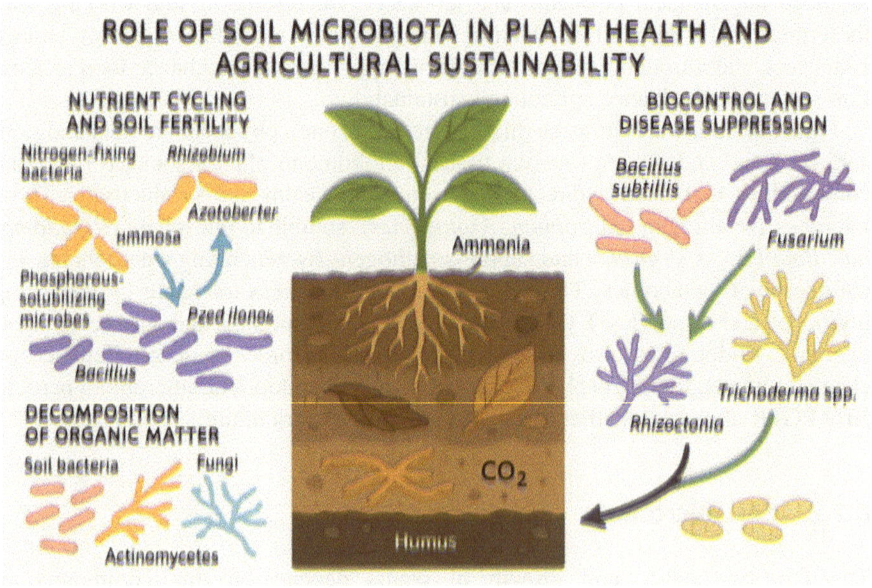

Fig. 8.3 Soil microbiota and their contributions to agriculture

compounds like sugars, amino acids, and organic acids. These decomposed materials enhance soil structure, boost microbial diversity, enhance water retention, and enrich the soil with humus. Furthermore, by releasing carbon dioxide into the atmosphere, the breakdown of organic matter affects soil carbon sequestration processes and adds to the global carbon cycle.

8.2.2.3 Biocontrol and Disease Suppression

The soil microbiota is essential to biocontrol because it facilitates suppressing pests and plant diseases. Beneficial microbes in the soil successfully reduce the prevalence of soil-borne diseases by outcompeting dangerous pathogens for nutrients and space. For example, some soil bacteria such as *Bacillus subtilis* produce antimicrobial compounds that stop pathogens like *Fusarium* and *Rhizoctonia* from growing (Lugtenberg and Kamilova 2009). It is well known that fungi like *Trichoderma* spp. parasitize and control harmful fungal infections. These bio-control systems promote more ecologically friendly farming practices by reducing the need for synthetic chemical pesticides. Certain soil microbes have the capacity to also make plants develop systemic resistance, which boosts immunity and provides long-term disease protection (Pieterse et al. 2014). The soil microbiota aids in soil fertility, nutrient cycling, organic matter decomposition, and biocontrol. By boosting nutrient availability, fortifying soil structure, and controlling pathogens, these microbial communities contribute to a strong, sustainable, and healthy agricultural system. One has the capacity to enhance soil health, reduce dependency on chemical inputs, and advance environmentally friendly farming practices by understanding and managing soil microbiota.

Figure 8.3 illustrates plant growth, soil fertility, and the sustainability of ecosystems depend on the soil microbiota (bacteria, fungi, archaea and protozoa). By solubilizing phosphorus (*Bacillus, Pseudomonas*), fixing atmospheric nitrogen (*Rhizobium, Azotobacter*), and releasing essential elements during the breakdown of organic matter, they control the cycling of nutrients. Microbes break down complex residues like proteins, cellulose, and lignin into simpler compounds which enriches soils with humus enhances water retention and aids in carbon sequestration. Beneficial organisms that produce antimicrobials, parasitize dangerous fungi, and cause systemic resistance in plants such as Bacillus subtilis and Trichoderma spp., suppress pathogens in biocontrol. When combined, these processes strengthen soil structure, lessen reliance on chemical inputs, and promote resilient, sustainable farming systems.

8.2.3 Animal Gut Microbiota

Immunological response, digestion, and overall health depend on the community of bacteria, fungi, viruses, and archaea that comprise an animal's gut. The microbiotas of many animal species vary greatly, especially those of monogastrics (pigs, chickens, and humans) and ruminants (cattle, sheep, and goats). Understanding how the gut microbiota of these animals differs in composition and function has the capacity to enhance feed conversion, immune control, and overall productivity (Fig. 8.4).

8.2.3.1 Ruminants and Monogastrics

A ruminant's stomach is divided into four separate sections: the omasum, reticulum, abomasum, and rumen. In the rumen, a diverse microbial community performs primary fermentation, which transforms complex plant fibers like cellulose into volatile fatty acids and gases like methane (Russell and Rychlik 2001). Fermentation also produces essential nutrients such as vitamin B and amino acids, which ruminants absorb and utilize for growth and milk production. A monogastric patient, on the other hand, has a single stomach chamber. The gut microbiota of monogastric animals, which is usually more species rich than that of ruminants, is essential for the breakdown of proteins, carbs and fats. In monogastrics, the microbiota ferments fiber to produce short-chain fatty acids which are an essential source of energy for the host. In contrast to ruminants, monogastrics mainly ferment microorganisms in the large intestine especially the cecum and different microbial populations have distinct functions in immune control and nutrient absorption.

8.2.3.2 Contribution to Feed Conversion, Immune Regulation, and Productivity

Feed conversion efficiency, a measure of an animal's capacity to convert feed into body mass, is significantly impacted by the gut microbiota. The fermentation process

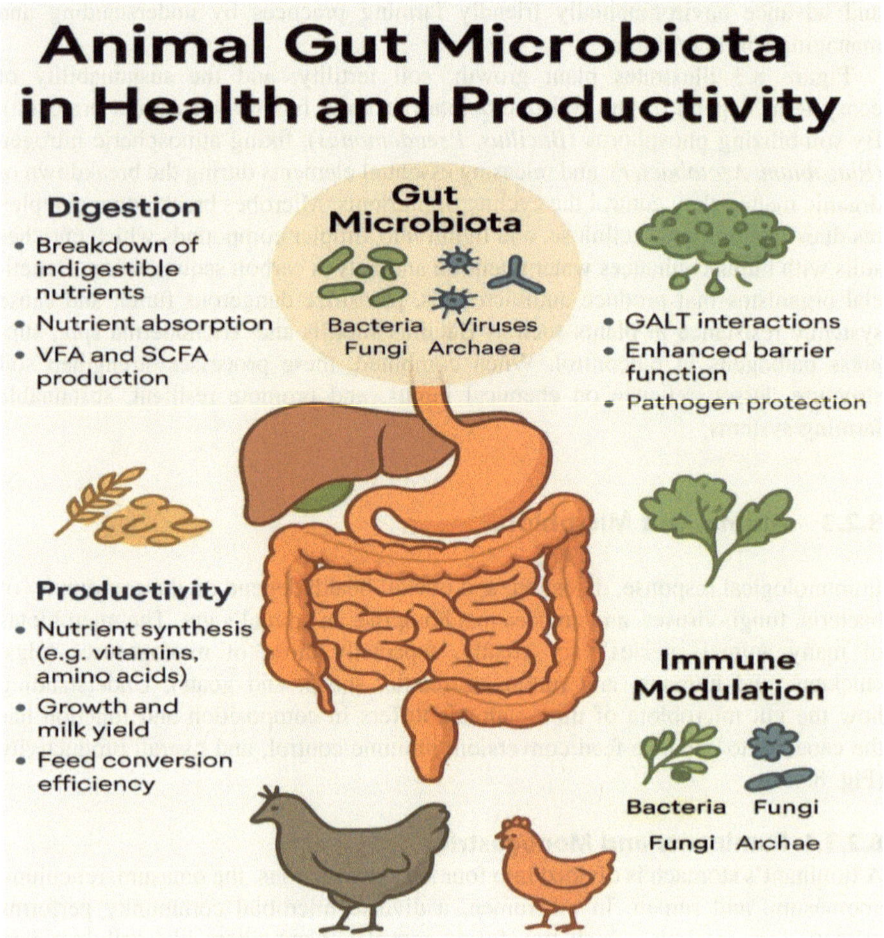

Fig. 8.4 Animal gut-microbiota in health and productivity

in ruminant's rumen has the capacity to break down complex plant fibers, improving nutrient accessibility. The efficient conversion of plant materials into energy-dense products such as Volatile Fatty Acids (VFAs) enhances milk production, feed utilization, and animal growth. In monogastrics, the gut microbiota aids in the breakdown of nutrients that the host cannot digest, enhancing energy and nutrient absorption and increasing feed efficiency and growth rates (Jami et al. 2013). Immune regulation is another crucial function of the gut microbiota. The gut contains the majority of the body's immune cells and the microbiota interacts with these cells to regulate immune responses. Bäckhed et al. (2005) state that gut microbes in monogastric and ruminant animals control systemic immune responses, protect against pathogenic bacteria, and preserve the integrity of the gut barrier. A balanced gut microbiota has the capacity to enhance an animal's overall health, immunity to infections, and prevention of immune-related diseases. The gut microbiota also affects gut-associated lymphoid tissue an integral component of the immune system (Round and Mazmanian 2009). Furthermore, gut microbiota has the capacity to directly affect productivity by influencing metabolic processes. Microbes in the rumen have the capacity to synthesize essential nutrients like vitamin B and amino acids which are critical for ruminant growth and productivity. Monogastric animals are more productive in terms of growth, egg production, and milk yield because gut microbes help break down and absorb nutrients. Thus, altering the gut microbiota through nutrition, probiotics, or prebiotics may enhance livestock productivity and health. Thus, the gut microbiota of an animal is one of the primary determinants of its health, productivity, and feed efficiency. It is essential for the breakdown of fibrous plant materials and the synthesis of essential nutrients in ruminants, even though it facilitates monogastrics to digest a wider range of nutrients and supports overall gut health. Both types of animals rely on their gut microbiota for optimal feed conversion, immunological regulation, and productivity. Making use of the potential of these microbial communities through microbiota management offers opportunities to reduce the use of antibiotics, promote more ecologically friendly farming practices, and elevate livestock productivity.

Figure 8.4 illustrates gut microbiota (bacteria, fungi, viruses, and archaea) is essential for immunity, animal performance, digestion, and nutrient absorption. Gut microorganisms enhance feed conversion efficiency, energy utilization, growth, and milk yield by decomposing complex feed materials and generating metabolites like volatile and short-chain fatty acids. Furthermore, they work with gut-associated lymphoid tissue to control immune responses, preserve the integrity of the gut barrier, and fend off infections. Microbial communities directly support productivity by synthesizing vital nutrients like vitamins and amino acids in addition to aiding in digestion and immunity. Utilizing probiotics, prebiotics, and nutrition techniques to manage these microbial populations has the capacity to enhance livestock health, lessen reliance on antibiotics, and support environmentally friendly agricultural methods.

8.2.4 Aquaculture Microbiota Specific

A range of microbial communities associated with the sediment, water column, biofilms, and the digestive tracts of shrimp and fish that are found in aquatic farming systems. Enhancing nutrient recycling, maintaining water quality, and boosting host immunity all depend on this microbiota. In modern aquaculture, especially intensive systems, microbial management strategies like probiotic supplementation and bio-floc technology have gained popularity as sustainable alternatives to chemical treatments and antibiotics (Fig. 8.5).

8.2.4.1 Probiotics and Biofloc in Fish and Shrimp Farming

A novel method known as Biofloc Technology uses a high carbon-to-nitrogen (C: N) ratio to control the microbial communities in aquaculture tanks or ponds. This

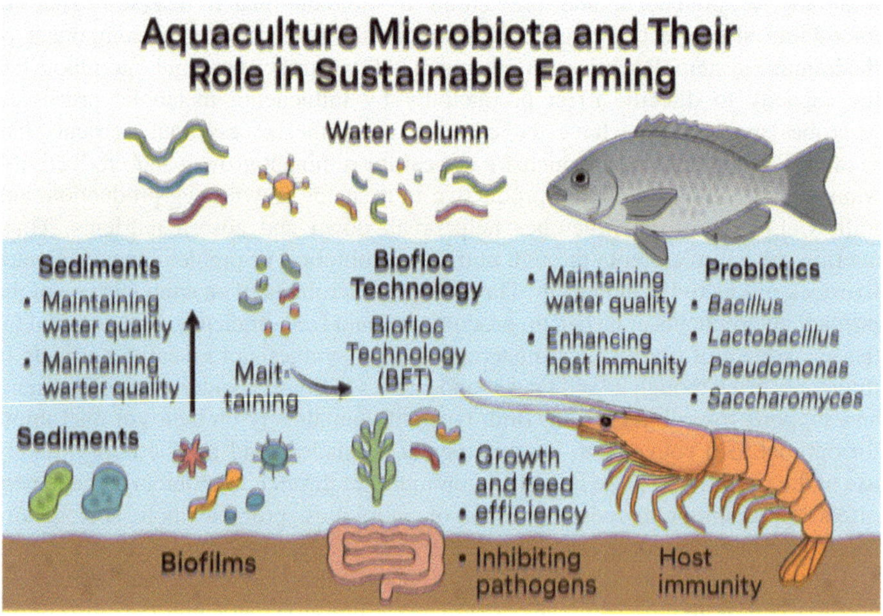

Fig. 8.5 Aquaculture microbiota in sustainable farming

encourages the growth of heterotrophic bacteria which utilize nitrogenous waste (ammonia and nitrites) to produce microbial biomass. The resultant biofloc gives shrimp and fish a nutrient-rich supplemental feed in addition to purifying the water (Crab et al. 2012). Bioflocs which are rich in proteins, lipids, vitamins, and good bacteria enhance growth performance and feed conversion ratios in species such as shrimp (*Litopenaeus vannamei*) and tilapia (*Oreochromis niloticus*). In order to enhance environmental conditions and host health, probiotics (live, healthy microorganisms) are administered through feed or water. Probiotic strains commonly used in aquaculture include *Bacillus, Lactobacillus, Pseudomonas*, and *Saccharomyces* species. These probiotics enhance digestion, change the immune system of the aquatic host, and inhibit pathogenic bacteria by generating antimicrobials and competitively excluding harmful bacteria (Hai 2015). Probiotic utilization has been associated with improved growth, higher survival rates, and fewer disease outbreaks in freshwater and marine aquaculture systems.

8.2.4.2 Impact on Disease and Water Quality Opposition

The water's quality which is directly tied to the microbial community in aquaculture systems determines the productivity and health of aquatic organisms. The accumulation of uneaten feed, feces, and organic matter has the capacity to lead to deteriorated water quality and the development of harmful pathogens. Microbiota-managed systems like BFT help maintain optimal water parameters by stabilizing pH and oxygen levels and reducing toxic nitrogen compounds (Yadav et al. 2025).

Furthermore, a balanced microbiota whether present in the host's gut or in the surrounding environment enhances disease resistance. Beneficial microbes have the capacity to produce antimicrobial compounds, stimulate the innate immune system, and create biofilms that stop pathogens from colonizing. For example, probiotics have been shown to provide protection against the white spot syndrome virus and Vibrio-related infections in shrimp farming (Zokaeifar et al. 2012). These microbial techniques reduce the need for antibiotics which aligns with sustainable and ecologically conscious aquaculture practices. The utilization of probiotics and biofloc to manage the microbiota has revolutionized aquaculture by highlighting the role of beneficial microbes in improving water quality, increasing immunity, and increasing overall productivity. Microbiota-based approaches are essential for food safety and security, environmental impact reduction, and sustainable aquaculture development.

Figure 8.5 illustrates water quality, host immunity, and nutrient recycling are all improved by microbial communities found in sediments, water, biofilms, and the guts of fish and shrimp. While probiotics like *Bacillus* and *Lactobacillus* aid in digestion, immunity, and pathogen suppression, Biofloc technology transforms nitrogenous waste into nutrient-rich biomass that enhances growth and feed efficiency. These methods lessen the use of antibiotics, prevent illnesses like Vibrio and WSSV, stabilize water quality, and support sustainable aquaculture.

8.3 Tools and Techniques for Microbiota Engineering

Our understanding to regulate microbial communities for enhanced host health, disease resistance, and ecosystem productivity has been transformed by the advent of microbiota engineering. This multidisciplinary field integrates ecological modeling, synthetic biology, biotechnological applications, and omics technologies to modify or enhance the structure and function of the microbiome in food systems, agriculture, and animal and aquaculture health. Engineering microbial communities creates new opportunities for climate resilience, biocontrol and sustainable food production because of their close relationship to the operation of biological and environmental systems (Fig. 8.6; Table 8.2).

8.3.1 Metagenomics and Amplicon Sequencing

These two methods are crucial for describing microbial communities. These culture-independent methods enable thorough profiling of microbial diversity, metabolic capabilities and ecological interactions. Shotgun metagenomics has the capacity to provide details about the microbial taxa and their genetic roles by sequencing all of the DNA molecules in a sample (Qin et al. 2010). It enables the discovery of new microbes in unculturable communities, the reconstruction of metabolic pathways, and the profiling of resistance genes. Amplicon sequencing, focusing on the 16S rRNA gene (bacteria), 18S rRNA (eukaryotes), and ITS region (fungi) is used for

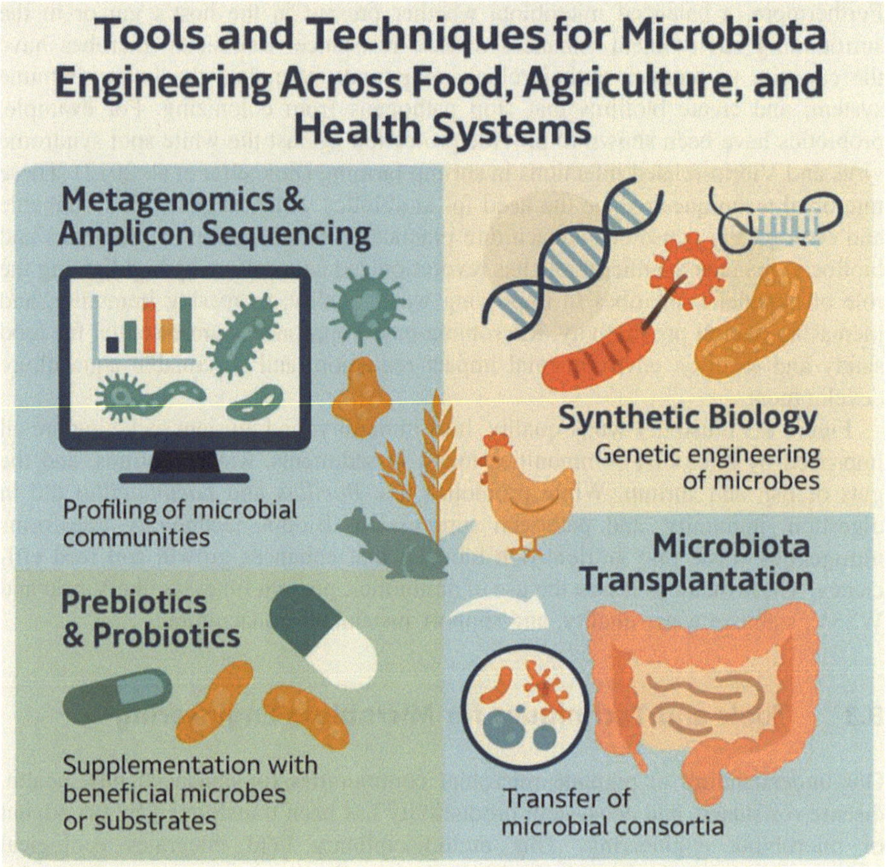

Fig. 8.6 Tools and techniques for microbiota engineering

rapid taxonomic identification and community composition analysis (Caporaso et al. 2011). These techniques have revolutionized the study of microbiome in soil, plant rhizospheres, animal guts and aquaculture systems. These tools are crucial for microbiota engineering because they help identify microbial species or genes that has the capacity to be changed to produce desired characteristics.

8.3.2 Synthetic Biology Methods

Synthetic biology offers powerful tools for generating microbial strains or consortia with predictable and programmable functions. This comprises: (1) Building a genetic circuit: It is possible to train engineered microorganisms to detect environmental cues, like pH or metabolites, and respond by producing the required products, like enzymes, signaling molecules, and antimicrobials (Brenner et al. 2008).

Table 8.2 Tools and techniques for microbiota engineering

Tool/ Technique	Description	Key examples/applications	References
Amplicon sequencing and metagenomics	Methods for characterizing the diversity and function of microorganisms those are not dependent on culture	16S/18S rRNA and ITS sequencing for taxonomic profiling in soil, plant, and gut microbiomes; shotgun metagenomics for functional gene analysis	Qin et al. (2010) and Caporaso et al. (2011)
Synthetic biology	creating microorganisms with specific, programmable characteristics	Gene editing using CRISPR-Cas; responsive production using synthetic circuits; and modulation of soil and gut microbiota through engineered consortia	Brenner et al. (2008) and Mimee et al. (2015)
Prebiotics and probiotics	Modifying the microbiota with selective substrates or live beneficial microorganisms	Animal feeds containing resistant starch, *Lactobacillus*, *Bifidobacterium*, inulin, GOS and functional foods for immunity and gut health	Hill et al. (2014) and Gibson et al. (2017)
Transplanting microbiota	Microbial consortia are transferred between people and environments	Transplanting fecal microbiota in livestock and aquaculture; stabilizing biofloc; and restoring the soil microbiome after chemical exposure	Youngster et al. (2014)
Modification of diet and food additive	Influencing microbiota with feed additives and food ingredients	High fiber diets, organic acids, enzymes, phytobiotics, and microbial boosters are used in aquaculture to improve water quality and biofloc formation	Flint et al. (2012) and Torok et al. (2011)

(2) CRISPR-Cas genome editing: According to Mimee et al. (2015), this technology enables precise genome modification of microorganisms, enabling the elimination of harmful genes or the integration of beneficial biosynthetic pathways. (3) Custom consortia engineering: Sometimes, synthetic biologists build multispecies consortia that perform complementary metabolic tasks, mimicking but surpassing natural synergy. (4) The food and agricultural industries are developing synthetic microbes to fight infections, enhance animal nutrition, break down environmental pollutants, and encourage plant growth.

8.3.3 Prebiotic and Probiotic Supplementation

One of the easiest and most popular methods to change the microbiota is to utilize probiotics and prebiotics: According to Hill et al. (2014), probiotics are live microorganisms that enhance the health of the host when given in sufficient quantities. Aquaculture, animal feed, and functional foods all make extensive use of them. *Bacillus, Saccharomyces, Bifidobacterius* and *Lactobacillus* are common genera. Prebiotics are fibers or compounds that utilize selective fermentation to promote the growth of good bacteria. GOS, inulin, and resistant starches are a few examples (Gibson et al. 2017). Their specific application has the capacity to enhance mucosal immunity, metabolite production (short-chain fatty acids) and microbial balance. In order to achieve synergistic microbiome modulation, food systems are increasingly using synbiotics or a combination of the two. Growth performance, immunity, gut health, pathogen resistance, and nutrient digestibility are all impacted by these formulations.

8.3.4 Microbiota Transplantation

This technique aims to restore or modify the recipient's microbiome by purposefully transferring microbial consortia from one environment or individual to another: Fecal microbiota transplantation is being researched in aquaculture and livestock to enhance gut health and disease resistance. It is a proven clinical practice for human recurrent *Clostridium difficile* infections (Youngster et al. 2014). After extensive chemical use, soil microbiome transplants are used in agriculture to revitalize degraded soils or restore microbial balance. Microbial inoculation from well-established systems or robust bioflocs is advantageous for aquaculture because it prevents pathogen blooms and stabilizes water quality. When the natural microbiome has been disrupted or when quick microbial colonization is required for system stability, this method works especially well.

8.3.5 Modification of the Dietary and Additive Microbiome

One of the best levers for microbiota engineering across systems is diet: By affecting the gut microbiota in humans and animals' dietary factors impact immunity, disease susceptibility, and host metabolism. For instance, whereas diets high in protein or fat may result in dysbiosis, diets high in fiber favor bacteria that produce butyrate (Flint et al. 2012). In order to change the microbial balance in livestock in favor of beneficial species, feed additives like enzymes, phytobiotics, organic acids, and essential oils are utilized. Torok et al. (2011) revealed that this increases growth rates, enhances nutrient absorption, and lessens the need for antibiotics. Microbial boosters and carbon sources are added to water in aquaculture to encourage the growth of beneficial bioflocs, which recycle nutrients and compete with pathogens. By adjusting dietary inputs to specifically enrich advantageous microbes and suppress undesirable ones, microbiome has the capacity to be directed toward their ideal functional states.

Figure 8.6 illustrates a variety of techniques used in microbiota engineering to mold microbial communities for sustainability, productivity, and health. Probiotics, prebiotics, and synbiotics are integral tools for enhancing gut health and pathogen resistance; metagenomics and amplicon sequencing are used for microbe profiling; and synthetic biology and CRISPR are used to create functional strains or consortia. Additional strategies include dietary or feed additives to control microbial composition and microbiota transplantation to restore equilibrium in soils, animals, or aquaculture. When combined, these tactics allow for targeted microbiome modification in food systems, agriculture, and health.

8.4 Applications in Agriculture and Livestock

Microbiota engineering is revolutionizing agriculture and livestock production by tailoring beneficial microbial consortia to enhance crop yield, animal health and food quality. It supports sustainable development goals and reduces environmental & health risks by replacing chemical inputs and antibiotics (Fig. 8.7).

8.4.1 Impact of Biopesticides and Biofertilizers on Crop Yield

This microbiota based biofertilizers enhance nutrient availability, soil fertility, and plant health. They consist of nitrogen fixers (*Rhizobium, Azospirillum,* and

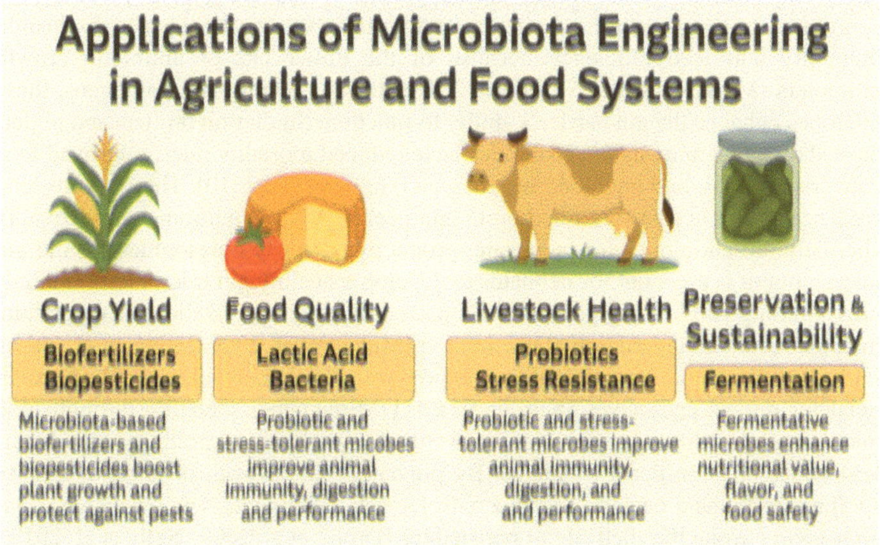

Fig. 8.7 Applications of microbiota, engineering in agriculture and livestock: this figure illustrates microbiota engineering mediated improved crop yield, livestock health, and food quality while reducing reliance on chemicals and antibiotics. Biofertilizers and biopesticides enhance nutrient cycling and pest control; stress-resistant microbes help crops adapt to drought and salinity, and probiotics enhance animal growth, immunity, and feed efficiency. In food systems, lactic acid bacteria and biocontrol yeasts extend shelf life, while fermentation employing particular microbial cultures enhances nutrition, safety, and flavor

Azotobacter) and phosphate solubilizers (*Pseudomonas* and *Bacillus*) (Bhattacharyya and Jha 2012; Compant et al. 2019). Because of their functions in plant-microbe symbiosis and nutrient cycling, they are crucial to low-input sustainable farming systems. Two examples of biopesticides that work by producing insecticidal or antifungal compounds and triggering immune responses in plants are *Bacillus thuringiensis* and *Trichoderma harzianum* (Chandler et al. 2008). These materials significantly reduce environmental contamination and the need for chemical pesticides.

8.4.2 Stress-Resistant Microbiota

Resilient microbial strains, either naturally occurring or artificially produced, aid crops in adapting to abiotic stressors. For example, aminocyclopropane-1-carboxylate (ACC) deaminase-active microbes alter ethylene levels to lessen growth inhibition brought on by stress (Glick 2014). Certain microorganisms help plants survive drought and salinity by improving water retention, ion balance, and the induction of antioxidant enzyme systems (Yang et al. 2009; Marasco et al. 2012).

8.4.3 Livestock Health and Growth

Probiotics such as *Lactobacillus* species, *Bacillus* species, and *Saccharomyces cerevisiae* are being used more often as antibiotic substitutes to enhance gut health, immunity and feed efficiency because of the drawbacks of antibiotic growth promoters (Markowiak and Śliżewska 2018). By competing with pathogens, these microbes enhance the gut barrier's ability to function. Studies on broilers and piglets have shown that animals fed probiotics have reduced mortality rates, improved feed conversion ratios, and increased weight gain (Gaggia et al. 2010). These results lend credence to the notion that microbiota engineering is an environmentally friendly alternative to antibiotics for increasing productivity. Early development of the gut microbiota: it is essential for neonates to develop a healthy gut microbiota. Immune programming and lifelong growth are impacted by early colonization via colostrum, maternal transfer, and microbial inoculation (Mulder et al. 2009; Milani et al. 2017). Early-life microbiota transplantation and prebiotic supplementation have been beneficial for piglets and calves (Kim et al. 2011). Enhancing shelf life through food quality and post-harvest microbial interventions: worldwide, spoiling and postharvest losses are significant problems. By preventing the growth of fungi that cause spoiling, lactic acid bacteria and biocontrol yeasts (*Metschnikowia fructicola, Candida sake*) elevate the shelf life of perishables (Droby et al. 2009; Nally et al. 2013). These microbes produce ethanol, mycocins, and volatile organic compounds, all of which have antimicrobial properties. Furthermore, by creating barriers against oxidation and moisture loss, microbial coatings and films containing *Lactobacillus plantarum* and *Pediococcus* species have the capacity to be used to preserve fresh meat and produce.

8.4.4 Fermentation of Microbes for Nutrition, Safety, and Taste

Fermentation is among the oldest forms of microbial food engineering. Fermentative bacteria and yeasts (*Lactobacillus, Leuconostoc,* and *Saccharomyces*) enhance food by adding organic acids, vitamins, and bioactive peptides, as well as preventing infections (Tamang et al. 2016). More recently, the creation of specialized starter cultures with enhanced metabolic qualities has made it feasible to precisely regulate flavor and health characteristics (Smid and Kleerebezem 2014). This is especially advantageous for soy, dairy, and cereal-based fermentations.

8.5 Effect on the Environment and Socioeconomics

Microbiota engineering has the capacity to boost environmental and socioeconomic resilience as well as agricultural productivity. By harnessing the functional capacities of microbial communities, we have the capacity to achieve sustainability in several domains, such as enhanced food security, reduced dependence on agrochemicals, climate mitigation, and rural empowerment (Fig. 8.8).

8.5.1 Carbon Sequestration and Climate Resilience

Because soil microbiota aid in the breakdown of organic matter and stabilize soil organic carbon, they are crucial for carbon sequestration. The conversion of plant waste into humic materials is one instance of a microbially mediated process that enhances carbon retention and fosters long term climate resilience (Lal 2004). Particularly in conservation tillage or agroforestry settings, applying mycorrhizal fungi and plant growth-promoting rhizobacteria has the capacity to elevate root biomass and promote carbon flow into soil systems (van der Heijden et al. 2008). Furthermore, microbiota that boost crops' resistance to drought and salinity indirectly aid climate adaptation, particularly in marginal environments (Glick 2014).

8.5.2 Reduction of Chemical Pesticides and Fertilizers

The adoption of biopesticides and fertilizers considerably lowers the usage of synthetic agrochemicals which are frequently connected to soil degradation, water pollution, and biodiversity loss (Malusá and Vassilev 2014). In certain systems, 50–70% of chemical fertilizers have the capacity to be substituted with nitrogen-fixing bacteria (*Rhizobium* and *Azospirillum*), phosphate-solubilizing *Bacillus* strains and potassium-mobilizing bacteria without reducing yield (Bhattacharyya and Jha 2012; Vessey 2003). Trichoderma spp. and *Pseudomonas fluorescens* are two examples of bio-pesticides that control diseases and pests while preserving ecological balance. This change reduces input costs, promotes integrated pest management, and lessens environmental risks (Chandler et al. 2008).

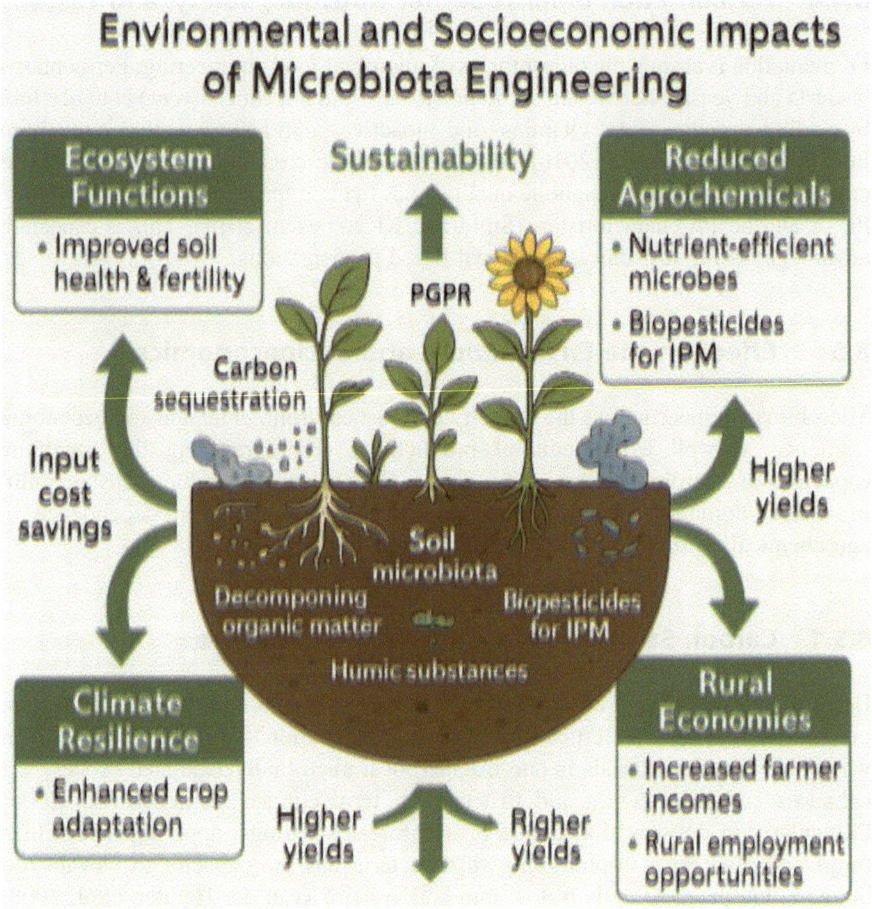

Fig. 8.8 Environmental and socioeconomic impacts of microbiota engineering

8.5.3 Increased Crop Resilience and Yields

Microbiota engineering directly contributes to food security by increasing crop resilience and yields, especially in low-input or climate-vulnerable areas. Improving output and quality increases farmers' income, and lowering input costs further boosts profit margin. Furthermore, the expansion of regional bio-inoculant industries and local microbial composting facilities has the capacity to enhance rural livelihoods and generate employment. Training farmers in microbiota-based technologies fosters knowledge sovereignty and capacity building by enabling rural communities to function as agents of sustainable change (Pretty et al. 2011).

Figure 8.8 illustrates by increasing soil fertility, increasing crop resilience to drought and salinity and improving carbon sequestration, microbiota engineering promotes sustainability. It lowers pollution, input costs, and ecological risks by substituting biofertilizers and biopesticides for chemical fertilizers and pesticides. These methods boost yields, raise farmer incomes, empower communities toward sustainable agriculture, and generate jobs in rural areas through regional bio-inoculant industries.

8.6 Challenges and Ethical Concerns

Despite the potential of microbiota engineering for sustainable food and agricultural systems, there are still a number of challenges and ethical concerns. These include biosafety concerns, complex regulatory frameworks, and public mistrust of genetically modified microorganisms. These barriers need to be eliminated in order for microbial technologies to be applied in a responsible and equitable manner.

8.6.1 Regulatory Barriers to Engineered Microbe Release

The release of artificial or modified microbes into the environment is tightly regulated due to the potential ecological consequences. Regulatory agencies such as the US Environmental Protection Agency, the Genetic Engineering Appraisal Committee of India, and the European Food Safety Authority enforce stringent guidelines prior to permitting the utilization of genetically modified organisms in agriculture. These include assessments of environmental persistence, gene stability, and non-target effects (Timmis et al. 2019). Furthermore, international regulatory frameworks are often not harmonized, which hinders transboundary research and commercialization financially and delays. This regulatory ambiguity stifles innovation and poses a serious problem for startups and developers in the public sector.

8.6.2 Microbiota Engineering, Horizontal Gene Transfer

HGT or the sharing of genetic material between microbial species is a significant biosafety concern since it has the capacity to lead to the unintended spread of altered traits (Heinemann and Traavik 2004). For example, pathogenic microorganisms may pose serious ecological or public health risks if they inherit genes that produce toxins or confer antibiotic resistance. Moreover, some altered microorganisms have the potential to displace native microbial populations, disrupting existing microbiome and ecological balances (Moe-Behrens et al. 2013). To reduce such risks, scientists are investigating bio-containment strategies such as synthetic auxotrophy and kill switch mechanisms, though they are not yet perfect (Wright et al. 2013).

8.6.3 Ethics and Public Perception

The way that people perceive engineered microorganisms, particularly those that employ genetic modification or synthetic biology, is one of the main issues confronting society. Corporate control over living things, environmental contamination, and food safety are common public concerns. These anxieties are made worse by a lack of candid communication and community involvement. Ethical questions are also raised by benefit sharing, especially when the microbiota originates from indigenous lands or traditional agricultural systems. Concerns about ownership, consent, and equitable access to the resulting biotechnologies necessitate the development of inclusive policies and participatory governance models (Wynberg and Laird 2020).

8.7 Future Trends

As microbiota research advances, new paradigms are emerging at the intersection of data science, biology, and sustainable innovation. Over the next decade, new methods and technologies that leverage microbes to enhance the health of humans, animals, plants, and ecosystems will proliferate. Three substantial areas that are influencing the direction of microbiota research: the development of microbial data banks and open access platforms; AI-guided microbiome engineering and customized microbiota solutions for farms.

8.7.1 Microbiome Engineering Under AI's Direction

Artificial intelligence and machine learning are revolutionizing our ability to understand, model, and design complex microbial ecosystems. Traditional statistical tools often cannot handle the massive amounts of complex data gathered from metagenomic, metabolomic, and transcriptomic investigations. AI bridges this gap by offering unprecedented accuracy in predictive modeling, network analysis, and the development of artificial communities. One recent advancement is MICOM, a tool that models metabolic interactions within microbial communities to forecast the effects of compositional changes (Diener et al. 2020). Similarly, disease states are predicted from microbiome signatures using autoencoder-based classifiers and deep learning frameworks like DeepMicro. By allowing researchers to model and fine-tune microbial consortia in silico before field application, these techniques reduce trial and error in both laboratory and industrial settings. Artificial intelligence is being used in agriculture to identify beneficial microorganisms that help crops that are being damaged by drought, salinity, or pests. To create formulations tailored to a particular crop or area, AI-powered systems integrate environmental factors and microbiome datasets. Precision microbial engineering may boost sustainability and productivity. By offering instruments to both understand and rewire those communities, artificial intelligence is revolutionizing the study of microbiome from observation to orchestration.

8.7.2 Tailored Solutions for Farm Microbiota

Farming systems differ significantly in terms of climate, microbial legacy, soil composition, crop rotation, and irrigation methods. These factors impact the native microbial communities and determine the success of any external microbial intervention. Personalized microbiota solutions are formulations tailored to these local conditions to optimize colonization, functionality, and benefit. To elevate yield and decrease disease, for instance, bioinoculants containing strains of *Bacillus, Pseudomonas*, and *Azospirillum* that are appropriate for specific pH and nutrient conditions have been shown to be more effective than general-purpose inoculants. Host-specific probiotics have also been demonstrated to enhance gut colonization and growth in high altitude-adapted poultry in animal agriculture. Precision agriculture, which makes decisions based on site-specific data, is comparable to this approach. Advances in environmental sensors, remote sensing, and soil microbiome profiling enable real-time insights into the microbial status of the farm ecosystem. These data inform the design of microbiota-based inputs, which offer a sustainable alternative to synthetic antibiotics and fertilizers.

8.7.3 Open-Source Microbiome Platforms and Microbial Data Banks

As microbiome research generates petabytes of sequencing and functional data, there is a growing need for standardized, searchable, and structured data repositories. Open-access microbial databases foster innovation and reproducibility by acting as knowledge bases for industry, academia, and regulatory bodies. By cataloguing the microbial diversity from over 27,000 environmental samples in 43 countries, notable projects like the Earth Microbiome Project have produced data on biogeography, function, and evolution (Thompson et al. 2017). Researchers have the capacity to utilize the MGnify platform from EMBL-EBI and Qiita to upload, analyze, and share microbiome data and associated metadata. New platforms that utilize blockchain technology for data provenance, access tracking, and collaborative research reward sharing are starting to appear in addition to static repositories. This ensures transparency, confidence, and equitable involvement in the global microbiome research. Microbiota data banks are the gene vaults of the twenty-first century, containing not only DNA but also the blueprint for ecological intelligence. The revival of microbiota research is being impacted by open science, artificial intelligence, and personalization. These recent advancements could elevate agricultural and environmental resilience and democratize access to microbial technology. These guidelines must be followed by researchers, practitioners, and policymakers in order to fully realize the potential of microbiota-based innovations.

8.8 Conclusion

Examining the transformative potential of microbiota in modern food systems reveals an intriguing intersection of biology, technology, and sustainability. Through the utilization of plant growth-promoting Rhizobacteria to boost crop yields and customized microbial consortia to enhance livestock productivity and health, microbiota engineering is revolutionizing our understanding of food security and agricultural resilience. It has the capacity to be used for food processing, preservation, and even waste reduction and it has a wide impact on the entire value chain. Nevertheless, the success of microbiota-driven innovations cannot be guaranteed by scientific advancements alone. By uniting environmental scientists, biotechnologists, agronomists, microbiologists, and lawmakers, it promotes interdisciplinary collaboration. Only by cooperating has the capacity to we get past moral dilemmas, legal restrictions, and scaling up challenges. Cultivating this synergy will be essential in the future to fully realize the potential of microbiota for a more robust, nutrient-dense, and sustainable global food system.

References

Bäckhed F, Ley RE, Sonnenburg JL, Peterson DA, Gordon JI (2005) Host–bacterial mutualism in the human intestine. Science 307(5717):1915–1920

Bhattacharyya PN, Jha DK (2012) Plant growth-promoting rhizobacteria (PGPR): emergence in agriculture. World J Microbiol Biotechnol 28(4):1327–1350

Brenner K, You L, Arnold FH (2008) Engineering microbial consortia: a new frontier in synthetic biology. Trends Biotechnol 26(9):483–489

Caporaso JG, Lauber CL, Walters WA et al (2011) Global patterns of 16S rRNA diversity at a depth of millions of sequences per sample. Proc Natl Acad Sci USA 108(Suppl 1):4516–4522

Chandler D, Davidson G, Grant W, Greaves J, Tatchell M (2008) Microbial biopesticides for integrated crop management: an assessment of environmental and regulatory sustainability. Trends Food Sci Technol 19(5):275–283. https://doi.org/10.1016/j.tifs.2007.12.009

Cheng X, Wang M, Yuan MM, Li J, Xiong W (2023) Editorial: rhizosphere microbiome engineering for crop cultivation. Front Bioeng Biotechnol 11:1267442. https://doi.org/10.3389/fbioe.2023.1267442

Compant S, Samad A, Faist H, Sessitsch A (2019) A review on the plant microbiome: ecology, functions, and emerging trends in microbial application. J Adv Res 19:29–37

Crab R, Defoirdt T, Bossier P, Verstraete W (2012) Biofloc technology in aquaculture: beneficial effects and future challenges. Aquaculture 356–357:351–356

Dey S, Talukdar A, Bhattacharya S (2025) Microbial degradation and valorization of food wastes: waste to wealth approaches towards sustainability. Discov Chem 2:147. https://doi.org/10.1007/s44371-025-00217-9

Diener C, Gibbons SM, Resendis-Antonio OR (2020) MICOM: metagenome-scale modeling to infer metabolic interactions in the gut microbiota. mSystems 5(1):e00606–e00619

Droby S, Wisniewski M, Macarisin D, Wilson C (2009) Twenty years of postharvest biocontrol research: is it time for a new paradigm? Postharvest Biol Technol 52(2):137–145

Fierer N (2017) Embracing the unknown: disentangling the complexities of the soil microbiome. Nat Rev Microbiol 15(8):508–514

Flint HJ, Scott KP, Duncan SH et al (2012) Microbial degradation of complex carbohydrates in the gut. Gut Microbes 3(4):289–306

Gaggia F, Mattarelli P, Biavati B (2010) Probiotics and prebiotics in animal feeding for safe food production. Int J Food Microbiol 141(Suppl 1):S15–S23

Gibson G, Hutkins R, Sanders M et al (2017) Expert consensus document: The International Scientific Association for Probiotics and Prebiotics (ISAPP) consensus statement on the definition and scope of prebiotics. Nat Rev Gastroenterol Hepatol 14:491–502. https://doi.org/10.1038/nrgastro.2017.75

Glick BR (2012) Plant growth-promoting bacteria: mechanisms and applications. Sci Prog 95(4): 462–472

Glick BR (2014) Bacteria with ACC deaminase has the capacity to promote plant growth and help plants tolerate abiotic stress. Microbiol Res 169(1):30–39

Hai NV (2015) The utilize of probiotics in aquaculture. J Appl Microbiol 119(4):917–935

Heinemann JA, Traavik T (2004) Problems in monitoring horizontal gene transfer in field trials of transgenic plants. Nat Biotechnol 22(9):1105–1109

Hill C, Guarner F, Reid G, Gibson GR, Merenstein DJ, Pot B, Morelli L, Canani RB, Flint HJ, Salminen S, Calder PC, Sanders ME (2014) Expert consensus document. The International Scientific Association for Probiotics and Prebiotics consensus statement on the scope and appropriate utilize of the term probiotic. Nat Rev Gastroenterol Hepatol 11(8):506–514. https://doi.org/10.1038/nrgastro.2014.66

Jami E, Israel A, Kotser A, Mizrahi I (2013) Exploring the bovine rumen bacterial community from birth to adulthood. ISME J 7(6):1069–1079. https://doi.org/10.1038/ismej.2013.2

Kim HB, Borewicz K, White BA et al (2011) Longitudinal investigation of the age-related bacterial diversity in the feces of commercial pigs. Vet Microbiol 153(1–2):124–133

Lal R (2004) Soil carbon sequestration impacts on global climate change and food security. Science 304(5677):1623–1627

Lavilla M, Amárita F (2025) Advances and future challenges in microbial food safety: volume I. Foods 14(13):2231. https://doi.org/10.3390/foods14132231

Lei Y, Kuai Y, Guo M, Zhang H, Yuan Y, Hong H (2025) Phosphate-solubilizing microorganisms for soil health and ecosystem sustainability: a forty-year scientometric analysis (1984–2024). Front Microbiol 16:1546852. https://doi.org/10.3389/fmicb.2025.1546852

Lugtenberg B, Kamilova F (2009) Plant-growth-promoting rhizobacteria. Ann Rev Microbiol 63: 541–556

Malusá E, Vassilev N (2014) A contribution to set a legal framework for biofertilisers. Appl Microbiol Biotechnol 98(15):6599–6607

Marasco R, Rolli E, Ettoumi B et al (2012) A drought resistance-promoting microbiome is selected by root system under desert farming. PLoS One 7(10):e48479

Markowiak P, Śliżewska K (2018) The role of probiotics, prebiotics and synbiotics in animal nutrition. Gut Pathog 10(1):21

Milani C, Duranti S, Bottacini F, Casey E, Turroni F, Mahony J, Belzer C, Delgado Palacio S, Arboleya Montes S, Mancabelli L, Lugli GA, Rodriguez JM, Bode L, de Vos W, Gueimonde M, Margolles A, van Sinderen D, Ventura M (2017) The first microbial colonizers of the human gut: composition, activities, and health implications of the infant gut microbiota. Microbiol Mol Biol Rev 81(4):e00036–e00017. https://doi.org/10.1128/MMBR.00036-17

Mimee M, Citorik RJ, Lu TK (2015) Microbiome therapeutics—advances and challenges. Adv Drug Deliv Rev 105(Pt A):44–54

Moe-Behrens GH, Davis R, Haynes KA (2013) Preparing synthetic biology for the world. Front Microbiol 4:5

Mulder IE et al (2009) Environmentally-acquired bacteria influence microbial diversity and natural innate immune responses at gut surfaces. BMC Biol 7:79

Nally MC, Pesce VM, Maturano YP et al (2013) Biocontrol of Botrytis cinerea in table grapes by Candida species. Postharvest Biol Technol 78:57–63

Pandey K, Saharan BS (2025) Soil microbiomes: a promising strategy for boosting crop yield and advancing sustainable agriculture. Discov Agric 3:54. https://doi.org/10.1007/s44279-025-00208-5

Pflughoeft KJ, Versalovic J (2012) Human microbiome in health and disease. Annu Rev Pathol 7: 99–122. https://doi.org/10.1146/annurev-pathol-011811-132421
Pieterse CM, Zamioudis C, Berendsen RL (2014) Induced systemic resistance by beneficial microbes. Annu Rev Phytopathol 52:347–373
Pretty J, Toulmin C, Williams S (2011) Sustainable intensification in African agriculture. Int J Agric Sustain 9:5–24. https://doi.org/10.3763/ijas.2010.0583
Qin J, Li R, Raes J, Arumugam M, Burgdorf KS, Manichanh C, Nielsen T, Pons N, Levenez F, Yamada T et al (2010) A human gut microbial gene catalogue established by metagenomic sequencing. Nature 464(7285):59–65. https://doi.org/10.1038/nature08821
Round JL, Mazmanian SK (2009) The gut microbiota shapes intestinal immune responses. J Immunol 182(1):59–63
Russell JB, Rychlik JL (2001) Factors that alter rumen microbial ecology. Science 292(5519): 1119–1122
Smid EJ, Kleerebezem M (2014) Production of aroma compounds in lactic fermentations. Annu Rev Food Sci Technol 5:313–326
Tamang JP, Watanabe K, Holzapfel WH (2016) Diversity of microorganisms in global fermented foods and beverages. Front Microbiol 7:377
Tan C, Kalhoro MT, Faqir Y, Ma J, Osei MD, Khaliq G (2022) Climate-resilient microbial biotechnology: a perspective on sustainable agriculture. Sustainability 14:5574. https://doi.org/10.3390/su14095574
Thompson LR, Sanders JG, McDonald D et al (2017) A communal catalogue reveals Earth's multiscale microbial diversity. Nature 551(7681):457–463
Timmis K, de Lorenzo V, Verstraete W et al (2019) The contribution of microbial biotechnology to sustainable development goals. Microb Biotechnol 12(5):995–1000
Torok VA, Hughes RJ, Ophel-Keller K, Ali M, Macalpine R (2011) Influence of litter on cecal microbiota in broilers. Poult Sci 90(2):435–444
van der Heijden MGA, Hartmann M (2016) Networking in the plant microbiome. PLoS Biol 14(2): e1002378. https://doi.org/10.1371/journal.pbio.1002378
van der Heijden MGA, Bardgett RD, van Straalen NM (2008) The unseen majority: soil microbes as drivers of plant diversity and productivity in terrestrial ecosystems. Ecol Lett 11(3):296–310
Vessey JK (2003) Plant growth promoting rhizobacteria as biofertilizers. Plant Soil 255(2):571–586
Vorholt JA (2012) Microbial life in the phyllosphere. Nat Rev Microbiol 10(12):828–840
Wright O, Stan GB, Ellis T (2013) Building-in biosafety for synthetic biology. Microbiology 159 (Pt 7):1221–1235
Wynberg R, Laird S (2020) A community perspective on the origin and benefit-sharing of genetic resources. Plants People Planet 2(6):587–600
Yadav NK, Paul S, Patel AB et al (2025) The role of biofloc technology in sustainable aquaculture: nutritional insights and system efficiency. Blue Biotechnol 2:7. https://doi.org/10.1186/s44315-025-00025-x
Yang J, Kloepper JW, Ryu CM (2009) Rhizosphere bacteria help plants tolerate abiotic stress. Trends Plant Sci 14(1):1–4
Youngster I, Russell GH, Pindar C, Ziv-Baran T, Sauk J, Hohmann EL (2014) Oral, capsulized, frozen fecal microbiota transplantation for relapsing Clostridium difficile infection. JAMA 312(17):1772–1778. https://doi.org/10.1001/jama.2014.13875
Zokaeifar H, Balcázar JL, Saad CR et al (2012) Effects of Bacillus subtilis on the growth performance, digestive enzymes, immune gene expression, and disease resistance of white shrimp (Litopenaeus vannamei). Fish Shellfish Immunol 33(4):683–689

Synthetic Biology

9

Abstract

Synthetic biology is a new collaborative field which combines engineering principles to biological systems to design and construct new functions, metabolic pathways, and organisms to fulfil human needs. Synthetic biology is different from genetic engineering; it involves construction of biological parts from scratch, with the help of systematic components like BioBricks to create modular and programmable systems. This chapter describes the essential principles of synthetic biology modularity, standardization, abstraction, and iterative stepwise plan that facilitate systematic biological engineering. These advancements support a vast range of applications in HEA agriculture, healthcare, industry, and environmental remediation. Special focus is kept on the design of artificial microbial consortia that reflect natural consortia and also facilitate cooperative metabolic interactions to achieve maximized production, as demonstrated through case studies on microbial biosynthesis of paracetamol and vitamin C. The chapter also descents deep into the dynamic function of microbiomes in personalized medicine, and how engineered microbes can potentially be used for diagnostics and targeted therapies. Although the potential of synthetic biology is vast, safety, regulatory, and ethical issues are still crucial. In general, synthetic biology is a strong strategy for creating robust and innovative biological systems.

Keywords

Synthetic biology · Biological engineering principles · BioBricks · Modular biological systems · Artificial microbial consortia · Cooperative metabolism · Microbial biosynthesis · Personalized medicine

R. C. Sobti et al., *Microbiota Unveiled: Science, Health, and Future Therapies*,
https://doi.org/10.1007/978-981-95-9031-5_9

9.1 Introduction

"Synthetic biology" is a new multidisciplinary field that uses concepts from biology, chemistry, engineering, computer science and technology to construct new biological tools, devices, and systems. It also redesigns new natural systems in contrast to the conventional genetic engineering, which involves manipulation of existing genes, synthetic biology enables scientists to construct completely new genetic constructs, often using standardized biological building blocks called BioBricks (Knight 2003). BioBricks is a collection of standardized DNA constructs encoding a specific biological function (e.g., promoters, ribosome binding sites, coding sequences, terminators). BioBricks can be constructed using simple restriction enzyme-based cloning methodology. To make biological systems controlled, modular, and predictable, molecular biology and engineering work together and develop it in the early 2000s (Endy 2005). Synthetic biology is the engineering of biological systems using engineering principles such as abstraction, standardization, and decoupling. This enables the design of complex systems such as biosensors, programmable cells, and genetic circuits. Numerous industries such as industrial biotechnology, medicine, agriculture, and environmental decontamination can witness radical transformation with the help of synthetic biology. For example, microbes are already genetically engineered to make biofuels, drugs (such as artemisinin to treat malaria), and bioplastics (Keasling 2010). Effective treatments and diagnosis in medicine have been made possible with the identification of synthetic circuits that can sense and respond to disease biomarkers in the human body (Huang et al. 2025). Molecular engineering and design concepts converge with molecular biology in the emerging field of synthetic biology to engineer and design novel biological functions and systems. Microbiological synthetic biology, particularly in bacteria, is enabling vast potential for the design of organisms to carry out a variety of functions, ranging from sensing and remediation in the environment to drug production (Yook et al. 2025). Bacteria are the best hosts for synthetic biology since they are fast-growing, have well-established genomes, and can be easily genetically engineered. Model organisms like *Escherichia coli* and *Bacillus subtilis* have long been employed as "biological chassis" for synthetic circuits and metabolic engineering (Nielsen and Keasling 2016). Using artificial networks of genes, these engineered bacteria can be programmed to produce useful chemicals, perform logical operations, and even communicate with one another. One of the most common uses of bacterial synthetic biology is the manufacture of drugs. *Escherichia coli* was the first bacterium to be employed in synthetic biology and it was used to construct the Repressilator (Elowitz and Leibler 2000) and the genetic toggle switch (Gardner et al. 2000). These pioneering studies demonstrated that synthetic circuits could be utilized to program bacterial cells in synthetic biology research. For instance, efforts have been made to design the synthesis of paracetamol (acetaminophen) in *E. coli* by utilizing genes that are involved in the conversion of 4-aminophenol and chorismate, a crucial step (Galanie et al. 2015). It is a cheap and renewable process in comparison to the traditional chemical synthesis. Also, the area of synthetic biology is more and more engaged in engineering bacteria that are associated with humans. Presently, gut-engineered bacteria can identify signs of illness, manage the immune system, or release therapeutic molecules into the gut directly (Charbonneau et al. 2020). They are the key to personalized treatment and second-generation probiotics.

Furthermore, on demand cell-free biomolecule production, minimal genome design, and biocomputing are mostly dependent on synthetic biology. The design build test learn (DBTL) cycle has been the driving force for discovery in the field due to the improvement in CRISPR-Cas systems, next-generation sequencing, and DNA-synthesis technologies. Though its potential is vast, synthetic biology also poses vast safety, ethical, and regulatory issues, particularly in the area of biosecurity and the release of genetically modified organisms (GMOs) into the environment. Synthetic biology is highly promising, but raises several safety, ethical, and regulatory issues, especially for biosecurity and the environmental release of genetically modified organisms (GMOs). Ethical research, innovation, and public debate are necessary to make this technology develop in a sustainable way.

9.2 Core Principles of Synthetic Biology

Novel biological systems with reliable and practical functions can be designed and built using the fundamental basis of synthetic biology. These bases set synthetic biology apart from conventional genetic engineering and establish it as a genuinely cross-disciplinary field that combines chemistry, biology, engineering, and computer science (Fig. 9.1).

9.2.1 Modularity

Modularity is a fundamental concept in synthetic biology in which biologists can build intricate biological systems with the help of standard and interchangeable parts. The biologists and engineers assemble genetic "modules" that consist of individual parts like promoters, ribosome binding sites, and regulatory sequence. The simpler modules are then tested and optimized to form more intricate biological

Fig. 9.1 Fundamental principles of synthetic biology

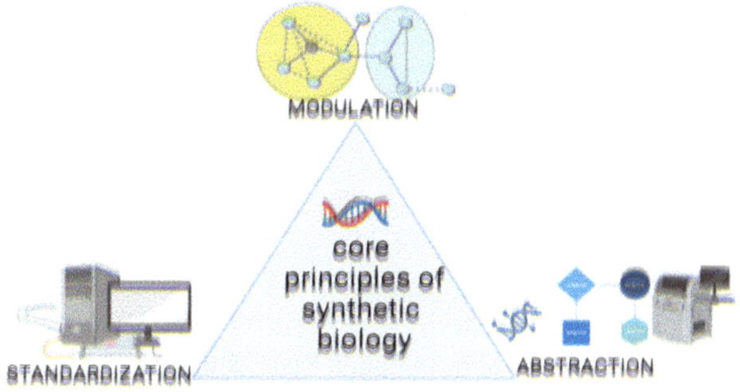

systems. This increases synthetic biology efficiency and predictability. Designing a module is the critical step that can be utilized in numerous applications. The central dogma explains how the genetic information flows within a cell which includes transcription, translation, and post-translational level. Similarly, modularity operates at several hierarchical degrees of biological system.

At level of transcription—synthetic promoters that are of different transcription initiation rates can regulate gene expression. Computational studies developed a model that can determine the strength of the sigma 70 promoter sequence. This model helps in designing of promoters with unique initiation rates of transcription (Fig. 9.2).

Control of gene expression can also be done using transcription factors. These TF affects the joining of the RNA polymerase and sigma factors at the promoter sequence. Recently, it was found that the transcription factor QF present within eukaryotes regulates the gene activation in *E. coli* BL21 by the help of Q- system upstream activating sequence (Chang et al. 2024) (Fig. 9.3).

At the level of translation, synthesis and implications of ribosome-binding sites (RBS) with different strength control synthesis of protein. Various computational tools are used to estimate the strength of these RBS and design them accordingly. Riboswitches are the cis-regulatory RNA sequence which are inserted at the 5' end of the target genes of interest. These sequences control the fine tuning of the gene

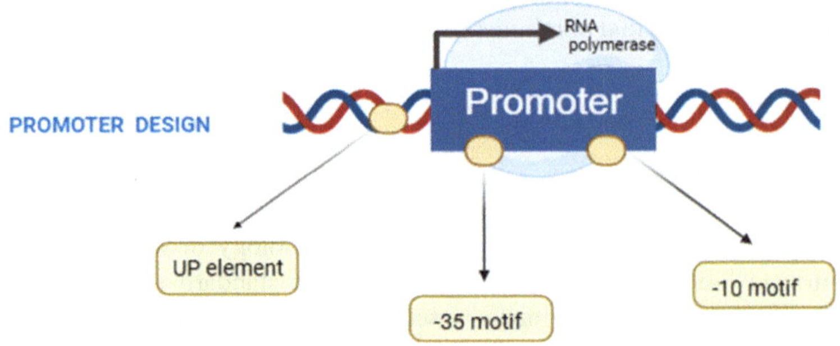

Fig. 9.2 Strategies for designing promoter sequence

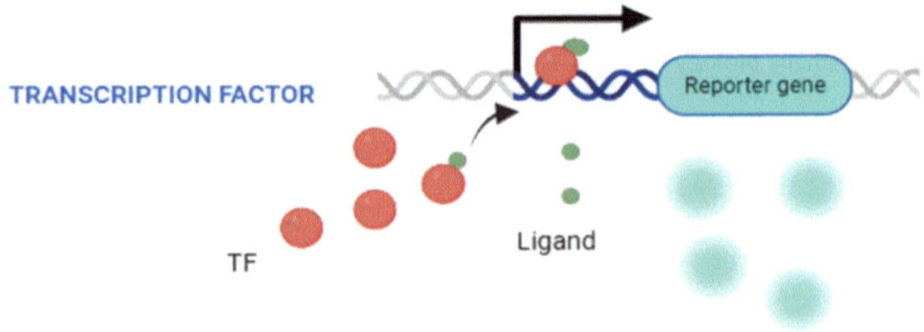

Fig. 9.3 Strategies for transcription factor designing

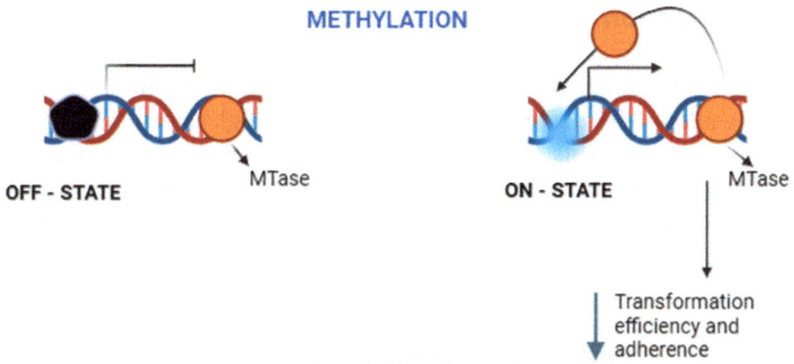

Fig. 9.4 Strategies for DNA methylation-based modulation

expression. A common example of these riboswitches is the toehold switch. In these switches, the hairpin structures are present on both sides of the RBS. Upon binding of the small partial complementary sequence of RNA, these switches open to activate the expression of the gene.

Post-translationally—beyond transcription and translation, gene expression can also be regulated at post translational level. In epigenetics, DNA methylation is found to be the key mechanism which can controls gene expression. In *Helicobacter pylori*, DNA methyltransferase methylates the first cytosine of TCTTC motif to 4-methylcytosine. This methylation alters the expression of the 102 genes which lead to a modified phenotype like reduced transformation efficiency and reduced adherence to gastric adenocarcinoma cells in human. Computational tools play a key role by modeling designs before they are actually built which improves success rates and minimizes experimental workload. Modularity has various applications such as smart therapeutics, environmental biosensors, and synthetic cells (Fig. 9.4).

9.2.2 Standardization

Standardization is one of the fundamentals of synthetic biology that allows biologists to design, construct, and exchange biological components and systems with enhanced efficiency. Standardization comprises creating biological parts such as sequence of DNA, proteins, and enzymes and established standard protocols (Decoene et al. 2018). This enables biologists to reuse checked parts, cooperate more effectively and enhance development by exploiting existing work rather than starting from scratch. One important example of standardization is the Bio Bricks Registry which manages a collection of DNA components which are built to be assemble like building blocks. These standardized components can be assembled to intricate biological devices.

9.2.3 Abstraction

Abstraction focuses on intricate systems and reduces them using high-level concepts. It allows the disassembly of intricate biological networks into simpler modular components. This hierarchical organization enables systematic analysis, manipulation, and designing of biological systems by enabling biologists to operate at various levels of complexity without requiring much knowledge of every molecular interaction.

9.2.4 Design–Build–Test–Learn (DBTL) Cycle

The Design–Build–Test–Learn (DBTL) cycle represents a core framework in synthetic biology which is inspired by basic key principles to engineer complex biological systems. During the design phase, biologists define the required biological function, which may include the biosynthesis of a target product or the control of certain cellular behavior. To identify suitable genetic components, metabolic reactions, and host organisms, different computational tools, genome-scale metabolic models, and manually curated biological databases are used. During the build phase, the actual implementation of the manual design is applied. This can be done using molecular cloning, gene synthesis, DNA assembly, or genome editing tools such as CRISPR-Cas systems, which enable precise construction of artificial circuits or metabolic pathways. During the test phase, the synthesized systems are examined under appropriate experimental and environmental conditions. They utilize biochemical assays, some omics-based methods, and quantitative measurements to determine system efficiency and identify any deviations from predicted results. The learn phase emphasizes on the analysis and interpretation of test data. Computational tools and statistical models are used to reveal functional insights, identify failures, and extract design principles. These observations help in enabling continuous refinement and optimization of the system (Kitano et al. 2023). Through multiple DBTL cycles, synthetic biologists can continuously improve system robustness, functionality, and efficiency. It replaces traditional trial and error experimental approaches in biology with a rational, engineering-driven methodology (Fig. 9.5).

9.3 Molecular and Advanced Computational Tools for Synthetic Biology

The evolution of advanced molecular and computational technology has been a primary driving force for the growth of synthetic biology as a field. These technologies allow for high-fidelity engineering of biological systems at the DNA, cell, and network levels. Four fundamental tools and technologies that are essential to synthetic biology: DNA synthesis and assembly, CRISPR and Genome Editing, Gene circuits and Logic Gates and Biosensors and Biocomputing systems.

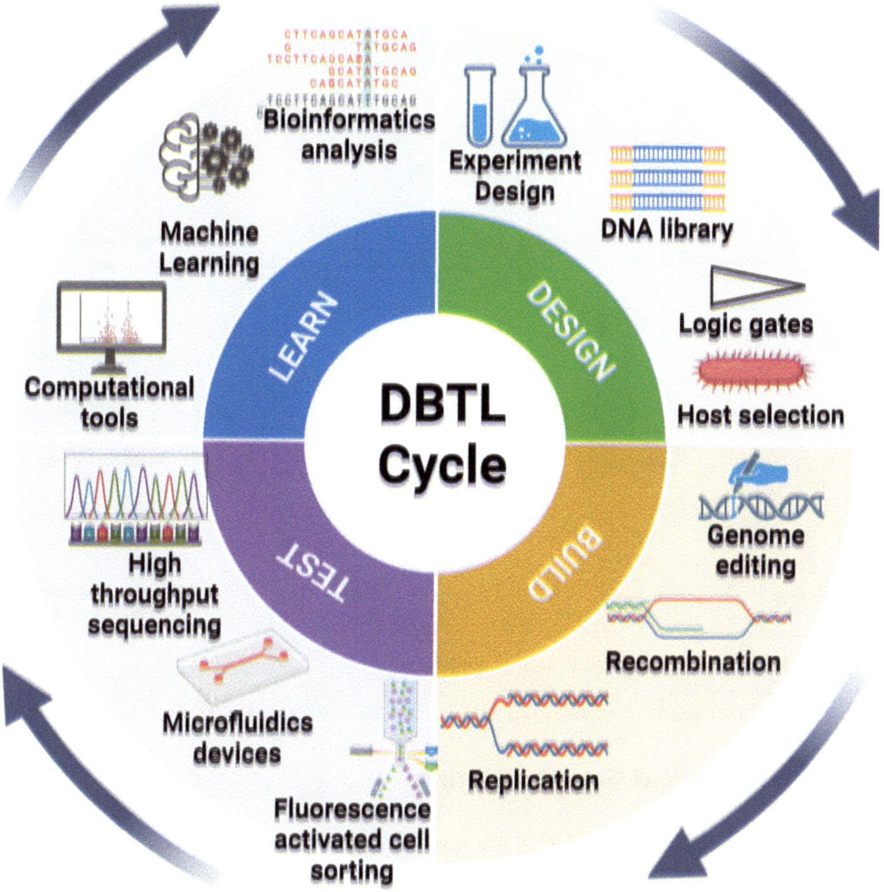

Fig. 9.5 Design–Build–Test–Learn (DBTL) cycle

9.3.1 DNA Synthesis and Assembly

The basis for synthetic biology is the ability to design new DNA sequences rather than relying on natural template DNA; scientists can design synthetic genetic structures through de novo DNA synthesis. When combined with assembly technologies like Gibson Assembly, Golden Gate Cloning, and yeast-based homologous recombination, it becomes possible to construct complex pathways and whole synthetic genomes efficiently (Gibson et al. 2009; James et al. 2025). These strategies allow for the high-throughput, module-based construction of metabolic pathways and genomic circuits. Synthetic biology has been driven worldwide by advances in the accuracy and decreasing cost of commercial DNA synthesis.

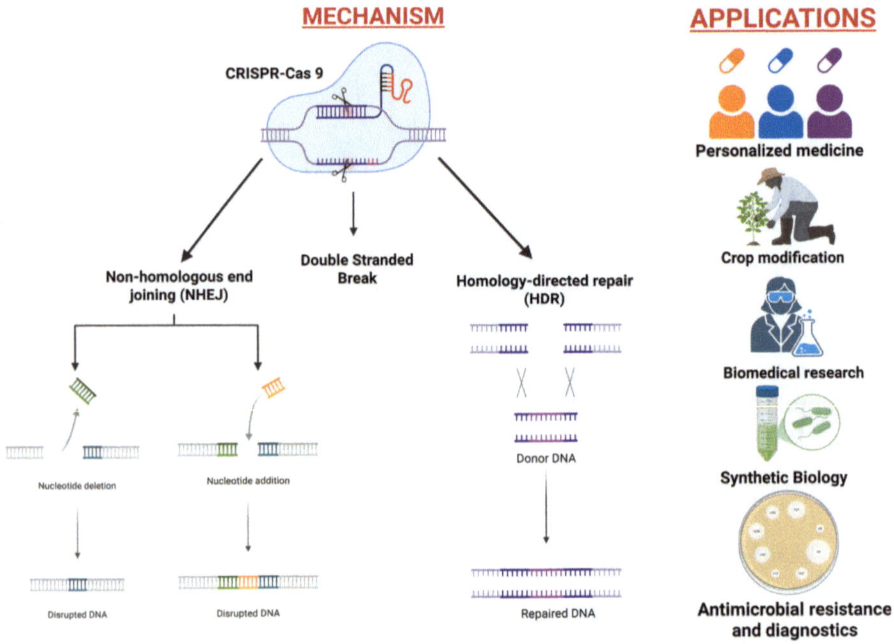

Fig. 9.6 Mechanism of CRISPER and its applications

9.3.2 CRISPR and Genome Editing

In synthetic biology, CRISPR-Cas is a novel technology, which provides a programmable and targeted genome editing at specific sites. CRISPR facilitates gene knock-in and knock-out, as well as gene silencing or activation (CRISPRi and CRISPRa), and the delivery of targeted mutations for directed evolution by retargeting the Cas nuclease to a target locus using guide RNA sequences. The system also enables multiplexed genome editing that can edit numerous gene targets simultaneously in a single cell. These features have facilitated the engineering of microbial chassis like *Saccharomyces cerevisiae* and *Escherichia coli* for producing highly optimized strains for uses like biological computation, environmental sensing, and biosynthesis. CRISPR has the core ability in synthetic biology that facilitates innovation in fundamental and applied research fields (Jinek et al. 2012) (Fig. 9.6).

9.3.3 Gene Circuits and Logic Gates

Synthetic biology synthesizes gene circuits made of regulatory components like promoters, repressors, and transcription factors in logic-gated networks based on electrical engineering to develop programmable living systems. These circuits utilize

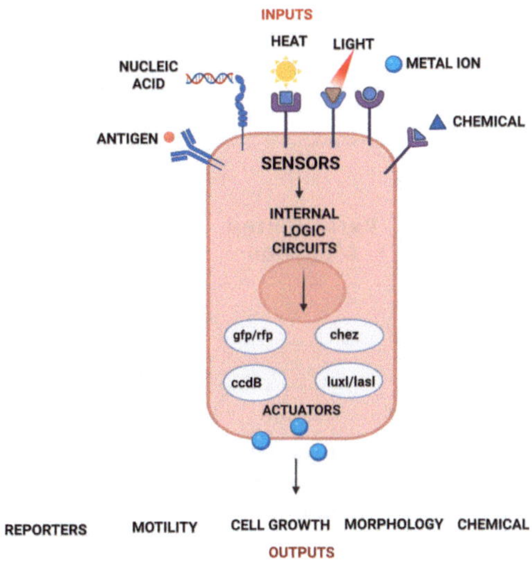

Fig. 9.7 Gene circuits and Logic Gates

biological logical gates like AND, OR, and NOT that are realized with elements like CRISPR-based regulators, riboswitches, or transcriptional repressors. Cells possessing such structures can perform complex functions such as oscillatory activity, signal integration, and decision-making. To enhance accuracy in artificial pathways, advanced gene circuits have been used to dynamically control gene expression based on certain cellular or environmental inputs. "Repressilator" built by Elowitz and Leibler (2000). It showed how gene circuit engineering can regulate temporal patterns of gene expression and introduced the first synthetic genetic oscillator (Fig. 9.7).

9.3.4 Biosensors and Biocomputing Systems

Biosensors are genetically modified machines that are applied in synthetic biology and that can respond to particular chemicals or environmental signals by generating an output that can be measured, e.g., fluorescence, color, or metabolic response. These biosensors play an essential role in variety of application that ranges from disease diagnosis. Lately, the world has faced notable challenges with the coronavirus disease 2019 COVID-19 pandemic, as well as other infectious diseases such as Zika and Ebola. Moreover, the rapid rise of noncommunicable diseases such as diabetes, heart disease, and cancer has placed a huge strain on healthcare resources and systems. Synthetic biology, a revolutionary field in modern science, shows considerable potential in advancing drug development and disease treatment and environmental monitoring, and in vivo metabolic regulation of engineered microorganisms. Scientists have integrated biosensors and artificial gene circuits to

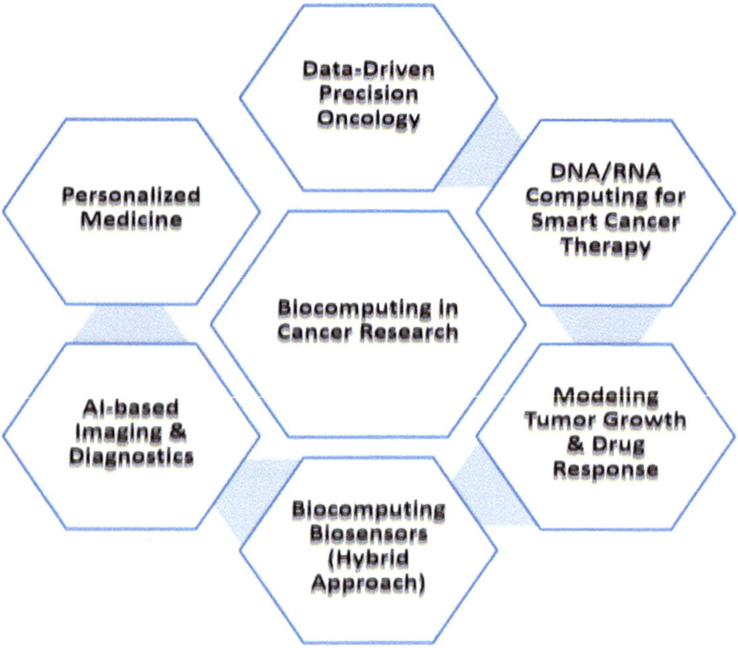

Fig. 9.8 Biocomputing in cancer research

create biocomputing devices that can analyze biological inputs and render logic-based decisions in living cells (Feng et al. 2024) (Fig. 9.8).

Hydrogel-based biosensors are a promising biosensor platform for cancer diagnosis because of their highwater content, biocompatibility, and tunable mechanical strength physicochemical properties. Recent advances highlight its potential for integration with responsive polymers, nanomaterials, and biomolecular recognition units to facilitate sensitive and selective detection of cancer biomarkers. Hydrogels can be engineered to structurally or optically respond to tumor-related signals, producing real-time and noninvasive diagnostic signals. Furthermore, integration of conductive nanoparticles has greatly improved the electrochemical performance of hydrogel sensors, with high-speed signal transduction and improved detection limits. These advances position hydrogel-based biosensors as a potential for early diagnosis of cancer, personalized therapy monitoring, and point-of-care diagnostics, bridging the laboratory and clinical gap (Sun and Chen 2024) (Table 9.1).

9.4 Application of Synthetic Biology

The field of synthetic biology has the potential to transform many industries by enabling the rapid engineering of biological systems for specific applications. The complexity and breadth of synthetic biology applications have increased

Table 9.1 Role of biosensors in cancer detection and therapeutic evaluation

Biosensors in cancer research		
Early detection and diagnosis	Detect circulating tumor DNA, microRNAs, or cancer biomarkers	
Real-time monitoring	Track tumor progression by measuring biomarker levels in blood or tissue fluids, monitor cancer	
Point of care testing	Portable biosensors can mimic tumor microenvironments to test new anticancer drugs	
Drug screening	Biosensors can mimic tumor microenvironments to test new anticancer drugs' effects	
Nano biosensors	Use nano materials, gold nanoparticles, carbon nanotubes, graphene for ultrasensitive detection of single cancer cells or exosomes	

significantly, from energy and environmental sustainability to medicine and agriculture. This section lists the major area of application enabled by current research and technical developments.

9.4.1 Agriculture

Synthetic biology has been a transformative catalyst in recent years of offering solutions for some of the most pressing issues facing contemporary agriculture, providing novel tools and approaches to enhancing crop productivity, sustainability, nutritional value, and use efficiency precritical considerations in light of the global imperative to grow food production by around 70% by 2050 with a rising population, climate stresses, and shrinking arable land. Contrary to traditional biotechnology, which tends to concentrate on manipulating single pathways or genes, synthetic biology draws on principles of modularity, standardization, and rational design paradigms well known in electronics and systems engineering to construct intricate, frequently multigenic improvements in plants and their related systems. The discipline has also seen the creation of standardized genetic components and modular cloning suites, with the backing of computational modeling and high-throughput analyses, that all enhance the speed of innovation and deployment in crops, although synthetic biology in plants remains behind advances experienced in microbes and mammalian systems (Roell and Zurbriggen 2020).

Synthetic biology in agriculture can be collectively placed under four integrated categories: maximizing photosynthetic and metabolic efficiency to increase yield, minimizing dependence on agrochemicals through enhanced nutrient utilization and plant-microbe interactions, increasing the nutritional content and safety of food crops, and using plants and photoautotrophs as green bioreactors for pharmaceuticals, biofuels, and high-value compounds. At the center of responding to global food security is the challenge of raising yields of crops within ecological and resource constraints. The agricultural yield is essentially constrained by how efficiently sunlight is captured, the conversion of light energy into biomass, and how much of the biomass is harvestable. Synthetic biology also provides new means of increasing the nutritional content and health safety of crops, addressing the problems of global malnutrition and healthier diets directly. The Golden Rice project is one of the milestones, as it employed gene technology to introduce phytoene synthase and carotene desaturase, two carotenoid biosynthetic enzymes, into the endosperm of rice, making it store provitamin A (beta-carotene) and giving one a means to combat vitamin A deficiency in monotonous rice-based diets. But the true game-changer is the modular, multigene engineering pipeline afforded by synthetic biology that can produce not just vitamins, but a range of value-added nutrients and health-improving metabolites. This comprises routes for carotenoids (other than beta-carotene, e.g., xanthophylls and astaxanthin, valuable for their antioxidant, immune-stimulatory, and market demand), anthocyanins (conferring color and antioxidant activity), and health-enhancing very long chain polyunsaturated fatty acids (VLC-PUFAs) like Eicosapentaenoic acid (EPA) and Docosahexaenoic acid (DHA), which are not present in crops but are crucial for brain and cardiovascular wellness. Such achievements have been in the form of transgenic Camelina and canola with seed-specific biosynthetic gene expression for PUFA, achieving up to 12–15% VLC-PUFA levels, which are sustainable plant sources that are likely to obtain regulatory clearance very soon. In addition to crop improvement, synthetic biology allows plants and other photoautotrophs to be utilized as scalable and sustainable bioproduction platforms for drugs, vaccines, specialty chemicals, biofuels, and industrial proteins.

All these developments are critically dependent on modeling and computational tools that allow in silico design, optimization, and prediction of synthetic routes prior to labor-intensive and costly experimental implementation. Although there has been tremendous advancement, there are still challenges ahead before the full potential of synthetic biology in agriculture is achieved. Plant transformation is still more difficult and less controllable than similar procedures in microbial or mammalian systems. Already, certain synthetic microbiota products that will cut fertilizer use are coming on stream, and crops designed for nutrient enhancement or for certain desirable qualities are moving through approval pipelines. With ongoing research and good governance, synthetic biology will revolutionize agriculture providing yield and resource use efficiency gains, enhancing resistance to environmental stress and pests, lowering ecological footprint, and making future food systems healthier and more sustainable (Roell and Zurbriggen 2020) (Fig. 9.9).

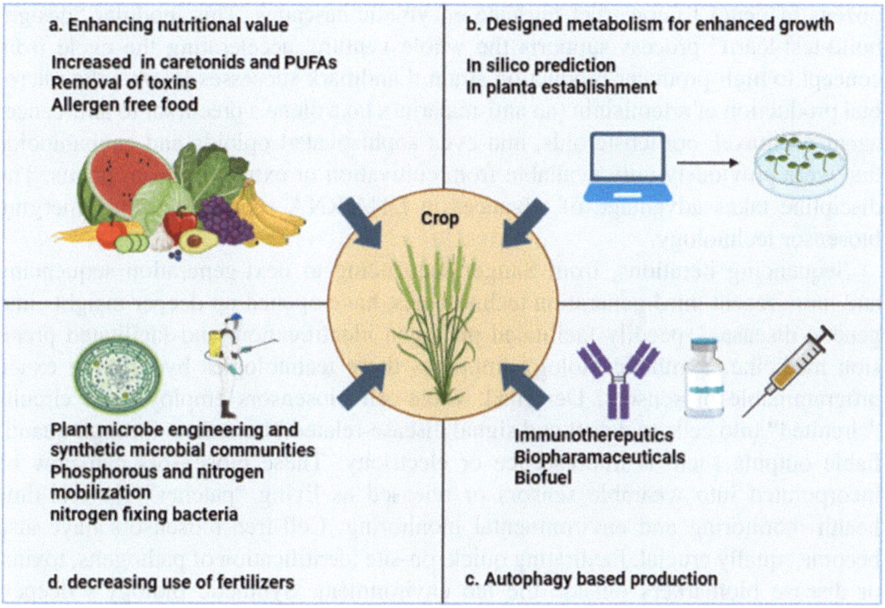

Fig. 9.9 Synthetic biology can support future food production through plant improvement of plant nutrition, growth augmentation through synthetic pathways, and minimization of fertilizer application through engineered plant microbe interactions. It also makes photoautotroph-based platforms for vaccines, therapeutics, and biofuels possible

9.4.2 Healthcare and Pharmaceutical

Synthetic biology is a fast-evolving discipline that combines biological approaches, engineering tools, computational science, and chemistry to create, assemble, and improve new and useful biological systems, components, and organisms with novel functions not previously seen in nature. In healthcare, its revolutionary potential is redefining how we go about discovering drugs, diagnosing diseases, and treating them. The past few years have underscored the need for medical innovation pressures of worldwide pandemics such as COVID-19 and the wave of noncommunicable diseases like cancer, diabetes, and cardiovascular disease have uncovered loopholes in conventional therapeutic development. One of the greatest uses of synthetic biology for medicine is creating engineered "cell factories" for pharmaceutical production. Synthetic biology utilizes microorganisms like *E. coli* and *Saccharomyces cerevisiae*, and other hosts like *Aspergillus niger*, *Bacillus subtilis*, and mammalian Chinese hamster ovary (CHO) cells (Feng et al. 2024). The procedure involves choosing an appropriate host (chassis) depending on considerations such as growth rate and genetic manipulability, optimizing the host for high-efficiency expression, and designing and refining metabolic pathways. New methods like CRISPR-Cas9, BAC-cloning-based transformation, and transformation-associated recombination allow engineers to insert, alter, or control dozens of genes to construct intricate enzymatic cascades. This modular "design-build-test-learn" process supports the whole venture, accelerating the cycle from concept to high-producer production strain. Landmark successes involve the microbial production of artemisinin (an anti-malarial), taxa diene a precursor to anti-cancer agent paclitaxel, corticosteroids, and even sophisticated opioids and cannabinoids that were previously only available from cultivation or extraction from plants. The discipline takes advantage of advances in DNA/RNA sequencing and emerging biosensor technology.

Sequencing iterations, from Sanger sequencing to next-generation sequencing and more recent third-generation technologies, have opened up deeper insights into genetic diseases, speedily facilitated pathogen identification, and facilitated precision medicine. Synthetic biology improves these technologies by offering exact, programmable biosensors. Designed whole-cell biosensors employ gene circuits "circuited" into cells to detect and signal disease-related biomarkers through quantifiable outputs such as fluorescence or electricity. These biosensors can now be incorporated into wearable sensors or released as living "patches" for real-time health monitoring and environmental monitoring. Cell-free biosensors have also become equally crucial, facilitating quick, on-site identification of pathogens, toxins, or disease biomarkers outside the lab environment. Synthetic biology's deepest influence perhaps lies in redesigning the logic of disease treatment. As opposed to conventional one-size-fits-all therapies, it holds the promise of personalized and programmable therapies. At its center is genome editing. The arrival of tools like CRISPR-Cas9, base editors, and prime editors which support accurate modifications, including deletions and insertions, now permits the correction of disease-inducing mutations at their origin. Synthetic biology also enables the design of viruses and bacteriophages as programmable therapeutics or delivery agents. New live attenuated or virus-like particle (VLP) vaccines are manufactured by codon re-engineering or de novo assembly, speeding vaccine development in response to emerging threats. Engineered viral vectors deliver gene therapies, genome editors, or therapeutic proteins with high specificity and better safety profiles. Designer bacteria, such as programmable probiotics and live therapeutic strains, are able to selectively colonize certain tissues, perceive microenvironmental signals, or deliver drugs locally. Certain bacteria are designed to immunomodulate, detoxify toxins, or restore a healthy balance of microbiome. Bacterial vesicles are new vaccine platforms or immunomodulators. The convergence of synthetic biology with technologies like artificial intelligence, computer-aided protein design, and organoid/organ-on-chip platforms promises a future in which disease modeling, drug screening, and patient-specific therapeutics are streamlined and significantly accelerated. The vision is to integrate engineered systems with native biology in a seamless manner, create smart living medicine, and achieve efficient, personalized health care solutions in spite of tremendous promise, there are still challenges bioethical issues, safety, cost, and compatibility with intricate endogenous networks (Feng et al. 2024) (Fig. 9.10).

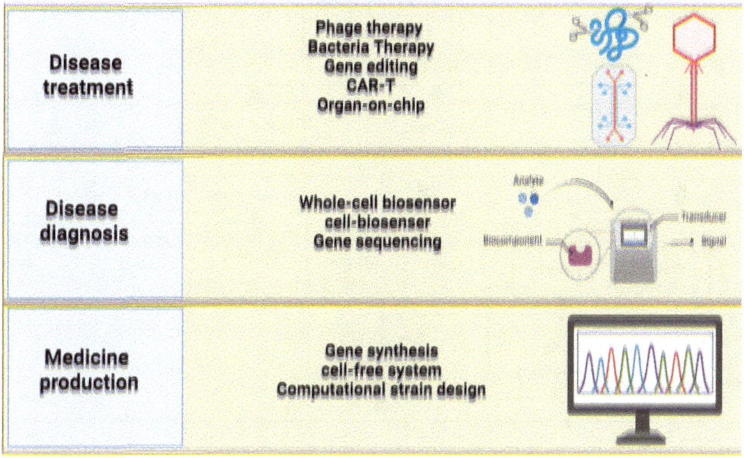

Fig. 9.10 Three parts of synthetic biology intersecting with medicine

9.4.3 Environment

Synthetic biology is revolutionizing how we approach environmental protection by enabling the precise engineering of microorganisms and plants for pollution sensing, monitoring, and remediation. Traditionally, the field focused on creating cost-effective cell factories for green chemical production, thereby indirectly reducing environmental impact by lowering reliance on fossil resources. However, recent advances have shifted the paradigm: synthetic biology is now being directly harnessed to address pollution, develop biosensors for toxic chemicals, engineer robust bioremediation agents, and leverage computational approaches for designing efficient interventions. This chapter evaluates key developments in synthetic biology as they relate to environmental monitoring, pollutant degradation, and the challenges and future directions of these technologies (Aminian-Dehkordi et al. 2023). Modern industrialization, urbanization, and intensive agriculture have led to the release of a wide variety of pollutants from heavy metals and persistent organic pollutants (POPs) to plastics and hydrocarbons into terrestrial and aquatic environments. Many of these are either widely dispersed, difficult to detect, or lack cost-effective remediation options. According to the WHO, environmental risk factors contribute to roughly a quarter of global human deaths annually. There is an urgent need for robust, scalable, and economical solutions for pollution management an area where synthetic biology offers promising new tools. Biosensors constitute one of the oldest and most significant applications. Genetically modified bacteria and yeast function as living sensors, designed with circuits to produce signals, e.g., fluorescence, luminescence upon encountering particular toxins such as heavy metals or organic pollutants. Contemporary designs prioritize specificity, modularity, and deployability in the field. In addition to living cells, cell-free systems employ only biological extracts and genetic circuits, providing safer and more portable solutions. Computational design and machine learning increasingly

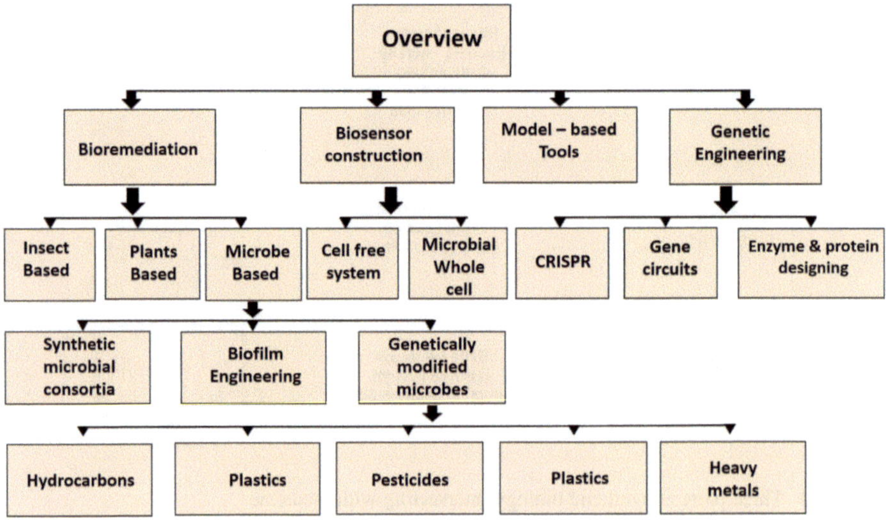

Fig. 9.11 An overview of hierarchy of various methods

assist in designing more sensitive and stronger biosensor (Aminian-Dehkordi et al. 2023). Bioremediation utilizes designer microbes sometimes in mixed populations to degrade pollutants. The designer strains are constructed with sophisticated genetic tools like CRISPR to enhance metabolic pathways that cleave pollutants such as hydrocarbons, plastics, pesticides, and heavy metals. Breakthroughs involve the use of unconventional microbial "chassis" with inherent robustness, the optimization of critical enzymes through directed evolution and metagenomics, and the design of synthetic consortia that partition labor to tackle complicated waste management. Biofilm engineering enhances the ability of bacteria to immobilize and sequester metals or break down pollutants more effectively. Engineered microbial communities can even convert waste into useful products like bioplastics or biofuels. Synthetic biology is also advancing the boundaries by making insects express pollutant-degrading enzymes, or changing plants for more efficient phytoremediation removing toxins like metals from soil by enhancing the natural uptake and detoxification processes in the plant. Computational methods are central, enabling researchers to simulate metabolic processes, foresee the effects of genetic alterations, and determine the most effective engineering targets. Machine learning and simulation software facilitate iterative design loops and optimize both biosensor circuits and metabolic engineering strategies (Aminian-Dehkordi et al. 2023) (Fig. 9.11).

A number of microbial whole-cell biosensors (MWCBs), mostly bacteria, have been constructed to detect environmental pollution. Perhaps one of the first biosensors targeted naphthalene, widely present in water as well as soil. For this biosystem, *Pseudomonas fluorescens* was genetically modified by the introduction of lux genes, which encode for the bioluminescence depicting enzyme luciferase, into the *nahG* locus and encodes salicylate hydroxylase, which allowed the detection

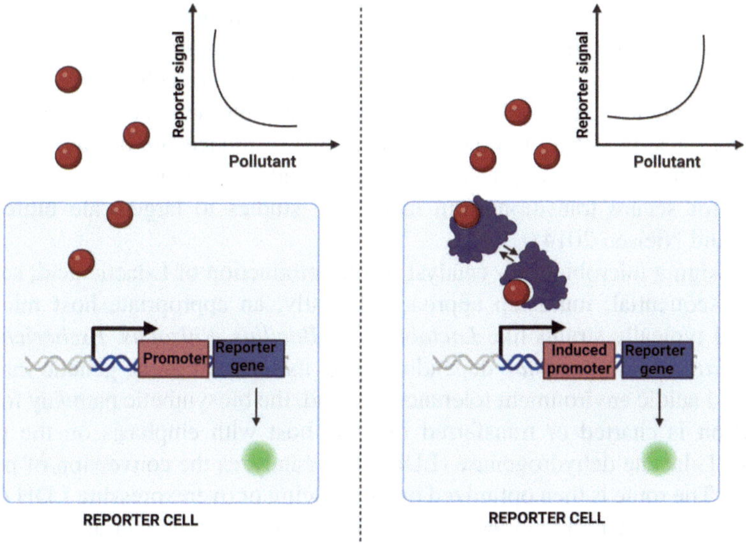

Fig. 9.12 Microbial whole-cell biosensors (MWCBs) employ reporter expression: constitutive MWCBs emit high signals that reduce with increasing toxin levels, whereas inducible MWCBs induce reporters through analyte-controlled proteins, emitting signals only if the promoter is activated

of aromatic hydrocarbon contamination through light emission. Later on, this method was enhanced with the use of *Acinetobacter* ADPWH lux, wherein the *nahAD* operon (naphthalene degradation) was borne on an episomal plasmid to generate salicylate when in contact with naphthalene. Salicylate in turn initiated the *salAR* and lux CDABE operons, which generated a bioluminescent signal. This improved biosensor showed greater specificity being insensitive to other PAHs and impressive sensitivity, being able to detect naphthalene at 0.01 mM concentrations. Because salicylate is a key intermediate in the breakdown of the majority of PAHs, the system enables the development of biosensors for other PAHs which are replaced by nahAD plasmid by the corresponding PAH-degrading operon (Aminian-Dehkordi et al. 2023) (Fig. 9.12).

9.4.4 Industry and Energy

Synthetic biology reshapes classical industrial microbiology by making it possible to design microbes that are tailored to have improved or new functionalities. Progress in bioinformatics, genetic tools, and metabolic engineering made possible the creation of new biosynthetic routes for renewable fuels, bioplastics monomers, long-chain alcohols, and other chemicals such as isoprenoids, e.g., dihydroartemisinic acid from engineered yeast. In addition to advancing conventional microbial platforms such as *E. coli* and *Saccharomyces cerevisiae*. Synthetic biology also covers designing synthetic microbial communities for tough industrial applications like biomining and bioremediation, and probing chemical synthetic biology to develop novel molecules. Innovations in technologies, including biosensor-regulated feedback loops to counteract biofuel toxicity and real-time evolution tracking (VERT) to increase microbial resilience, are driving the field. In addition to these breakthroughs, risk mitigation and biocontainment are addressed to provide for secure transition from laboratory studies to large-scale biorefineries (Zhang and Nielsen 2014).

To design a microbial bio- catalyst for the production of L-lactic acid, scientists adopt a sequential, multistep approach. Initially, an appropriate host microbe is identified typically strains like *Lactobacillus, Bacillus, Rhizopus, Escherichia coli*, or *Saccharomyces cerevisiae* depending upon its inherent traits, genetic manipulability, and acidic environment tolerance. Second, the biosynthetic pathway for L-LA production is charted or transferred into the host with emphasis on the primary enzyme, L-lactate dehydrogenase (LDH), that catalyzes the conversion of pyruvate to L-LA. The route is then optimized by introducing or overexpressing LDH (usually from effective external sources) to increase L-LA synthesis, and at the same time knocking out genes that channel carbon flow toward undesirable by-products, like ethanol or glycerol, thus providing increased metabolic flux toward L-LA. Genome editing precision tools, such as CRISPR/Cas9, are employed to add, remove, or refine genes in the microbial genome for maximum pathway performance. The process is finally optimized and scaled up, often using low-cost, renewable raw materials like agricultural waste, and advanced fermentation technologies to ensure maximal industrial utility and sustainability. This integrated strategy has made it possible to develop high-performance microbial strains with the ability to produce L-lactic acid sustainably at industrial levels (Liu et al. 2022) (Fig. 9.13).

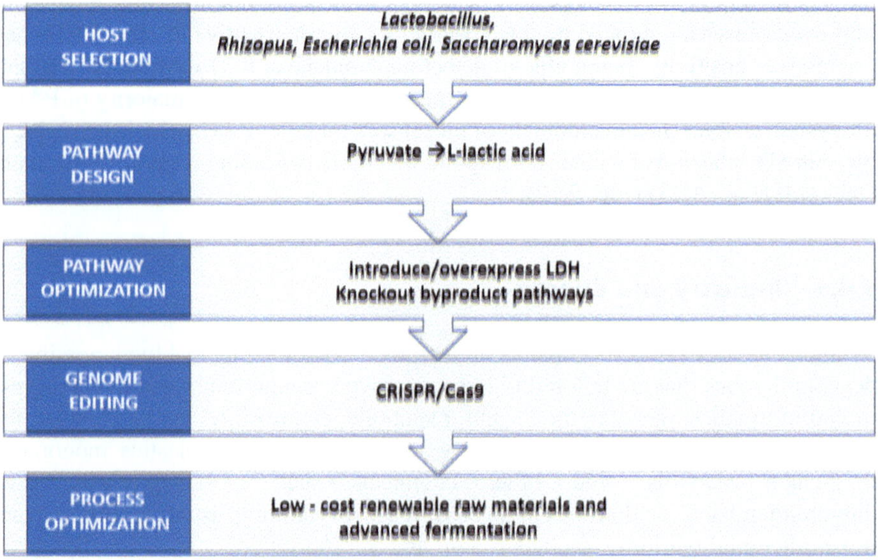

Fig. 9.13 Work flow to design a microbial cell factory for the synthesis of L-lactic acid to produce it sustainably at industrial levels

9.5 Designing Synthetic Microbial Communities

Synthetic biology is an upcoming research field that combines biology with engineering. Instead of altering genetic sequences like traditional genetic engineering, in this gene and circuits are generated from scratch, like we build machines. The aim is to create systems in living organisms that can be predicted, controlled, modulated, and improved to do specific tasks, like synthesis of chemicals or responding to certain signals. Scientists have also started redesigning entire metabolic pathways which helps in the production of useful products used in chemical industry and medicine which is beneficial for pharmaceutical (Song et al. 2014). Synthetic biology not a single cell based. Now the focus has shifted towards creating engineered microbial communities, also called synthetic consortia. This is important because in nature, microbes rarely live alone they are part of a group where they share resources or signals and interact with each other. Some are even dependent on each other for survival. So, by designing how microbes interact, whether they help, harm, or compete with each other and understanding their behavior scientists can now build complex systems that mimic natural communities but with specific, useful functions which can be beneficial to mankind (Bober et al. 2018).

9.5.1 Engineering Interspecies Communication

Microorganisms constantly interact and communicate with each other, and these interactions are very important because they affect how each microbe survives, evolves, and behaves in a group. These interactions also decide how stable or active the whole microbial community is. By learning all this, scientists can design new and useful combinations of microbes (called microbial consortia). Effects of these interactions can be neutral, positive, or negative for the organisms involved. Understanding these interactions helps scientists know how microbial communities work as a system in some cases is neutralism, where there is no interaction at all between the two microbes. For example, if two microbes use completely different types of nutrients, they may not affect each other's growth. This is a rare event, because microbes usually at least compete for space or some resources. There are six main types of interactions between two microbes called binary or pairwise interactions (Table 9.2).

Table 9.2 Types of binary interactions in microbial ecosystems and their effects on interacting species

Binary interaction model in microbial ecosystem	
Binary interaction models	Nature of interaction
Mutualism	+/+
Commensalism	+/0
Parasitism or predation	+/−
Competition	−/−
Amensalism	0/−
Neutralism	0/0

0 (null: neutral); + (positive; benefit or win); − (negative: detrimental or loss)

9.5.2 Molecular Mechanisms to Show Interactions in Microbial Consortia

Microorganisms living in communities usually interact via sharing signals and molecules. These interactions help them behave as a group, almost like a single, organized system. Learning about these molecular-level interactions helps understand both natural microbial behavior and also design new artificial microbial communities with required traits. Microbes can interact in two main ways:

9.5.2.1 Contact-Independent Interactions

When there is release of chemicals into the environment which are detected and used by other microbes without direct contact. One most common contact-independent interaction is Quorum Sensing (QS). In this microbe release, the signaling molecules are released into their environment. When the concentration of these signals reaches a certain level, the microbes start responding collectively leading to changing gene expression and behavior. Quorum sensing helps microbes coordinate things like biofilm formation, movement, virulence (ability to cause disease), and cooperation. Disruption of these signals is called quorum quenching this can help fight harmful bacteria by blocking their communication. Apart from this, microbes also release small and large molecules such as gases like hydrogen or methane and proteins. Quorum sensing (QS) is a widespread process by which bacteria sense cell density and adapt gene expression to it. Most bacteria employ regulator proteins that, upon ligand binding, regulate transcription from QS promoters. Of all the QS modes, LuxR/LuxI systems with homoserine lactone ligands are particularly convenient due to their versatility and simplicity. QS has been used in synthetic biology for pathogen detection, pattern generation, synchronized oscillators, biofilm formation, and constructing synthetic ecosystems. In metabolic engineering, QS facilitates timed chemical production at optimized cell density, enhancing efficiency (Scott and Hasty 2016). Natural microbial populations organize group behavior through quorum sensing (QS), a cell–cell communication process including production, secretion, and accumulation of soluble signal molecules referred to as autoinducers. Such molecules tend to be acyl-homoserine lactones (AHLs) in Gram-negative bacteria, diffusing unhindered across membranes and binding to specific receptors to initiate cascades of gene expression, often involving positive feedback loops to magnify signal production. Initially identified in *Vibrio fischeri* to regulate bioluminescence, AHL-dependent QS is currently recognized to govern a wide range of behavior including biofilm formation, pathogenicity, metabolite production, and competence, frequently combining several autoinducer signals to communicate intra- and interspecies. With inspiration drawn from these biological systems, synthetic biology developed toolkits for the synthesis and detection of AHL for engineered microbial populations to coordinate activities such as oscillating signals, differential gene expression, culture density regulation, and cooperation. In a recent breakthrough, researchers introduced three QS pathways Lux, Las, and Tra into the cyanobacterium *Synechococcus elongatus* PCC 7942, while AHL synthases were expressed in sucrose-grown *E. coli*. Connecting sucrose secretion in *S. elongatus* into *E. coli* cell

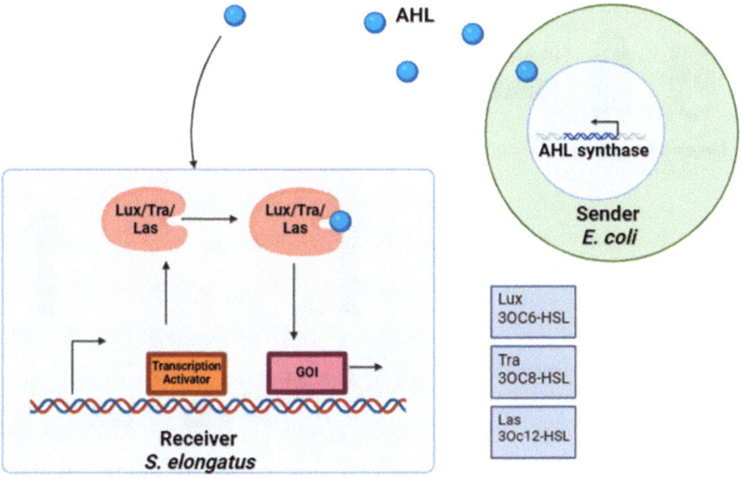

Fig. 9.14 Diagram illustrates quorum sensing genetic circuit: sender cell *E. coli* secretes AHL molecules diffusing into receiver cell *S. elongatus*, where activators LuxR/LasR/TraR bind and induce target gene expression through promoter

density showed, for the first time, the application of QS modules in cyanobacteria for adjusting gene expression and cross-species communication, with implications for coordinated light-driven bioproduction (Kokarakis et al. 2023). *E. coli* sender strains were co-cultured with *Synechococcus elongatus* receiver strains in co-culture experiments to verify the performance of the developed quorum-sensing system. Efficient interspecies communication was validated by the induction of cyanobacterial reporter genes and operable modules, including sucrose secretion, upon receiving AHL signals from *E. coli*. The engineered Lux and Las circuits behaved as expected and could be properly tuned within the mixed-species environment, validating the resilience of the system to coordinate gene expression across species (Fig. 9.14).

9.5.2.2 Contact-Dependent Interactions

In this type of interaction, cells need to physically touch each other to share proteins, DNA, or electrons. Microbe also interact by direct physical contact. In some cases, they directly exchange electrons; this is known as direct interspecies electron transfer. Certain materials such as activated carbon or magnetite can help speed up this electron transfer by acting like tiny wires. Some bacteria can also form nanotubes which are small bridges that connect neighboring cells. Through which materials like plasmids, proteins, or small molecules are passed. Systems biology helps by giving a complete view of what's happening inside and between cells, such as which genes are active, which proteins are produced, and what chemical reactions are taking place. Systems biology Genomics (studying all the genes), Transcriptomics (studying which genes are being used), Proteomics (studying the

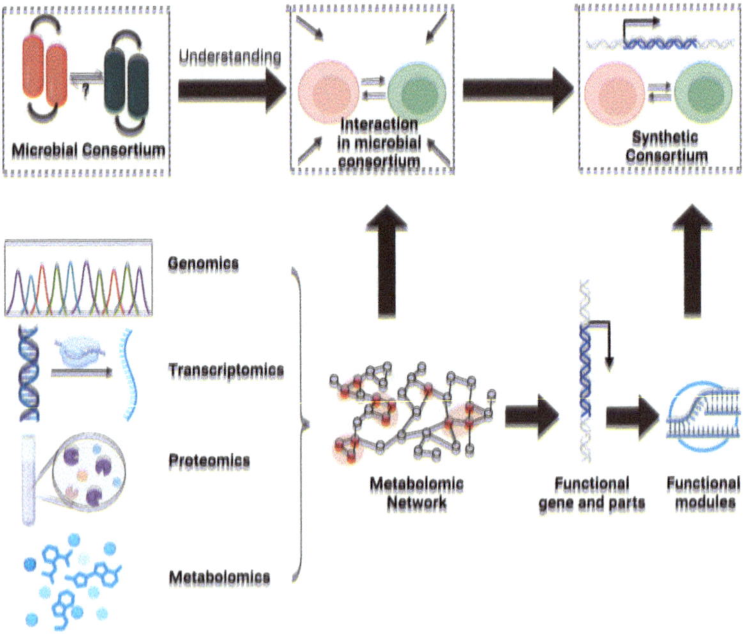

Fig. 9.15 Molecular mechanisms of interactions in microbial consortia

proteins made), and Metabolomics (studying the small molecules involved in metabolism). These tools help scientists figure out how microbial behaves and interacts. By applying these techniques, researchers can analyze both natural microbial communities (like those found in oceans, soil, or the human gut) and compare them with artificial ones built in the lab. Hence, systems biology gives the blueprint to create better-functioning microbial ecosystems (Fig. 9.15).

9.5.3 Example of Synthetic Microbial Community Designing

9.5.3.1 Vitamin C

Vitamin C which is also known as L-ascorbic acid, is a water-soluble vitamin that is naturally present in some foods and is available as a dietary supplement. Microbes *Bacillus megaterium* and *Ketogulonicigenium vulgare* are cultured together. Scientists used tools like genomics, proteomics, metabolomics, and trans-omics to study how these two microbes exchange metabolites. It was found that *B. megaterium* helps *K. vulgare* grow and survive better by supporting its energy and nutrient pathways, which boosts the production of 2-keto-gulonic acid (2-KLG) which is an important ingredient for making vitamin C. Interestingly, *K. vulgare* can also affect *B. megaterium* in both positive and negative ways. It may provide useful amino acids but can also hinder its growth by producing too much 2-KLG. More

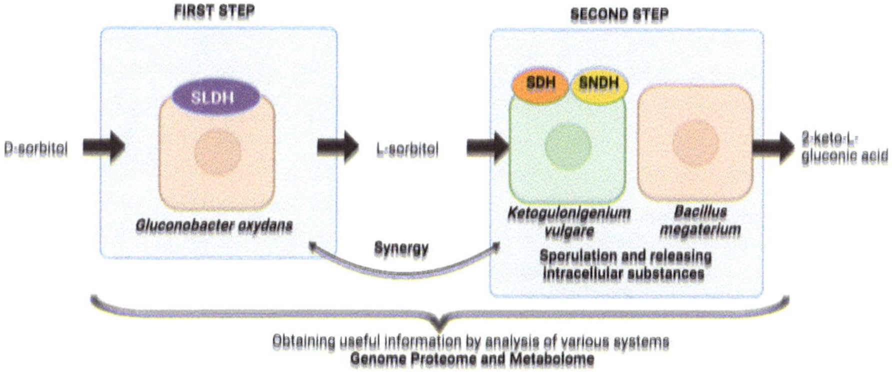

Fig. 9.16 Vitamin C is synthesized through a two-step fermentation in which *Gluconobacter oxydans* reduces sorbitol to sorbose, and *Ketogulonicigenium vulgare* and *Bacillus megaterium* co-cultured convert sorbose to 2-keto-gulonic acid (vitamin—C precursor). B. *megaterium* promotes *K. vulgare* by providing nutrients, mitigating stress, and maintaining redox balance, and systems biology tools like genomics, proteomics, and metabolomics are used to optimize this synthetic microbial consortium

detailed studies showed that *B. megaterium*, even after cell death, releases molecules that support *K. vulgare's* growth and 2-KLG production. Using this deep understanding, scientists were able to genetically modify the microbes, enhancing certain genes to boost the production of 2-KLG by 20%. The complete genetic maps of these microbes made such targeted improvements possible (Song et al. 2014) (Fig. 9.16).

9.5.3.2 Microbial Production of Paracetamol

Among the most commonly used analgesic and antipyretic drugs globally is paracetamol, or acetaminophen, or N-acetyl-p-aminophenol (APAP). Traditionally, it is produced in a multistep chemical synthesis from petroleum-derived benzene followed by subsequent acetylation, reduction, and nitration reactions. The process is dubious in regards to sustainability and environmental responsibility with the use of nonrenewable feedstocks, production of toxic byproducts, and the need for harsh reaction condition. Synthetic biology approaches have been investigated as a more sustainable and environmentally benign route to the production of paracetamol from genetically modified microorganisms like *Escherichia coli*. By a series of heterologous enzymatic transformations, scientists have shown the potential of *E. coli* to be engineered for the production of 4-aminophenol from simple carbon feedstocks, the key intermediate in the synthesis of APAP. Acetylation is subsequently used to produce the final product. Enzymes such as arylamine N-acetyltransferase and 4-aminobenzoate hydrolase, developed for expression and activity in the bacterial host, are used in this microbial platform for reconstructing biosynthetic pathways. The process has been established in two forms: one as a fully integrated de novo

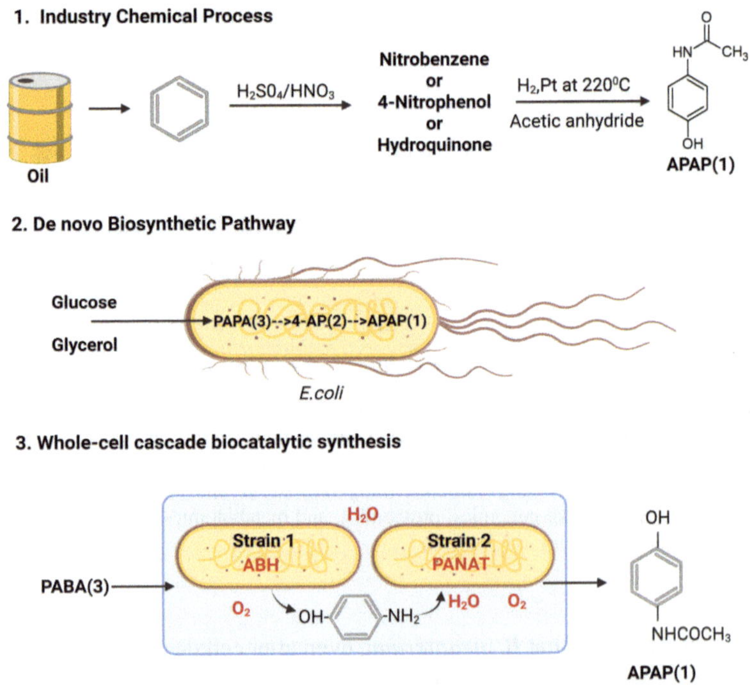

Fig. 9.17 Designed biotransformation process and chemical route for the manufacture of paracetamol (APAP, 1). (1) Industrial process for the production of 1 via multistep chemical oxidation, reduction, nitration, and acetylation from benzene. (2) De novo biosynthetic process of (1). (3) Whole-cell cascade biocatalytic process for the synthesis of 1 from para-aminobenzoate (PABA, 3)

biosynthetic pathway where glucose is the sole carbon source, and another as a whole-cell biocatalytic process where para-aminobenzoic acid (PABA) is supplied externally and converted to APAP within the engineered *E. coli* cells. This innovation is part of a broader shift in pharmaceutical manufacturing, aiming to reduce dependence on fossil fuels, lower carbon footprints, and meet the rising global health demands with minimal environmental impact. Cutting-edge metabolic engineering and dynamic pathway control technologies have been used to maximize product yield, cofactor balance, and precursor supply. Challenges remain, however, such as low titers constraining commercial scale-up, enzyme instability, and route toxicity (Hou et al. 2021). Enhancing strain strength, increasing bioreactor yields, and applying computer modeling for optimization of pathway design are the goals of future initiatives. This proof of concept illustrates the possibility of synthetic biology to decrease reliance on conventional petrochemical-based processes and enable ecologically sustainable production of medicines (Fig. 9.17).

9.6 Microbiome Community Dynamics and Their Role in Personalized Medicine

The human microbiome is a complex and dynamic population of microorganisms whose organization, function, and relationship with the host are now understood to play critical roles in governing health and disease states. Alterations in the balance of microbes a state known as dysbiosis have been associated with a variety of disorders, such as metabolic, inflammatory, neurological, and gastrointestinal diseases. Whereas the human genome is static, the microbiome is exceedingly plastic and malleable. Its structure is not only extremely individualized but also able to change within the same individual over the years due to age, geography, circadian cycles, environmental exposures, diet, and medication. Plasticity may account for why two people with the same genetic risk of a disease may have vastly different clinical manifestations. Consequently, it is essential to recognize the regulators of microbiome diversity and complexity in order to optimize personalized strategies for the prevention, diagnosis, and treatment of disease. Classic precision medicine approaches largely focus on genomic and biochemical profiling, with especial emphasis in oncology, where these technologies have become routine in-patient treatment. Yet, these methods do not always succeed well to address variability in the manifestation of disease and treatment responses at the individual level. Thus, there is increasing enthusiasm to harness microbiome-derived information as an additional source of physiological information (Ratiner et al. 2024). An advantage of microbiome data is that intraindividual stability samples drawn consistently from the same person often produce reproducible patterns, and it is therefore a valuable resource in diagnostic applications like forensic identification. Broader clinical application is presently limited by low reproducibility between studies, primarily because they are often retrospective in nature and limited to homogeneous study populations. To resolve this will need to involve large-scale, heterogeneous cohort studies coupled with integration with computational platforms, including big data analysis, to delineate the complex interactions between host and microbiome in a more globally applicable format. One of the important advancements in the field of personalized medicine is probiotics. Probiotics can be designed from single-strain products by synthetic multispecies microbial consortia specifically engineered to restore gut microbiota balance in the context of dysbiosis and disease. Which overcomes the limitations of traditional probiotics and fecal microbiota transplantation, emphasizing new approaches that use computational "in silico" and laboratory "in vitro" models to predict, assemble, and validate multistrain bacterial cocktails. Central to these designs are advanced genetic and metabolic analyses, including omics-based profiling, ecological network modeling, and genome-scale metabolic reconstruction, which together identify and optimize beneficial interactions between microbial species. Suitable communities of microbes are tested in laboratory gut simulators and cell-based models for stability, resilience, metabolic output, and host interactions key criteria before advancing to clinical trials. To understand the importance of functionally validating these consortia for their ability to produce health-promoting metabolites, restore gut diversity, displace pathogens, and support

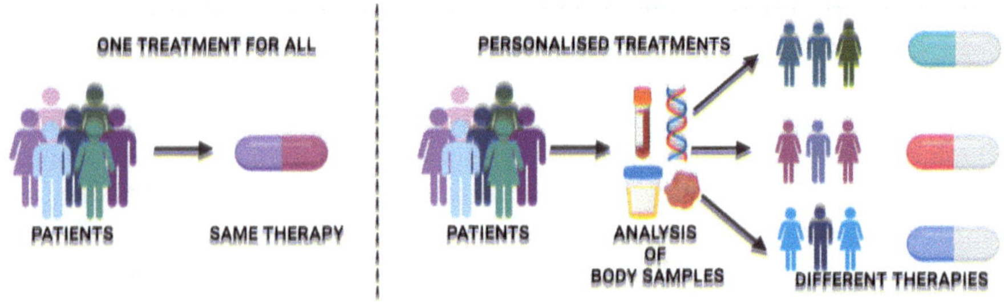

Fig. 9.18 Microbial community and their role in personalized medicine

immune and barrier function. Hence, synthetic microbial consortia act as a promising, customizable, and scientifically rigorous strategy to treat microbiota-associated diseases, paving the way for next-generation biotherapeutics targeting gut health (Vázquez-Castellanos et al. 2019) (Fig. 9.18).

9.7 Conclusion

Synthetic biology has become a revolutionary area that transforms our ability to engineer life by integrating biological sciences, engineering, computation-based design, and systems-level thinking. Unlike traditional genetic engineering, which changes existing genes, synthetic biology builds new functions from standardized, modular, and programmable biological building blocks, with the possibility of obtaining predictable and scale-up designs. This has already led to revolutionary breakthroughs in health care, agriculture, industry, and environmental sciences. From the design of probiotics to regulate human immunity and microbiota, to nutrient-enhanced plants, to microbes that make medicines and renewable fuels, synthetic biology has expanded the boundaries of what is technologically and biologically achievable. The chapter demonstrated how the core principles modularity, standardization, abstraction, and the design–build–test–learn (DBTL) cycle are the cornerstones of the design of resilient and reproducible complex biological systems. These principles, combined with advanced tools such as DNA synthesis and assembly, CRISPR-based genome editing, artificial gene circuits, biosensors, and computational modeling, enable researchers to not only replicate but also surpass the efficiency of natural processes. Examples such as microbial biosynthesis of vitamin C and paracetamol demonstrate how engineered microbial consortia and pathway engineering rationalization can replace traditional petrochemical-based processes, making the production pathways more sustainable, cost-effective, and eco-friendly. After integration with artificial intelligence, machine learning, multiomics, and systems biology, the design and predictive aspects of the discipline are enhanced further. The intersection allows high-throughput optimization of strains, real-time modeling of microbial communities, and dynamic genetic circuit control. Significantly, the emerging ability to design synthetic microbial biofilms and engineer microbiomes promises to open the way for personalized medicine, where in engineered consortia and bacteria diagnose, treat, and even prevent disease in a patient-specific manner. In short, synthetic biology is not a collection of tools but a revolutionary paradigm for reinventing how we engineer living systems for human and planetary good. With the potential to design genomes, cells, and microbial consortia with precision, it is a blueprint to innovations that can solve major challenges in medicine, food security, environmental sustainability, and renewable energy. In the future, the discipline will more depend on interdisciplinary teamwork, integrative computational platforms, and ethically guided practices to unlock its full potential. In the end, synthetic biology is leading a new biological revolution, ready to bring sustainable, programmable, and life-saving solutions that will shape the future of science, technology, and society.

References

Aminian-Dehkordi J, Rahimi S, Golzar-Ahmadi M, Singh A, Lopez J, Ledesma-Amaro R, Mijakovic I (2023) Synthetic biology tools for environmental protection. Biotechnol Adv 68: 108239. https://doi.org/10.1016/j.biotechadv.2023.108239

Bober JR, Beisel CL, Nair NU (2018) Synthetic biology approaches to engineer probiotics and members of the human microbiota for biomedical applications. Annu Rev Biomed Eng 20(1): 277–300. https://doi.org/10.1146/annurev-bioeng-062117-121019

Chang M, Ahn SJ, Han T, Yang D (2024) Gene expression modulation tools for bacterial synthetic biology. Biotechnol Sustain Mater 1(1):6. https://doi.org/10.1186/s44316-024-00005-y

Charbonneau MR, Isabella VM, Li N, Kurtz CB (2020) Developing a new class of engineered live bacterial therapeutics to treat human diseases. Nat Commun 11(1). https://doi.org/10.1038/s41467-020-15508-1

Decoene T, De Paepe B, Maertens J, Coussement P, Peters G, De Maeseneire SL, De Mey M (2018) Standardization in synthetic biology: an engineering discipline coming of age. Crit Rev Biotechnol 38(5):647–656. https://doi.org/10.1080/07388551.2017.1380600

Elowitz MB, Leibler S (2000) A synthetic oscillatory network of transcriptional regulators. Nature 403(6767):335–338. https://doi.org/10.1038/35002125

Endy D (2005) Foundations for engineering biology. Nature 438(7067):449–453. https://doi.org/10.1038/nature04342

Feng Y, Su C, Mao G, Sun B, Cai Y, Dai J, Ma Y (2024) When synthetic biology meets medicine. Life Med 3(1):lnae010. https://doi.org/10.1093/lifemedi/lnae010

Galanie S, Thodey K, Trenchard IJ, Filsinger Interrante M, Smolke CD (2015) Complete biosynthesis of opioids in yeast. Science 349(6252):1095–1100. https://doi.org/10.1126/science.aac9373

Gardner TS, Cantor CR, Collins JJ (2000) Construction of a genetic toggle switch in Escherichia coli. Nature 403(6767):339–342. https://doi.org/10.1038/35002131

Gibson DG, Young L, Chuang R-Y, Venter JC, Hutchison CA, Smith HO (2009) Enzymatic assembly of DNA molecules up to several hundred kilobases. Nat Methods 6(5):343–345. https://doi.org/10.1038/nmeth.1318

Hou F, Xian M, Huang W (2021) *De novo* biosynthesis and whole-cell catalytic production of paracetamol on a gram scale in *Escherichia coli*. Green Chem 23(20):8280–8289. https://doi.org/10.1039/d1gc02591k

Huang R, Kravchik V, Zaatry R, Habib M, Geva-Zatorsky N, Daniel R (2025) Engineering coupled consortia-based biosensors for diagnostic. Nat Commun 16(1):3761. https://doi.org/10.1038/s41467-025-58996-9

James JS, Dai J, Chew WL, Cai Y (2025) The design and engineering of synthetic genomes. Nat Rev Genet 26(5):298–319. https://doi.org/10.1038/s41576-024-00786-y

Jinek M, Chylinski K, Fonfara I, Hauer M, Doudna JA, Charpentier E (2012) A programmable dual-RNA–guided DNA endonuclease in adaptive bacterial immunity. Science *337*(6096): 816–821. https://doi.org/10.1126/science.1225829

Keasling JD (2010) Manufacturing molecules through metabolic engineering. Science 330(6009): 1355–1358. https://doi.org/10.1126/science.1193990

Kitano S, Lin C, Foo JL, Chang MW (2023) Synthetic biology: learning the way toward high-precision biological design. PLoS Biol 21(4):e3002116. https://doi.org/10.1371/journal.pbio.3002116

Knight T (2003) Idempotent vector design for standard assembly of Biobricks (No. MIT Synthetic Biology Working Group0; MIT Synthetic Biology Working Group0). Massachusetts Institute of Technology. Department of Biological Engineering. http://hdl.handle.net/1721.1/21168

Kokarakis EJ, Rillema R, Ducat DC, Sakkos JK (2023) Developing cyanobacterial quorum sensing toolkits: toward interspecies coordination in mixed autotroph/heterotroph communities. ACS Synth Biol 12(1):265–276. https://doi.org/10.1021/acssynbio.2c00527

Liu T, Xu X, Liu Y, Li J, Du G, Lv X, Liu L (2022) Engineered microbial cell factories for sustainable production of L-lactic acid: a critical review. Fermentation 8(6):279. https://doi.org/10.3390/fermentation8060279

Nielsen J, Keasling JD (2016) Engineering cellular metabolism. Cell 164(6):1185–1197. https://doi.org/10.1016/j.cell.2016.02.004

Ratiner K, Ciocan D, Abdeen SK, Elinav E (2024) Utilization of the microbiome in personalized medicine. Nat Rev Microbiol 22(5):291–308. https://doi.org/10.1038/s41579-023-00998-9

Roell M-S, Zurbriggen MD (2020) The impact of synthetic biology for future agriculture and nutrition. Curr Opin Biotechnol 61:102–109. https://doi.org/10.1016/j.copbio.2019.10.004

Scott SR, Hasty J (2016) Quorum sensing communication modules for microbial consortia. ACS Synth Biol 5(9):969–977. https://doi.org/10.1021/acssynbio.5b00286

Song H, Ding M-Z, Jia X-Q, Ma Q, Yuan Y-J (2014) Synthetic microbial consortia: from systematic analysis to construction and applications. Chem Soc Rev 43(20):6954–6981. https://doi.org/10.1039/c4cs00114a

Sun S, Chen J (2024) Recent advances in hydrogel-based biosensors for cancer detection. ACS Appl Mater Interfaces 16(36):46988–47002. https://doi.org/10.1021/acsami.4c02317

Vázquez-Castellanos JF, Biclot A, Vrancken G, Huys GR, Raes J (2019) Design of synthetic microbial consortia for gut microbiota modulation. Curr Opin Pharmacol 49:52–59. https://doi.org/10.1016/j.coph.2019.07.005

Yook G, Nam J, Jo Y, Yoon H, Yang D (2025) Metabolic engineering approaches for the biosynthesis of antibiotics. Microb Cell Factories 24(1):35. https://doi.org/10.1186/s12934-024-02628-2

Zhang W, Nielsen DR (2014) Synthetic biology applications in industrial microbiology. Front Microbiol 5. https://doi.org/10.3389/fmicb.2014.00451

Microbiota-Based Vaccine Alternatives 10

Abstract

Vaccination continues to be one of the most successful public health measures; however, the traditional vaccine platforms face constant hurdles, such as the efficacy of vaccines being variable in different populations, poor mucosal immunity, the need for cold chain storage, and reduced immunogenicity in the setting of immune compromise or microbiome dysbiosis. Over the past decade, there has been an increasing body of evidence that has revealed the human microbiota to be a critical regulator of innate and adaptive immunity, significantly influencing susceptibility to infection and vaccine response. This chapter explores the concept of microbiota-based vaccine alternatives as a revolutionary shift in the field of immunoprophylaxis. It critically examines the role of commensal microbes, microbial metabolites, and microbiota-derived materials as endogenous adjuvants, antigen carriers, or independent immunomodulators. The fundamental mechanisms explored include pattern recognition receptor activation, immune modulation by microbial metabolites, antigen presentation enhancement, and the generation of mucosal and systemic immune memory. This chapter also focuses on the importance of probiotics, prebiotics, postbiotics, genetically engineered microbial vectors, outer membrane vesicles, and fecal microbiota transfer in enhancing or replacing traditional vaccine approaches. The ethical, biosafety, and regulatory issues associated with live microbial therapy are also critically discussed, in addition to the issues of interindividual microbiome diversity and translational limitations. Finally, future directions are outlined, incorporating the concepts of synthetic biology, next-generation probiotics, and personalized microbiome engineering, thus positioning microbiota-based vaccine alternatives as a rapidly advancing area of research for the development of safer, more effective, and context-specific immunotherapeutic approaches.

R. C. Sobti et al., *Microbiota Unveiled: Science, Health, and Future Therapies*,
https://doi.org/10.1007/978-981-95-9031-5_10

Keywords

Vaccine alternatives · Microbiota-based vaccines · Gut microbiome · Natural adjuvants · Prebiotics · Probiotics · Postbiotics · Engineered postbiotics · Fecal microbiota transplant (FMT) · Mucosal immunity · Outer membrane vesicles (OMVs) · Next-generation probiotics · Dysbiosis

10.1 Introduction

One of the most successful public health initiatives for a long time, vaccination has dramatically decreased the prevalence of infectious illnesses and the deaths that go along with them globally. Inactivated pathogens, live attenuated organisms, toxoids, or subunit antigens are examples of traditional vaccination techniques that have produced broad protection against a variety of microbial dangers. Nevertheless, despite these achievements, conventional vaccinations still have a number of drawbacks, including as low effectiveness in immunocompromised people, pathogen antigenic diversity, cold chain dependencies, and uncommon but severe adverse side effects (Pulendran and Ahmed 2006). Furthermore, new infections, antibiotic resistance, and an increasing desire for customized and mucosal immunization techniques necessitate creative solutions that transcend traditional paradigms. The recent years have seen various vaccine-based options arrive to overcome the shortcomings of conventional methods. These are DNA and mRNA vaccines, viral-vectored recombinant vaccines, peptide- and protein-based vaccines, and nanoparticle platforms. These have allowed for the rapid design of antigens, scale-up of manufacturing, and targeting of highly mutable pathogens. The COVID-19 pandemic showed how mRNA vaccines can be quickly adapted to new variants, while viral-vectored vaccines (e.g., adenovirus-based systems) demonstrated their ability to induce humoral and cellular immune responses. There are still challenges, however, despite these developments, like variability in efficacy in diverse populations, safety concerns with repeated dosing, and reliance on cold-chain systems. This has prompted consideration of even more radical paradigms, like microbiota-based vaccine alternatives.

The past decade has seen human microbiota act as an important modulator of host immunity. It has the capability to shape the function as well as development of adaptive and innate immunity. The gut microbiota, in fact, has a subtle balance between immune tolerance and activation. Perturbation of microbial composition, i.e., dysbiosis, has been demonstrated to play a role in increased susceptibility to infection, autoimmune disease, and even vaccine failure (Belkaid and Hand 2014; Lynn et al. 2022). This has created growing interest in the potential to harness the immunomodulatory role of the microbiota to enhance or replace conventional vaccine responses.

Microbiota-based vaccine methods are a new field in immune-prophylaxis. Such methods attempt to utilize commensal bacteria, microbial metabolites, genetically-engineered bacterial vectors as adjuvants or as stand-alone immunogens to enhance immune efficacy. Probiotics, postbiotics as well as bacterial outer membrane vesicles (OMVs) are explored for their ability to act as an antigen delivery vehicles, stimulate mucosal immunity, and create long-term immunity without the use of inactivated or live pathogen components (Ahmed et al. 2023; Van Niel et al. 2018; Wells et al. 2017). Besides, technological advances in synthetic biology, systems immunology, and microbiome engineering enable one's own individual microbial signature to be utilized when crafting immunization schedules. This is specifically relevant for populations whose microbiota profiles are modified in one manner or another by variables such as age, geography, diet, antimicrobial exposure, or the presence of chronic disease.

In addition to reviewing recent experimental and clinical advancements in microbial delivery systems, this chapter considers the potential of prebiotics, probiotics, and postbiotics as immunological adjuvants and addresses the theoretical and practical aspects of vaccine substitutes derived from microbiota. It also discusses the main immunological processes by which microbiota modulates vaccine response. Additionally, the difficulties, safety considerations, and legal implications of integrating microbiota into the field of vaccination are covered. With growing demand of world for faster, safer, and more effective vaccination approaches to combat global health issues, microbiota represents an innovative and unexplored field for vaccine development (Fig. 10.1).

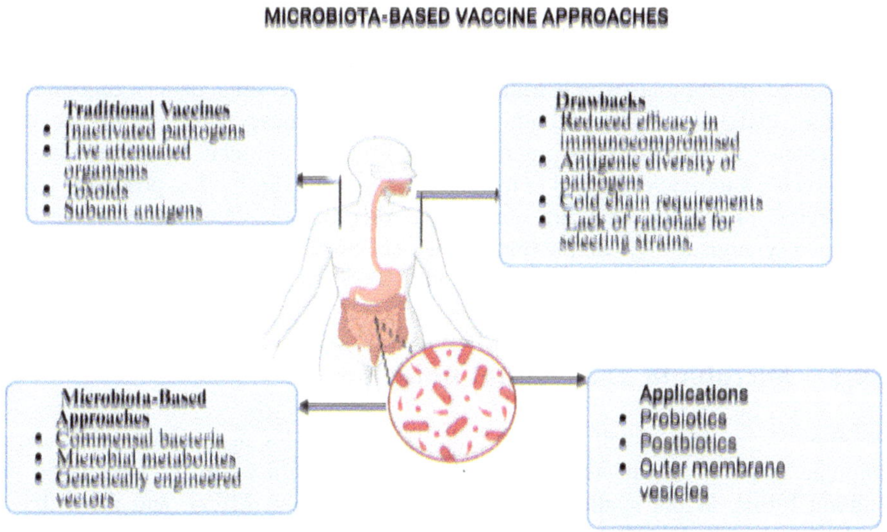

Fig. 10.1 Microbiota-based vaccine strategies influence commensal bacteria and metabolites to overcome limitations of traditional vaccines, enhancing immune modulation with applications in probiotics and engineered vectors

10.2 Dysbiosis and Its Impact on Vaccine Alternatives

Dysbiosis, or commensal diversity or abundance disturbance, has been associated with increased susceptibility to enteric infection, such as chronic inflammation and compromised vaccine responsiveness (Belkaid and Hand 2014). The commensal loss induced by antibiotics can lead to pathogenic overgrowth, such as boom of *Clostridioides difficile* and barrier failure, with consequent systemic microbial inflammation and translocation. Dysbiosis has also been shown to be involved with disrupted responsiveness to oral and parenteral vaccines, especially in low-income settings where microbiome diversity and enteropathy are prevalent (Lynn et al. 2022).

Destruction of microbial signals interrupts pattern recognition receptor (PRR) activation and antigen presentation pathways, which leads to suboptimal B and T cell activation. In addition, dysbiosis patterns have been reported in autoimmunity, allergy, and metabolic disorders, suggesting the broad-reaching influence of microbial imbalance on systemic immunity. This hints at the need for maintaining or restoring microbial homeostasis as a requirement for optimal immune instruction and vaccine efficacy.

Apart from enhancing vulnerability to enteric infection, dysbiosis, or commensal variety and abundance change, also impacts the effectiveness of conventional and next-generation vaccinology. As a case in point, suboptimal efficacy has been attributed to chronic enteropathy in low-resource settings despite the introduction of advanced vaccine platforms like viral vectors and mRNA vaccines, whilst microbial depletion by antibiotics diminishes the efficacy of oral and parenteral immunizations. Full potential of new vaccine technologies is held back by an imbalanced microbiota, which influences antigen processing, pattern recognition receptor (PRR) signaling, and mucosal immunity. These findings indicate that maintenance or restoration of microbial equilibrium is crucial to the creation of new vaccine alternatives as well as to general well-being. Probiotics, postbiotics, and engineered microbial vectors are some of the microbiota-targeted interventions that present a synergistic approach in optimizing vaccine efficacy, especially in high-risk individuals where dysbiosis is the norm (Fig. 10.2).

10.3 Microbiota as a Natural Adjuvant

Traditional vaccines usually require adjuvants molecules that enhance the host's immune response to an antigen in order to work, particularly in inactivated or subunit forms. While synthetic adjuvants such as alum, MF59, and AS01 have been used widely, they have been weak in efficacy in certain populations and have had adverse reactions (O'Hagan et al. 2011). For this purpose, the microbiota and its components have been suggested as a potential alternative or addition to conventional adjuvants. Their intrinsic immunomodulatory activity, which is physiologically compatible and achieved through host immune system co-evolution, offers a natural mechanism to synergize immune responses. To enhance antigen presentation

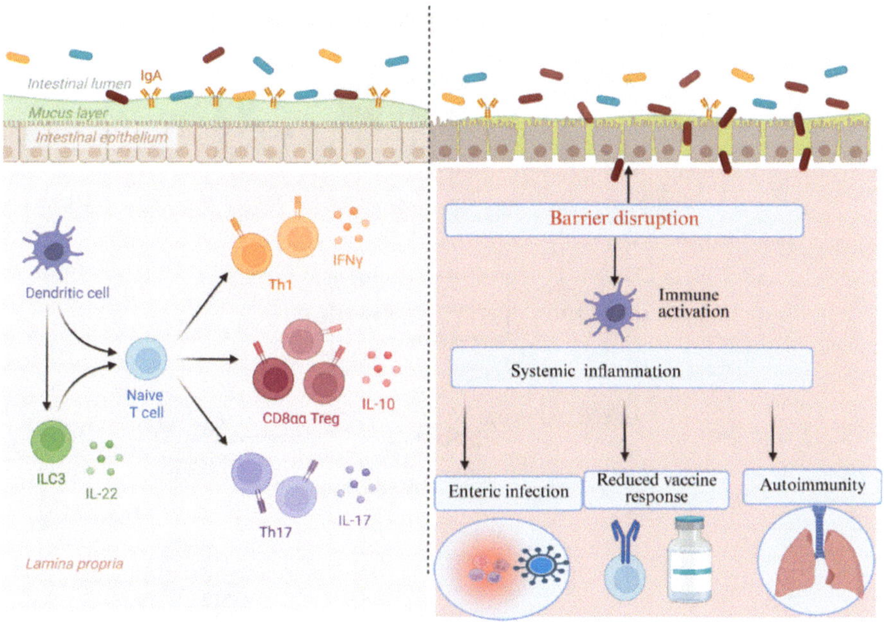

Fig. 10.2 Dysbiosis leads to systemic inflammation and barrier disruption, increasing susceptibility to enteric infections, autoimmunity, and reduced vaccine responses

and antibody responses, alum does so primarily by establishing a depot at the site of injection and triggering NLRP3 inflammasome activation, leading to the release of IL-1β and IL-18, respectively. Through induction of the recruitment of monocytes, dendritic cells, and granulocytes to the site of injection and promotion of antigen uptake, MF59, an oil-in-water formulation of squalene emulsion, enhances vaccine performance. Through induction of NF-κB activation, AS01, a liposome-based adjuvant system formulation with MPL (a TLR4 agonist) and QS-21 saponin, enhances innate immune mechanisms and enhances humoral and cellular immunity. This has prompted the possibility that the microbiota and its components might serve as alternatives or additives to conventional adjuvants. They possess a physiologically compatible intrinsic immunomodulatory activity as a natural mechanism of coordinating immune responses, through host immune system coevolution.

10.3.1 Microbial Metabolites and Immunomodulation

Short-chain fatty acids (SCFAs), such as acetate, butyrate, and propionate, are significant microbial metabolites produced when intestinal microbes break down food fibers. To operate, the host immune system needs SCFAs. For instance, butyrate suppresses regulatory T cell histone deacetylases (HDACs), resulting in an anti-inflammatory phenotype and increased mucosal tolerance (Furusawa et al.

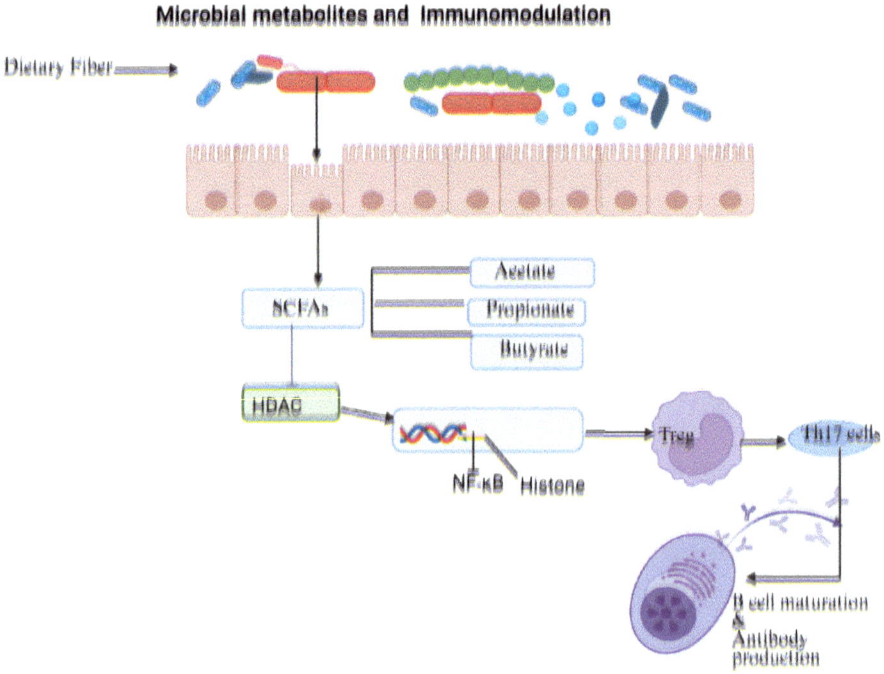

Fig. 10.3 Microbial metabolites from dietary fiber [butyrate and acetate], modulate host immunity via receptors like GPCRs and AhR, promoting Treg and Th17 responses for antibody production and immune tolerance

2013). SCFAs can, however, improve vaccine-induced humoral immunity through supporting B cell maturation and antibody formation (Park et al. 2015).

Secondary bile acids and tryptophan metabolites further interact with host receptors, such as the aryl hydrocarbon receptor (AhR) and G-protein coupled receptors (GPCRs), and modulate the Th17/Treg ratio and dendritic cell function. These may be potential targets for microbiota-derived adjuvant therapy to modulate the immune response to a particular vaccine regimen (Fig. 10.3).

10.3.2 Pattern Recognition Receptors (PRRs) and Microbial Signaling

The innate immune system uses pattern recognition receptors (PRRs), such as NOD-like receptors (NLRs), Toll-like receptors (TLRs), as well as C-type lectin receptors, to recognize conserved microbial compounds, sometimes referred to as pathogen-associated molecular patterns (PAMPs). Remarkably, most of the molecules of commensal microbes act as physiological ligands of such receptors. They generate regulated immune responses without triggering inflammation (Medzhitov 2007). For example, *Lactobacillus peptidoglycans* and lipoteichoic

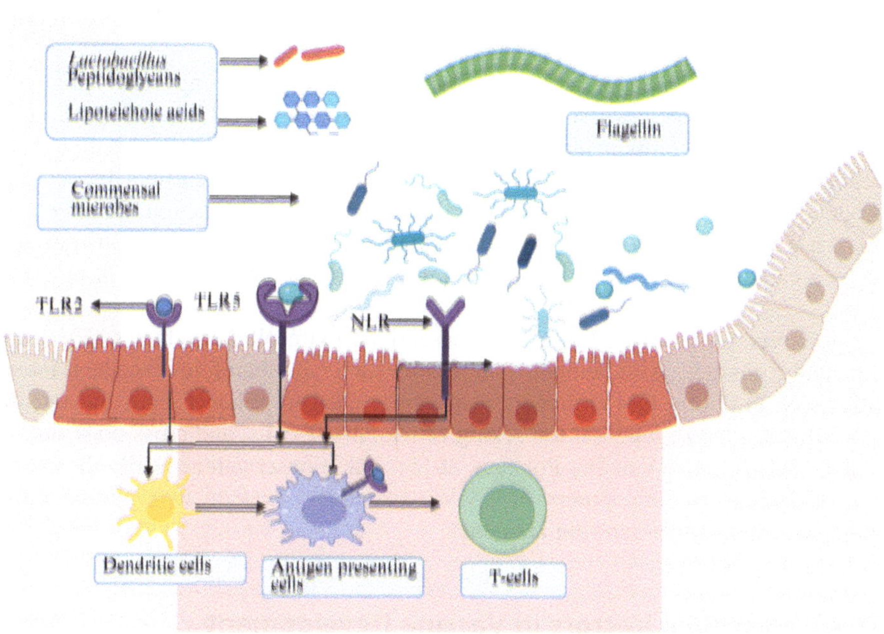

Fig. 10.4 Commensal microbes activate innate receptors (TLR2-5, NLR) on dendritic cells, promoting antigen presentation and T cell activation

acids can induce activation of TLR2, resulting in antigen presentation and maturation of dendritic cells. Commensal flagellin, a potent TLR5 agonist, can promote mucosal immunity and induce a strong CD4+ and CD8+ T cell response, augmenting vaccine immunogenicity (Camacho et al. 2011). Thus, it is reasonable to selectively use or manipulate these microbial molecules to mimic naturally occurring immune priming mechanisms in vaccine formulations (Fig. 10.4).

10.3.3 Enhancement of Antigen Presentation

Effective antigen presentation by antigen-presenting cells (APCs), such as B cells, dendritic cells, and macrophages, is necessary for successful vaccination responses. APC recruitment, maturation, and function at mucosal surfaces, such as Peyer's patches and the mesenteric lymph nodes, are regulated by microbiota. For example, *Bacteroides fragilis*, a commensal produce capsular polysaccharides that induce expression of MHC class II and costimulatory molecules on dendritic cells (Mazmanian et al. 2005).

Excretion of outer membrane vesicles (OMVs) containing pathogen-associated molecular patterns (PAMPs) by other Gram-negative commensals, e.g., *Escherichia coli* Nissle 1917, triggers Toll-like receptors (TLRs) on APC and facilitates cross-presentation to CD8+ T lymphocytes. Similarly, it was demonstrated that

Helicobacter pylori OMVs trigger dendritic cell activation and IL-12 secretion, which skews immunological responses toward Th1 immunity.

Gram-positive bacteria can augment antigen processing and augment co-stimulatory molecules (CD80 or CD86) by activating TLR2 on dendritic cells with cell wall components such as lipoteichoic acids and *Lactobacillus peptidoglycans*. By producing butyrate, certain *Clostridium* clusters induce the production of regulatory T cells, indirectly modifying APC function. *Lactococcus lactis* has also been genetically engineered to deliver viral antigens, demonstrating it to be a candidate live vector for delivering antigens directly to APCs in mucosal sites.

Moreover, similar to OMVs, microbe-derived extracellular vehicles (EVs) are inherently charged with antigen and PAMP and are therefore excellent nano-adjuvants. Probiotic OMVs have already been used to enhance cytokine release (e.g., IL-12, IFN-γ), mucosal trafficking of memory T cells, and cross-presentation to the CD8+ T cells (Van Der Pol et al. 2015). These observations justify the use of gene-modified probiotic strains, or OMVs, as delivery systems that immunize the immune system while carrying antigens.

10.4 Microbial Vectors in Vaccine Development

New advances in microbiome science and biotechnology towards the investigation of microbiota-derived structures and microbes as vaccine delivery systems have opened a potential avenue to explore. These biologically active molecules can potentially act as immunogenic platforms, antigen delivery systems, or immune adjuvants and thus offer an alternative to traditional vaccines. Their host compatibility due to their natural origin, mucosa-targeting ability and adjuvant-like action render them highly promising entities for future vaccinology.

Microbial vectors have matured from concept to bedside, and Gram-negative and Gram-positive-derived systems are now utilized as immunogenic scaffolds, antigen-delivery vehicles, and self-adjuvants. Classic bacterins (killed whole cells) remain pertinent models: oral inactivated *Vibrio cholerae* vaccines protect and mitigate disease in endemic settings through extensive presentation of PAMPs engaging PRRs on APCs, precipitating strong mucosal IgA and systemic immunity (TLR4/TLR2/NOD signaling). Recent studies confirm prolonged protection and detail waning dynamics and underscore their public-health significance. In parallel, outer-membrane vesicles (OMVs) from Gram-negative bacteria display nano-sized antigen/adjuvant complexes that promote APC uptake, DC maturation, and cross-presentation to CD8+ T cells; licensed OMV-based MenB vaccines not only prevent meningococcal disease but are shown to display measurable cross-protection against gonorrhea, illustrating OMVs' translational depth. OMVs are being bioengineered to display heterologous antigens (e.g., *Shigella*, *Klebsiella*, viral epitopes) and even to function as adjuvants that enhance acellular pertussis antigens, broadening their application across mucosal and systemic indications. Lactic-acid bacteria, including *Lactococcus lactis* and *Lactiplantibacillus*/*Lactobacillus* species, along with

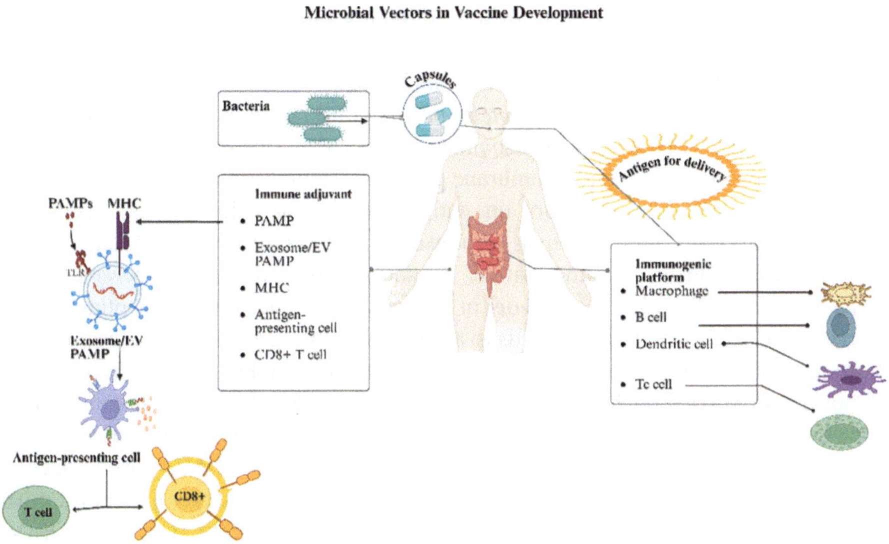

Fig. 10.5 Microbial vectors deliver antigens to mucosal sites, activating dendritic cells as well as antigen-presenting cells to drive B and T cell responses for vaccine development

spore-formers, are being developed for oral or nasal use on the Gram-positive side. Surface-displayed or secreted antigens, often combined with responsive genetic circuits, trigger strong mucosal IgA while also offering good safety and manufacturing profiles. Genetically engineered LAB are noted as flexible mucosal carriers in 2025 updates, and recent studies show that *L. lactis* can express full-length viral antigens for oral vaccines. Bacteriocins, such as nisin and colicins, act as microbiota-modulating, immune-active peptides. They work alongside these vaccine vectors to improve vaccine uptake and function as adjuvant-like components by reducing competing pathobionts and adjusting communication in epithelial and innate immune systems (Fig. 10.5).

10.4.1 Bacterial Outer Membrane Vesicles (OMVs)

Gram-negative bacteria secrete spherical, bilayered lipid structures called outer membrane vesicles (OMVs). Outer membrane proteins, lipopolysaccharides (LPS), peptidoglycans, nucleic acids, and other periplasmic components make up OMVs, which range in diameter from 20 to 250 nm (Kuehn and Kesty 2005; Zhang et al. 2025b). OMVs are naturally immunogenic and are recognized by innate immune receptors, such as TLR4, TLR2, and NOI, which induce strong antigen-presenting cell activation and release of pro-inflammatory cytokines (Ahmed et al. 2023; Ellis and Kuehn 2010).

Recent advancements in microbiome science and biotechnology reveal how microbiota-derived forms and bacterial strains may be employed as vaccine delivery systems, immunogenic platforms, or self-adjuvants. One of the oldest applications is the use of bacterins, which are inactivated whole-bacterial preparations traditionally employed against infectious diseases. Gram-negative bacterins, like inactivated *Vibrio cholerae* for cholera or *Bordetella pertussis* for pertussis, target by presenting a wide range of pathogen-associated molecular patterns (PAMPs), like lipopolysaccharides and outer membrane proteins, which activate Toll-like receptors (TLRs) and NOD-like receptors on antigen-presenting cells (APCs), triggering antibody responses and T-helper responses (Higgs et al. 2012; Montero et al. 2023). Analogously, Gram-positive bacterins like those of *Streptococcus pneumoniae* or *Clostridium chauvoei* in veterinary vaccines employ peptidoglycan, lipoteichoic acids, and teichoic acids to activate dendritic cells and induce protective immunity. Following this concept, outer membrane vesicles (OMVs) released by Gram-negative bacteria like *Neisseria meningitidis*, *Shigella*, and *Salmonella* provide natural nano-sized carriers of antigens and adjuvants, which can activate CD4+ and CD8+ T-cell responses (Sangiorgio et al. 2024). Indeed, OMV-based vaccines are already licensed against meningococcal disease and are in development for other enteric and respiratory pathogens, with recent advances showing engineered OMVs presenting SARS-CoV-2 and Klebsiella pneumoniae antigens with strong mucosal protection.

In Gram-positive species, spore-based bacteria, such as *Bacillus subtilis* spores and engineered vesicles, have been defined as stable oral vehicles with the ability to survive harsh gastrointestinal conditions and deliver heterologous antigen effectively to mucosal surfaces. In addition, live-attenuated and commensal pathogens such as *Salmonella typhi* Ty21a, *Listeria monocytogenes*, *Lactococcus lactis*, and *Lactobacillus* spp. have been modified to produce antigens secreted transiently from mucosal tissue on colonization and elicit robust IgA responses and T-cell priming (Cho and Chung 2020; Riedel et al. 2019). Engineered probiotics are particularly promising, with *Lactobacillus plantarum* and *Escherichia coli* Nissle 1917 strains recently engineered to present viral epitopes or tumor antigens, with consequent augmentation of systemic and mucosal immunity in preclinical models (Yang et al. 2025; Zhang et al. 2025a). Another useful category comprises bacteriocins, ribosomally synthesized antimicrobial peptides such as nisin from *Lactococcus lactis* or colicins from *Escherichia coli*, which, in addition to shaping the microbial community to augment vaccine responsiveness, also exert direct immunomodulatory effects on epithelial and immune cells. Together, these Gram-positive and Gram-negative microbial vehicles exhibit a range of mechanisms ranging from the activation of PRRs, maturation of APC, as well as cross-presentation to the modulation of mucosal microbiota by which bacterial systems can enhance vaccine immunogenicity. Through the integration of natural immunogenicity and engineered precision, these microbial vehicles form a rapidly evolving frontier in vaccinology, with applications ranging from cholera and typhoid prevention to experimental immunotherapies for cancer (Lahiri et al. 2023; Ruhal and Kataria 2021).

For enhanced immunogenicity and multivalent presentation, OMVs can be engineered to exhibit heterologous antigens on their surface. OMVs of *Neisseria meningitidis* are the primary ingredient of the approved meningococcal B vaccine

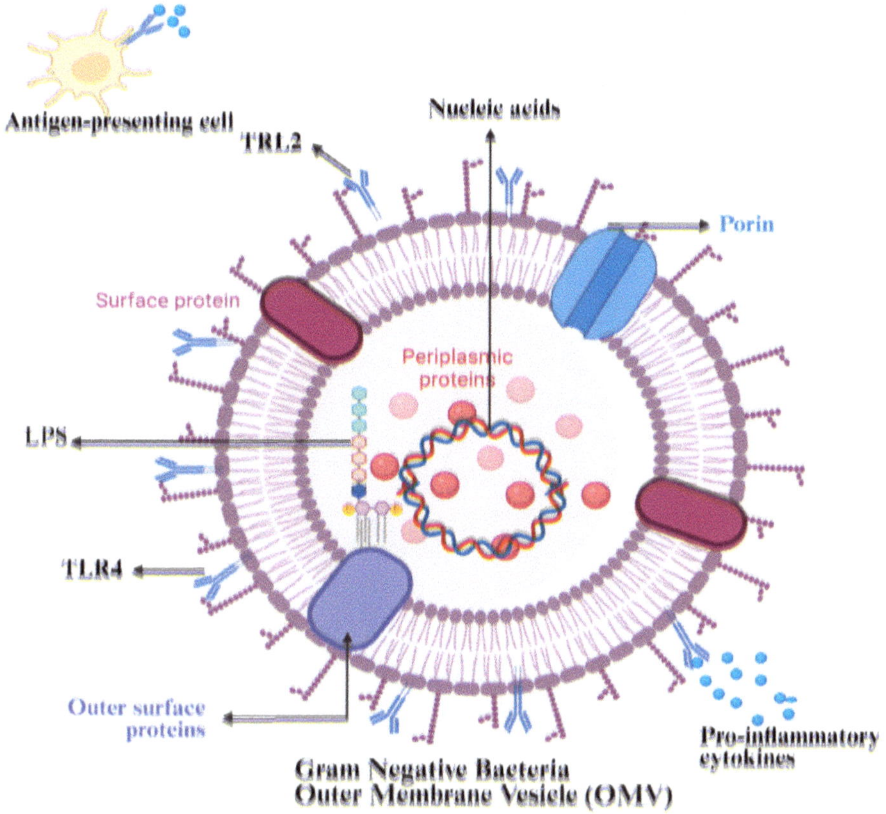

Fig. 10.6 Bacterial outer membrane vesicles (OMVs) carry proteins, nucleic acids, and lipopolysaccharides to antigen-presenting cells, triggering innate immune receptors (TLR2, TLR4, and NOD) then induces proinflammatory cytokines

(Bexsero®), which serves as one such example (Giuliani et al. 2006). They are highly useful for oral or intranasal vaccine administration due to their nano-range, mucosal permeability, and tolerance to the gastrointestinal environment, particularly in low-resource environments where cold-chain reliance is a constraint (Fig. 10.6).

10.4.2 Commensal Bacteria as Live Vectors

Certain probiotic and commensal bacteria such as *Escherichia coli* Nissle 1917, *Bifidobacterium*, and *Lactobacillus* have been successfully utilized as live vectors for the presentation of recombinant antigens to mucosal surfaces (Han et al. 2024). These microbes can interact with mucosal immune cells such intraepithelial lymphocytes, dendritic cells, and M cells after temporarily colonizing the gastrointestinal or urogenital tract (Wells et al. 2011). With the lowest risk of live attenuated

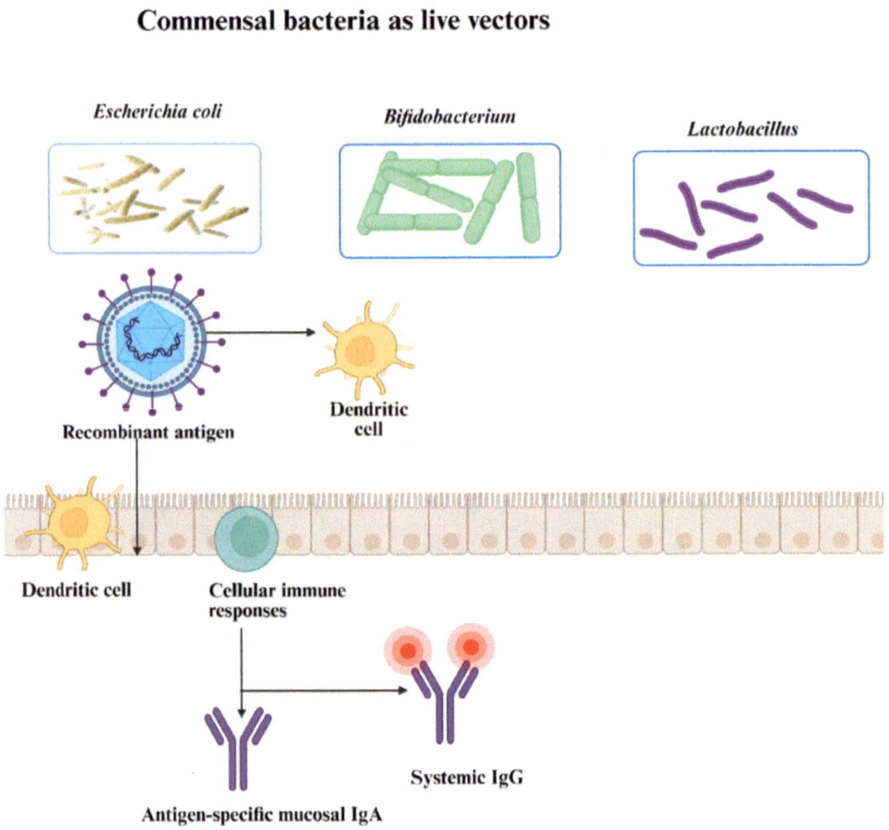

Fig. 10.7 Commensal bacteria [*E. coli* Nissle 1917, *Bifidobacterium*, *Lactobacillus*] deliver recombinant antigens to dendritic cells in gut, inducing mucosal IgA and systemic IgG responses

pathogens, recombinant strains that produce viral, bacterial, or parasite antigens can elicit antigen-specific mucosal IgA, systemic IgG, and cellular immune responses. For instance, *Lactobacillus plantarum* has been genetically modified to generate SARS-CoV-2 spike protein or influenza hemagglutinin, and in preclinical models, it has demonstrated immunogenicity (Hwang et al. 2023). Furthermore, the strains may be administered orally or could also be lyophilized, which offers a flexible and patient-responsive approach of vaccine administration (Fig. 10.7).

10.4.3 Engineered Probiotics for Antigen Delivery

Synthetic biology has enabled the design of probiotics with non-natural functions such as functions other than native antigen expression, enabling the construction of microbial platforms that are both immune modulators and antigen carriers. Engineered microbes can be programmed to respond to host or environmental cues, regulating their own antigen release or replication in a spatially and temporally controlled manner (Contreras-Salgado et al. 2024). Genetic circuits that are responsive to physiological states such as pH, bile acids, or inflammatory cytokines have been incorporated to trigger antigen production only within the targeted niches, hence enhancing safety and specificity. Surface display technologies such as sortase-mediated LPXTG anchoring or fusion to cell-wall binding domains facilitate stable antigen presentation to mucosal immune cells (Xia et al. 2019). In addition to single antigens, probiotics have been designed to codeliver adjuvants such as flagellin, cytokines (e.g., IL-10, IL-12), or GM-CSF, both enhancing innate and adaptive responses. Besides single antigens, probiotics have been designed to co-deliver adjuvants like flagellin, cytokines (e.g., IL-10, IL-12), or GM-CSF to enhance innate and adaptive immunity.

Recent developments in synthetic biology and genome editing tools, such as CRISPR/Cas9 and recombination-based tools, have opened up new avenues for the engineering of probiotics (Jeong et al. 2023). For example, *Lactobacillus plantarum* and *Lactococcus lactis* were engineered to be capable of expressing viral antigens like influenza hemagglutinin and SARS-CoV-2 spike epitopes, inducing protective mucosal IgA and systemic IgG immune responses in preclinical models (Tobias et al. 2025). Likewise, *E. coli* Nissle 1917 was engineered to be capable of delivering tumor-associated antigens, and it showed anti-tumor effect through increased CD8+ T-cell activation in murine models (Zhang et al. 2025a).

Engineered probiotics with additional functions than natural antigen expression have been made possible by the processes of synthetic biology. The modified microbes may be taught to recognize environmental or host cues, which in turn helps them govern their own self-replication, or distribute antigens in a much more regulated and localized manner (Mimee et al. 2015). Context-specific activation can be achieved by integrating genetic circuits that trigger antigen production in response to changes in pH, bile acid, or host immunological stimuli. In addition, the surface features of probiotic bacteria can be chemically or genetically modified to promote the presentation of antigens by anchoring motifs (e.g., LPXTG motifs or sortase-conjugated anchoring), which permits for a stable presentation as well as host recognition (Xu et al. 2013). In order to improve the potency of the immunological response, bacterial strains have also been modified to co-deliver the adjuvants such as flagellin or interleukins. The potential of modified probiotics is hampered by biosafety concerns about strain management, and regulatory compliance. These research gaps are rapidly being filled by recent advancements in genome editing methods like as recombination, CRISPR/Cas9, whole-genome synthesis, and others, which are advancing microbiota-based vaccines (Zhang et al. 2025b).

Together, these modified microorganisms show several different ways to work, including targeted immunotherapy, mucosal adjuvant distribution, direct antigen presentation, and local cytokine network modification. Their uses are rapidly expanding beyond infectious diseases to include metabolic disorders, autoimmune control, and cancer immunotherapy. Although biosafety and regulatory issues still

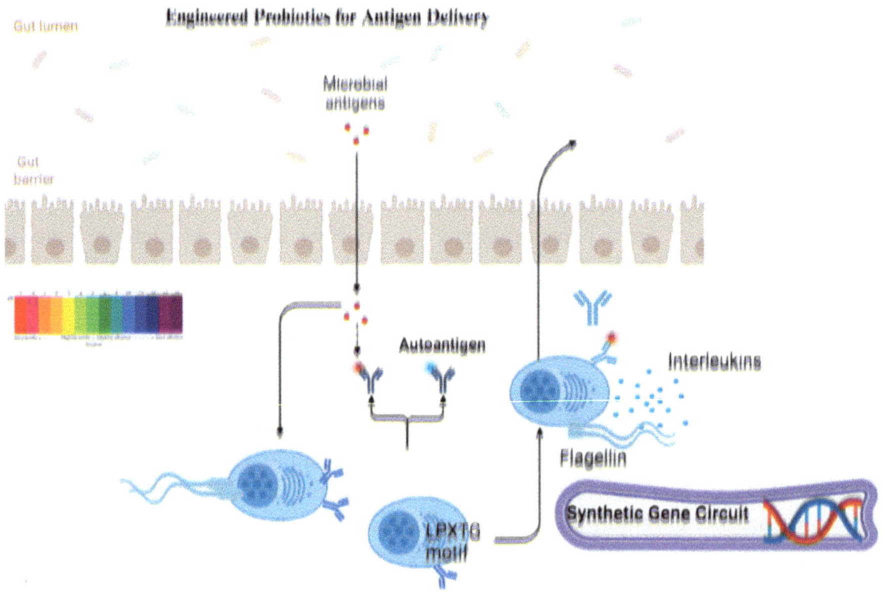

Fig. 10.8 Engineered probiotics detect gut signals, exhibit antigens, and codeliver adjuvants to strengthen the mucosal immunity

exist, ongoing progress in synthetic biology is driving the development of next-generation probiotics as flexible and personalized vaccine delivery vehicles (Fig. 10.8).

10.5 Fecal Microbiota Transplantation (FMT) and Immunity

10.5.1 History of Fecal Microbiota Transplantation

In the year 1957, Stanley Falkow, a distinguished American microbiologist, and a trailblazer in domain of molecular microbial pathogenesis, encouraged the administration of capsules containing a patient's own stool before surgery in order to treat antibiotic-associated diarrhea. In 1958, Ben Eiseman reported in the United States that fecal enemas act as an effective therapy for pseudomembranous enterocolitis, which is commonly known as *C. difficile* colitis and this infection has become more frequent and difficult to treat due to the widespread use of strong, broad-spectrum antibiotics. These medicines can cause dysbiosis in the gut, allowing *C. difficile* to multiply, spread more easily, and become resistant to treatment. In order to address this, doctors now use a technique called fecal microbiota transplantation (FMT), in which stool from healthy donors is introduced into a patient's intestine to restore microbial balance and improve health (Thanush et al. 2023).

10.5.2 FMT in Restoring Immune Homeostasis

Faecal Microbiota Transplantation is defined as a medical procedure in which the complete fecal microbial content is introduced into the gastrointestinal tract of a recipient, typically via enema from a healthy individual. The main aim is to re-establish the natural proportion of the gut microbiome in individuals experiencing dysbiosis by facilitating the restoration of a healthy gut microbiota composition, a state often referred to as *normobiosis*. The United States Food and Drug Administration (FDA) has recognized FMT as an investigational new drug (IND). The procedure begins with rigorous donor screening to rule out bacterial, viral, or protozoan infections. A healthy donor is selected based on the absence of metabolic, autoimmune, or cancer-related conditions in their personal and family history.

FMT is considered a robust and comprehensive method for manipulating the gut microbiome. By reconstituting the entire microbial population, it facilitates stable engraftment of beneficial microbes. This stable colonization is strongly associated with maintaining remission and supporting long-term gut health in several gastrointestinal disorders. Initially, FMT demonstrated significant success in controlling recurrent *Clostridium difficile* infections (CDIs), where it effectively restored a healthy intestinal microbiome. This success has sparked considerable interest in its potential utility for other chronic gastrointestinal disorders, such as inflammatory bowel disease. FMT contributes towards the restoration of immune homeostasis by introducing beneficial microbial communities that can initiate the formation of short-chain fatty acids, specifically butyrate, propionate, and acetate. These SCFAs are primarily made by microbial fermentation process with the help of gut-resident anaerobes and are critical for regulating not only innate but also adaptive immune responses. The immunomodulatory effects of SCFAs operate through multiple mechanisms. These mechanisms involve activation of Guanine nucleotide-binding protein-coupled receptors (GPR41, GPR43), inhibition of histone deacetylase (HDAC) and their integration into cellular metabolism via acetyl-CoA. Among SCFAs, butyrate is particularly important in so far as it facilitates the differentiation of regulatory T cells (Tregs) through increased expression of FoxP3, and repression of inflammatory cytokines such as tumor necrosis factor-alpha (TNF-α), interleukin-6 (IL-6) and interferon-gamma (IFN-γ). In terms of innate immunity, short-chain fatty acids (SCFAs) drive macrophage differentiation toward the M2 subtype, known for its anti-inflammatory properties. They also modulate the activity of innate immune components such as neutrophils, dendritic cells, and various subsets of innate lymphoid cells. Additionally, SCFAs contribute to gut barrier integrity by upregulating mucin proteins (MUC1, MUC4), strengthening tight junctions, and inducing antimicrobial peptides including CAMP and S100A8. FMT also restores bile acid metabolism, enhancing the secondary bile acids production such as deoxycholic acid, which inhibits *Clostridioides difficile* growth and supports Treg-mediated immune tolerance. Experimental studies provide further evidence for these effects. FMT has been shown to rebalance cytokine profiles in CD4+ T cells notably IL-10, IL-17, and IFN-γ following dysbiosis. Importantly, blocking IL-10 signaling reduces the therapeutic efficacy of FMT, highlighting its pivotal role in immune regulation.

10.5.3 FMT for Enhancing Vaccine Responses

Creation of effective vaccines is a complicated, time-consuming, and expensive process. In most cases, vaccines do not advance to the stage of clinical trials. Many vaccine candidates that perform well in laboratory and animal experiments often fail to protect humans, may cause harmful side effects, or even intensify the very illness they are intended to prevent. It is a novel method that helps improve the effectiveness of vaccines by restoring the variety and balance of helpful microbes in the gut. The organisms lives in gut significantly play an important role in the body's immune system under the condition of dysbiosis, the immune system becomes weak and may not respond well to vaccines. During dysbiosis, important groups of beneficial microbes such as *Bifidobacteria*, *Clostridia*, and *Ruminococcaceae* are often reduced. This weakens the body's ability to fight against infections.

In FMT, the transfer of fecal microbial content from a healthy donor to the gastrointestinal tract of a recipient is reported in order to restore eubiosis, which finally enhances vaccine responses by producing SCFAs that help to maintain the intestinal barrier through stimulating the formation of mucin proteins such as MUC1 and MUC4. This restoration helps in initiation and maturation of essential immune cells, like, T lymphocytes, dendritic cells and B-lymphocytes, thereby improving humoral responses such as antibody generation. Notably, increased levels of immunoglobulin Y (IgY) have been documented in FMT-treated and vaccinated avian models, underscoring the potential of this intervention in enhancing vaccine efficacy. Moreover, FMT promotes the production of volatile fatty acids, including, acetate, butyrate and propionate which work as immunomodulatory metabolites. These SCFAs influence cytokine signaling pathways by regulating molecules such as interleukin-10 (IL-10),interferon-gamma (IFN-γ), interleukin-6 (IL-6), and tumor necrosis factor-alpha (TNF-α). These cytokines work in concert to balance proinflammatory and anti-inflammatory responses, facilitating a more robust and sustained immunological response to vaccination (Fig. 10.9).

10.5.4 Ethical and Regulatory Concerns in FMT

10.5.4.1 Suitable Healthy Donor

Identification of an appropriate, healthy donor is one of the most complex ethical issues in fecal microbiota transplantation (FMT) even though, stool is considered as widely available raw material, finding an appropriate donor is extremely challenging task because there is no clear-cut definition of a "suitable donor" or an "optimal microbiota profile." Donor selection is based on very strict criteria, including blood and stool tests, medical history, and physical examinations, all of which greatly decrease the number of eligible donors. However, there is a lack of uniformity in

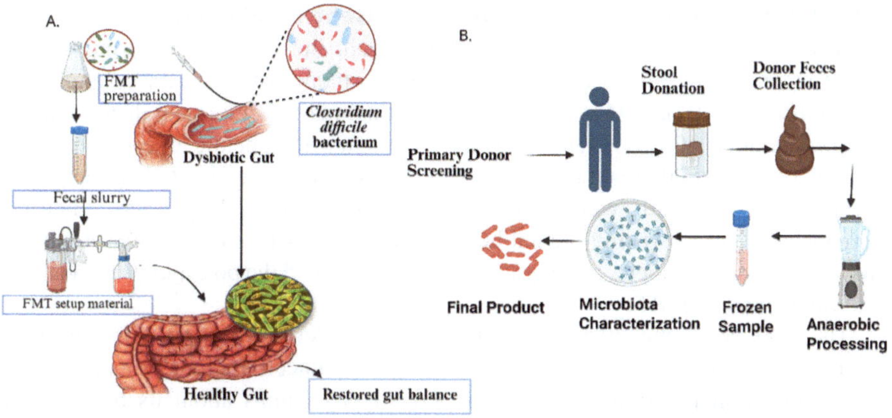

Fig. 10.9 Workflow and mechanism of fecal microbiota transplantation (FMT). (**a**) Instillation of fecal matter from a healthy donor which introduces beneficial microbes into the patient's gut, restoring microbial balance, displacing *Clostridioides difficile*, and thereby resolving symptoms. (**b**) The preparation process involves donor screening, feces collection, anaerobic processing, freeze drying, and microbiota characterization before generating the final transplant

screening protocols across different countries in terms of test requirements, timing and interval frequency. Some additional uncertainties are also linked with the donor profile. For example, it is not clear whether age restrictions should apply to donors; whether children, woman in gestational phase, or breastfeeding women should be excluded from donating; and whether the donor's religious or cultural background is associated with dietary customs, it should be considered when allocating fecal microbiota to recipients Now, questions also arise as to whether the donor and recipient should be of the same gender, or whether cross-gender transplants could affect a recipient's sexual identity or fertility. According to a study, serum testosterone levels in young female mice significantly increased when the gut microbiota is transferred to immature females from adult male mice. These findings suggest that gender-related effects may need to be carefully considered in the context of human fecal microbiota transplantation. There are also ethical concerns regarding the privacy and protection of donors, especially if screening reveals a serious illness or if a donor is dishonest due to financial incentives. Practical difficulties include limited availability of FMT facilities, the need to travel long distances, and the heavy workload on doctors. To address these challenges, some countries have set up stool banks (NDFB, OpenBiome, Taymount Clinic, etc.) but even there, the selection process remains difficult for example, in a study from Australia, only 12 out of 116 potential donors were eligible. Considering all these factors, identifying a healthy donor becomes not only a medical but also a deeply rooted ethical and social issue. (Thanush et al. 2023)

10.5.4.2 Vulnerability and Voluntary Consent of IBD Patients

FMT is considered a novel therapeutic approach, but it raises important ethical concerns regarding informed consent, especially among patients with inflammatory bowel disease (IBD). IBD involves chronic inflammation, emotional distress, and a reduced quality of life. Many patients experience therapeutic desperation after several conventional treatments fail. This desperation can affect their ability to make voluntary and well-informed decisions. Financial pressure, nutrient deficiencies, and coexisting mental health challenges such as depression or anxiety may cloud their judgment, making it harder to understand the possible risks, benefits, and limitations of FMT.

The risks are higher because FMT is still considered experimental, except for treating *Clostridium difficile* infection. There is also uncertainty about its long-term effects on brain function and behavior. Overhyped claims about its benefits can mislead patients and reduce the reliability of their consent. Ethical guidelines, such as those in the Declaration of Helsinki, allow untested treatments for serious illnesses only when strong protections are in place. Doctors must be honest and not influenced by money or institutional pressure. Consent should be clear, compassionate, and guided by what is best for the patient.

Regulatory authorities like the FDA have emphasized that FMT remains investigational for most medical uses. Therefore, cautious and well-informed decision-making in clinical settings is essential. To uphold ethical standards and trust, it is necessary to improve informed consent procedures and conduct more thorough long-term studies on the safety and broader applications of FMT (Faintuch and Faintuch 2019).

10.5.4.3 Risks, Safety, and Privacy Concerning FMT

Generally, FMT is considered safe, although the method of administration may have certain drawbacks. The most common and prominent side effects include gas, constipation, bloating, headache, and diarrhea. Some patients may experience cramping, fever, or elevated C-reactive protein levels within a few hours to a few days after the procedure. These reactions can potentially lead to new infections, increased severity of illness, and interference with current treatments. Such physiological disturbances may amplify the overall risk profile and raise ethical and clinical concerns. Because of its application, there is a potential for privacy violations, especially for donors who have more knowledge about a person than their human DNA. It has been shown that individual microbiomes include enough differentiating traits to distinguish a person over time with recent advancements in human microbiome research.

10.5.5 Regulatory Concerns

FMT faces several regulatory concerns because there is no universally accepted definition for FMT. Regulatory concerns vary from country to country. FMT is regulated under varying classifications. In United States, the Food and Drug

Administration (FDA) regulates FMT which is derived from human feces and classifies it as both a biological product and a drug. FMT is also considered as a therapy that has not received full regulatory approval under existing biologics standards. Normally, this is a strict regulatory category, but the FDA has made a special exception for treating recurrent *Clostridium difficile* infection (rCDI), provided that informed consent is obtained from the patient. This is known as an enforcement discretion policy. It shows how regulators face challenges in balancing safety rules with making treatment available to patients.

The first and foremost randomized clinical trial on FMT was published in 2013 by Dutch infectious diseases specialist, demonstrating its superior efficacy in treating recurrent *Clostridioides difficile* infection (rCDI), which helped increase interest in FMT within the United States. That same year, the U.S. Food and Drug Administration (FDA) conducted an open public workshop to discuss scientific, and regulatory issues related to fecal microbiota transplantation. At the time, when FDA stated that FMT was not FDA-approved and that doctors had to submit an Investigational New Drug (IND) application before offering it to any patient. It could only be given during clinical trials or emergencies. This requirement worried physicians, researchers, and advocacy groups, especially for patients suffering from rCDI.

In July 2013, the FDA announced a temporary discretionary enforcement approach. The agency explained this was not a permanent rule, and further study was needed. Over time, the FDA released several draft guidelines to get public feedback, but none were finalized. In 2014, the FDA proposed changes, including requiring that patients and doctors know who the stool donor is, and that all donor screening must be supervised by the doctor doing the procedure. This policy would allow hospitals to run their own stool banks, and let family or friends donate to public stool banks like OpenBiome in Massachusetts. However, some patient advocates and doctors were concerned about how these changes might reduce access to treatment. In 2016, another draft guideline was released suggesting that for rCDI, clinicians should use public stool banks under an IND. No final decisions were made after the public comment period ended.

Many experts suggest that classifying FMT as a human cell, tissue, or cellular- or tissue-based product (HCT/P) could help balance safety and access. Under this classification, treatments are regulated differently depending on how they are used. But making FMT available while keeping it safe continues to require careful planning. It should be explained clearly to the patient that FMT is under experimental phase and its long-term consequences are not yet completely understood. If FMT is to be employed in the treatment of any other condition (such as IBD or obesity), an IND application is mandatory.

In Canada, FMT is classified as the novel biologic medicine, which is under experimental phase so any research involving FMT requires a Clinical Trial Application (CTA). According to Health Canada the person struggling from *Clostridioides difficile* infection (rCDI) under strict interim guidelines. The guidelines in Canada are stricter compared to United States, and require doctors to obtain informed consent, maintain detailed medical records, ensure donors test negative for specific infections and microbiome-related illnesses, and confirm that either the patient or their treating healthcare provider personally knows the donor. If this criterion is not met, FMT for rCDI patients must only be performed under an approved Clinical Trial Application (CTA).

In Australia, the Therapeutic Goods Administration (TGA) has deemed fecal microbiota transplantation (FMT) to be an unapproved medicinal product. There are four main ways for patients to obtain FMT: (1) the special access scheme, which allow doctors to prescribe it in rare, necessary cases; (2) the Authorized Prescriber Scheme, which permits physicians to prescribe FMT for a specific indication to a specific patient group; (3) personal importation, which necessitates direct TGA approval if a patient wants to import FMT material for their own use; and (4) clinical trials involving TGA-sanctioned FMT protocols. Along with reviewing current regulations, the TGA met with stakeholders in October 2018 to discuss ways to guarantee that FMT can continue to be safely used for therapeutic purposes in the future.

10.6 Prebiotics, Probiotics, and Postbiotics in Vaccine Enhancement

10.6.1 Probiotics

The word probiotic is derived from the Latin word "pro," that is, "for," and the Greek word "bio," that is, "Life." Probiotics are live microorganisms that, when administered in adequate amounts, confer a health benefit on the host. This term was originally coined by Werner Kollath in 1953 to denote biologically active entities. The use of probiotics dates back thousands of years, but it was not until 1907 that the concept of beneficial microorganisms, now known as "probiotics," was formally introduced. He studied the effect of consumption of fermented products, such as yogurt, on human health. He also conceptualized that the food we consume influences our intestinal microflora, and changes in our diet can help replace pathogenic microbes in our gut with beneficial ones (Fig. 10.10).

10.6.2 Probiotics as Immune Modulators

Probiotics are widely recognized agents for influencing gut microbiota and enhancing host health. These probiotic microorganisms operate through various mechanisms that frequently work in concert. The primary mechanisms include the modulation of the immune system, resistance to colonization, enhancement of the intestinal barrier, and the production of metabolites that function both locally (such as antimicrobials, enzymes, and organic acids) and systemically (including neurochemicals and hormones). There is substantial evidence supporting the safety and effectiveness of several probiotics, including *Lactobacillus* spp., *Bifidobacterium* spp., and *Saccharomyces* spp. Additionally, other promising candidates such as *Roseburia* spp. and *Faecalibacterium* spp. are emerging.

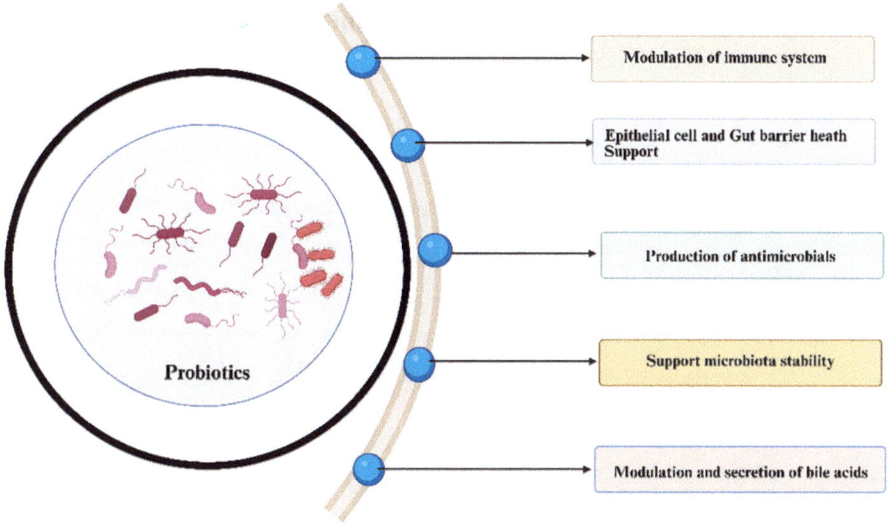

Fig. 10.10 Probiotics promote host health by changing the immune system, improving gut barrier and epithelial integrity, and producing antimicrobial substances. They also influence host metabolism and help stabilize the gut flora by modulating and secreting bile acids

Increasing evidence suggests that probiotics administered orally may assist patients suffering from health issues such as diabetes, cardiovascular diseases, or neurological disorders. Nevertheless, high-quality randomized controlled clinical trials are essential to gain a clearer understanding of which strains, formulations, and dosing regimens are beneficial for specific conditions. Furthermore, the administration of probiotics by mothers during or after pregnancy also alters the microbiota composition of both the fetus and newborns, respectively. These microbe-based substances are mainly found in food items or dietary supplements that provide nutritive benefits. Critical physical and chemical barriers that maintain immunological homeostasis in the intestines, mucosal surfaces are the initial point of contact between the immune system and the microbial flora of the gut. The mucus layer, produced by goblet cells (GCs), serves to keep the gut microbiota from contacting the intestinal epithelium, creating spatial segregation while sustaining microbial surveillance, and is an example of such a barrier. The intestinal epithelial cells have different PRRs, which can recognize MAMPs on the bacterial surface, triggering a series of responses, such as the formation of AMPs, which inhibit bacterial proliferation and maintain mucosal integrity.

TGF-β and RA, two substances secreted by the IECs, promote the development of tolerogenic dendritic cells (DCs) in the lamina propria. Immune tolerance is produced by these DCs, which act as antigen-presenting intermediaries and promote T cell development into regulatory T cells (Treg). B cells produce IgA after differentiating into plasma cells (PCs). In order to link innate and adaptive immune responses, IgA subsequently moves through the epithelium and into the mucous

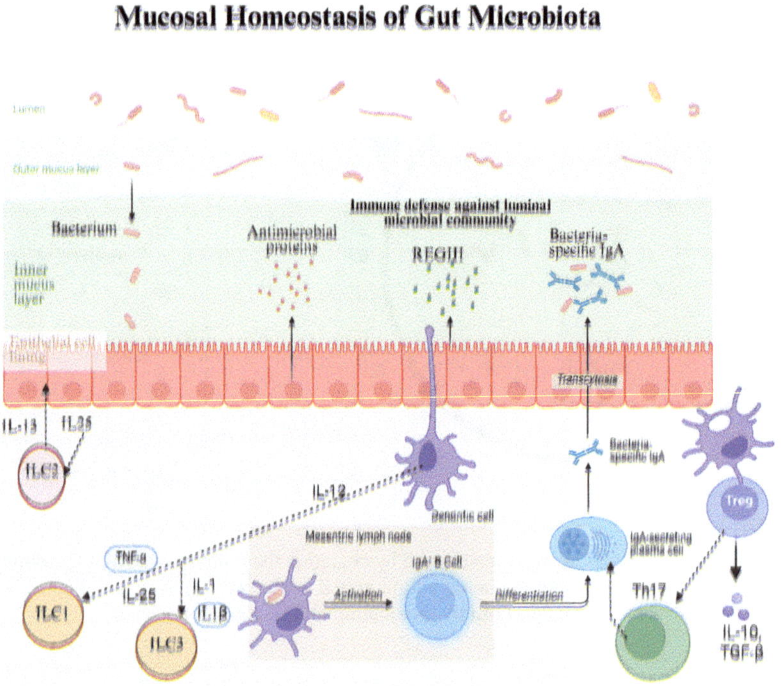

Fig. 10.11 The early defense against luminal microbes is regulated by epithelial cells, antimicrobial proteins, and bacterium-specific IgA. Regulation of B cell activation and IgA production is controlled by dendritic cells and innate lymphoid cells, while protective immunity and tolerance are regulated by Th17 and Treg cells. All these coordinated mechanisms regulate gut microbial and immune homeostasis

layer, where it controls bacterial adhesion to host tissues. Activated by signals such as flagellin, macrophages release IL-23, which activates ILC3 to secrete IL-22. In response to IL-22, IECs secrete and release the antimicrobial peptide RegIIIγ, further supporting local defense against microbial invasion. ILC2 also helps regulate mucus by secreting IL-13 a cytokine that stimulates intestinal epithelial stem cells to differentiate into goblet cells, which then produce mucin glycoproteins, again strengthening the mucus barrier. There are several types of probiotics that used as an alternative for vaccine such as probiotic-derived polyphosphate, etc. (Ciabattini et al. 2019) (Fig. 10.11).

10.6.2.1 Probiotic-Derived Polyphosphates

Probiotic-derived polyphosphate is a multifunctional molecule produced by *Lactobacillus brevis* that plays an essential role in strengthening the intestinal epithelial barrier. It increases the production of the protective protein HSP27 in epithelial cells through integrin p38 MAPK signaling, thereby enhancing mucosal integrity and maintaining epithelial homeostasis. Polyphosphate reduces oxidative stress-induced damage to the gut barrier, stabilizes the cytoskeleton, and suppresses inflammation.

These combined effects help protect the gut lining against environmental stressors and potential pathogens, particularly under gut inflammatory conditions. Thus, probiotic-derived polyphosphate is a promising alternative to vaccines for supporting mucosal health and defense, on the other hand, traditional probiotics also referred to as first-generation probiotics, which is mainly obtained from fermented foods, such as kefir, yogurt, sauerkraut, kimchi, and miso. These foods naturally contain a variety of bacteria that result from fermentation. *Lactobacillus* and *Bifidobacterium* strains are among the most widely employed in dietary supplements and functional foods due to their well-established safety and health benefits. For example, *Lactobacillus rhamnosus* GG has been shown to enhance mucosal immunity, a critical component in the defense against respiratory and enteric infections.

Traditional probiotic bacteria like *Lactobacillus* and *Bifidobacterium* have a crucial function in gut-related immune-responses by interacting with the gut-associated lymphoid tissue (GALT). This interaction modulates not only innate but also adaptive immune mechanisms. These microbes stimulate immune cells like dendritic cells and macrophages, and augment the release of proinflammatory cytokines like IL-12, IL-6, and TNF-α, leading to stronger mucosal immunity. These probiotics compete with pathogens for adherence to the intestinal walls. This competitive exclusion inhibits pathogen colonization and toxin production, thereby playing a vital role in preventing infections and post-vaccination gut health maintenance (Fig. 10.12).

10.6.2.2 Next-Generation Probiotics

Many next-generation probiotics (NGPs), like *F. prausnitzii*, lead to the formation of important short-chain fatty acids such as butyrate, acetate, and propionate. These SCFAs promote immune balance and intestinal well-being. SCFAs can directly inhibit pathogenic growth by lowering ph. They also provide anti-inflammatory effects by blocking histone deacetylases and stimulating G-protein coupled receptors in intestinal epithelial and immune cells (Al-Fakhrany and Elekhnawy 2024).

10.6.3 Prebiotics and Gut-Brain-Immune Axis

Prebiotics

Prebiotics were first introduced in 1995. They are fermented by gut microbes and help improve the activity and composition of gut microbiota. This modification in the gut microbial community is beneficial for the health of the host. SCFAs and lactic acid are produced during the fermentation process and enhance gut health.

Impacts of Probiotics and Prebiotics on the CNS

Galactooligosaccharides promote the growth of *Bifidobacteria* and *Lactobacilli*. Additionally, GOS encourages the proliferation of Bacteroidetes, Firmicutes, and Enterobacteria. A specific type of starch, known as resistant starch (RS), is capable of withstanding digestion in the upper gut. During the fermentation of prebiotics by

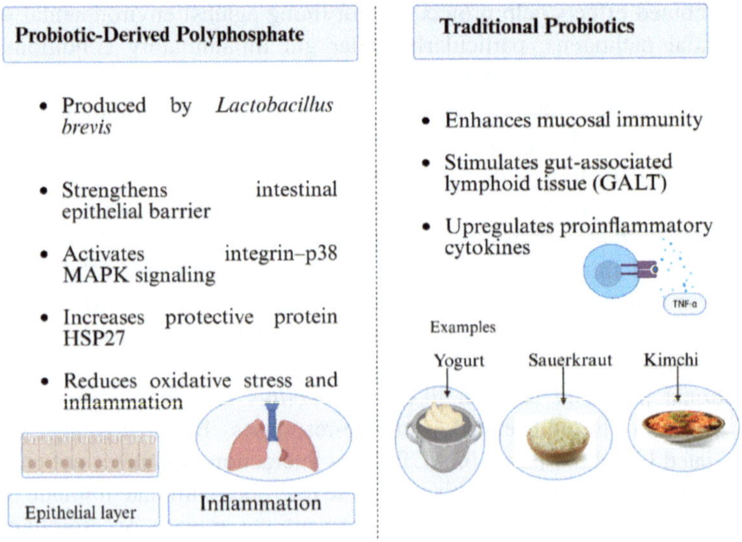

Fig. 10.12 Probiotic-derived polyphosphate versus traditional probiotics. *Lactobacillus brevis* polyphosphate strengthens epithelial barrier function, initiates MAPK signaling, enhances HSP27, and inhibits oxidative stress and inflammation. Traditional probiotics such as those found in yogurt, sauerkraut, and kimchi act mainly by the stimulation of gut-associated lymphoid tissue (GALT), enhancing mucosal immune function, and upregulating proinflammatory cytokines. (Corrales et al. 2025)

gut microbiota, RS generates significant amounts of short-chain fatty acids (SCFAs), like acetic acid, butyric acid, and propionic acid. These SCFAs are crucial for preserving the health of both the gut and the brain. One of their vital roles involves the activation of G-protein-coupled receptors (GPCRs), such as GPR41, GPR43, and GPR109A. These receptors are located in both the enteric nervous system (ENS) and the central nervous system (CNS). Particularly in the striatum, GPCRs assist in regulating immune responses by sustaining T cell homeostasis and mitigating inflammation within the brain. Thus, the presence of prebiotics enhances SCFA production, which subsequently stimulates the activation of GPCRs, which in turn supports metabolic and immune balance in the body (Ansari et al. 2023).

Gut-Brain-Immune Axis

The gut, microbiota, and brain are connected through a network of signals which involves hormonal (endocrine), humoral, and neural pathways. Along with these pathways, there is a distinctive nervous system within the gut, known as the enteric nervous system (ENS) that is commonly called as the "second brain," which maintains a direct and constant link with the central nervous system through blood flow and vagus nerves. This complex neuronal network is embedded in the walls of the gastrointestinal tract and remains separated from gut microbes by a protective layer of mucous-producing cells, preventing direct microbial contact. However, the

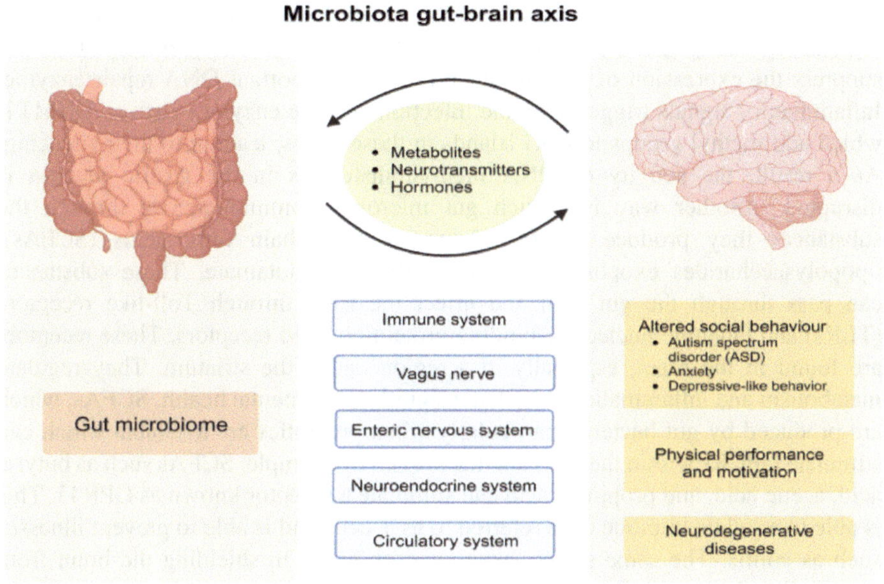

Fig. 10.13 The microbiota gut—brain axis. Gut microbiota communicates to the brain through metabolites, neurotransmitters, and hormones, influencing many pathways including the immune system, vagus nerve, enteric nervous system, neuroendocrine system, and circulation. These interactions affect social behavior, mood disorders, physical performance, motivation, and lead to neurodegenerative disease development

intestinal microbiota can still communicate indirectly with the ENS by transmitting signals from the gut lumen to the lamina propria, often via microfold cells or dendritic cells. Direct microbial invasion into this region can damage intestinal tissues, leading to ulceration or perforation (Ansari et al. 2023) (Fig. 10.13).

A system for bi-directional communication which establishes a connection between, the gut and the brain which involves the interaction of the gut microbiota, the ENS, the immune system, and the enteroendocrine system. This interaction fosters the process of signal transmission to the central nervous system via blood flow and vagus nerves. The equilibrium of this communication is maintained by diverse molecules such as neurotransmitters, hormones, and cytokines and the whole process is performed with the assistance of gut microbiota which is a vital constituent of the gut-brain axis (GBA). This axis communicates through both direct and indirect pathways, which can influence gene expression without changing the DNA sequence with the help of a process known as epigenetic reprogramming and it is maintained through modification of histone tail, methylation of DNA and noncoding RNA. All such changes in the composition of the microbiota can trigger epigenetic regulation which profoundly impacts behavior. For example, according to the study conducted by Sepulveda (2013) *Helicobacter pylori* is an infectious disease caused by the bacterium *H. pylori*, which induces gastritis through epigenetic alterations. One such change is the excessive methylation of DNA in gastric epithelial cells, particularly in regions regulating genes responsible for DNA repair. This hypermethylation can suppress the expression of the MGMT protein, an important DNA repair enzyme. Inflammatory signals triggered by the infection activate enzymes such as DNMT1, which add methyl groups to CpG islands in these genes, leading to their silencing. As a result, the activity of DNA methyltransferases in the gastric mucosa is disrupted. Another way by which gut microbes communicate is through the substances they produce. These substances short-chain fatty acids (SCFAs), lipopolysaccharides exopolysaccharides (EPS) and glutamate. These substances can pass through the gut wall and affect the ENS through Toll-like receptors (TLRs) and Guanine nucleotide-binding protein-coupled receptors. These receptors are found in the brain, especially in a region called the striatum. They regulate metabolism and inflammation, which are significant in mental health. SCFAs, which are produced by gut bacteria, particularly when prebiotics are available which can stimulate GPCRs within the ENS and the brain. For example, SCFAs such as butyric acid, acetic acid, and propionic acid can stimulate a receptor known as GPR43. This is able to regulate immune cells referred to as T cells and is able to prevent illnesses such as colitis. The same effects have been observed in shielding the brain from mental disorders and inflammation (Ansari et al. 2023).

TLRs are receptors that react to chemicals such as EPS and LPS produced by gut microbes. Upon activation, they assist the immune system and ENS in releasing beneficial chemicals such as cytokines and neurohormones. These fortify both the gut and the brain. For instance, when TLR2 is stimulated, it ensures that the ENS stays healthy by causing more glial cells and neurons to be present. It also facilitates the release of a unique protein known as glial cell line-derived neurotrophic factor, which aids in the growth and survival of neurons.

Studies have found a correlation between the gut microbiome and gene expression, especially in regions of the brain that control mood and brain-related disorders. For example, when mice were given *Lactobacillus rhamnosus* as a dietary supplement, the transcription and gene expression of GABA subunits across certain brain regions could be regulated. Prebiotic supplements can lead to specific modifications in the composition and activity of the gut microbiota, helping to improve neurobehavioral abnormalities through epigenetic changes as a result, phenotypic attenuation has been reported.

10.6.4 Postbiotics and Their Role in Mucosal Immunity

Postbiotics are biologically active compounds produced by probiotics during fermentation and bacterial metabolism and they provide significant health benefits without the presence of live microorganisms. Postbiotics is defined as "crude cell extracts that benefit the human or animal consumer when administered orally or topically in sufficient amounts". Postbiotic ingredients derived from probiotics are thought to be promising alternative supplements to eradicate the risks associated with live attenuated vaccines or probiotic. For example, Bacteriocins are small,

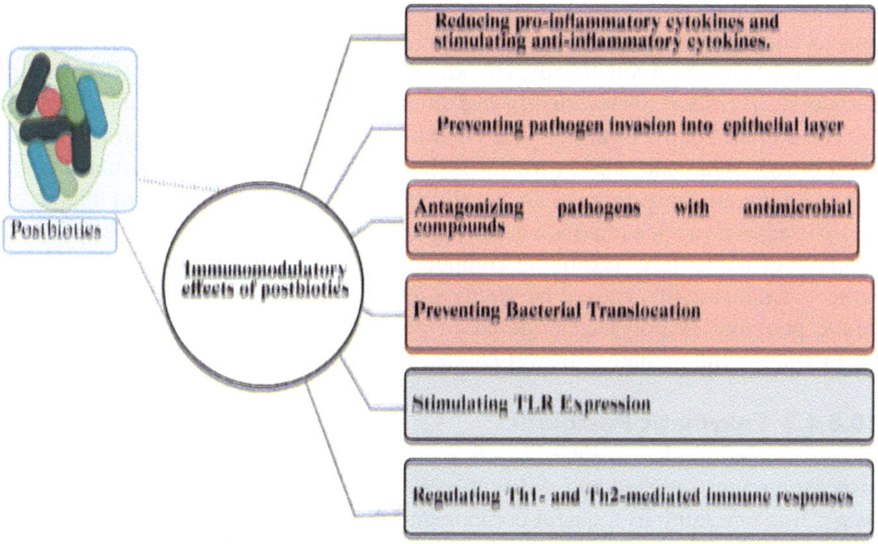

Fig. 10.14 Postbiotics' immunomodulatory action. By inhibiting pro-inflammatory cytokines, augmenting anti-inflammatory responses, and stemming bacterial or pathogen transmigration, postbiotics exert immune-regulatory effects. They augment host defense and mucosal homeostasis via action by antimicrobial substances, Toll-like receptor (TLR) expression induction, and Th1/Th2-mediated regulation of immune response

proteinaceous, ribosomally synthesized antimicrobial substances produced by bacteria which kills or limit microorganisms of the same species or related ones. They are produced by indigenous microbiota, such as the probiotic bacterium *Lactobacillus plantarum* (Mojgani et al. 2025).

Bacteriocins mainly target the cell membranes or cell wall synthesis of harmful bacteria, thereby killing cells without influencing useful microbiota. These peptides evince strong antimicrobial activity against both Gram-positive as well as Gram negative pathogens. Some strains of commensal *Escherichia coli* are capable of producing bacteriocins, which directly inhibit pathogenic enterohorrhagic *Escherichia coli* (EHEC) (Fig. 10.14).

10.6.4.1 Polysaccharides-Based Postbiotics

Polysaccharides are long chains of sugars made of repeating units of monosaccharides, disaccharides and they are mainly present in a variety of dietary sources, like fruits, vegetables, mushrooms, algae, and fungi. For example, cellulose is a polymeric polysaccharide made up of only one type of monosaccharide, which is glucose. Some commonly found monosaccharides are glyceraldehyde, sialic acid, galactose, mannose, fructose, arabinose, galacturonic acid, etc. Disaccharides are formed by the cluster of these simple-sugar molecules. Polysaccharides, which may be linear or branched, can contain hundreds to thousands of repeating sugar units, most of which are six-carbon sugars (hexoses).

One notable polysaccharide-based postbiotic is *Fucoidan,* a sulphated polysaccharides-based postbiotics rich in L-fucose, present in algal cell wall. Fucoidan exists in various forms, including *fucoidan extract fraction* (FeF) and its derivative, *fucoidan enzymatically hydrolyzed modified polysaccharide* (FeHMP) (Krylova et al. 2020). Both FeF and FeHMP are bioactive polysaccharides derived from brown algae and are known for their immunomodulatory properties and can stimulate immune responses in a natural way by interacting with the gut microbiota hence used as a vaccine alternative. Polysaccharide A from *Bacteroides fragilis* is an example of a structural postbiotic compound with anti-inflammatory properties. It protects against colitis caused by *Helicobacter hepaticus*, demonstrating the therapeutic potential of polysaccharide-based postbiotics in managing gastrointestinal disorders like Inflammatory Bowel Disease.

10.6.4.2 Enzymatic Postbiotics

Enzymatic postbiotics include bioactive enzymes such as *Lactocepin*, is a serine protease enzyme secreted by LAB such as *Lactobacillus paracasei* and *Lactococcus lactis*. It is emerging as a useful component in microbiota-based vaccine strategies due to its ability to degrade specific pro-inflammatory chemokines like IP-10 (CXCL10). This action helps reduce intestinal inflammation and offers a precise way to regulate immune responses in chronic diseases, just like Inflammatory Bowel Disease (IBD). By cleaving chemokines such as IP-10 and Eotaxin (CCL11), lactocepin prevents excessive recruitment of immune cells to inflamed tissues. In disorders like IBD, the overexpression of these chemokines leads to uncontrolled immune activity, causing tissue damage and delayed healing. Lactocepin targets these molecules at the site of inflammation, helping restore immune balance without affecting the overall immune system. Purified lactocepin or its active fragments can be developed into postbiotic formulations and used alongside oral vaccines. These formulations function as mucosal adjuvants, which means they enhance the immune response at mucosal surfaces. By regulating local immunity and reducing side effects, lactocepin helps improve the efficacy and safety of vaccines administered through the gastrointestinal route.

10.6.4.3 S-Layer Proteins

In many bacteria and nearly all archaea, the outermost surface is covered with a proteinaceous layer, which is commonly referred to as Surface-layer. This layer consists of one or more glycol-proteins, which are called S-layer proteins (SLPs and these self-assemble to present a regularly spaced array on the cell surface. *Bacillus anthracis* and *Clostridioides difficile* have good examples of S-layer proteins and their encoding genes that produce them are found in *Bacillus anthracis and Clostridioides difficile* Surface-layer proteins (Slps) from *Lactobacillus acidophilus* NCFM are biologically active molecules that can be administered in controlled doses. Unlike the administration of entire microbial communities, which may disturb the normal balance of bacteria in the body, Slps are specific and safe which modulate immune responses by interacting with dendritic cells and T lymphocytes. These proteins suggest a mechanism for probiotics to influence host immunity without colonization.

However, Slps have been shown to reduce inflammation induced by lipopolysaccharide (LPS), a bacterial component known to trigger immune responses. This anti-inflammatory effect is achieved with inhibition and production of proinflammatory cytokines such as reactive oxygen species (ROS), nitric oxide (NO), IL-1β, prostaglandin E2 (PGE2), and TNF-α. S-layer proteins act by inhibiting key intracellular signaling pathways, including MAPK and NF-κB, which are central to the activation of inflammation-related genes.

These properties make Slps as promising candidates for microbiota-based vaccine alternatives. Although the carbohydrates fermentation mediated through gut microbes produces short-chain fatty acids such as butyrate and acetate. These SCFAs help control the growth and spread of pathogens like *Salmonella enterica,* serovar Enteritidis and *Salmonella typhimurium.* They also modulate inflammation by suppressing proinflammatory cytokines and reinforcing epithelial defense.

10.6.4.4 Secretion of Antimicrobial Peptides and Toxins

Indigenous microbes produce substances known as *bacteriocins* and *microcins,* which are antimicrobial peptides that specifically target and inhibit other closely related bacterial strains. For example, Escherichia coli Nissle secretes microcins that exploit iron-uptake receptors on pathogens (a "Trojan horse" strategy) to deliver lethal toxins. Additionally, commensals may express *Type VI Secretion Systems (T6SS)* to inject toxins into competitors (Fig. 10.15).

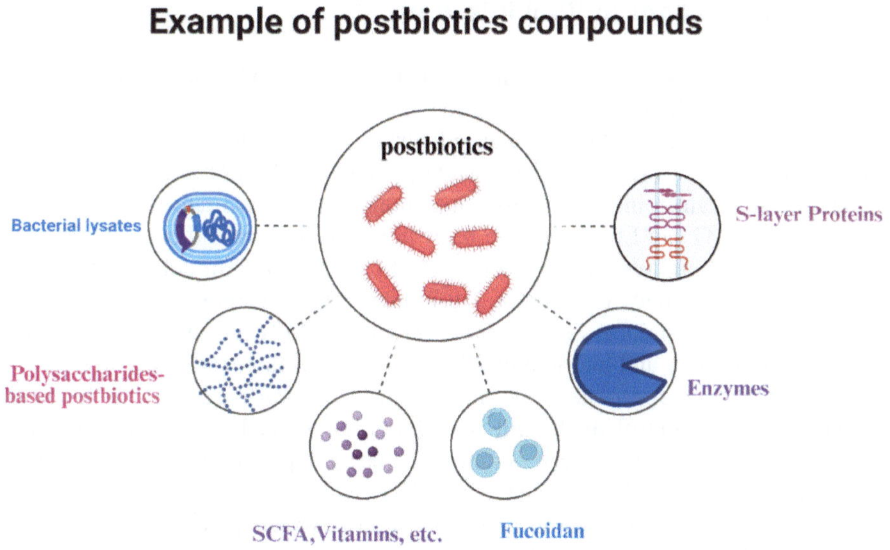

Fig. 10.15 Postbiotic compound examples. Bacterial lysates, polysaccharide-rich fractions, short-chain fatty acids (SCFAs), vitamins, fucoidan, S-layer proteins, and enzymes are a few of the bioactive molecules that are classified as postbiotics. These molecules enhance host health through the modulation of immunity, barrier function enhancement, and antibacterial and metabolic activity

10.6.4.5 Enhancing Innate Immunity

Beneficial microbes stimulate immune cells to produce: antimicrobial peptides like REG3γ, REG3β, and α-defensins (from Paneth cells), which inhibit pathogen colonization. Cytokines like IL-1β and IL-22, promoting neutrophil recruitment and epithelial defense. SCFA-mediated signaling, which interacts with epithelial receptors and enhances epithelial tight junctions and immune signaling. For example, *Lactobacillus reuteri* produces indole-3-aldehyde from tryptophan, which activates the aryl hydrocarbon receptor (AHR) on innate lymphoid cells (ILCs), stimulating IL-22 production and antimicrobial activity.

10.6.4.6 Promoting Adaptive Immunity

Microbes play an important role in training the adaptive immune system: Th17 cells induced via segmented filamentous bacteria (SFB), these cells produce IL-17 and IL-22 and enhance resistance against enteric pathogens. Regulatory T cells (Tregs): promoted by *Clostridium spp.* and *Bacteroides fragilis*, these cells produce IL-10 to maintain immune tolerance and reduce inflammation. Secretory IgA (sIgA): commensals trigger dendritic cells and B cells to produce sIgA, which coats microbes and limits their access to epithelial surfaces and sIgA also maintains gut homeostasis and microbiota composition.

10.7 Microbiota-Based Strategies in Infectious Disease Prevention

10.7.1 Applications in Viral Infections

Microbiota is an essential function in maintaining health and a balanced microbiota that is important in infection prevention, particularly in premature infants and immunocompromised children. Balanced status is needed for the development of the immune system. Probiotics *Bifidobacteria* and *Lactobacillus* have shown beneficial effects against influenza virus infections of the respiratory tract. Strains *Lactobacillus* GG and Lb. casei have been shown to prevent and control microbial infections in gut and respiratory tracts caused by influenza. Bifidobacterium breve has been shown in mice to increase anti-influenza IgG, thereby protecting against respiratory tract infections. *Lactobacillus rhamnosus* M21 has also been shown to increase interferon-γ and interleukin-2 levels in mice with pneumonia.

Systemic immune responses and distant mucosal surfaces such as the lungs are under significant control of the gut microbiota. Of special interest, the probiotic *Lactobacillus plantarum* DR7 strain has been reported to inhibit harmful pro-inflammatory cytokines such as IFN-γ and TNF-α and increase anti-inflammatory cytokines such as IL-4 and IL-10, which potentially protect the lungs in viral infections such as COVID-19. Oral probiotics may be of tremendous importance in the majority of COVID-19 patients. These strains increase the integrity of the tight junction by increasing short-chain fatty acids (SCFAs) such as butyrate, which is theoretically capable of blocking the entry of SARS-CoV-2.

10.7.2 Bacterial Translocation and Alterations of Gut Microbiota in HIV Infection

The gastrointestinal tract plays a central role in HIV replication, regardless of how the virus is acquired. A large concentration of CD4+ T lymphocytes is located in the gut-associated lymphoid tissue (GALT). Unlike the peripheral blood, these gut-resident T cells are memory cells that remain in an activated state, making them highly vulnerable to HIV. Consequently, HIV infection is often regarded as a disease of the gut, as it causes substantial depletion of CD4+ T cells from mucosal surfaces, especially within the GALT. The early stage of HIV infection is characterized by the selective loss of CD4+ coreceptor-bearing cells CCR5 and CXCR4 within the gut, but it later involves all body CD4+ T cells. After antiretroviral therapy (ART), significant CD4+ cell recovery occurs in the peripheral blood, which is different from the slow and partial CD4+ cell recovery within the gut. The main cause of inadequate immune restoration in the gut during HIV infection remains unclear. One contributing factor is the infection of macrophages within the intestinal mucosa, which facilitates viral spread through direct cell-to-cell contact. Additionally, cytokines secreted by infected macrophages attract and activate CD4+ T lymphocytes, making them highly susceptible to further infection thereby intensifying the viral impact at the gut level.

HIV also significantly reduces the number of mucosal Th17 cells, which are vital for antimicrobial defense. The depletion of gastrointestinal CD4+ T lymphocytes lead to structural disruptions in lymphoid tissues and compromises both the integrity and functionality of the mucosal barrier. This loss of immune protection allows microbial molecules especially lipopolysaccharides (LPS) to migrate across the epithelial barrier into the lamina propria and, eventually, into systemic circulation. This phenomenon, known as bacterial translocation, was first documented in HIV-infected individuals in 2006 and is now recognized as a persistent feature throughout the course of infection.

Bacterial products such as LPS trigger an immune response by engaging toll-like receptors (TLRs), promoting proinflammatory cytokine production. Other receptors including macrophage scavenger receptors (MSR), potassium (K^+) channels, and CD11/CD18 complexes also identify LPS and contribute to inflammation. Binding of LPS to these receptors activates kinase pathways and leads to the mobilization of NF-κB, a transcription factor that drives the expression of inflammatory genes such as TNF-α, IL-1, and IL-6.

10.8 Challenges and Limitations

Research on the microbiome has revealed significant links between disease states and microbial ecosystems. However, a number of challenges and barriers in the field make it challenging to interpret results, apply them widely, and use them in clinical settings. These problems are mostly caused by the intricate and ever-changing relationships that exist between microbial communities and their human hosts.

10.8.1 Individual Microbiome Variability

The human foetus grows in a sterile environment, during birth it comes in contact with microbes. The mode of delivery is also vital in shaping the initial microbial communities that colonize the gut. For example, vaginally delivered infants possess *Lactobacillus, Prevotella*, or *Sneathia* species-dominant bacterial communities on their skin, oronasopharyngeal mucosae, and meconium, which are comparable to those of the mothers. This phenomenon is foundational to our understanding of microbiome development but presents significant challenges in study design and participant selection. The route of delivery is a major predictor variable that must be carefully considered and controlled for in microbiome research, as it introduces early and substantial differences in microbial colonization. Failure to account for these initial patterns can lead to confounding factors that complicate comparisons across study participants. Variability in methodologies employed is the major challenging task. Both 16S rRNA gene sequencing and WGS techniques have limitations that are greatly slowing the progression of basic research to therapeutic application.

10.8.2 Stability and Colonization of Microbial Strains

Microbial colonization and stability in the human body are shaped by a number of biological and environmental factors. *Bacteroides fragilis* is one of the species that exhibits colonization resistance in the gut; once a germ-free mouse is colonized with a single strain, it cannot be colonized by the same species again. By using the Commensal Colonization Factors (CCF) system- a molecular toolkit that permits stable attachment to gut surfaces, particularly colonic crypts- *Bacteroides* overcome this restriction. In contrast, CCF-deficient mutants fail to occupy these niches, showing poor colonization and increased vulnerability to environmental stress. Furthermore, in hosts with complex microbial ecosystems, such as specific pathogen-free mice, introducing new strains becomes more difficult due to competition with resident microbes. Similarly, in the vaginal microbiome, microbial composition and stability vary widely across individuals and life stages. In healthy women of reproductive age, dominance by *Lactobacillus* species such as *L. crispatus* or *L. iners*, helps maintain vaginal health by creating an acidic environment that inhibits pathogen growth. However, this pattern differs by ethnicity; only about 50% of Black or Hispanic women exhibit *Lactobacillus* dominance, highlighting population-level variability. Moreover, factors like menstruation, sexual activity, and pregnancy introduce fluctuations in microbial stability. Some women maintain consistent microbial profiles, while others experience frequent shifts not directly linked to disease. For example, bacterial vaginosis (BV), marked by the depletion of *Lactobacilli* and overgrowth of anaerobes like *Gardnerella vaginalis*, is associated with increased susceptibility to sexually transmitted infections and adverse pregnancy outcomes. However, treatment with antibiotics does not always prevent complications.

Together, these examples from the gut and vaginal microbiomes illustrate that microbial colonization and resistance are influenced by niche specificity, host environment, and individual variation. Molecular mechanisms like the CCF system and host factors like hormonal cycles or immune responses must be considered to understand microbial dynamics and design effective microbiota-based therapies.

10.8.3 Biosafety Concerns of Live Bacterial Vectors

Despite lactic acid bacteria (LAB) being generally recognized as safe (GRAS status) for humans and animals, live vaccine strains should not be considered avirulent. Biosafety assessments are crucial and include:

Antibiotic Resistance and Horizontal Gene Transfer There are concerns about the detection and potential horizontal transfer of antibiotic resistance genes between microbes, even though some resistances (like vancomycin resistance in *Lactobacillus*) are chromosomally located and less easily transmitted. The widespread use of antibiotic selection markers in genetic modification further complicates this, necessitating the development of alternative food-grade marker.

10.8.4 Strain-Specific Variability

The efficacy of microbiota-based strategies is highly dependent on the specific disease etiology and the probiotic strain used. Different LAB strains exhibit varied immunomodulatory characteristics, affecting cytokine profiles and immune responses. Optimal genetic sequences for antigen production can differ significantly even between strains. It is unlikely that a single strain will be an "ideal vector" for delivering antigenic molecules to different hosts or even different mucosal cavities within the same host.

10.8.5 Limited Research on Parasitic Infections

Most investigations into postbiotics have historically focused on infections of bacterial origin, rather than those of parasitic origin, indicating a gap in current knowledge and application for protozoa like *Entamoeba histolytica.*

10.8.6 Considerations for Clinical Development and Trial Design

Developing microbiome treatments is complicated because a person's microbiome can vary when disrupted, and the specific mix of bacteria delivered clinically (e.g., via fecal microbiota transplantation or FMT) cannot be predicted. These studies have several factors that are difficult to control. Moreover, selecting an appropriate patient group and accurately identifying eligible participants for clinical trials can be challenging.

10.8.7 Regulatory Frameworks

Policy infrastructure is still changing to keep up with microbiome-based medicines, and there is still no clear way to test these therapies. The U.S. FDA has put out draft guidelines and said it is open to working with the industry, but it is still unclear if this will lead to the creation of a new set of rules just for microbiome-based treatments. The FDA has given orphan drug status to two microbiome therapies that are currently in phase III trials for recurrent *Chloridoids difficile* infection (rCDI). Manufacturers may charge high prices to make up for the money they spent on these complicated and expensive treatments.

10.9 Future Perspectives and Innovations

10.9.1 Synthetic Biology and Next-Generation Probiotics

Synthetic biology is a transformative discipline that combines principles of engineering with biological sciences, applying concepts such as abstraction, modular design, and standardization to create and construct a new biological system. Biology incorporates a wide range of scientific techniques and approaches. Through synthetic biology, scientists can create engineered systems that produce useful products such as ethanol, pharmaceuticals, and even complete synthetic organisms. For example, complex bacteria can be designed to digest and neutralize toxic chemicals, offering solutions for environmental cleanup and industrial sustainability.

10.9.1.1 Applications of Synthetic Biology

Synthetic biology provides emerging technologies for bioremediation, biosensing, and ecosystem recovery in the environmental arena. Cleanup processes can be precise and effective by directing engineered microbes to break down toxic substances such as plastics, heavy metals, and oil-based contaminants. Moreover, environmental pollutants such as nitrates or arsenic can be detected in real-time, in situ, by integrating synthetic biosensors with microbial system. Synthetic biology offers tools to rebalance microbiome dysbiosis, helping prevent or treat various diseases. Genetically engineered bacteria are being developed for delivering targeted therapies in conditions like cancer, inflammatory bowel disease (IBD), diabetes, and HIV.

10.9.1.2 Next-Generation Probiotics

The term of next-generation probiotics (NGPs) was suggested by O'Toole PW in 2017. These are live bacteria identified through comparative analysis of the human gut microbiota, and when administered in appropriate amounts, they offer health benefits. Unlike vaccines, the live microbes present in biological formulations are designed for disease prevention in humans. Notable examples include *Akkermansia muciniphila*, *Faecalibacterium prausnitzii*, *Bacteroides* species, *Eubacterium hallii*, and another species of *Clostridium* cluster IV. Since these organisms originate from the human gut, thorough safety evaluation is essential before their use. Furthermore, the development of advanced techniques to incorporate these NGPs into functional food products is an urgent priority (O'Toole et al. 2017).

10.9.1.3 Role of Next-Generation Probiotics (NGPs)

Next-generation probiotics (NGPs) are emerging as powerful agents in cancer prevention and therapy, with documented potential against colorectal, gastric, hepatocellular, breast, and cervical cancers. Their mechanisms include enhancing epithelial barrier integrity, modulating inflammatory signaling, inducing cell death in cancer cells, regulating cytokine profiles, and producing immunomodulatory metabolites like acetate, butyrate and propionate etc. Specific strains such as *Lactobacillus reuteri*, *L. acidophilus*, *L. plantarum*, *L. rhamnosus*, *L. casei*, *Streptococcus thermophilus* and *Bifidobacterium bifidum* have demonstrated anticancer effects both in vitro and in vivo by lowering colonic pH, inhibiting carcinogenic enzymes, suppressing growth of tumor, and influencing oncogene expression. Species like *Akkermansia muciniphila*, *Faecalibacterium prausnitzii*, *Bacteroides thetaiotaomicron*, and *Christensenella minuta* show promise in improving immunotherapy responses through T-cell modulation. *Butyricicoccus pullicaecorum* is noted for its production of butyrate, which regulates tumor suppressor genes and reduces inflammation-linked colorectal cancer risk. Additionally, *L. crispatus* and *L. acidophilus* have shown antiproliferative activity against breast cancer cells, while *L. casei* may be linked to lower breast cancer incidence with long-term exposure. Novel approaches even explore the use of *Salmonella* species to target tumors, deliver cancer-specific toxins or antigens, and convert prodrugs within the tumor microenvironment. Although these findings are encouraging, well-designed clinical trials remain essential to establish the safety, underlying mechanisms, and therapeutic effectiveness of microbial interventions in treating human cancers (Abouelela and Helmy 2024).

10.9.2 Personalized Probiotic Therapies

The field of next-generation probiotics (NGPs) is progressing with the help of new and advanced molecular strategies to improve probiotic therapies. One important area is the creation of personalized probiotics, based on the fact that each person's gut microbiome is different. Using techniques like metagenomic sequencing, scientists try to find specific bacterial strains that can give the good health benefits to each individual. *Multiomics* approaches integrate genomic, transcriptomic, metabolomic, and proteomic data to understand how microbial interactions effect the host physiology. Another important method is the use of synthetic biology and gene editing to make probiotics work better. These new technologies help in designing probiotics that can survive longer in the digestive system, reach the right part of the gut, and even produce helpful substances. Scientists are also studying how probiotics can work better when used along with prebiotics, postbiotics, and regular medicines. At the same time, new delivery systems like nanoparticles and time-release capsules are being developed to make sure that probiotics reach the right place in the gut and stay active for a long period of time. These developments are expected to create special types of NGPs that can help treat disease condition like inflammatory bowel disease (IBD), allergies, and disorders related to metabolism. This shows a big step forward in using personalized and accurate methods to improve gut health.

10.9.3 Integration with mRNA and Nanoparticle Technologies

The integration of mRNA and nanoparticle technologies offers promising directions for bacterial vaccine development, especially since many bacteria evade both innate and adaptive immunity. Lipid nanoparticles (LNPs) are the primary delivery systems, while newer carriers like NLCs, CNEs, CAFs, LION™, and QTAP improve cellular uptake and stability. Future advancements focus on optimizing mRNA platforms using modified nucleotides (e.g., N1-methylpseudouridine), cap analogs, codon usage, untranslated region (UTR) selection, and secondary structure analysis. Novel RNA formats such as self-amplifying RNA (saRNA) and circular RNA have shown effectiveness in vaccines against *Mycobacterium tuberculosis*, *Streptococcus pyogenes*, *Chlamydia trachomatis*, and *Yersinia pestis*. Routes of administration like intramuscular, intranasal, and intradermal are being explored to enhance immune responses. Antigen design now includes multiepitope and multivalent constructs, supported by in silico immune informatics and AI tools to identify safe and immunogenic epitopes. Linker and adjuvant sequences are being refined, and cross-reactive vaccines targeting conserved pathogen motifs are under development. Incorporating epitopes for unconventional MHC molecules such as MHC-E may further improve efficacy. Strategic approaches like heterologous prime–boost regimens and boosting innate immunity are gaining attention. Although preclinical data are encouraging, mRNA vaccines for bacterial infections are still in early stages, with few clinical candidates. Future research must focus on optimizing antigen and adjuvant sequences, validating computational predictions through in vivo studies (including nonhuman primates), and exploring therapeutic combinations with antibiotics.

10.10 Conclusion

The interface interaction between our immune system and microbiota is a new area for vaccines. Instead of relying on traditional vaccinations, which may be less effective for roughly a group of individuals and require advanced delivery or storage systems, microbiota-directed strategies try to enhance immunity by stimulating the body's own good bacteria and their derivatives. To enhance the immune function and allow for safer, more targeted vaccination strategies, they employ probiotics, postbiotics, genetically modified microorganisms, and even bacterial vesicles. There are, though, issues in this new discipline, because no two individuals possess the identical gut flora, the variety of the human microbiome may be the most significant hurdle to universal microbiota-based therapies. The long-term effect of gut microorganism modification, regulatory ambiguity, and safety issues surrounding genetically modified organisms are other issues to be researched carefully and controlled. For all lab-effective solutions to be established beyond reasonable doubt, there also must be large-scale clinical trials. Vaccinations could be made more accessible, personalized, and natural in the future. Science has achieved this with advancements in next-generation probiotics and synthetic biology to develop vaccines that are not only highly effective but also customized to every individual's distinct microbial signature. Microbiota-based options can drive infectious disease prevention and treatment to the next level of investment, sharper regulation, and interdisciplinary coalitions. This would lead us closer to wiser, safer, and human-centered vaccinations.

References

Abouelela ME, Helmy YA (2024) Next-generation probiotics as novel therapeutics for improving human health: current trends and future perspectives. Microorganisms 12(3):430. https://doi.org/10.3390/microorganisms12030430

Ahmed AAQ, Besio R, Xiao L, Forlino A (2023) Outer membrane vesicles (OMVs) as biomedical tools and their relevance as immune-modulating agents against H. pylori infections: current status and future prospects. Int J Mol Sci 24(10):8542. https://doi.org/10.3390/ijms24108542

Al-Fakhrany OM, Elekhnawy E (2024) Next-generation probiotics: the upcoming biotherapeutics. Mol Biol Rep 51(1):505. https://doi.org/10.1007/s11033-024-09398-5

Ansari F, Neshat M, Pourjafar H, Jafari SM, Samakkhah SA, Mirzakhani E (2023) The role of probiotics and prebiotics in modulating of the gut-brain axis. Front Nutr 10:1173660. https://doi.org/10.3389/fnut.2023.1173660

Belkaid Y, Hand TW (2014) Role of the microbiota in immunity and inflammation. Cell 157(1): 121–141. https://doi.org/10.1016/j.cell.2014.03.011

Camacho AGA, Teixeira LH, Bargieri DY, Boscardin SB, Soares IDS, Nussenzweig RS, Nussenzweig V, Rodrigues MM (2011) TLR5-dependent immunogenicity of a recombinant fusion protein containing an immunodominant epitope of malarial circumsporozoite protein and the FliC flagellin of Salmonella Typhimurium. Mem Inst Oswaldo Cruz 106(suppl 1):167–171. https://doi.org/10.1590/s0074-02762011000900021

Cho W-I, Chung M-S (2020) Bacillus spores: a review of their properties and inactivation processing technologies. Food Sci Biotechnol 29(11):1447–1461. https://doi.org/10.1007/s10068-020-00809-4

Ciabattini A, Olivieri R, Lazzeri E, Medaglini D (2019) Role of the microbiota in the modulation of vaccine immune responses. Front Microbiol 10:1305. https://doi.org/10.3389/fmicb.2019.01305

Contreras-Salgado EA, Sánchez-Morán AG, Rodríguez-Preciado SY, Sifuentes-Franco S, Rodríguez-Rodríguez R, Macías-Barragán J, Díaz-Zaragoza M (2024) Multifaceted applications of synthetic microbial communities: advances in biomedicine, bioremediation, and industry. Microbiol Res 15(3):1709–1727. https://doi.org/10.3390/microbiolres15030113

Corrales D, Alcántara C, Monedero V, Zúñiga M (2025) Polyphosphate from lactic acid bacteria: a functional molecule for food and health applications. Foods 14(13):2211. https://doi.org/10.3390/foods14132211

Ellis TN, Kuehn MJ (2010) Virulence and immunomodulatory roles of bacterial outer membrane vesicles. Microbiol Mol Biol Rev 74(1):81–94. https://doi.org/10.1128/mmbr.00031-09

Faintuch J, Faintuch S (eds) (2019) Microbiome and metabolome in diagnosis, therapy, and other strategic applications. Elsevier/Academic

Furusawa Y, Obata Y, Fukuda S, Endo TA, Nakato G, Takahashi D, Nakanishi Y, Uetake C, Kato K, Kato T, Takahashi M, Fukuda NN, Murakami S, Miyauchi E, Hino S, Atarashi K, Onawa S, Fujimura Y, Lockett T et al (2013) Commensal microbe-derived butyrate induces the differentiation of colonic regulatory T cells. Nature 504(7480):446–450. https://doi.org/10.1038/nature12721

Giuliani MM, Adu-Bobie J, Comanducci M, Aricò B, Savino S, Santini L, Brunelli B, Bambini S, Biolchi A, Capecchi B, Cartocci E, Ciucchi L, Di Marcello F, Ferlicca F, Galli B, Luzzi E, Masignani V, Serruto D, Veggi D et al (2006) A universal vaccine for serogroup B meningococcus. Proc Natl Acad Sci 103(29):10834–10839. https://doi.org/10.1073/pnas.0603940103

Han L, Liu X-W, Li W-X, Zang T, Li Z-P, Deng Y, Tan L-L, Liang D-S, Zhang X-J, Shen X-F, Tang S-Q, Liao X-P, Ren H, Zhang C, Jiang G-B, Sun J (2024) Innovative probiotic Escherichia coli nissle 1917-derived outer membrane vesicles coupled microspheres demonstrate anti-inflammatory and barrier-enhancing properties in colitis. Chem Eng J 487:150711. https://doi.org/10.1016/j.cej.2024.150711

Higgs R, Higgins SC, Ross PJ, Mills KHG (2012) Immunity to the respiratory pathogen Bordetella pertussis. Mucosal Immunol 5(5):485–500. https://doi.org/10.1038/mi.2012.54

Hwang I-C, Valeriano VD, Song JH, Pereira M, Oh JK, Han K, Engstrand L, Kang D-K (2023) Mucosal immunization with Lactiplantibacillus plantarum-displaying recombinant SARS-CoV-2 epitopes on the surface induces humoral and mucosal immune responses in mice. Microb Cell Factories 22(1). https://doi.org/10.1186/s12934-023-02100-7

Jeong SH, Lee HJ, Lee SJ (2023) Recent advances in CRISPR-Cas technologies for synthetic biology. J Microbiol 61(1):13–36. https://doi.org/10.1007/s12275-022-00005-5

Krylova NV, Ermakova SP, Lavrov VF, Leneva IA, Kompanets GG, Iunikhina OV, Nosik MN, Ebralidze LK, Falynskova IN, Silchenko AS, Zaporozhets TS (2020) The comparative analysis of antiviral activity of native and modified Fucoidans from brown algae Fucus evanescens in vitro and in vivo. Marine Drugs 18(4):224. https://doi.org/10.3390/md18040224

Kuehn MJ, Kesty NC (2005) Bacterial outer membrane vesicles and the host–pathogen interaction. Genes Dev 19(22):2645–2655. https://doi.org/10.1101/gad.1299905

Lahiri D, Nag M, Ghosh S, Dey A, Dash S, Ray RR (2023) Microbial communications in Gram-positive and Gram-negative bacteria. In: Emerging technologies in applied and environmental microbiology. Elsevier, pp 29–44. https://doi.org/10.1016/B978-0-323-99895-6.00008-3

Lynn DJ, Benson SC, Lynn MA, Pulendran B (2022) Modulation of immune responses to vaccination by the microbiota: implications and potential mechanisms. Nat Rev Immunol 22(1):33–46. https://doi.org/10.1038/s41577-021-00554-7

Mazmanian SK, Liu CH, Tzianabos AO, Kasper DL (2005) An immunomodulatory molecule of symbiotic bacteria directs maturation of the host immune system. Cell 122(1):107–118. https://doi.org/10.1016/j.cell.2005.05.007

Medzhitov R (2007) Recognition of microorganisms and activation of the immune response. Nature 449(7164):819–826. https://doi.org/10.1038/nature06246

Mimee M, Tucker AC, Voigt CA, Lu TK (2015) Programming a human commensal bacterium, Bacteroides thetaiotaomicron, to sense and respond to stimuli in the murine gut microbiota. Cell Syst 1(1):62–71. https://doi.org/10.1016/j.cels.2015.06.001

Mojgani N, Ashique S, Moradi M, Bagheri M, Garg A, Kaushik M, Hussain MS, Yasmin S, Ansari MY (2025) Gut microbiota and postbiotic metabolites: biotic intervention for enhancing vaccine responses and personalized medicine for disease prevention. Probiot Antimicrob Proteins. https://doi.org/10.1007/s12602-025-10477-7

Montero DA, Vidal RM, Velasco J, George S, Lucero Y, Gómez LA, Carreño LJ, García-Betancourt R, O'Ryan M (2023) Vibrio cholerae, classification, pathogenesis, immune response, and trends in vaccine development. Front Med 10:1155751. https://doi.org/10.3389/fmed.2023.1155751

O'Hagan DT, Rappuoli R, De Gregorio E, Tsai T, Del Giudice G (2011) MF59 adjuvant: the best insurance against influenza strain diversity. Expert Rev Vaccines 10(4):447–462. https://doi.org/10.1586/erv.11.23

O'Toole PW, Marchesi JR, Hill C (2017) Next-generation probiotics: the spectrum from probiotics to live biotherapeutics. Nat Microbiol 2(5):17057. https://doi.org/10.1038/nmicrobiol.2017.57

Park J, Kim M, Kang SG, Jannasch AH, Cooper B, Patterson J, Kim CH (2015) Short-chain fatty acids induce both effector and regulatory T cells by suppression of histone deacetylases and regulation of the mTOR–S6K pathway. Mucosal Immunol 8(1):80–93. https://doi.org/10.1038/mi.2014.44

Pulendran B, Ahmed R (2006) Translating innate immunity into immunological memory: implications for vaccine development. Cell 124(4):849–863. https://doi.org/10.1016/j.cell.2006.02.019

Riedel S, Hobden JA, Miller S, Morse SA, Mietzner TA, Detrick B, Mitchell TG, Sakanari JA, Hotez P, Mejia R (2019) Spore-forming gram-positive bacilli: bacillus and clostridium species. In: Jawetz, Melnick, & Adelberg's Medical Microbiology, 28e, vol 1–Book, Section. McGraw-Hill Education. https://accessmedicine.mhmedical.com/content.aspx?aid=1163279314

Ruhal R, Kataria R (2021) Biofilm patterns in gram-positive and gram-negative bacteria. Microbiol Res 251:126829. https://doi.org/10.1016/j.micres.2021.126829

Sangiorgio G, Nicitra E, Bivona D, Bonomo C, Bonacci P, Santagati M, Musso N, Bongiorno D, Stefani S (2024) Interactions of gram-positive bacterial membrane vesicles and hosts: updates and future directions. Int J Mol Sci 25(5):2904. https://doi.org/10.3390/ijms25052904

Sepulveda AR (2013) Helicobacter, inflammation, and gastric cancer. Curr Pathobiol Rep 1(1):9–18. https://doi.org/10.1007/s40139-013-0009-8

Thanush D, Basavaraj HC, Gowrav MP (2023) Current regulation and initial considerations for successful development and commercialization of microbiome therapies. Adv Gut Microbiome Res 2023:1–11. https://doi.org/10.1155/2023/6657515

Tobias J, Heinl S, Dendinovic K, Ramić A, Schmid A, Daniel C, Wiedermann U (2025) The benefits of Lactiplantibacillus plantarum: from immunomodulator to vaccine vector. Immunol Lett 272:106971. https://doi.org/10.1016/j.imlet.2025.106971

Van Der Pol L, Stork M, Van Der Ley P (2015) Outer membrane vesicles as platform vaccine technology. Biotechnol J 10(11):1689–1706. https://doi.org/10.1002/biot.201400395

Van Niel G, D'Angelo G, Raposo G (2018) Shedding light on the cell biology of extracellular vesicles. Nat Rev Mol Cell Biol 19(4):213–228. https://doi.org/10.1038/nrm.2017.125

Wells JM, Rossi O, Meijerink M, Van Baarlen P (2011) Epithelial crosstalk at the microbiota–mucosal interface. Proc Natl Acad Sci 108(supplement_1):4607–4614. https://doi.org/10.1073/pnas.1000092107

Wells JM, Brummer RJ, Derrien M, MacDonald TT, Troost F, Cani PD, Theodorou V, Dekker J, Méheust A, De Vos WM, Mercenier A, Nauta A, Garcia-Rodenas CL (2017) Homeostasis of the gut barrier and potential biomarkers. Am J Physiol-Gastrointest Liver Physiol 312(3):G171–G193. https://doi.org/10.1152/ajpgi.00048.2015

Xia P-F, Ling H, Foo JL, Chang MW (2019) Synthetic genetic circuits for programmable biological functionalities. Biotechnol Adv 37(6):107393. https://doi.org/10.1016/j.biotechadv.2019.04.015

Xu W, Zhang Y, Wang X, Huang M, Gao X, Zhang D, Li C, Ai Q (2013) Display of heterologous protein on the surface of Lactobacillus plantarum by using the CspI anchor protein. Ann Microbiol 63(3):855–861. https://doi.org/10.1007/s13213-012-0536-2

Yang Y, Yu P, Huang Y, Zhang W, Nie Y, Gao C (2025) Metabolic engineering of Lactobacilli spp. For disease treatment. Microb Cell Factories 24(1):53. https://doi.org/10.1186/s12934-025-02682-4

Zhang L, Chen N, Chen H, Tang C, Wang J, Wang Y, Zhang Y, Guo H, Yuan J (2025a) Recent advances of engineered probiotics for therapeutic applications. BioDesign Res 7(3):100039. https://doi.org/10.1016/j.bidere.2025.100039

Zhang X, Wang YEQ, Naveed M, Wang X, Liu Y, Li M (2025b) The biological activity and potential of probiotics-derived extracellular vesicles as postbiotics in modulating microbiota-host communication. J Nanobiotechnol 23(1):349. https://doi.org/10.1186/s12951-025-03435-6

Safety, Standardization, and Policy Development

11

Abstract

Strong frameworks for safety, standardization, and policy development are required due to the expanding utilize of microbiota in the food, agricultural, and therapeutic sectors. This chapter provides a comprehensive analysis of the difficulties and methods currently used to assess the safety of products derived from microbiota such as toxigenicity assessment, clinical safety validation, microbial strain selection, and antibiotic resistance profiling. The fact that sampling, sequencing techniques, bioinformatics pipelines and metadata reporting are still applied differently across studies and geographical locations emphasizes the need for uniform and repeatable standards in microbiota research. A substantial research gap is the absence of widely accepted criteria for the efficacy of microbiota-based treatments with respect to strains, dosage regularity and long-term impact assessment. Furthermore, the lack of a comprehensive international regulatory consensus complicates the classification and approval processes for medications and agricultural inputs derived from microbiota. Examples of ethical, legal and social problems that highlight the necessity of inclusive and open policy development include benefit sharing, bio-prospecting and consumer misinformation. Future perspectives suggest that the development of globally standardized standards has the capacity to be aided by interdisciplinary collaborations, flexible regulatory frameworks that has the capacity to accommodate new microbiome technologies and the utilize of digital tools such as blockchain, artificial intelligence and omics-based surveillance for real time safety and quality control. The chapter concludes with a call for evidence-based policymaking and improved public private partnerships to support the responsible development, marketing and governance of microbiota-based solutions globally.

R. C. Sobti et al., *Microbiota Unveiled: Science, Health, and Future Therapies*,
https://doi.org/10.1007/978-981-95-9031-5_11

Keywords

Microbiota safety assessment · Standardization in microbiome research · Antibiotic resistance profiling · Microbial strain selection · Sequencing and bioinformatics · Microbiota based therapeutics

11.1 Introduction

Policy, uniformity, and safety are critical for microbiota research and application. Research on the microbiota has completely changed our knowledge of the complex ecosystems that surround and coexist with living things. The skin, mucosal surfaces, and gut are home to trillions of microorganisms that make up the human microbiota. According to Sommer and Bäckhed (2013) and Zheng et al. (2020), the microbiota is essential for preserving health because it affects immune system function, metabolism, and even neurological processes. According to Berg et al. (2019), crop productivity, nutrient cycling and sustainable agriculture all depend on the microbiomes of plants and soil. Research on everything from engineered probiotics and microbiome manipulation in plants to boost yield and stress resistance to fecal microbiota transplantation for recurrent *Clostridium difficile* infections has been accelerated by metagenomics, bioinformatics and high-throughput sequencing (Gilbert et al. 2018). These applications have the potential to revolutionize environmental sustainability, personalized medicine, and global food security. Nevertheless, there are complex safety concerns associated with these advancements. Instead of using conventional drugs or biologics, microbiota-based interventions utilize live, dynamic microbial communities that have the capacity to adapt, interact with hosts, and change over time. This complexity increases the likelihood of unintended negative outcomes such as pathogen overgrowth, immunological dysregulation, horizontal transfer of antibiotic resistance genes or ecological imbalance (Schmidt et al. 2018; Olesen and Alm 2016). Strict quality control, continuous monitoring, and comprehensive risk assessments are necessary for safety in order to safeguard individuals and populations. Standardization is necessary due to the high variability in microbiota composition caused by variables such as geography, diet, genetics, and disease status. Without standardized procedures for sampling, sequencing, microbial identification, and therapeutic formulation, it is challenging to compare and validate research findings which hinder clinical translation and regulatory approval (Knight et al. 2018a, b). Products based on microbiota must also be produced with consistent potency, purity, and stability to ensure safety and efficacy. Policy frameworks provide the regulatory foundation for microbiota research and applications. Current regulations are often inadequate or unclear due to the novelty of microbiome-based products and treatments, resulting in a regulatory grey area. Effective policies must take ethical factors like informed consent, data privacy, and equitable access into account in order to strike a balance between innovation and public safety. Furthermore, standardized international regulations and guidelines are necessary due to the global scope of microbiota research and commercialization.

Specific challenges with therapies based on microbiota: The complexity and novelty of microbiota interventions present a number of particular challenges:

1. Biological complexity and individual differences: The complex ecosystem known as the microbiota includes bacteria, viruses, fungi, and archaea. It has the capacity to speak to its host as well as to itself. The highly personalized microbial composition is influenced by a variety of factors such as age, diet, environment, and health status (Turnbaugh et al. 2007). This variability complicates clinical efficacy studies and safety evaluations and makes it more challenging to determine healthy microbiota baselines and predict individual responses to microbiota-based therapies (Sommer and Bäckhed 2013).
2. Use of live and dynamic organisms: Smillie et al. (2011) state that live organisms with the capacity for gene transfer, replication, and environmental adaptation are commonly employed in microbiota-based therapies. These traits raise concerns about microbial persistence, potential pathogenicity, and the horizontal spread of antibiotic resistance or virulence genes, especially when using fecal transplants or genetically modified strains (Johnson and Grossman 2015). Unlike traditional drugs that target specific molecules or pathways, the microbiota influences host health in multiple ways including immune modulation, metabolic regulation, and barrier function enhancement. These complex and often indirect effects complicate safety and efficacy evaluations and make it difficult to determine causal relationships and forecast long-term outcomes (Zheng et al. 2020). Because different studies utilize various sequencing platforms, bioinformatic pipelines, DNA extraction protocols, and microbiota sampling techniques (estool/mucosal biopsies), the absence of standard methods leads to inconsistent results and interpretations across studies (Knight et al. 2018a, b). Moreover, it has the capacity to be challenging to maintain batch-to-batch consistency, stability, and viability when creating live microbial products.
3. Privacy and ethical concerns: Microbiome data are increasingly recognized as sensitive and private personal health information. Among the ethical concerns that arise are informed consent, data ownership, potential discriminatory or commercial use of microbiome data, and fair access to treatments (Berg et al. 2019).
4. Changing regulatory environment: Generally speaking, products related to microbiota are exempt from the standard regulatory frameworks that apply to drugs, devices, and biologics. To assist regulatory agencies in striking a balance between patient access to novel therapies and the requirement for robust safety and efficacy data, flexible and adaptive policy models are required.

Goals and Purview of the Chapter In the quickly developing field of microbiota research and applications, this chapter aims to offer a comprehensive analysis of the various facets of safety, standardization and policy development. The chapter specifically seeks to: (1) Determine and examine the main safety concerns related to microbiota based interventions in the fields of agriculture, human health and the environment. (2) Outline current safety evaluation strategies and new approaches such as post market surveillance and preclinical and clinical evaluation. (3) Highlight initiatives to create global standards and best practices while discussing the vital significance of standardization in microbiota research protocols, analytical methods and therapeutic product development. (4) Examine the current regulatory environment that oversees microbiota products, clarifying any moral, legal or societal ramifications. (5) Use pertinent case studies such as fecal microbiota transplantation and engineered probiotics to demonstrate problems and solutions. (6) Examine future viewpoints, highlighting the necessity of flexible, evidence based regulations and global cooperation to promote microbiota based innovations that are safe and efficient.

11.2 Safety Concerns Regarding Microbiota Treatments

The power of intricate microbial ecosystems is harnessed by microbiota-based therapies to boost immunity, promote agricultural productivity, or restore health. Despite their great potential, these interventions present particular safety risks because microbiomes are dynamic, adaptive, and complex. The main safety risks related to microbiota modulation are examined in this section with an emphasis on current knowledge and lingering issues (Fig. 11.1).

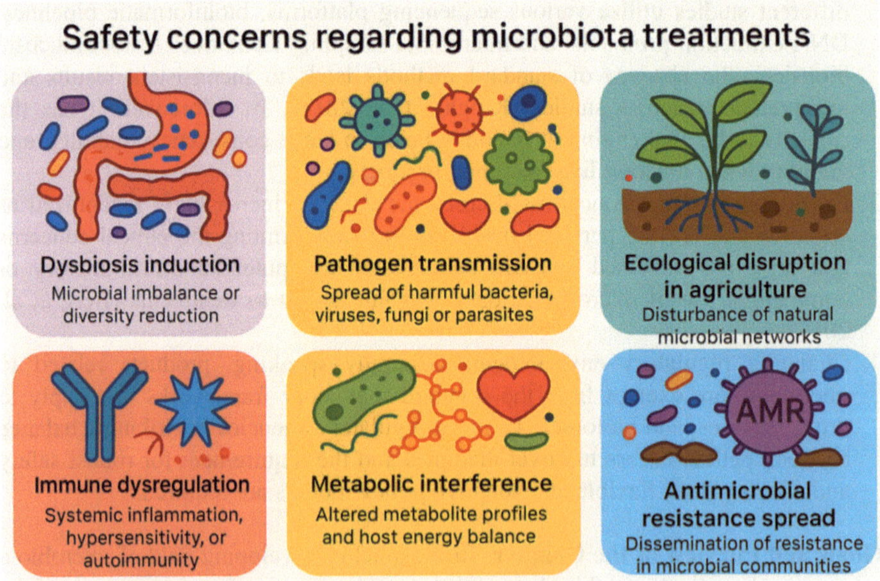

Fig. 11.1 Safety concerns regarding microbiota treatments

11.2.1 Negative Impacts of Microbiota Change

Restoring the balance of the microbial community to a healthy state is the main objective of microbiota therapies. Nevertheless, because microbial ecosystems are complex, interventions may occasionally have unanticipated and harmful effects:

1. Dysbiosis induction: Dysbiosis, i.e., microbial imbalance or microbial diversity reduction, overgrowth of pathogenic species, or the loss of beneficial microbes. Unintentionally upsetting native microbial communities through the therapeutic introduction of microbes or prebiotics has the capacity to result in inflammation, impaired barrier function, or metabolic dysfunction (Becattini et al. 2016). For example, some patients have developed new onset conditions like obesity or autoimmune disorders following fecal microbiota transplantation, indicating that transferred microbiota may have unpredictable systemic effects (Alang and Kelly 2015).
2. Pathogen transmission: There is a chance of spreading harmful bacteria, viruses, fungi, or parasites when donor-derived microbiota is used, especially in FMT. Multidrug-resistant *Escherichia coli* and other pathogens have been known to spread despite strict donor screening resulting in serious infections and even death (DeFilipp et al. 2019). This emphasizes the need for strict manufacturing controls and standardized thorough screening procedures.
3. Ecological disruption in agriculture: When microbial inoculants are introduced into plant and soil microbiome engineering, they may upset the natural microbial networks, which may result in a decline in soil biodiversity or the growth of opportunistic pathogens. According to Mendes et al. (2013), these disturbances may decrease soil fertility, elevate crop susceptibility to disease, or compromise ecosystem services like nitrogen fixation.
4. Risks to the immune system and metabolism: The host immune system and metabolic pathways are significantly impacted by microbiota. Therapy that modifies these interactions may have detrimental effects on the immune system or metabolism.
5. Immune dysregulation: The microbiota informs and regulates immune responses abrupt or unwarranted changes in microbial populations may result in systemic inflammation, hypersensitivity or autoimmunity. For instance, although generally safe, probiotics containing strains of *Lactobacillus* or *Bifidobacterium* have occasionally caused bacteremia or sepsis, especially in patients with weakened immune systems, indwelling devices or serious illnesses (Verna and Lucak 2010). Furthermore, transplanted microbiota may trigger immune pathways that have been dormant, aggravating graft versus host disease and inflammatory bowel disease.
6. Metabolic interference: The microbiota plays a foundational role in a role in vitamin synthesis, xenobiotic metabolism, and nutrient absorption. Short-chain fatty acids which are metabolized by microorganisms alter the immune system and host energy balance. Metabolite profiles may change as a result of changing microbiota composition sometimes in a negative way. For example, elevated

levels of gut microbe-produced trimethylamine-N-oxide or specific bile acids have been associated with an increased risk of cardiovascular disease (Krautkramer et al. 2021). Because these metabolic changes have systemic and indirect effects that go beyond microbial viability, they make safety evaluations more difficult.

7. Concerns about antimicrobial resistance and horizontal gene transfer: The horizontal gene transfer (HGT), a natural process by which microbes exchange genetic material, including genes conferring virulence or antibiotic resistance is one of the most concerning risks connected to microbiota-based interventions.
8. Antimicrobial resistance spread: Resistance genes are stored in microbiota especially in the gut. Mobile genetic elements carrying AMR genes such as plasmids, transposons, or integrative conjugative elements may be introduced or spread by microbial therapies. The effectiveness of antibiotics and public health may be jeopardized if commensal bacteria undergo such gene exchange and develop into multidrug-resistant pathogens (Smillie et al. 2011; Johnson and Grossman 2015).
9. Environmental dissemination: Derived from the microbiota, like soil inoculants or animal probiotics may add to the larger environmental resistance when they are released into the environment. This has the capacity to make containment challenging by promoting the spread of AMR genes among microbial communities in soil, water, and animals (Wellington et al. 2013).

Figure 11.1 illustrates substantial risks associated with microbiota-based therapies: (1) Immune dysregulation (systemic inflammation, hypersensitivity/autoimmunity). (2) Pathogen transmission (through bacteria, viruses, fungi, or parasites). (3) Ecological disruption in agriculture (disruption of natural microbial networks). (4) Metabolic interference (altered metabolite profiles and disrupted host energy balance). (5) Antimicrobial resistance (spread through dissemination of resistant traits in microbial communities) and (6) Dysbiosis induction which results in microbial imbalance or diversification.

11.2.2 Techniques to Reduce HGT Risk

Current methods for lowering HGT risks include using defined microbial consortia with well-characterized genomes, genetic engineering to eliminate or minimize mobile elements and rigorously screening donor material for AMR genes. Furthermore, it is advised to conduct ongoing post treatment surveillance in order to identify any new resistance patterns. Other things to think about are: (1) Lack of long term safety data: Since many microbiota treatments are still in their infancy, there is a dearth of longitudinal data on their long term impacts such as potential late-onset adverse events, microbiome stability, and chronic immune modulation. (2) Responses particular to the patient: personalized safety evaluations are necessary because therapy outcomes and risks are influenced by an individual's immune status, diet, genetics, and microbiota composition. (3) Manufacturing and quality control: standardized production and validation procedures are necessary because variations in microbial viability, purity, and product formulation have the capacity to impact safety and efficacy.

11.2.3 Safety Assessments in Clinical and Preclinical Contexts

To ensure the safety of microbiota-based treatments, extensive preclinical and clinical evaluations tailored to the unique biological complexity of host microbe interactions are required. Unlike conventional pharmaceuticals, microbial interventions has the capacity to involve living things and have dynamic, often unpredictable effects on human health, necessitating the use of multilayered safety protocols and advanced monitoring equipment.

11.2.4 Techniques for Assessing Safety in Human and Animal Trials

1. Animal models for preclinical safety testing: For the preliminary assessment of microbiota-based interventions, animal models are essential. The dynamics of the human microbiome are commonly simulated under controlled conditions using gnostic and germ-free rodent models (e.g., mice and rats). These models have the capacity to be used to assess metabolic alterations, host immune responses, microbial colonization, and potential off-target effects.
2. Common endpoint assessments consist of biochemical markers of inflammation and stress; toxicity indicators (histological organ damage or weight loss); Gut permeability and the barrier function, the presence of genes associated with horizontal gene transfer or antimicrobial resistance (Zhang et al. 2015). Large animal models like pigs and nonhuman primates are also used when a higher degree of anatomical and physiological resemblance to humans is needed (Liu et al. 2018). The fact that interspecies differences in microbiota composition limit the full translational value of preclinical data emphasizes the importance of additional clinical research.
3. Clinical safety assessments: Clinical trials of microbiota-based treatments follow standard Phase I to III structures and incorporate microbiota-specific endpoints. Phase I trials assess immediate safety and tolerability in small groups of healthy patients or volunteers. Microbial colonization efficiency, gastrointestinal symptoms, and immune biomarkers (CRP and cytokine profiles) are some of the parameters (Feuerstadt et al. 2022). Phase II and III trials expand to include larger patient populations and focus on long-term safety and efficacy. To monitor negative consequences such as immunological imbalances, infection risk, and dysbiosis; microbiome sequencing and host response profiling are employed (Aroniadis and Brandt 2013). FMT-specific protocols require microbial product traceability, consistent preparation methods, and stringent donor screening. The U.S. FDA views fecal microbiota transplantation as a biologic under the Investigational New Drug (IND) framework, which necessitates thorough safety reporting and adherence to GMP (FDA 2013a, b; Cammarota et al. 2017).

11.2.5 Monitoring Tools and Biomarkers for Safety Evaluation

The development of sensitive and accurate biomarkers is crucial for identifying adverse effects of microbiota modification at an early stage. Inflammatory indicators such as C-reactive protein (CRP), tumor necrosis factor alpha (TNF-α), and interleukins (IL-6, IL-1β) are used to quantify systemic immune activation (Liu et al. 2018). Zonulin, calprotectin, and lactulose mannitol ratios are indicators of gut integrity that shed light on inflammation and mucosal barrier function. To evaluate broader metabolic impact, metabolic biomarkers like glucose, lipid levels, trimethylamine-N-oxide and short-chain fatty acids are employed (Zhang et al. 2015). (1) Microbial biomarkers and omics tools (2) Next-generation sequencing techniques like 16S rRNA gene profiling and shotgun metagenomics are used to monitor shifts in microbial diversity, the quantity of pathogenic taxa, and the appearance of AMR genes. With digital PCR and quantitative PCR (qPCR), integral microbial genes such as those encoding toxins or resistance have the capacity to be monitored in high resolution (Pamer 2016). (3) Transcriptomics, metabolomics, and proteomics provide a systems-level understanding of host microbe interactions by highlighting the functional implications of changes in the microbiome. (4) Real-time monitoring and digital surveillance: wearable health trackers, digital symptom diaries, and AI-assisted microbiome monitoring platforms are among the new technologies being incorporated into trial designs to enhance continuous safety assessment in outpatient settings (Zhang et al. 2015). In conclusion, safety evaluations in microbiota interventions must combine cutting-edge molecular tools with traditional toxicology in order to manage the unique risks posed by live microbial products. Preclinical animal models help identify risks early, while biomarker-based monitoring and human trials ensure long-term patient safety. We must develop our diagnostic methods and regulatory frameworks in tandem with microbiota research if we are to fully realize the therapeutic potential of the microbiome.

11.2.6 Pharmacovigilance and Postmarket Surveillance

Following their release into the market, microbiota-based therapies must be continuously monitored for safety and effectiveness. Preclinical and clinical trials often have a small patient pool and controlled environments, despite their importance. In order to identify rare side effects, chronic adverse events, and unforeseen interactions across broader populations and time periods, pharmacovigilance and post market surveillance are crucial. Microbiota-based treatments, as opposed to chemical medications, make use of living things that are capable of growth, modification, and dynamic interaction with their surroundings and host biology. This enables long-term security and the tracking of unforeseen consequences. This inherent complexity makes it especially challenging to keep an eye on their long-term safety.

11.2.7 Key Elements of PMS for Products Based on Microbiota

After the trial period, active surveillance systems are used to collect safety data, especially for populations with chronic illnesses or immunocompromised conditions (Cammarota et al. 2017). Longitudinal microbiome profiling uses methods such as 16S rRNA sequencing and shotgun metagenomics to monitor microbiome changes that may be connected to unanticipated outcomes like dysbiosis, metabolic alterations, or autoimmune triggers (Zhang et al. 2015). Wearable biosensors, digital health platforms, and patient registries are used to collect real-world evidence in order to assess safety in various clinical and demographic contexts (Feuerstadt et al. 2022). Horizontal gene transfer and the spread of antimicrobial resistance genes should be closely monitored, particularly in probiotic or FMT products as these factors may contribute to the development of antibiotic resistance in the gut (Pamer 2016).

Microbiota-based product safety examples issues: several real-world incidents highlight the importance of robust pharmacovigilance systems for microbiota-based interventions:

Case study 1: Bacteremia associated with FMT: In 2019, the U.S. FDA reported two cases of severe infections following fecal microbiota transplantation. *Escherichia coli* which generates extended-spectrum beta-lactamases was the cause of the infections. One patient died as a result (FDA 2019a, b). The incident led to increased awareness of the possibility of pathogen transmission even with screened material and modifications to FDA regulations requiring strict multidrug-resistant organism screening for donors.

Case study 2: Fungemia linked to probiotics: using probiotics has been linked to systemic infections such as *Saccharomyces fungemia* or *Lactobacillus bacteremia* especially in patients with severe illnesses or weakened immune systems. A commonly reported case of a patient with ulcerative colitis who developed a fatal *Saccharomyces boulardii* fungemia serves as an illustration of the importance of risk stratification (Zhang et al. 2015).

Case study 3: Delayed immune responses: furthermore, new research indicates that treatments that alter the immune signaling pathways in the gut may cause or worsen autoimmune diseases in vulnerable people. Despite being uncommon, these risks require long-term surveillance because they are hard to predict during early-phase trials. To sum up, postmarket surveillance of microbiota-based interventions needs to incorporate real-world data analytics, molecular profiling, microbiological monitoring and the conventional tracking of adverse events. Case studies demonstrate how in practical settings even tried and true therapies have the capacity to have unanticipated adverse effects. To advance microbiota therapies and protect public health, it will be essential to establish centralized registries, international safety databases, and multidisciplinary pharmacovigilance frameworks.

11.3 Research and Products on Microbiota

11.3.1 Consistency Is Essential

Research on microbiota is a fast growing field that could transform biotechnology, agriculture, and medical care. Nevertheless, the absence of uniform protocols across studies, goods, and clinical applications is one of the primary obstacles to its widespread utilization and regulatory approval. For safe, efficient product development as well as reproducibility and interstudy comparability standardization is crucial.

11.3.1.1 Disparities in the Microbiota Function and Make Up Across Populations

Numerous factors such as geographic location, diet, age, genetics, lifestyle, and disease status, influence the naturally dynamic and diverse compositions of human and environmental microbiota. For example, compared to people who live in urbanized, industrialized settings, members of rural agrarian communities generally have a more diverse gut microbiome (Yatsunenko et al. 2012). Probiotics, prebiotics, and fecal microbiota transplantation are among the interventions whose predictability and generalizability are impacted by this interindividual and interpopulation variability. Furthermore, it complicates the search for healthy reference microbiomes or universal therapeutic targets (Zhao et al. 2018). It restricts the reproducibility of cross-study research especially when data analysis pipelines, sample handling, or sequencing are conducted on different platforms (Knight et al. 2018a, b). Standardizing sampling strategies, storage settings, and analytical procedures is essential to producing meaningful, comparable datasets that have the capacity to guide clinical and regulatory decision making.

11.3.1.2 Value of Comparability, Reproducibility, and Regulatory Approval

Standardization makes reproducibility possible in microbiota research, which is essential for confirming results and turning them into evidence-based products and treatments. The following critical areas require standardization: (1) Sample collection and storage: differences in how soil or fecal samples are preserved have the capacity to result in inaccurate readings and changes in the microbial community (Costea et al. 2017). (2) DNA extraction and sequencing platforms: the taxa discovered have the capacity to vary significantly depending on the kits and bioinformatics tools used (Pollock et al. 2018). (3) Production of microbiome-based products: Suez et al. (2019) demonstrated microbial consortia must be produced with consistent potency, purity, and viability metrics under stringent GMP conditions. The significance of consistent, well-characterized products in microbiota-based treatments is emphasized by regulatory bodies like the U.S. Food and Drug Administration (FDA) and the European Medicines Agency (EMA). For example, Live Bio-therapeutic Products (LBPs) need to have defined mechanisms of action, batch-to-batch consistency and no contaminants in addition to efficacy (FDA 2016). Due to variations in donor stool material, FMT is still in the investigational stage. Without standardization, it is impossible to ensure product safety, efficacy, and scalability which hinders commercialization and the global harmonization of regulatory pathways. In summary, standardization is not only a technical requirement but also a scientific and regulatory one for microbiota research and product development. Standardizing processes for sample handling, data processing, and microbial formulation will significantly enhance the dependability and translational potential of microbiota science. Furthermore, international collaborative efforts and regulatory frameworks must be supported in order to develop consensus standards that bridge the gap between research and application.

11.3.2 Standardization of Microbiota Analysis and Sampling

The validity of microbiota-based research and products depends on the reliability, consistency, and reproducibility of microbial data. Nonetheless, methodological variability continues to be a significant barrier in the field. From sample collection to computational analysis, standard operating procedures must be adhered to in order to minimize technical biases and enhance interstudy comparability (Fig. 11.2).

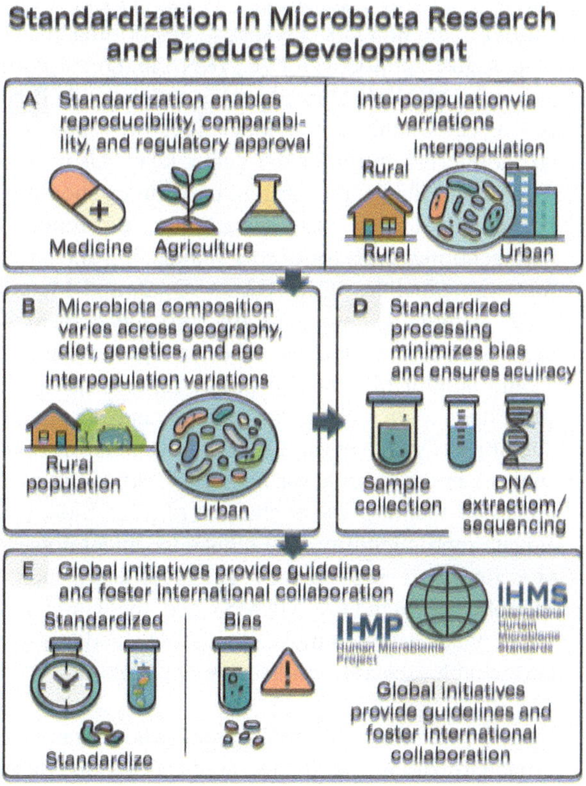

Fig. 11.2 Standardization in microbiota research and product development

1. Instructions for gathering, preserving, and handling samples: One of the most integral sources of bias in microbiome research is variation in sample handling. Microbial community profiles have the capacity to be significantly changed by variables like processing time, storage temperature, and preservation media. The following are integral standardization measures: sample collection: sterile, DNA-free kits must be used to collect stool, oral, skin, or environmental samples. Stool is the most representative sample type for studies of the gut microbiome. When feasible, collection should take place in a consistent environment and with a consistent diet (Sinha et al. 2016).
2. Preservation and storage: Stabilizing buffers such as RNAlater or OMNIgene, GUT, or immediate freezing at −80 °C prevent microbial integrity during transportation and processing delays (Costea et al. 2017).
3. Processing timeframes: Anaerobic degradation and the proliferation of aerotolerant bacteria have the capacity to be encouraged by postponed processing. Research indicates that microbial diversity measurements have the capacity to be distorted by even 24-h delays at room temperature (Song et al. 2016).
4. Sequencing and bioinformatics pipelines standard operating procedures: Variability is also introduced by downstream DNA extraction, sequencing, and bioinformatics techniques. Cross-study comparisons are frequently unreliable due to a lack of standardized SOPs. (1) Extracting DNA: the effectiveness of various extraction kits in lysing spore-forming and Gram positive bacteria varies. Standardized protocols for 16S rRNA and metagenomic investigations frequently utilize the MO BIO UltraClean Kit and the QIAGEN PowerSoil Kit (Pollock et al. 2018). (2) Methods of Sequencing: (a). 16S rRNA gene sequencing: Offers taxonomic resolution at the genus level. Using consistent primer sets (V3–V4 region) and library preparation procedures should be part of standardization (b). Higher taxonomic and functional resolution is provided by Shotgun Metagenomics calls for deeper sequencing and stricter quality control (c). Bioinformatics pipelines: QIIME2, DADA2, and MetaPhlAn are widely recognized tools. Although these platforms provide curated workflows, output may vary depending on the parameter selections made (Knight et al. 2018a, b). Researchers should submit raw data to public repositories like MG-RAST or EMBL-EBI, share open-source code, and register analysis pipelines in order to elevate reproducibility (Quince et al. 2017). The standardization of microbiota sampling and analysis is the foundation for the validity of microbiome science. Standardized processes facilitate the evaluation of microbiota-based products for regulations, elevate reproducibility, and reduce technical noise. Guidelines for standardizing procedures are provided by global projects such as the International Human Microbiome Standards (IHMS) and the Human Microbiome Project (HMP). Industry and research must adopt these standards to transform microbiota science into reliable, scalable, and secure applications.

Figure 11.2 illustrates main areas where standardization is essential such as (1) Standardization facilitates applications in biotechnology, agriculture, and medicine by enabling reproducibility, comparability, and regulatory approval. (2) Age, diet, genetics, and geographic location all have an impact on the microbiota composition with notable differences between rural and urban areas. (3) Technical areas for standardization include sample collection, storage (−80 °C), and DNA extraction/sequencing protocols because delays or incorrect handling have the capacity to distort results. (4) Standardized processing ensures accuracy and minimizes bias. (5) International initiatives to standardize procedures such as the Human Microbiome Project and IHMS provide guidelines and promote international collaboration.

11.3.3 Microbiota to Standardize Probiotics and Treatments

The clinical and commercial success of microbiota-based interventions such as probiotics live bio-therapeutic products (LBPs) and fecal microbiota transplantation depends on strict standardization. Identifying microbial strains, ensuring viability, optimizing dosages, and incorporating quality control into production processes are all part of this. Without such standardization, regulatory approval, efficacy, and reproducibility remain elusive (Fig. 11.3).

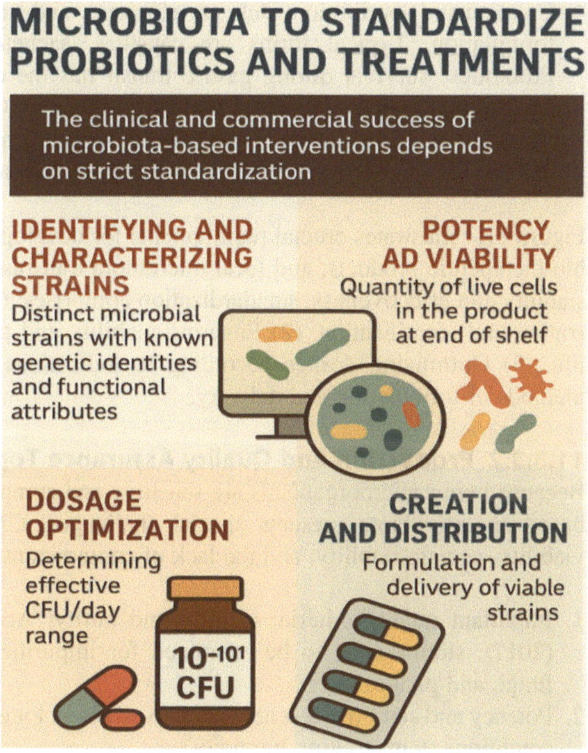

Fig. 11.3 Standardization of microbiota-based interventions

11.3.3.1 Developing Dosage, Formulation, Viability, and Strains Guidelines

1. Identifying and characterizing strains: Every microbiota-based product must contain distinct microbial strains with known genetic identities, metabolic characteristics, and functional attributes. Because strain level differences have the capacity to have a substantial impact on host responses, accurate identification through whole genome sequencing or molecular typing techniques is essential (Hill et al. 2014; Brodmann et al. 2017). The International Scientific Association for Probiotics and Prebiotics (ISAPP) states that since health benefits must be strain specific, unproven strains should not be marketed as probiotics (Hill et al. 2014).
2. Potency and viability: One of the most integral quality indicators is viability or the quantity of live cells in the product. Colony-forming units (CFUs) per dose should be disclosed by products at the end of their shelf life not just when they were manufactured. Due to poor formulation or insufficient storage, many commercial probiotics have been found to fall short of their label claims.
3. Dosage optimization: Microbiota-based treatments do not have a fixed dosage. Nonetheless, research indicates that depending on the strain and indication, the usual range of effective probiotic dosages is 10^8–10^{11} CFU/day (Sanders et al. 2019). Extensive dose finding trials are used to determine strain-specific therapeutic doses for LBPs such as SER-109 (Feuerstadt et al. 2022).
4. Creation and distribution: Delivery efficiency and strain viability are impacted by formulation. Typical forms are capsules, sachets, or lyophilized powders. Probiotics' survival during gastric transit has the capacity to be improved by encasing them in coatings that are resistant to acid (Ouwehand et al. 2021). FMT formulations have changed over time, moving from fresh or frozen stool to oral delivery of specific microbial consortia (Wilcox et al. 2017).

Figure 11.3 illustrates crucial requirements for developing effective probiotics, live bio-therapeutic products, and fecal microbiota transplantation. In order to preserve stability and effectiveness, standardization comprises: (1) Accurate strain characterization and identification. (2) Ensuring viability and potency throughout the shelf life. (3) Optimizing dosage according to indications and strains. (4) Improving methods of formulation and delivery.

11.3.3.2 Production and Quality Assurance Top Techniques

Because living microorganisms are sensitive and complex, manufacturing products based on microbiota presents special challenges. GMP must include microbial viability, genetic stability, and the lack of contaminants (FDA 2016).

1. Important quality criteria: identity and purity: According to Brodmann et al. (2017), strains need to be examined for impurities such as harmful bacteria, fungi, and phages.
2. Potency and stability: It is necessary to establish long-term stability under specific conditions (temperature, humidity).

3. Batch consistency: According to Morelli and Capurso (2012), live microbial products' cell count, makeup, and functionality need to be consistent across batches.
4. Regulatory viewpoints: For LBPs and other medications based on microbiota, the FDA and EMA in the US demand comprehensive Chemistry, Manufacturing and Controls (CMC) documentation (FDA 2016). Current gaps in standardization particularly with regard to donor variability, preparation techniques, and storage protocols are reflected in the investigational status of FMT (Sanders et al. 2019). Since their safety profiles and reproducibility have gained regulatory acceptance, emerging companies are now manufacturing next-generation microbiota-based medications as defined consortia (Feuerstadt et al. 2022; Wilcox et al. 2017). Thus, standardization of the development of microbiota-based therapies is necessary to guarantee their efficacy, safety, and adherence to global regulations. Vigorous manufacturing procedures, strain-specific characterization, dose optimization, viability assurance, and robust formulation are all necessary for this. Adopting consistent standards and regulatory frameworks will enable scalable, dependable microbiota-based health solutions.

11.3.4 Challenges in Developing Standards

Creating standardized processes for microbiota research and product formulation is extremely difficult due to the field's complexity and rapid change. Because microbiota-based interventions involve living organisms with dynamic, context-dependent effects unlike traditional pharmaceuticals uniform standardization is especially difficult.

11.3.4.1 Diverse Microbial Ecosystems and Rapid Scientific Progress

The ongoing development of microbiota science both in terms of fundamental understanding and technological capacity is one of the primary obstacles. Our knowledge of the functions, relationships and impacts of microbes in health and illness is continually evolving due to new discoveries. Consequently, current regulations may soon become obsolete (McDonald et al. 2018). Microbial ecosystems are also inherently complex and context dependent. Among hosts, geographical locations, environments, diets, and even anatomical locations within the same organism they vary significantly (Yatsunenko et al. 2012; Lloyd-Price et al. 2016). Because of this diversity, developing universal reference standards for the makeup or function of microorganisms is difficult. For example, a healthy gut microbiome in one population might be quite different from one in another. A medicinal strain that is effective in one environment (*Lactobacillus rhamnosus* GG) may not be effective at all or even work against you in another due to ecological interactions (Zmora et al. 2018). Therefore, the fluidity of microbial definitions, functional redundancy, and cross-domain interactions complicates the definition of core microbiota or targetable therapeutic pathways all of which need to be taken into account in standardization efforts.

11.3.4.2 Balancing Regulatory Requirements with Innovation

Finding a balance between the rigorous deadlines and regulations of regulatory agencies and the quick progress of science is a substantial challenge. The regulatory frameworks for microbiota-based therapies are still in their infancy and often draw inspiration from traditional pharmaceutical models that may not be fully applicable to engineered probiotics, live biotherapeutic products or fecal microbiota transplantation (Khanna et al. 2022). Among the main regulatory bottlenecks are the following: (1) Lack of clearly defined endpoints: Surrogate markers, like microbiome diversity or composition, are frequently used to evaluate the clinical effectiveness of microbiota-based interventions, even though they are not always directly related to clinical outcomes (NIH Human Microbiome Portfolio Analysis Team 2019). (2) One regulatory issue with FMT is donor-to-donor variability. There is ongoing discussion regarding the best practices for screening, standardizing or qualifying donor stool (Bafeta et al. 2017). (3) No generally recognized standards exist: Regarding microbiota products, different countries have adopted differing positions. For instance, FMT is regulated as a drug in the United States but as a tissue or biological product in other jurisdictions which complicates multinational trials and commercialization (Petrof et al. 2013). Furthermore, novel methods of microbiota engineering such as CRISPR-based modifications or synthetic microbial consortia make safety assessment more difficult and necessitate the development of new evaluation models and ethical considerations (Sheth et al. 2016). (4) Developing robust and flexible standards for microbiota-based research and therapies presents numerous scientific and legal obstacles. The dynamic nature of microbial ecosystems, the contextual variability of host microbe interactions and the lack of widely accepted regulatory frameworks make standardization difficult. Nevertheless, collaborative international efforts, adaptable regulatory frameworks, and iterative standard updates in accordance with new research are essential to bridging the gap between discovery and application.

11.4 Creating Policies for Microbiota Applications

The growing field of microbiota research and its utilization in consumer products, diagnostics, and therapies necessitate the development of robust policy frameworks. These frameworks need to balance rapid innovation with stringent safety and ethical oversight in order to advance science, preserve public trust, and guarantee equitable access (Fig. 11.4).

11.4.1 Environment of Regulation

A summary of the laws that are currently in effect regarding microbiota-based products and treatments. Microbiota-based interventions come in a wide variety of forms, from dietary supplements like probiotics and prebiotics to complex biologics like fecal microbiota transplantation and live bio-therapeutic products. Foods,

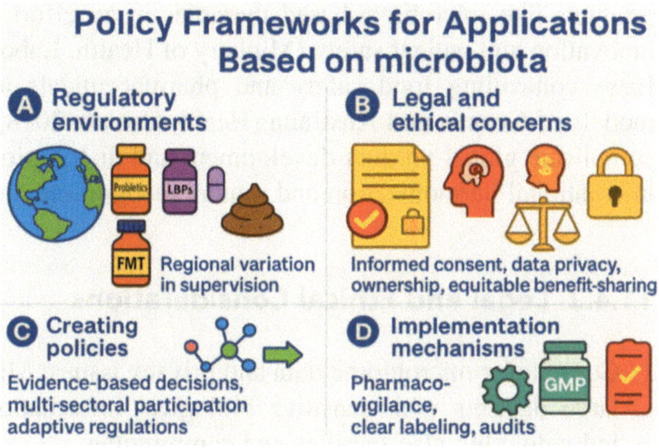

Fig. 11.4 Policy frameworks for applications based on microbiota

biologics, and pharmaceuticals are often difficult to distinguish from one another, posing a challenge to traditional regulatory classifications. Since probiotics and prebiotics are typically governed as foods or dietary supplements with health claims, regulators have adjusted their oversight protocols accordingly. The primary focuses of oversight are safety (the US's Generally Recognized as Safe (GRAS) designation) and proper labeling. Nonetheless, assertions of therapeutic benefits or efficacy are often not supported by additional research and analysis (Sanders et al. 2019). (1) Live bio-therapeutic products or LBPs: According to the FDA, they are biological products intended to prevent, treat, or cure diseases and contain living organisms. These comprehensive drug approval procedures require rigorous manufacturing controls, controlled clinical trials, and preclinical research (FDA 2016). At the moment, investigational new drug (IND) applications regulate the practice of fecal microbiota transplantation. Although the FDA has enforcement discretion over recurrent *Clostridium difficile* infections that do not enhance with standard treatment broader applications are still being investigated clinically (FDA 2019a, b). (2) Microbiome-derived diagnostics: to validate multiparametric tests, regulators need to create new frameworks that have the capacity to manage complex bioinformatics as the role of the microbiome in disease diagnosis increases. Regional differences in regulatory strategies: international regulatory bodies' approaches show both convergence and divergence: (1) United States (FDA): employs a risk-based, product-specific framework. While LBPs undergo rigorous clinical testing, probiotic supplements, for example, are not as tightly regulated. The FDA's guidance documents emphasize consistent manufacturing quality, safety profiling, and efficacy data (FDA 2020). (2) The European Union's (EMA) primary goal is the centralized evaluation of medicinal microbiota products. The EMA requires adherence to GMP standards and enforces advanced therapy medicinal product regulations when necessary. Furthermore, it facilitates collaborations between academia and industry and manages harmonization initiatives via the Innovative Medicines Initiative (EMA 2019). (3) Asia Pacific and other regions: countries like Japan have set up accelerated approval procedures for regenerative medicine products like microbiota-based therapies in an effort to strike a balance between innovation and patient safety (Ministry of Health, Labour and Welfare Japan 2018). Laws controlling food safety and pharmaceuticals are combined in the hybrid models of Canada and Australia (Health Canada 2018). These regional differences complicate global product development and distribution, underscoring the need for international harmonization and mutual recognition agreements.

11.4.2 Legal and Ethical Considerations

1. Consent for microbiome data and privacy issues: Microbiome research generates large datasets with sensitive biological information that affects not only the individual but also families and communities.
2. Among the moral issues are: Informed consent-participants must be aware of the potential uses of their microbiome data such as future research or commercialization. Because they provide continuous participant engagement and control over data usage, dynamic consent models are becoming more and more popular.
3. Privacy and confidentiality: Due to the fact that microbiome profiles have the capacity to identify people or disclose medical conditions, there is a higher risk of reidentification, data breaches, or abuse by employers or insurers. To protect participants, strong data governance procedures, encryption, and restricted access are required (Knoppers and Chadwick 2019).
4. Data sharing: While open science and data sharing speed up research, they also necessitate striking a balance between confidentiality and transparency. Models for responsible, secure data sharing are offered by frameworks such as the Global Alliance for Genomics and Health (Knoppers and Chadwick 2019).
5. Microbiota resource ownership and commercialization—Microbiota sample ownership and benefit sharing are still open questions; ownership of the sample, unlike human tissue microbiota is thought to be a component of an individual's biological makeup, but it also represents ecosystems in the environment and society. Uncertainty surrounds the legal rights of microbiota samples donated for bio-banking or research.
6. Benefit sharing: Ethical guidelines support equitable distribution of commercialization-related benefits such as profit sharing or access to the resulting treatments, particularly for vulnerable or indigenous groups whose microbiota may exhibit distinctive traits.
7. Intellectual property: Bioprospecting, drug accessibility, and incentives for innovation are among the ethical conundrums raised by the patenting of microbial strains, synthetic consortia, or engineered microbes. Policy must balance encouraging innovation with avoiding monopolization or bio-piracy.

11.4.3 Structures for Developing Policies-Engaging Stakeholders

Researchers, physicians, businesses, patients, and regulators: (1) Multisectoral involvement is necessary for effective policymaking in order to guarantee that policies are clinically relevant, socially acceptable, commercially viable, and scientifically sound. Scientists and researchers offer integral insights into the biology of the microbiome, safety issues, and technological capabilities. (2) Clinicians offer insights on practical efficacy, patient safety, and therapeutic application. (3) Industry partners provide expertise in manufacturing, market dynamics, and product development. (4) Regulatory authorities create legally binding guidelines in line with public health objectives, patients and public representatives guarantee that policies address issues of access, consent, and equity. (5) Strategies for engaging stakeholders include workshops, public consultations, advisory boards, and continuous feedback loops. (6) Flexible regulatory models and evidence-based policymaking. Because microbiota science is still in its infancy and is developing quickly, regulatory frameworks need to be flexible. (7) Evidence-based methods: strong scientific data from postmarket surveillance, epidemiological research, and clinical trials should serve as the foundation for policies. Systematic reviews and meta-analyses inform risk-benefit analyses (Hoffmann et al. 2021). (8) Adaptive regulations: to enable prompt policy changes, regulatory bodies are putting in place adaptable procedures that collect empirical data on safety and efficacy and provide early access to promising treatments. (9) Risk-based stratification: this separates low-risk products (probiotics) from higher risk treatments (engineered microbial consortia) to enable appropriate regulatory scrutiny (Turnbaugh et al. 2007).

11.4.4 Execution and Conformance

Ensuring adherence to safety and quality standards requires effective policies and robust mechanisms for implementation. (1) GMP: adhering to GMP ensures product potency, consistency, purity and absence of contaminants all crucial components of microbiota-based treatments (Suez et al. 2019). (2) Pharmacovigilance: long-term safety and systems to monitor adverse events are crucial especially for novel microbiota interventions involving complex host interactions. (3) Consumer information and labeling: when products are properly labeled with details on dosage, storage conditions, strain identification, and contraindications consumers are better able to make informed decisions (Hill et al. 2014). (4) Audits and inspections: regular regulatory inspections are conducted at production facilities, clinical trial sites and supply chains to guarantee compliance.

Figure 11.4 illustrates pivotal components required for microbiota research and treatment governance such as (1) Regulatory environments: there are regional variations in the supervision of probiotics, prebiotics, live biotherapeutic products, and fecal microbiota transplantation in Asia and other regions. (2) Legal and ethical concerns: among the primary concerns are informed consent, data privacy, ownership of microbiota resources, and equitable benefit sharing. (3) Creating policies:

effective frameworks require evidence-based decision-making, multisectoral participation, and adaptable regulations that have the capacity to evolve in response to scientific advancements. (4) Implementation mechanisms: pharmacovigilance, clear labeling, adherence to GMP standards, and regular audits are essential for ensuring product quality, safety, and public trust.

11.5 Importance of International Collaboration and Harmonizing Policies

The worldwide scope of microbiota research necessitates cross-border cooperation: Worldwide Guidelines: Organizations such as the WHO, Codex Alimentarius and ICH create standardized guidelines to help with product approval and market access. (1) Data sharing and regulatory convergence: clinical data and best practices are shared to reduce duplication and speed up regulatory decision making. (2) Building capacity: promoting the establishment of regulatory frameworks in emerging economies enables equitable access to microbiota-based innovations (WHO 2021). (3) Ethical harmonization: global frameworks ensure consistent protections for donor rights, privacy, and benefit sharing. This section looks at real-world examples of microbiota-based treatments and research standardization efforts to highlight regulatory challenges, successes, and adaptable policy solutions. Future developments and the governance of microbiota applications depend on learning from these experiences (Fig. 11.5; Table 11.1).

11.5.1 Microbiota Treatments

Regulatory acceptance and challenges of fecal microbiota transplantation: recurrent *Clostridium difficile* infection is the primary condition treated with FMT, one of the most widely used microbiota-based therapies. Although it has been clinically successful, the regulatory path has been challenging: (1) Regulatory status: The US

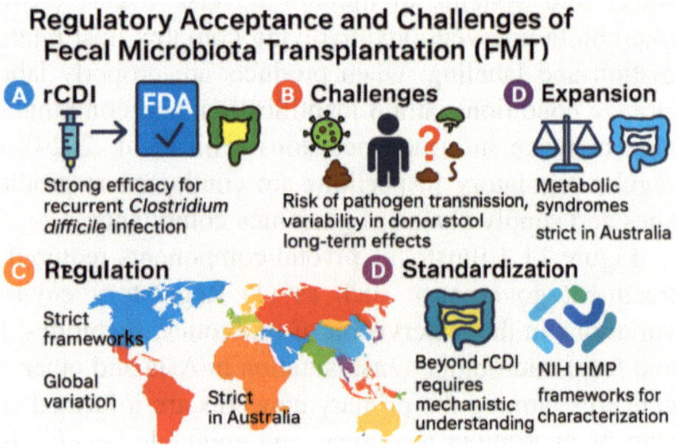

Fig. 11.5 Regulatory acceptance and challenges of fecal microbiota transplantation

Table 11.1 Research and applications of microbiota: safety, standardization, and policy development

Topic	Key points	Challenges	Recommendations and solutions	References
Safety considerations	Complex microbial ecosystems; possible dangers include resistance, pathogenicity, and contamination	Unknown long-term consequences Live therapy safety	Thorough safety testing postmarket monitoring, and risk assessment models	Suez et al. (2019) and Hill et al. (2014)
Standardization needs	Comparability and reproducibility Coordinated data analysis, sequencing, and sampling Manufacturing GMP	Variability between methods and populations The absence of standardized protocols	SOPs, international consensus standards (IHMS), and benchmarks for quality control	Knight et al. (2018a, b), Costea et al. (2017), and Quince et al. (2017)
Microbiota therapeutics standardization	Specifying strains, dosage, and viability Quality control and manufacturing best practices	Uncertainty in regulations and batch variability	Clear regulatory guidelines; GMP production; defined potency and purity standards	FDA (2016), Hill et al. (2014), and Suez et al. (2019)
Regulatory landscape	Various strategies (FDA, EMA) Changing policy frameworks	Inconsistencies across borders and regulatory lag	Risk-based, flexible regulation; attempts at harmonization Participation of stakeholders	FDA (2016), EMA (2019), and Suez et al. (2019)
Ethical and legal considerations	Consent and data privacy Microbiota resource ownership	Privacy violations and disputes over benefit-sharing	Strong consent procedures and open data governance Equitable benefit-sharing schemes	Knight et al. (2018a, b) and Sankar and Parker (2017)
Policy frameworks and implementation	The creation of multistakeholder policies Enforcing compliance International cooperation	Disjointed policies and difficulties with enforcement	Adaptive, evidence-based policies International networks for collaboration Standardized systems for monitoring	EMA (2019) and Sankar and Parker (2017)

(continued)

Table 11.1 (continued)

Topic	Key points	Challenges	Recommendations and solutions	References
Future directions	Integration of multiomics Policies that are dynamic Global governance and public education	Integration of multiomics Policies that are dynamic Global governance and public education	Multiomics protocols that are standardized Constantly updating policies Initiatives for public engagement	Quince et al. (2017) and Knight et al. (2018a, b)

Food and Drug Administration (FDA) initially classified FMT as an Investigational New Drug (IND), requiring strict clinical trial oversight (FDA 2013a, b). Nevertheless, because of clinical demand and efficacy, the FDA established an enforcement discretion policy for rCDI treatment allowing its use in some situations without an IND (FDA 2019a, b). (2) Challenges: safety issues are brought up by the possibility of pathogen transmission, long-term consequences, and donor material variability. Standardizing donor screening, stool processing, and delivery practices is still necessary to lower risks (Cammarota et al. 2017). (3) Global variation: different countries' regulatory positions, which range from strict frameworks (like Australia) to complete drug regulation (like Canada) make it more challenging to collaborate on clinical practice and research internationally. (4) Emerging applications: the expansion of FMT into other diseases (like metabolic syndrome and inflammatory bowel disease) is subject to stiffer regulatory scrutiny due to higher risk profiles and a lack of a clear understanding of the mechanisms, highlighting the need for strong clinical evidence. (5) Effective standardization efforts by the microbiome research consortium: an outline of the Human Microbiome Project (HMP): the innovative HMP consortium was formed by the NIH to employ standardized protocols to systematically characterize the human microbiome across individuals and body sites (Fig. 11.5; Table 11.1) (Turnbaugh et al. 2007).

Figure 11.5 illustrates FMT's remarkable clinical success in treating recurrent Clostridium difficile infection (rCDI) but obtaining regulatory approval remains a challenge. FMT was initially classified as an Investigational New Drug by the US FDA but it later adopted an enforcement discretion policy for the treatment of rCDI. Major challenges that highlight the need for standardized stool processing and donor screening include unknown long-term effects, variability in donor material, and pathogen transmission risks. Different regulatory approaches around the world ranging from strict oversight in Australia to drug-level regulation in Canada impede international harmonization. Initiatives such as the NIH Human Microbiome Project encourage standardization and provide frameworks for expanding FMT into new therapeutic areas under more stringent evidence requirements.

11.5.2 Attempts to Standardize

1. Sampling and data collection: Standardized DNA extraction, sequencing platforms, sample collection, and metadata annotation ensured cross-study comparability.
2. Data sharing: Open-access databases and bioinformatics pipelines accelerated reproducibility and secondary analyses.
3. Quality control: Data reliability was improved through the use of control samples and batch effect monitoring.
4. Impact: By establishing benchmarks for the caliber of microbiome research, HMP contributed to the development of new studies and products on a global scale according to the Integrative HMP Research Network Consortium (2019).
5. International Human Microbiome (IHMS) Standards: IHMS developed consensus protocols for processing, sequencing, and reporting data from microbiome samples in order to reduce methodological biases and enhance interlaboratory reproducibility (The Human Microbiome Standards Consortium 2012): many research groups and industry stakeholders have adopted these standards, which have improved the robustness and comparability of microbiota data worldwide (Table 11.1).

11.5.3 Policy Responses to Emerging Microbiota-Related Issues

1. Regulatory agencies are now carefully reviewing probiotic health claims due to exaggerated or unsupported marketing.
2. The European Food Safety Authority (EFSA) rejected several probiotic claims for lack of evidence, which prompted calls for comprehensive clinical substantiation and transparent communication (EFSA 2012). This position has improved consumer protection by pressuring manufacturers to embrace standardized labeling and higher scientific standards.
3. Keeping microbiome data private: As direct-to-consumer microbiome testing has increased, new regulations have been developed in response to worries about data security and misuse. The European Union's General Data Protection Regulation (GDPR), which requires express consent and limits secondary use, safeguards personal data, including microbiome data (Regulation (EU) 2016/679). Global adoption or consideration is underway for comparable privacy frameworks that uphold participant rights and ethical data stewardship.
4. Global harmonization initiatives: By focusing on equitable access to microbiota resources and the sharing of conservation benefits, programs like the Global Microbiome Conservancy influence policy discussions on microbial biobanking and biodiversity.
5. WHO, FDA, EMA, and other agencies are collaborating to standardize regulatory approaches for microbiota-based products in order to enable safer and more efficient market entry (WHO 2021) (Table 11.1).

11.6 Opportunities and Recommendations

In the rapidly evolving field of microbiota science, new methods are required to ensure safe, effective, and ethical applications. This section identifies pivotal areas for further research, policy formation, and public engagement to enhance the translational potential and societal impact of microbiota-based innovations.

11.6.1 Utilizing Multiomics Data to Enhance Safety and Standardization

Advances in high-throughput technologies such as metagenomics, metatranscriptomics, metabolomics, and proteomics enable thorough characterization of microbial communities and their functional dynamics (Hasin et al. 2017). The integration of these multiomics datasets has several benefits: (1) Improved safety profiles: by exposing microbial gene expression patterns, metabolic processes, and host-microbe interactions that single-layer analyses might miss, multiomics has the capacity to aid in the early detection of potential pathogenicity or adverse effects (Tierney et al. 2024). (2) Better standardization: multiomics integration has the capacity to produce functional signatures and trustworthy biomarkers that serve as repeatable standards for clinical monitoring and product quality assurance (Quince et al. 2017). (3) Personalized therapeutics are data-driven strategies that help create precision microbiota treatments that are tailored to each host's environment and genetic makeup reducing risks and boosting effectiveness. It is still challenging to harmonize multiomics data collection, analysis pipelines, and interpretation frameworks, underscoring the need for international consensus standards and shared computational resources (Heintz-Buschart et al. 2016).

11.6.2 Creating Adaptable Guidelines to Stay Up to Date with Microbiota Research

The increasing complexity and uniqueness of microbiota applications necessitate the need for adaptable regulatory frameworks that have the capacity to be updated to take into account new information and developments in technology. (1) Adaptive regulation models: while traditional stringent approval pathways may stifle innovation, adaptive models that leverage real-world data, phased approvals, and conditional licensing have the capacity to expedite access while ensuring safety (Turnbaugh et al. 2007). By categorizing products based on risk profiles (live bio-therapeutics/metabolites derived from microorganisms), risk-based approaches enable tailored regulatory oversight that is in line with potential hazards. (2) Continuous stakeholder engagement: it is crucial that scientists, regulators, industry, clinicians, and patients have continuous conversations to make sure that policy is in accordance with scientific findings and societal values (Yadegar et al. 2024). (3) World harmonization: coordinated international efforts have the capacity to reduce unnecessary testing, facilitate cross-border research, and ensure equitable access to microbiota innovations (WHO 2021).

11.6.3 Improving Education and Public Awareness

The effective utilization of microbiota-based products and treatments depends on public acceptance and awareness. (1) Transparent communication: transparent fact-based messaging regarding advantages, risks, and limitations has the capacity to also elevate trust and reduce misinformation. (2) Including microbiota science in curricula: future professionals will be better prepared for this multidisciplinary field if microbiome literacy is incorporated into medical, life science, and public health education (Cho and Blaser 2012). (3) Including communities: Citizen Science and participatory research projects empower people, promote moral data exchange, and guarantee cultural sensitivity.

11.6.4 Promoting Collaborative Global Governance Frameworks

Governance frameworks must cross national borders because microbial ecosystems and product markets are global in scope: (1) International consortia: organizations like the International Microbiome Standards Consortium have the capacity to lead initiatives to reach agreements on data exchange, protocols, and moral principles (The Human Microbiome Standards Consortium 2012). (2) Benefit sharing models: sustainable and just policies that guarantee the equitable distribution of the benefits of microbiota resources particularly those derived from biodiverse and indigenous environments. Global platforms for postmarket surveillance, adverse event reporting, and tracking antibiotic resistance enhance safety and provide information for regulatory updates through monitoring and surveillance networks.

11.7 Conclusion

There are unprecedented opportunities to advance human health, agriculture, and environmental sustainability through the study of microbiota and its applications at this pivotal moment in history. Nevertheless, achieving this potential will necessitate overcoming formidable obstacles in the areas of policy development, standardization, and safety. Strict evaluation frameworks that take into account the complexity and variability found in microbial communities are necessary to ensure the safety of products derived from microbiota. There is still a significant chance of negative effects, contamination or unforeseen ecological repercussions in the absence of thorough safety assessments. Comparability, reproducibility, and regulatory approval are all aided by standardization. Standardized processes for sampling, analysis, manufacturing, and quality control are essential for generating reliable data, substantiating therapeutic claims, and enabling the scalable production of microbiota-based interventions. Consensus guidelines and globally coordinated efforts will speed up progress and facilitate access to global markets. Even though laws governing the utilization of microbiota are evolving, more work must be done to keep up with new discoveries in science. Dynamic and flexible regulatory models founded on evidence-based decision making and multistakeholder engagement are necessary to strike a balance between innovation and public safety. Ethical concerns like data privacy, informed consent, and fair benefit sharing must also be carefully considered in order to develop public trust and social license. Importantly, to tackle these complicated problems, interdisciplinary methods combining clinical sciences, bioinformatics, microbiology, regulatory affairs, ethics and public participation are required. Networks of cooperation between academia, industry, regulators, and communities will be crucial for the ethical and successful application of microbiota science. Hence, the future of microbiota applications will be determined by a confluence of scientific rigor, technological innovation, regulatory foresight, and societal discourse. This kind of all-encompassing strategy promises to safeguard sustainability and health while optimizing the potential of microbiota to have transformative effects across industries.

References

Alang N, Kelly CR (2015) Weight gain after fecal microbiota transplantation. Open Forum Infect Dis 2(1):ofv004

Aroniadis OC, Brandt LJ (2013) Fecal microbiota transplantation: past, present and future. Curr Opin Gastroenterol 29(1):79–84

Bafeta A, Yavchitz A, Riveros C, Batista R, Ravaud P (2017) Methods and reporting studies assessing fecal microbiota transplantation: a systematic review. Ann Intern Med 167(1):34–39

Becattini S, Taur Y, Pamer EG (2016) Antibiotic-induced changes in the intestinal microbiota and disease. Trends Mol Med 22(6):458–478

Berg G, Rybakova D, Fischer D, Cernava T, Champomier Vergès MC et al (2019) Microbiome definition re-visited: old concepts and new challenges. Microbiome 7(1):103

Brodmann T, Endo A, Gueimonde M, Vinderola G, Salminen S (2017) Safety of novel microbes for human consumption: practical examples of assessment in the European Union. Front Microbiol 8:1725

Cammarota G et al (2017) European consensus conference on faecal microbiota transplantation in clinical practice. Gut 66(4):569–580

Cho I, Blaser MJ (2012) The human microbiome: at the interface of health and disease. Nat Rev Genet 13(4):260–270

Costea PI, Zeller G, Sunagawa S et al (2017) Towards standards for human fecal sample processing in metagenomic studies. Nat Biotechnol 35(11):1069–1076

DeFilipp Z, Bloom PP, Torres Soto M, Mansour MK, Sater MRA et al (2019) Drug-resistant E. coli bacteremia transmitted by fecal microbiota transplant. N Engl J Med 381(21):2043–2050

EFSA (2012) Scientific opinion on the substantiation of health claims related to probiotics. EFSA J 10(3):1–31

EMA (2019) Reflection paper on the regulatory requirements for probiotics as medicinal products [cited 2025 Jun 20]

FDA (2013a) Guidance for industry: early clinical trials with live biotherapeutic products [cited 2025 Jun 20]

FDA (2013b) Guidance for industry: enforcement policy regarding investigational new drug requirements for utilize of fecal microbiota for transplantation [cited 2025 Jun 20]

FDA (2016) Early clinical trials with live biotherapeutic products: chemistry, manufacturing, and control information. Guidance for Industry [cited 2025 Jun 20]

FDA (2019a) Enforcement policy regarding IND requirements for FMT to treat Clostridioides difficile infection [cited 2025 Jun 20]

FDA (2019b) Safety alert: serious adverse reactions reported following fecal microbiota for transplantation [cited 2025 Jun 20]

FDA (2020) Microbiome-based products: regulatory considerations. Draft guidance for industry [cited 2025 Jun 20]

Feuerstadt P, Louie TJ, Lashner B, Wang EEL, Diao L, Bryant JA, Sims M, Kraft CS, Cohen SH, Berenson CS, Korman LY, Ford CB, Litcofsky KD, Lombardo MJ, Wortman JR, Wu H, Auniņš JG, McChalicher CWJ, Winkler JA, McGovern BH, Trucksis M, Henn MR, von Moltke L (2022) SER-109, an oral microbiome therapy for recurrent clostridioides difficile infection. N Engl J Med 386(3):220–229. https://doi.org/10.1056/NEJMoa2106516. PMID: 35045228

Gilbert JA, Quinn RA, Debelius J, Xu ZZ, Morton J et al (2018) Microbiome-wide association studies link dynamic microbial consortia to disease. Nat Rev Microbiol 16(11):681–692

Hasin Y, Seldin M, Lusis A (2017) Multi-omics approaches to disease. Genome Biol 18:83

Health Canada (2018) Natural health products regulations: probiotics [cited 2025 Jun 20]

Heintz-Buschart A, May P, Laczny CC, Lebrun LA, Bellora C, Krishna A, Wampach L, Schneider JG, Hogan A, de Beaufort C, Wilmes P (2016) Integrated multi-omics of the human gut microbiome in a case study of familial type 1 diabetes. Nat Microbiol 2:16180. https://doi.org/10.1038/nmicrobiol.2016.180. Erratum in: Nat Microbiol. 2016 Oct 24;2:16227. https://doi.org/10.1038/nmicrobiol.2016.227. PMID: 27723761

Hill C, Guarner F, Reid G, Gibson GR, Merenstein DJ, Pot B et al (2014) Expert consensus document: the ISAPP consensus statement on the scope and appropriate utilize of the term probiotic. Nat Rev Gastroenterol Hepatol 11(8):506–514

Hoffmann TC et al (2021) Risk-based regulation of microbiota therapeutics: a pragmatic approach. Nat Rev Drug Discov 20(6):396–397

Johnson CM, Grossman AD (2015) Integrative and conjugative elements (ICEs): what they do and how they work. Annu Rev Genet 49:577–601

Khanna S, Sims M, Louie TJ, Fischer M, LaPlante K, Allegretti J, Hasson BR, Fonte AT, McChalicher C, Ege DS, Bryant JA, Straub TJ, Ford CB, Henn MR, Wang EEL, von Moltke L, Wilcox MH (2022) SER-109: an oral investigational microbiome therapeutic for patients with Recurrent Clostridioides difficile Infection (rCDI). Antibiotics (Basel) 11(9):1234. https://doi.org/10.3390/antibiotics11091234

Knight R, Callewaert C, Marotz C, Hyde ER, Debelius JW et al (2018a) The microbiome and human biology. Annu Rev Genomics Hum Genet 19:1–22

Knight R, Vrbanac A, Taylor BC et al (2018b) Best practices for analysing microbiomes. Nat Rev Microbiol 16(7):410–422

Knoppers BM, Chadwick R (2019) Human microbiome research: privacy and ethics. Science 363(6427):553–554

Krautkramer KA, Kreznar JH, Romano KA, Vivas EI, Barrett-Wilt GA et al (2021) Diet–microbiota interactions mediate global epigenetic programming in multiple host tissues. Mol Cell 78(5):1039–55.e5

Liu Y, Alookaran JJ, Rhoads JM (2018) Probiotics in autoimmune and inflammatory disorders. Nutrients 10(10):1537

Lloyd-Price J, Abu-Ali G, Huttenhower C (2016) The healthy human microbiome. Genome Med 8(1):51

McDonald D, Hyde E, Debelius JW, Morton JT, Gonzalez A, Ackermann G et al (2018) American Gut: an open platform for citizen science microbiome research. mSystems 3(3): e00031–e00018

Mendes R, Garbeva P, Raaijmakers JM (2013) The rhizosphere microbiome: significance of plant beneficial, plant pathogenic, and human pathogenic microorganisms. FEMS Microbiol Rev 37(5):634–663

Ministry of Health, Labour and Welfare Japan (2018) Regenerative medicine legislation and microbiota therapies [cited 2025 Jun 20]

Morelli L, Capurso L (2012) FAO/WHO guidelines on probiotics: 10 years later. J Clin Gastroenterol 46:S1–S2

NIH Human Microbiome Portfolio Analysis Team (2019) A review of 10 years of human microbiome research activities at the US National Institutes of Health, Fiscal Years 2007–2016. Microbiome 7(1):31. https://doi.org/10.1186/s40168-019-0620-y. PMID: 30808411; PMCID: PMC6391833

Olesen SW, Alm EJ (2016) Dysbiosis is not an answer. Nat Microbiol 1(5):16228

Ouwehand AC, Derrien M, de Vos WM, Tiihonen K, Burcelin R (2021) Prebiotics and other microbial substrates for gut functionality: a review. Br J Nutr 126(S1):S54–S65

Pamer EG (2016) Resurrecting the intestinal microbiota to combat antibiotic-resistant pathogens. Science 352(6285):535–538

Petrof EO, Gloor GB, Vanner SJ, Weese JS, Allen-Vercoe E (2013) Stool substitute transplant therapy for the eradication of Clostridium difficile infection: 'RePOOPulating' the gut. Microbiome 1(1):3

Pollock J, Glendinning L, Wisedchanwet T, Watson M (2018) The madness of microbiome: attempting to find consensus "best practice" for 16S microbiome studies. Appl Environ Microbiol 84(7):e02627-17

Quince C, Walker AW, Simpson JT, Loman NJ, Segata N (2017) Shotgun metagenomics, from sampling to analysis. Nat Biotechnol 35(9):833–844

Sanders ME, Merenstein DJ, Reid G, Gibson GR, Rastall RA (2019) Probiotics and prebiotics in intestinal health and disease: from biology to the clinic. Nat Rev Gastroenterol Hepatol 16(10): 605–616

Sankar PL, Parker LS (2017) The Precision Medicine Initiative's All of Us Research Program: an agenda for research on its ethical, legal, and social issues. Genet Med 19(7):743–750

Schmidt TSB, Raes J, Bork P (2018) The human gut microbiome: from association to modulation. Cell 172(6):1398–1414

Sheth RU, Cabral V, Chen SP, Wang HH (2016) Manipulating bacterial communities by in situ microbiome engineering. Trends Genet 32(4):189–200

Sinha R, Abnet CC, White O, Knight R, Huttenhower C (2016) The microbiome quality control project: baseline study design and future directions. Genome Biol 16:276

Smillie CS, Smith MB, Friedman J, Cordero OX, David LA, Alm EJ (2011) Ecology drives a global network of gene exchange connecting the human microbiome. Nature 480(7376):241–244

Sommer F, Bäckhed F (2013) The gut microbiota—masters of host development and physiology. Nat Rev Microbiol 11(4):227–238

Song SJ, Amir A, Metcalf JL et al (2016) Preservation methods differ in fecal microbiome stability, affecting suitability for field studies. mSystems 1(3):e00021–e00016

Suez J, Zmora N, Segal E, Elinav E (2019) The pros, cons, and many unknowns of probiotics. Nat Med 25(5):716–729

Tierney BT, Kim J, Overbey EG, Ryon KA, Foox J, Sierra MA, Bhattacharya C, Damle N, Najjar D, Park J, Garcia Medina JS, Houerbi N, Meydan C, Wain Hirschberg J, Qiu J, Kleinman AS, Al-Ghalith GA, MacKay M, Afshin EE, Dhir R, Borg J, Gatt C, Brereton N, Readhead BP, Beyaz S, Venkateswaran KJ, Wiseman K, Moreno J, Boddicker AM, Zhao J, Lajoie BR, Scott RT, Altomare A, Kruglyak S, Levy S, Church GM, Mason CE (2024) Longitudinal multi-omics analysis of host microbiome architecture and immune responses during short-term spaceflight. Nat Microbiol 9(7):1661–1675. https://doi.org/10.1038/s41564-024-01635-8. Epub 2024 Jun 11. PMID: 38862604; PMCID: PMC11222149

The Human Microbiome Project Consortium (2012) Structure, function and diversity of the healthy human microbiome. Nature 486:207–214. https://doi.org/10.1038/nature11234

The Integrative HMP (iHMP) Research Network Consortium (2019) The integrative human microbiome project. Nature 569:641–648. https://doi.org/10.1038/s41586-019-1238-8

Turnbaugh PJ et al (2007) The Human Microbiome Project. Nature 449(7164):804–810

Verna EC, Lucak S (2010) Use of probiotics in gastrointestinal disorders: what to recommend? Ther Adv Gastroenterol 3(5):307–319. https://doi.org/10.1177/1756283X10373814. PMID: 21180611; PMCID: PMC3002586

Wellington EM, Boxall AB, Cross P, Feil EJ, Gaze WH et al (2013) The role of the natural environment in the emergence of antibiotic resistance in gram-negative bacteria. Lancet Infect Dis 13(2):155–165

WHO (World Health Organization) (2021) Global policy framework for microbiome-related products

Wilcox MH, Gerding DN, Poxton IR, Kelly C, Nathan R, Birch T, Cornely OA, Rahav G, Bouza E, Lee C, Jenkin G, Jensen W, Kim YS, Yoshida J, Gabryelski L, Pedley A, Eves K, Tipping R, Guris D, Kartsonis N, Dorr MB, MODIFY I and MODIFY II Investigators (2017) Bezlotoxumab for prevention of recurrent clostridium difficile infection. N Engl J Med 376(4):305–317. https://doi.org/10.1056/NEJMoa1602615. PMID: 28121498

Yadegar A, Bar-Yoseph H, Monaghan TM, Pakpour S, Severino A, Kuijper EJ, Smits WK, Terveer EM, Neupane S, Nabavi-Rad A, Sadeghi J, Cammarota G, Ianiro G, Nap-Hill E, Leung D, Wong K, Kao D (2024) Fecal microbiota transplantation: current challenges and future landscapes. Clin Microbiol Rev 37(2):e0006022. https://doi.org/10.1128/cmr.00060-22. Epub 2024 May 8. PMID: 38717124; PMCID: PMC11325845

Yatsunenko T, Rey FE, Manary MJ, Trehan I, Dominguez-Bello MG, Contreras M et al (2012) Human gut microbiome viewed across age and geography. Nature 486(7402):222–227

Zhang YJ, Li S, Gan RY, Zhou T, Xu DP, Li HB (2015) Impacts of gut bacteria on human health and diseases. Int J Mol Sci 16(4):7493–7519

Zhao L, Zhang F, Ding X, Wu G, Lam YY, Wang X et al (2018) Gut bacteria selectively promoted by dietary fibers alleviate type 2 diabetes. Science 359(6380):1151–1156

Zheng D, Liwinski T, Elinav E (2020) Interaction between microbiota and immunity in health and disease. Cell Res 30(6):492–506

Zmora N, Zilberman-Schapira G, Suez J, Mor U, Dori-Bachash M, Bashiardes S, Kotler E, Zur M, Regev-Lehavi D, Brik RB, Federici S, Cohen Y, Linevsky R, Rothschild D, Moor AE, Ben-Moshe S, Harmelin A, Itzkovitz S, Maharshak N, Shibolet O, Shapiro H, Pevsner-Fischer-M, Sharon I, Halpern Z, Segal E, Elinav E (2018) Personalized gut mucosal colonization resistance to empiric probiotics is associated with unique host and microbiome features. Cell 174(6):1388–1405.e21. https://doi.org/10.1016/j.cell.2018.08.041

Challenges, Ethical Considerations, and Regulatory Frameworks in Microbiota-Based Medicine

12

Abstract

The intricate problems pertaining to the modification of human and environmental microbiota are critically examined in this chapter. The conflict between human rights and scientific advancement is exemplified by ethical concerns such as informed consent, data privacy, genetic editing and equitable access. The ongoing ambiguity and fragmentation of the regulatory environment including differing definitions of microbiota-based therapies such as foods, medications or biologics complicates approval procedures and market authorization. Numerous ethical and regulatory issues have arisen as a result of the quick development of microbiota research and its growing applications in biotechnology, healthcare and agriculture. These issues need to be carefully and quickly considered. This section's thorough analysis of these urgent issues lays the groundwork for future discussions and the development of legislation intended to promote moral innovation in microbiota research. Commercialization attempts are complicated by a number of issues including intellectual property rights, bio-prospecting, benefit sharing, public misconceptions, cultural sensitivities and the environmental impact of releasing engineered microbial strains.

Keywords

Biological complexity and dysbiosis · Microbiota-based therapies · Clinical translation challenges · Regulatory uncertainty · Ethical and societal implications · Safety and risk assessment

R. C. Sobti et al., *Microbiota Unveiled: Science, Health, and Future Therapies*,
https://doi.org/10.1007/978-981-95-9031-5_12

12.1 Introduction

Microbiota-based medical interventions: Among the most significant findings in modern biomedical science is the human microbiota, i.e., a complex ecosystem of microorganisms that interact dynamically with the host. These microbial communities influence immunity, digestion, metabolism, neurological functions and they have an impact on both health as well as illness. Advancements in culturomics, functional metagenomics, and sequencing technologies have never provided a better understanding of the diversity and roles of the microbiota (Suez et al. 2019). This knowledge has sparked the development of microbiota-based treatments including fecal microbiota transplantation (FMT), probiotics, prebiotics, and next-generation engineered microbial consortia. These techniques are now regarded as promising therapeutic modalities that can effectively enhance conventional medical treatments, treat disorders associated with dysbiosis and also restore microbial balance. The translational potential of microbiome research is therefore demonstrated by a variety of therapeutic modalities. Probiotics (live microorganisms that confer a health benefit to the host when administered in adequate amounts) have been added to functional foods and dietary supplements. They are commonly used to treat gastrointestinal conditions such as irritable bowel syndrome, antibiotic-associated diarrhea, and recurrent *Clostridium difficile* infections. FMT, which involves moving stool from a healthy donor into a patient's gastrointestinal tract, has shown cure rates of over 85% for antibiotic-resistant recurrent *C. difficile* infections (Fig. 12.1).

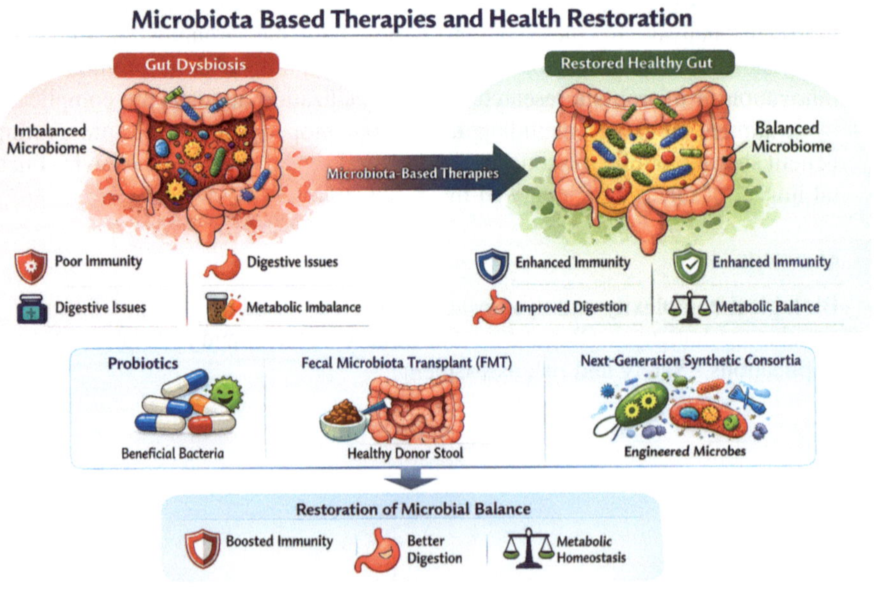

Fig. 12.1 Microbiota-based therapies and health restoration

Synthetic biology and systems biology techniques are enabling the engineering of microbial consortia with particular functional properties as a standardized and controlled alternative to donor-derived preparations. Despite these promising applications, the field still faces many challenges from the fields of biology, technology, medicine, ethics, and society. The intrinsic complexity, high inter-individual variability, and susceptibility to environmental influences of the microbiome complicate the standardization of therapeutic approaches. Regulatory frameworks, safety protocols, and large-scale manufacturing pipelines are still in their infancy. Concerns like equitable access, patient data privacy, and public perception of microbiome-based therapies must also be carefully taken into account in order to ensure the reliability of ethical implementation. This chapter brings these points of view together to form a coherent discussion that addresses therapeutic applications, present challenges, ethical and societal implications, real-world safety cases, and future research priorities. It aims to provide a comprehensive overview of the opportunities and challenges related to microbiota-based medical interventions, while also highlighting the critical areas that require coordinated scientific, clinical, and policy efforts to translate potential into safe, accessible, and effective therapies.

Figure 12.1 illustrates the conceptual framework showing how microbiota interventions can effectively correct gut dysbiosis and restore microbial balance. Different therapeutic approaches can effectively be used to target dysbiosis, which is characterized by altered microbial composition and function. These include (i) introducing beneficial live microorganisms (probiotics), (ii) transferring healthy donor stool (fecal microbiota transplantation), and (iii) engineering microbes with specific functions (next-generation synthetic consortia). By restoring a healthy gut microbiome, these tactics work together to enhance host immunity, digestion, and metabolic homeostasis.

12.2 Therapeutic Applications of Microbiota-Based Interventions

12.2.1 Gastrointestinal Disorders

The GI tract is the most studied site for microbiota-based interventions due to its dense microbial populations and intimate connection to dietary, environmental, and therapeutic exposures. Dysbiosis of the gut microbiota has been linked to a number of gastrointestinal conditions including inflammatory bowel disease (IBD), antibiotic-associated diarrhea, *Clostridium difficile* infection (CDI) and irritable bowel syndrome (IBS). So, in order to address microbial imbalances, probiotics, FMT, and engineered consortia have shown great therapeutic promise. FMT has revolutionized the treatment of recurrent CDI by achieving cure rates above 85%, even in refractory cases. According to Bakken et al. (2011) its efficacy is attributed to the restoration of a diverse microbial ecosystem that outcompetes pathogenic *C. difficile*, restores colonization resistance and reestablishes normal metabolic and immunological signaling. New research suggests that FMT may play a role in conditions other than CDI including Crohn's disease, ulcerative colitis and IBS though results are still uncertain due to host-specific factors and differences in donor material. Probiotics such as *Lactobacillus rhamnosus* GG, *Bifidobacterium longum* and *Saccharomyces boulardii* have been demonstrated to reduce antibiotic-associated diarrhea and prevent CDI relapse. They function by secreting antimicrobial metabolites, strengthening the epithelial barrier, competing with pathogens and altering the host's immune system. In addition, probiotic supplements have been tested for IBS and strains like *Bifidobacterium infantis* 35624 which have been shown to reduce irregular bowel habits, bloating, and abdominal pain. However, the outcomes of these trials remain mixed due to strain specificity and host variability. These findings also highlight the importance of the gut microbiota in maintaining gastrointestinal homeostasis and the therapeutic benefits of restoring microbial balance. However, the need for accurate strain identification, standardized donor selection, and standardized clinical trial protocols to ensure the reliability of safety and reproducibility is highlighted by the variability in results (Fig. 12.2).

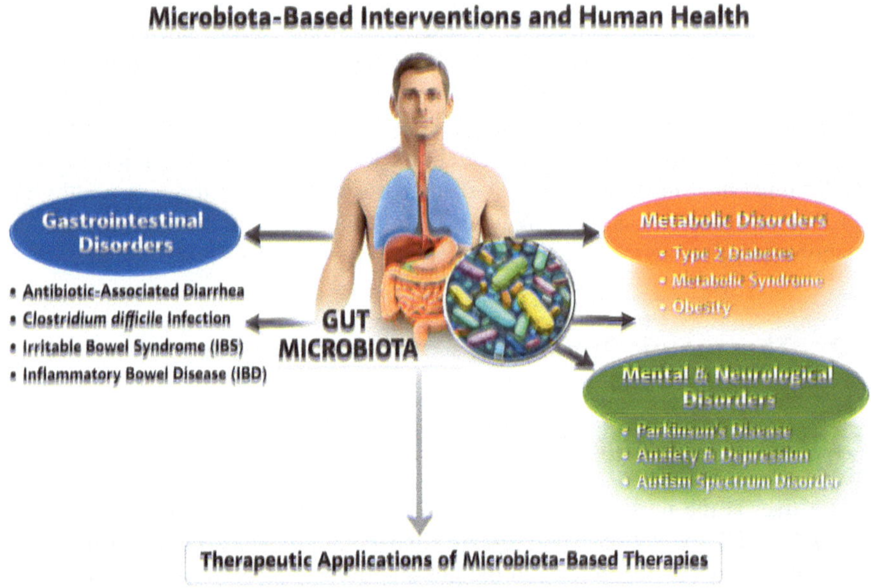

Fig. 12.2 Therapeutic applications of microbiota-based interventions

12.2.2 Metabolic Disorders

The human microbiota has a substantial influence on energy metabolism, nutrient absorption and the regulation of metabolic signaling pathways. Dysbiosis has been associated with obesity, metabolic syndrome, and type 2 diabetes mellitus (T2DM). Patients with type 2 diabetes and obesity frequently have altered Firmicute to Bacteroidete ratios and reduced microbial diversity which can effectively lead to altered bile acid metabolism, increased food energy harvest and chronic low-grade inflammation (Suez et al. 2019). The potential of FMT to enhance insulin sensitivity and metabolic profiles has been studied. According to a randomized controlled trial, lean donor FMT significantly improved peripheral insulin sensitivity in obese individuals with metabolic syndrome; however, the effects diminished over time suggesting that more dietary support or repeated administration is required. Probiotics and prebiotics are also being researched as modulators of metabolic health. For example, the next-generation probiotic *Akkermansia muciniphila* has shown promise in promoting glucose homeostasis and reducing weight gain by altering host-microbe signaling pathways and strengthening the gut barrier (Suez et al. 2019). Controlling metabolic health may also benefit from dietary interventions that encourage the growth of beneficial bacteria. Prebiotics like inulin and fructo-oligosaccharides elevate the production of short-chain fatty acids (SCFAs) which enhance satiety, reduce inflammation, and elevate metabolic flexibility by specifically promoting beneficial bacteria like Bifidobacterium. These strategies suggest that microbiota-targeted interventions may complement conventional pharmacological therapies for diabetes and obesity, despite the lack of long-term safety and efficacy data (Fig. 12.2).

12.2.3 Mental and Neurological Disorders

The two-way communication between the gut microbiota and the central nervous system known as the gut-brain axis has opened up new treatment options for neuropsychiatric and neurodegenerative diseases. Through immunological modulation, neurotransmitter synthesis, and vagus nerve signaling, there is mounting evidence that the gut microbiota can effectively influence mood, cognition as well as behavior. Preclinical and clinical studies support the role of microbiota in conditions such as autism spectrum disorders, anxiety, depression, and Parkinson's disease. By controlling systemic inflammation and lowering cortisol levels some probiotic strains such as *Bifidobacterium longum* R0175 and *Lactobacillus helveticus* R0052 have been shown to alleviate depressive symptoms. Improvements in gastrointestinal comorbidities and behavioral symptoms in Autism Spectrum Disorder (ASD) have been observed with FMT and probiotic treatments suggesting a potential therapeutic relationship mediated by gut microbial metabolites and immune responses. Recent research also suggests that neurodegenerative diseases may be influenced by gut microbiota. According to Suez et al. (2019) dysbiosis characterized by decreased SCFA production has been linked to increased α-synuclein aggregation whereas *Akkermansia muciniphila* and bacteria that produce SCFA are thought to protect against neuroinflammation in models of Parkinson's disease. Although their application to clinical practice is still in its infancy, these findings represent a paradigm shift in the way that neurological disorders are viewed through a microbial lens (Fig. 12.2).

Figure 12.2 illustrates (A) gastrointestinal disorders—probiotics, FMT and engineered consortia restore microbial balance, prevent *C. difficile* infection, and relieve IBS/IBD symptoms, (B) metabolic disorders—lean donor FMT, next-generation probiotics (e.g., *Akkermansia muciniphila*), and prebiotics enhance insulin sensitivity, weight control, and SCFA production, (C) mental and neurological disorders—gut-brain axis modulation via probiotics and FMT alleviates depression, autism symptoms, and may protect against neuro-inflammation in Parkinson's disease.

12.3 Barriers and Challenges in Microbiota-Based Medicine

The potential of human microbiota-based therapeutic interventions to transform treatment approaches for infections, neuropsychiatric disorders, gastrointestinal disorders, metabolic disorders, and even cancer has sparked a great deal of scientific and clinical interest. The road from discovery to bedside application is paved with challenges despite their potential. Biological complexity, methodological constraints, clinical variability, regulatory uncertainty, unresolved ethical and social issues are main causes of these difficulties. A thorough examination of these challenges clarifies why the translation of microbiota-based therapies remains a challenging task (Fig. 12.3).

12.3.1 Biological and Ecological Complexity

With trillions of microorganisms interacting with host genetics, immune responses, diet, and environment the human microbiota is one of the most known diverse and dynamic ecosystems. One of the main obstacles to creating standardized microbiota-based treatments is this inherent complexity. Significant interindividual variation exists in metabolite production, functional pathways and microbial composition even in healthy populations (Lloyd-Price et al. 2016). These differences make it more difficult to find universal therapeutic microorganisms and restrict the reproducibility of studies. Furthermore, because microbial communities are ecologically resilient, interventions might not produce predictable outcomes. For instance, probiotics may successfully colonize one individual but not establish or persist in another due to differences in niche availability, present microbial competitors, and host immune responses (Zmora et al. 2018). Similarly, although fecal microbiota transplantation has shown mixed results for recurrent Clostridium difficile infections, it has shown mixed results for other conditions such as inflammatory bowel disease likely due to ecological incompatibilities between donor and recipient

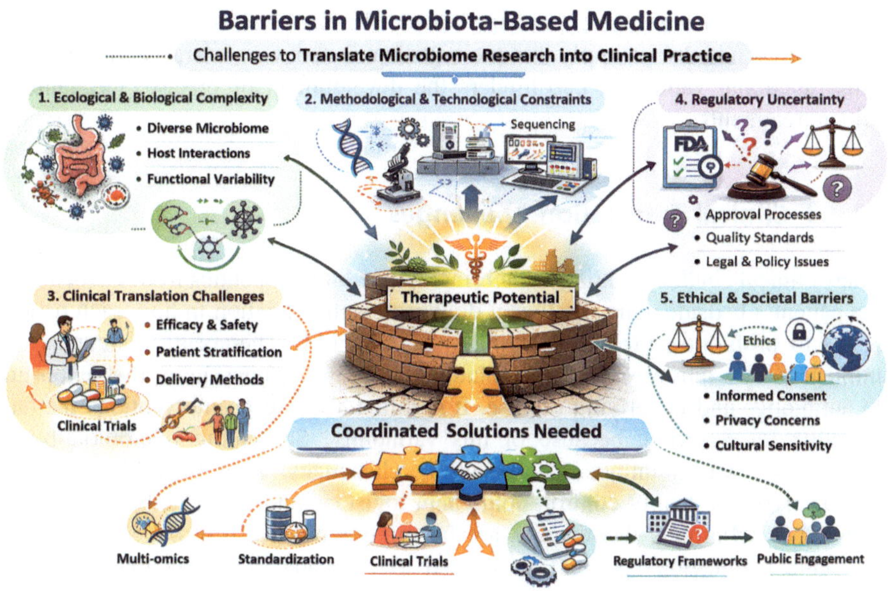

Fig. 12.3 Barriers in microbiota-based medicine

microbiomes. Complicating matters is the dynamic nature of microbial ecosystems which can effectively adapt to different selective pressures like diet, antibiotics or therapeutic interventions. Microbes may change metabolite production, develop resistance mechanisms or transition to opportunistic pathogenic states all of which can effectively compromise treatment efficacy and create new risks (Suez et al. 2019). These biological uncertainties continue to pose a significant obstacle to the safe and effective use of microbiota-based therapies.

12.3.2 Methodological and Technological Limitations

The study of microbiome is limited by technological challenges in bioinformatics analysis, sequencing, and culturing. Because a significant portion of gut microbes are unculturable by standard laboratory methods, functional characterization is therefore constrained. Even though advances in culturomics and organoid models are helping to close this gap, therapeutic development is hindered by an inadequate representation of the microbial community. While meta-genomics, meta-transcriptomics, metabolomics, and proteomics are examples of sequencing technologies that provide valuable insights, they also produce large and complex datasets that are difficult to integrate. The lack of harmonized bio-informatic pipelines and standardized analytical methods leads to poor comparability between studies and inconsistent results. Biases in reference databases further complicate interpretation especially for rare or poorly understood taxa. Moreover, functional inference from sequence data remains limited. The existence of a gene or microbial species does not always imply that the organism is active or interacts with its host. This functional validation gap hinders the development of precise, mechanism-driven treatments.

12.3.3 Clinical Translation Challenges

Many clinical challenges arise when bringing live biotherapeutics from the bench to the patient's bedside. One significant drawback is the variation in how each patient reacts to probiotics, prebiotics, synbiotics, and FMT. Therapeutic outcomes are influenced by a number of factors including genetics, immune status, diet, exposure to antibiotics and underlying disease. It is difficult to create interventions that are effective for everyone because of this variability. So scaling microbiota-based treatments presents unique manufacturing challenges. Live microbial products require rigorous quality control, cold chain logistics, and reproducible production techniques. Microbial therapeutics are far more expensive to produce than small molecule medications because they need to be made in a way that maintains viability, guarantees purity and prevents contamination. Adverse events have also been reported in clinical applications. For example, FMT has been linked to the spread of multidrug-resistant organisms and probiotics have occasionally caused bloodstream infections in immunocompromised patients (Suez et al. 2019). The necessity of careful strain selection, donor screening and patient risk assessment before clinical use is highlighted by these safety concerns.

12.3.4 Regulatory Uncertainty

The regulatory landscape pertaining to microbiota-based treatments is still evolving. Unlike conventional pharmaceuticals, these interventions make use of living organisms whose behavior is unpredictable. There are variations in the oversight and approval procedures because different jurisdictions classify products based on microbiota differently, whether as drugs, biologics, tissues or dietary supplements. Despite the Food and Drug Administration's (FDA) regulation of FMT under investigational new drug (IND) applications, the U.S. Food and Drug Administration's (FDA) enforcement discretion regarding recurrent *C. difficile* infections demonstrates a lack of standardized worldwide guidance. In addition, comprehensive clinical testing is often skipped by probiotics marketed as dietary supplements which raise concerns about unsubstantiated health claims and inconsistent product quality. The absence of standardized international regulatory frameworks puts patient safety at risk, impedes collaboration, and delays translation. Establishing globally accepted standards for microbial therapies is therefore essential.

12.3.5 Ethical and Societal Barriers

The transition from microbiota research to clinical practice raises critically important ethical and social questions. Fair access is a substantial issue because therapies like FMT and engineered microbial consortia are still expensive and logistically difficult. This is particularly true in LMICs which have the highest rates of chronic and infectious diseases. In the absence of proactive policies to lower costs and elevate capacity these innovations run the risk of making global health disparities worse. Another barrier is privacy and data ownership. Microbiome profiles which can effectively reveal lifestyle choices, health conditions, and even environmental exposures, raise questions about consent, abuse and potential discrimination. Unlike human genomic data, microbiome data is less protected by laws, making breaches or commercial exploitation more likely (Lloyd-Price et al. 2016). Finally, the hype and misinformation surrounding treatments based on microbiota exacerbate barriers. Commercial probiotics are commonly marketed with exaggerated or unverified claims, which can effectively lead to irrational expectations and potentially erode public trust. To guarantee informed decision making and safeguard consumers, effective science communication, patient education, and regulatory monitoring are required.

In conclusion, the potential of microbiota-based medicine is constrained by substantial challenges in the domains of biology, methodology, clinical practice, regulation, and society. Microbial adaptability, high inter-individual variability and a lack of mechanistic understanding complicate therapeutic design. Limitations in sequencing, culture, and bioinformatics hinder functional validation. Clinical translation is beset by problems with cost, scalability, and safety as well as regulatory frameworks remain fragmented. Ethical issues like disinformation, privacy, and equity make responsible application more challenging. Improved mechanistic research, standardized standards, interdisciplinary cooperation and proactive safety and accessibility laws are needed to overcome these obstacles. The full potential of microbiota-based medicine to enhance human health will only be realized at that point.

Figure 12.3 illustrates various obstacles that must be overcome in order to translate microbiome research into clinical practice: (1) Ecological and biological complexity. (2) Methodological and technological constraints. (3) Clinical translation challenges. (4) Regulatory uncertainty and (5) Ethical and societal barriers. To fully realize the therapeutic potential of microbiota-based interventions each domain needs coordinated solutions.

12.4 Ethical, Societal, and Safety Considerations

As microbiota-based medical interventions move from experimental research into routine clinical practice, they raise critically important ethical, societal, and safety concerns. These elements are essential to ensuring that these treatments are applied responsibly, inclusively, and fairly. This includes public trust, informed consent,

Fig. 12.4 Ethical, societal, and safety considerations in microbiota-based medical interventions

patient privacy, fair access and potential adverse outcomes. If these problems are not fixed, barriers to acceptance and widespread use may arise for even the most scientifically promising interventions (Fig. 12.4).

Figure 12.4 illustrates the key domains influencing responsible clinical implementation, including equitable access to treatment, informed consent processes, data security and patient privacy, public perception and trust, regulatory oversight, and potential adverse outcomes. Together, these interconnected factors determine the safe, fair, and sustainable integration of microbiota-based therapies into routine healthcare practice.

12.4.1 Equitable Access to Microbiota-Based Treatments

Equitable access is one of the main barriers to the broad adoption of microbiota-based therapies. Probiotics, fecal microbiota transplantation and engineered microbial consortia are examples of interventions that are often expensive to develop, produce, and administer. The availability is primarily restricted to healthcare systems with sufficient funding because of these costs and the need for specialized logistics like cold-chain transportation and advanced microbiology labs. However, accessibility is an issue in low- and middle-income countries which often have the highest rates of infectious and chronic illnesses. For instance, despite being a highly successful treatment for recurrent *Clostridium difficile* infections, FMT is still either unavailable or too costly in many LMICs. In a similar vein, advanced infrastructure, regulatory frameworks, and diagnostics are needed for next-generation microbial therapies and precision probiotics, all of which are commonly lacking in developing nations. By denying vulnerable groups access to treatments that could change their lives, this disparity could worsen already existing health disparities. The creation of standardized, reasonably priced formulations and lyophilized or synthetic microbial consortia with longer shelf lives and simpler distribution is two initiatives aimed at democratizing microbiome-based medicine. To enhance affordability and guarantee global equity, policy-level measures like government subsidies, health insurance coverage, and the addition of microbial therapies to essential drug lists are equally crucial. Without these safeguards, the advantages of microbiome research might be concentrated among wealthy populations, compromising its potential as a breakthrough in universal healthcare.

12.4.2 Data Security, Privacy, and Consent

A person's environment, lifestyle, identity, and health are all strongly correlated with their microbiome which is a complex layer of biological information. Unlike conventional medical data, microbiome profiles can effectively reveal dietary patterns, disease susceptibilities, and even behavioral traits. As microbiome sequencing becomes increasingly integrated with genomic, metabolomic, and electronic health record data the amount and sensitivity of personal data is rapidly increasing. Currently, human genomic data is governed by far more sophisticated regulatory frameworks than microbiome data. This regulatory gap raises urgent concerns about ownership, informed consent, and potential misuse of private data. Microbial signatures associated with obesity, infectious disease risk, or mental health conditions, for example, may be used for insurance discrimination, social stigma, or employment bias if privacy is not adequately protected (Lloyd-Price et al. 2016). The commercial microbiome industry, which collects massive amounts of consumer data through direct-to-consumer testing, usually lacks robust privacy protections. To reduce these risks, experts recommend dynamic consent models that provide participants ongoing control over the use of their microbiome data, clear data governance and strong cybersecurity. International cooperation is also crucial because microbiome research typically involves cross-border data sharing that must abide by a number of ethical and legal requirements.

12.4.3 Public Perception, Misinformation and Trust

The rapid advancement of microbiome science has captured the public's interest due to media coverage and commercial interest. Although this enthusiasm has raised awareness and funding, it has also resulted in misconceptions and unreasonable expectations. Commercial probiotics and microbiome supplements usually lack regulatory approval and strong scientific validation and their health claims are often exaggerated or unsubstantiated. These marketing tactics have the potential to mislead consumers into stopping evidence-based medical interventions or taking dangerous self-medication. Furthermore, early negative results could undermine public confidence if experimental microbiota-based treatments are not managed transparently. If exaggerated claims are followed by unanticipated issues, credibility in microbiome research may suffer over time. Reducing these risks requires effective science communication. For researchers, medical professionals and industry players, accurate communication of the potential and limitations of microbiota-based medicine is crucial. To encourage well-informed decision making, partnerships with patient advocacy organizations, open communication of clinical results and moral marketing techniques are essential tactics. Regulatory bodies are essential for keeping an eye on claims, making sure that goods fulfill safety and effectiveness requirements and shielding customers from false information.

12.4.4 Safety Concerns in Clinical Applications

Despite the great potential of microbiota-based treatments, real-world clinical experience has shown significant safety concerns that require careful observation. Two significant examples illustrate the risks:

12.4.4.1 Fecal Microbiota Transplantation Safety Risks

With cure rates of over 85% in refractory cases, FMT has revolutionized the treatment of recurrent *C. difficile* infections. However, safety warnings have been issued due to the spread of infections through improperly screened donor stool. The FDA issued a safety warning in 2019 after reports of severe infections caused by multidrug-resistant organisms including *Escherichia coli* that produces extended-spectrum beta-lactamase (ESBL) and led to patient deaths. The risks are increased by the inherent complexity of donor stool, the variation in screening standards between facilities and the lack of standardized manufacturing processes. Particularly in recipients with weakened immune systems, incomplete donor panels can effectively enable opportunistic pathogens to avoid detection. To lower these risks, the FDA and other authorities strongly emphasize adherence to good manufacturing practices and expanded pathogen screening. Researchers are also investigating safer alternatives such as synthetic microbial consortia and standardized FMT capsules to reduce reliance on raw donor stool.

12.4.4.2 Probiotic-Induced Sepsis in Immunocompromised Patients

Although immunocompromised individuals have encountered adverse side effects, probiotics are generally regarded as safe for healthy individuals. Case reports describe bloodstream infections and fungemia associated with probiotic organisms particularly Lactobacillus and Saccharomyces species (Suez et al. 2019). For example, *Lactobacillus rhamnosus* sepsis was reported to have occurred in a neutropenic patient receiving probiotics. Weakened mucosal barriers or contaminated central venous catheters are often linked to these infections. These findings emphasize that when administering probiotics to vulnerable groups, careful patient selection and a thorough risk-benefit analysis are essential. Clinical guidelines recommend strain-specific safety testing, tailored treatment options, and close observation to identify low-pathogenicity candidates.

In conclusion, the ethical, societal, and safety factors are crucial for the responsible development and application of microbiota-based therapies. Ensuring equitable access, safeguarding patient privacy, addressing misinformation and preserving clinical safety are not optional rather they are necessary for successful implementation. Ignoring these factors damages trust, increases health disparities and exposes patients to preventable harm. However, proactive measures to address these problems like equitable policy frameworks, robust data governance, transparent communication and stringent safety regulations can effectively facilitate microbiota-based medicine realize its transformative potential in a responsible and inclusive manner.

12.5 Case Studies and Clinical Experiences

Clinical experiences and real-world case studies provide valuable information about the potential and hazards of microbiota-based therapies. Although controlled clinical trials offer structured evidence, actual patient outcomes often reveal individual variability, long-term effects, and unexpected complications. Documented cases of both success and failure show the potential of microbiota therapies and emphasize the importance of strict safety monitoring and ethical conduct.

12.5.1 Fecal Microbiota Transplantation Successes and Failures

FMT is one of the best microbiota-based therapies particularly for recurrent *Clostridium difficile* infection. Numerous case series and randomized controlled trials have reported cure rates of over 85% for patients who did not respond to standard antibiotic regimens. In a groundbreaking study, van Nood et al. (2013) demonstrated that FMT was significantly more effective than vancomycin therapy, leading to both rapid symptom resolution and sustained remission. Case Example (Success): A 65-year-old patient with recurrent rCDI who was not responding to multiple antibiotic courses experienced complete remission following a single FMT via colonoscopy. The fact that the intervention not only eliminated the infection but also restored the diversity of gut microbes showed its therapeutic potential. However, there have also been documented instances of adverse events especially when donor screening and preparation are inadequate. Case Example (Failure/Complication): In 2019, the FDA reported two cases of severe infection caused by *Escherichia coli* that produced extended-spectrum beta-lactamases (ESBLs) and were transmitted by FMT. One patient's death brought attention to the risks of inadequate donor screening and the potential for life-threatening complications. This case led to a push for standardized, pathogen-screened microbial preparations and stronger FDA warnings. These incidents demonstrate the dual nature of FMT while it is highly

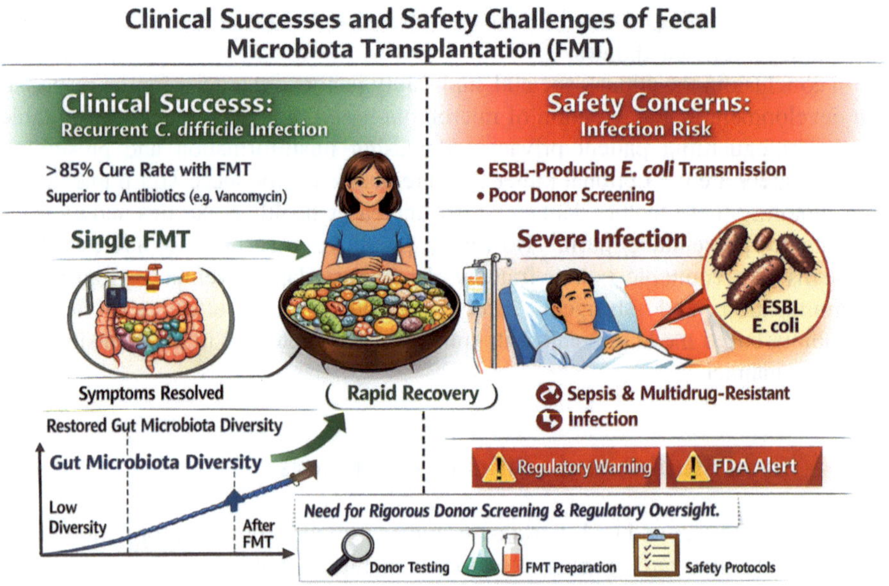

Fig. 12.5 Clinical successes and safety challenges of fecal microbiota transplantation (FMT)

effective in some circumstances; it requires stringent safety regulations and may not be yet ready for widespread, unregulated use (Fig. 12.5).

Figure 12.5 illustrates the high efficacy of FMT in recurrent *Clostridioides difficile* infection, including rapid symptom resolution and sustained remission, alongside documented safety concerns such as transmission of ESBL-producing *Escherichia coli* due to inadequate donor screening. It highlights the need for strict safety regulations and standardized protocols to ensure safe clinical use.

12.5.2 Probiotic-Associated Sepsis in Immunocompromised Patients

Even though probiotics are widely used and thought to be safe, actual cases demonstrate that they can effectively be detrimental to vulnerable populations. For example, when *Lactobacillus rhamnosus* GG was administered to a preterm newborn, the probiotic strain that caused the bacteremia was identified by blood culture to be the same. Similarly, fungemia associated with *Saccharomyces boulardii* has been reported in critically ill patients receiving central venous catheters. In another case, a patient with severe ulcerative colitis receiving immuno-suppressive treatment developed *Lactobacillus sepsis* as a result of taking probiotics. According to these studies, probiotics can effectively be very harmful for those with impaired immune systems or compromised mucosal barriers even though they are generally beneficial. Stricter safety reviews, strain-specific risk assessments, and tailored recommendations that restrict the use of probiotics in high-risk groups have all been demanded as a result of these incidents (Suez et al. 2019).

12.5.3 Engineered Microbial Therapies and Experimental Applications

Emerging case studies are beginning to draw attention to the potential of genetically modified microorganisms as therapeutic agents. Early clinical experiences exhibit both ambiguity and creativity. For example, Synlogic Inc. developed a strain of *Escherichia coli* Schultz and Burton (2017) designed to break down excess ammonia in hyperammonemia patients. Early trials of the treatment revealed a reduction in serum ammonia levels and metabolic activity. Despite promising outcomes, patient responses varied and additional research was required to demonstrate long term efficacy and safety. As an example of experimental cancer therapy, Chowdhury et al. (2019) reported the use of engineered bacteria in preclinical cancer therapy where the microbes were created to specifically colonize tumor tissue and deliver therapeutic payloads. Although these techniques have not yet undergone extensive human testing they show promise for microbiota-based medicine. These experimental uses underscore the need for robust ethical frameworks and careful patient consent because the risks of genetically modified microbes such as horizontal gene transfer and ecological disruption are still poorly understood (Table 12.1).

12.5.4 Long-Term Outcomes and Monitoring Gaps

Despite recent successes, there is still a lack of long-term follow-up data on microbiota-based interventions. Some case reports highlight long-term benefits, while others highlight late-onset complications. For example, patients treated with FMT for rCDI have shown years of sustained remission and some indication of a lower risk of recurrence when compared to antibiotics. However, other studies caution that alterations to the microbiome may have unanticipated repercussions such as causing recipients to become obese or changing their metabolism. The lack of standardized long-term monitoring frameworks means that many of these outcomes go unreported. These gaps underscore the importance of patient registries, post-market surveillance and international collaboration in capturing the full spectrum of risks and benefits.

In conclusion, the clinical case studies and real-world experiences with microbiota-based interventions demonstrate their immense therapeutic potential as well as their complex risks. Success stories demonstrate the potential to save lives particularly when using FMT to manage rCDI. However, adverse effects such as sepsis linked to probiotics and the spread of pathogens through FMT highlight flaws that require careful observation. As engineered microbial therapies enter clinical trials, safe innovation must be guided by lessons learned from the past. Strong patient consent protocols, standardized manufacturing and screening processes and

Table 12.1 Comprehensive overview of ethical, regulatory, safety, socioeconomic, environmental, public perception, and future directions in microbiota-based medicine

Aspect	Key points	Implications/ challenges	References
Ethical considerations	Potential stigmatization and discrimination; privacy and ownership of microbiome data; informed consent for microbiome research; ethical concerns in human microbiota transplantation (FMT/HMT); limitations of genetic modification in microbiota engineering from an ethical standpoint	Long-term uses might not be fully understood by participants; sensitive data could be misused; HMT necessitates careful donor selection and risk communication; and engineered microbes could have unanticipated ecological or health effects	Quinn et al. (2014), O'Doherty et al. (2012), Heeney et al. (2011), Ma et al. (2017), Weil and Hohmann (2015), Carlson (2020), and Ribeiro and Shapira (2019)
Regulatory challenges	Regulatory supervision of products based on microbiota; unclear product classification (food, drug, and probiotic); absence of standardized international standards; challenges in designing clinical trials; Long-term risk assessment and post-market monitoring	Global research and commercialization are hampered by regulatory fragmentation; approval and labeling are complicated by classification ambiguity; and live therapeutics are not intended for traditional regulatory frameworks. inadequate long-term safety surveillance	Mimee et al. (2016), Aguilar-Toalá et al. (2018), Rodriguez et al. (2025), Lloyd-Price et al. (2019), Langdon et al. (2016), and Bik (2020)
Commercialization and IP	Natural microbiota patentability, bioprospecting and benefit sharing, accessibility, and moral licensing	Unfair benefit sharing, monopolistic licensing, and patent legal ambiguity can all hinder innovation and restrict access to treatments	Jensen and Murray (2005), Sherkow (2015), CBD (2011), Kamau et al. (2010), So et al. (2008), and Kapczynski et al. (2005)
Public perception and misinformation	Media hype, unstated risks, and exaggerated claims; cultural sensitivities around microbiota interventions	Cultural taboos may prevent FMT and probiotics from being accepted, and false information may decrease trust and adoption	Bresalier and Chapkin (2020), de Vos et al. (2022), and Caulfield and Condit (2012)
Environmental & Ecological considerations	Gaps in environmental risk assessment; effects of probiotics or	Unintentional ecological disruptions,	Barbey and Latour (2024), De Lorenzo (2010), Jansson and

(continued)

Table 12.1 (continued)

Aspect	Key points	Implications/ challenges	References
	microbiome modulators on environmental microbiota; introduction of modified microorganisms into ecosystems	horizontal gene transfer, and inadequate environmental regulations	Hofmockel (2020), and Timmis et al. (2017)
Socioeconomic and Equity issues	Cost, insurance, and reimbursement; marginalization and health disparities; accessibility of microbiota-based treatments in low-income environments	Lack of insurance coverage encourages risky do-it-yourself methods; underrepresentation in research results in biased interventions; high cost and infrastructure constraints limit access	Yadav and Chauhan (2021), Subramanian et al. (2014), Sinha et al. (2015), Olesen and Alm (2016), and Baunwall et al. (2020)
Safety considerations	Safety issues with FMT; sepsis caused by probiotics in immunocompromised people	Pathogen transmission risk; requirement for strain safety testing and standardized donor screening; and customized risk assessment	Suez et al. (2019)
Future policy and research directions	Standardization of clinical and analytical procedures; customized microbiome treatments; multi-omics studies; social and ethical protections; international cooperation; and advancements in synthetic biology and culture	improves safety and reproducibility, permits precision medical techniques, lessens disparities, and guarantees responsible innovation	Lloyd-Price et al. (2019), and Barbey and Latour (2024)

long-term monitoring frameworks are all necessary for patient safety. In the end, empirical evidence will remain the cornerstone for advancing microbiota-based medicine, developing clinical guidelines and maintaining public trust.

12.6 Regulatory and Policy Frameworks

The rapid development of microbiota-based treatments, including fecal microbiota transplantation, probiotics, prebiotics, synbiotics and postbiotics has outpaced the establishment of robust regulatory frameworks. The variety of these interventions presents many challenges for regulators even though they have a lot of clinical promise. Unlike conventional drugs, microbiota products often include live or complex microbial consortia, donor-derived materials or engineered organisms all of these require special protocols to ensure the reliability of safety, efficacy and reproducibility.

12.6.1 Current Regulatory Landscape

12.6.1.1 United States

The Food and Drug Administration (FDA) in the United States regulates microbiota-related products under a number of different categories, depending on the product's intended use. The Dietary Supplement Health and Education Act, which requires manufacturers to ensure the reliability of safety but does not require premarket approval, apply to probiotics marketed as dietary supplements. However, because FMT and other microbiota-based treatments are categorized as biological products they can effectively be applied for as FDA Investigational New Drugs (INDs). The FDA has exercised its enforcement discretion to allow the use of FMT for recurrent Clostridium difficile infections that are not getting better with traditional treatments provided that informed consent is obtained (Cammarota et al. 2017). However, in response to recent adverse events such as pathogen transmission, the FDA has reinforced its guidelines for pathogen testing and donor screening.

12.6.1.2 European Union

The European Union (EU) depending on their classification has multiple regulatory frameworks that apply to microbiota-based products. Probiotics marketed as foods or supplements are governed by the European Food Safety Authority (EFSA) but therapeutic applications may be categorized as advanced therapy medicinal products by the European Medicines Agency (EMA). The EMA has published guidelines on microbiota-derived products that prioritize donor screening, quality control, and safety despite the fact that there is not yet a unified EU-wide regulation for FMT.

12.6.1.3 Asia and Other Regions

Regulations in Asia differ widely. For example, in Japan, the Foods for Specified Health Uses system which allows health claims after scientific review authorizes certain probiotics. Conversely, China and India struggle to keep up with commercialization due to their fragmented and evolving regulatory frameworks. This uneven environment hinders international standardization and cross-border clinical trials.

12.6.2 Challenges in Regulation

Several specific problems make regulating products based on microbiota more difficult: complexity of biological materials: unlike single-molecule drugs, microbiota therapies employ living consortia or donor-derived samples which are inevitably variable. Donor screening and standardization: differences in donor microbiota and potential pathogen transmission lead to safety concerns in FMT. Strain-specific effects: probiotics exhibit strain-specific clinical outcomes leading to inconsistent efficacy and mismatched claims despite the fact that many regulations generalize at the species level (Suez et al. 2019). International harmonization is made more challenging by classification ambiguities which occur when a product is regulated differently as a food, supplement or medication depending on the jurisdiction and intended use. Monitoring Long Term Safety: Most frameworks lack post-market surveillance, patient registries and adverse event reporting tools tailored to microbiota-based interventions.

12.6.3 Harmonization and Global Policy Needs

Due to the global nature of microbiota research and commercialization, harmonized frameworks are desperately needed. International organizations such as the World Health Organization (WHO) and the International Council for Harmonization of Technical Requirements for Pharmaceuticals for Human Use (ICH) could set global standards for manufacturing, safety, and quality control. Proposals for harmonization include: Good Manufacturing Practices (GMP) were developed especially for live biotherapeutics, the establishment of international donor registries for FMT, creation of strain-specific probiotic databases that link genetic identity and clinical efficacy, terms and definitions that are frequently used to distinguish between probiotics, synbiotics, postbiotics and live biotherapeutic products.

12.6.4 Ethical and Legal Considerations in Policy

Regulatory frameworks must also address broader ethical and legal concerns. These include obtaining informed consent for microbiota transplantation, the implications of genetically modified organisms, equitable access to treatments, and intellectual property rights over microbial strains. For instance, the commercialization of donor-derived microbiota raises ethical concerns about ownership and compensation. Similar to this, patents on engineered probiotics or microbial consortia may restrict accessibility, sparking debates between public health and proprietary rights.

In conclusion, the regulation of microbiota-based therapies remains fragmented due to the diversity and uniqueness of these treatments. The U.S. FDA, EU EMA, and regional regulators have taken steps to address safety and efficacy despite ongoing challenges with donor screening, strain-specific effects, standardization, and classification. Support from international organizations for worldwide harmonization will be crucial for promoting safe innovation, cross-border research,

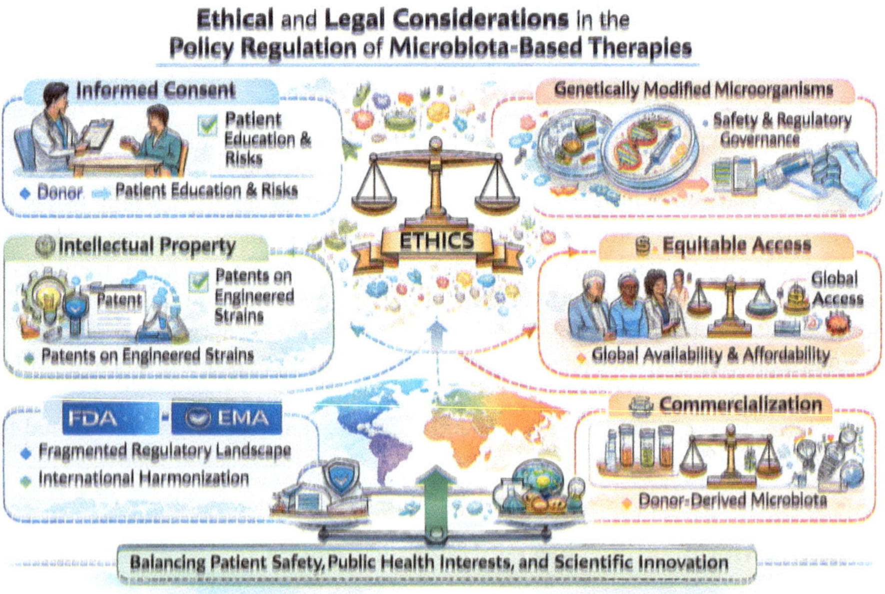

Fig. 12.6 Ethical and legal considerations in the policy regulation of microbiota-based therapies

and equitable access to microbiota-based therapies. As the field progresses, regulatory frameworks must adapt to balance patient safety and scientific advancement allowing microbiota-based interventions to fully realize their potential to transform healthcare (Fig. 12.6).

Figure 12.6 illustrates key regulatory domains including informed consent for microbiota transplantation, governance of genetically modified microorganisms, intellectual property rights over engineered strains, commercialization of donor-derived microbiota, and equitable access to treatment. It also highlights the fragmented regulatory landscape across agencies such as the U.S. Food and Drug Administration and the European Medicines Agency, emphasizing the need for international harmonization to balance patient safety, public health interests, and scientific innovation.

12.7 Future Perspectives and Emerging Trends

Future developments in the quickly developing field of microbiota-based interventions are probably going to change the way that healthcare is currently provided. Precision interventions are replacing descriptive studies in microbiota research thanks to the combination of sophisticated biotechnological tools, computational methods, and synthetic biology. Although established treatments such as fecal microbiota transplantation and probiotics have shown great promise, new approaches hold the promise of overcoming their drawbacks and opening up new therapeutic opportunities (Table 12.2).

Table 12.2 Microbiota-based therapeutic interventions and associated risks

Intervention	Disease/indication	Mechanism	Known risks/adverse events	References
Fecal microbiota transplantation (FMT)	Recurrent *Clostridium difficile* infection; experimental application in metabolic syndrome and IBD	By passing stool from a healthy donor to the patient, the gut microbial diversity and balance are restored; pathogenic bacteria are competitively inhibited, SCFA production is restored, and the immune response is modulated	Transmission of infections (ESBL-producing *E. coli*, MDROs); adverse immune responses; donor-dependent variable efficacy; reported deaths from contaminated stool; and unclear long-term effects	
Probiotics (e.g., *Lactobacillus*, *Bifidobacterium*, *Saccharomyces*)	IBS, *H. pylori* infection, antibiotic-associated diarrhea, and overall gut health	Competition with pathogens, secretion of antimicrobial compounds, colonization and temporary persistence in the gut, and immune response modulation	Immunocompromised patients' sepsis or fungemia brought on by probiotics; production contamination risk; translocation risk through compromised mucosa; and strain-specific effects	Suez et al. (2019)
Engineered microbial consortia	Experimental treatment for cancer, chronic inflammation, and metabolic diseases	Synthetic consortia for accurate microbial community manipulation; genetically modified microorganisms intended to generate medicinal compounds, break down toxins, or alter host immunity	Possible horizontal gene transfer, ecological disruption, off-target consequences, erratic host interactions, and long-term colonization safety issues	De Lorenzo (2010)
Synbiotic/designer probiotic formulations	Metabolic syndrome, inflammatory bowel disease, and targeted gut dysbiosis	Prebiotics and specific microbial strains combined to improve colonization and functional activity; immune pathway and microbial metabolite modulation	Inadequate clinical evidence, inconsistent effectiveness, possible negative immunological or metabolic effects, and production contamination risk	Lloyd-Price et al. (2019)
Human microbiota transplantation (HMT, non-fecal)	Experimental treatment for metabolic modulation and systemic disorders	Transferring a donor's defined microbiota to a recipient with the goal of reestablishing the structure and function of the microbial community	Similar to FMT, safety concerns exist; donor selection and screening are crucial; long-term effects are unknown; and vulnerable populations have ethical considerations	Weil and Hohmann (2015)

12.7.1 Personalized and Precision Microbiome Medicine

One of the most promising trends is the move toward personalized microbiota-based therapies that take into account an individual's genetic makeup, microbiome composition, diet, and lifestyle. There is growing evidence that interindividual variability in microbiome composition influences different therapeutic responses (Suez et al. 2019). For example, one patient may benefit from a specific strain of probiotics while another may not due to underlying differences in gut ecology. To overcome this variability, cutting-edge technologies such as metabolomics, machine learning algorithms and metagenomic sequencing are being used to identify patient-specific microbial biomarkers and predict therapeutic outcomes. With precise methods, clinicians might be able to select the best probiotic strains, form distinctive microbial consortia, and personalize the FMT donor selection. In line with the broader trend of precision medicine, this patient-centered approach considers microbiota as a critical component in assessing treatment efficacy (Fig. 12.7).

12.7.2 Next-Generation Probiotics and Engineered Strains

The future of probiotics goes beyond well-known genera like *Lactobacillus* and *Bifidobacterium*. Commensal bacteria like *Faecalibacterium prausnitzii*, *Roseburia*

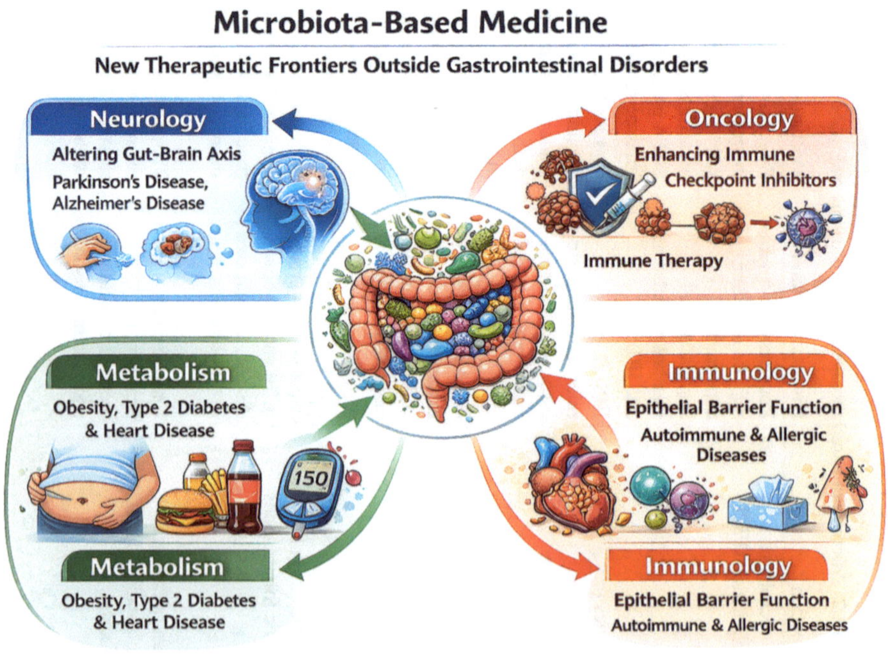

Fig. 12.7 Microbiota-based medicine

spp., and *Akkermansia muciniphila* that have been shown to have health-promoting qualities are included in next-generation probiotics. These species play critically important roles in gut homeostasis, metabolic regulation, and immune modulation making them attractive therapeutic candidates. At the same time, advancements in synthetic biology make it possible to produce microbial strains with specific characteristics. Engineered bacteria can effectively be programmed to release anti-inflammatory substances, degrade toxic substances or transport drugs to disease sites. For example, *Escherichia coli* Schultz and Burton (2017) strains that have been altered have been developed to directly produce therapeutic proteins in the gut environment. Beyond what naturally occurring strains can effectively achieve; these advancements may greatly expand the therapeutic range of microbiota interventions.

12.7.3 Synthetic Microbial Consortia

Another recent development is the formation of artificial microbial consortia, which are groups of multiple microbial species with complementary functions. Consortia can effectively be designed to replicate the complexity of healthy microbiome in contrast to traditional probiotics that usually contain one or a small number of strains. In illnesses associated with dysbiosis, this can effectively aid in reestablishing ecological balance. Synthetic consortia targeting *Clostridium difficile* infections are being developed as an alternative to FMT. Standardization, reproducibility and a decreased risk of pathogen transmission are benefits of these consortia. These consortia can effectively also be modified to provide specific therapeutic advantages like enhanced short-chain fatty acid synthesis, improved barrier integrity or focused pathogen suppression.

12.7.4 Postbiotics and Microbiota-Derived Metabolites

Many of the therapeutic effects of probiotics may be attributed to bioactive metabolites also referred to as postbiotics. These include short-chain fatty acids (SCFAs), bacteriocins, vitamins and extracellular vesicles (Aguilar-Toalá et al. 2018). Postbiotics are more stable, easier to standardize and less likely to cause infection in patients with weakened immune systems than live microorganisms. In order to encourage the body's natural production of healthy metabolites, future treatment plans may place more of an emphasis on administering postbiotics directly or on utilizing pharmacological or dietary modifications. These techniques could aid in bridging the gap between traditional probiotics and pharmaceutical-grade treatments.

12.7.5 Integration with Digital Health and Artificial Intelligence

Combining digital health technologies with artificial intelligence is a new field of study in microbiota. Mobile health apps, wearable technology, and digital food diaries can effectively all provide continuous data on a patient's symptoms, diet, and way of life. When paired with metabolomic profiling and microbiome sequencing, AI algorithms can effectively analyze massive datasets and generate predictive models of microbiota-host interactions. The intersection of microbiome science and digital health may enable real-time monitoring of therapeutic responses, adaptive treatment plans and predictive diagnostics. For example, customized dietary recommendations based on microbiota composition are already being researched to enhance weight management and glycemic control.

12.7.6 Expanding Therapeutic Frontiers Beyond Gastrointestinal Diseases

Current applications are primarily focused on gastrointestinal disorders, but it is expected that future microbiota-based interventions will spread into other therapeutic areas. Research on the roles of microbiota is ongoing in the following areas: Neurodegenerative diseases like Parkinson's and Alzheimer's may progress differently if the gut-brain axis is altered. Metabolic disorders like obesity, type 2 diabetes, and cardiovascular disease where the microbiota influences energy harvesting and inflammatory pathways. Immuno-checkpoint inhibitors are more effective in cancer treatment when microbiota is regulated. Autoimmune and allergic disorders can effectively be improved by altering immune tolerance and strengthening the epithelial barrier. These emerging frontiers have made the microbiota a substantial therapeutic target in many medical specialties.

Figure 12.7 illustrates new therapeutic frontiers outside of gastrointestinal disorders. Four main domains are connected to the central gut image: (1) neurology: altering the gut-brain axis in Parkinson's and Alzheimer's disease; (2) metabolism: focusing on pathways linked to obesity, type 2 diabetes, and heart disease; (3) oncology: increasing the effectiveness of immune checkpoint inhibitors; and (4) Immunology: boosting the function of the epithelial barrier and controlling immune tolerance in autoimmune and allergic diseases.

12.7.7 Global Collaboration and Policy Development

The success of microbiota-based therapies will depend on future scientific discoveries, global collaboration and wise policymaking. International donor registries, harmonized regulatory frameworks and standardized definitions are necessary to ensure the reliability of quality, safety, and reproducibility. Collaborations between academia, industry, and governments can effectively speed up clinical translation by facilitating open-access microbial databases, sharing anonymized patient data, and funding large-scale multicenter trials. Policies must also address ethical concerns like benefit-sharing, informed consent, and equitable access when it comes to the commercialization of microbiota-based products. In conclusion, the future directions in microbiota research have the potential to transform healthcare by moving beyond conventional probiotics and FMT and toward postbiotic therapies, engineered strains, next-generation probiotics, precision medicine, and synthetic consortia. By integrating AI, digital health technologies, and systems biology, predictive power and personalization will be enhanced. The microbiota's expanding applications in immunology, metabolism, oncology and neurology highlight how critically important it is to human health. To achieve this potential, however, standardized regulation, ethical oversight, and international cooperation are required. As science, technology and policy come together microbiota-based treatments have the potential to transform modern medicine by offering safer, more customized, and more effective treatments for a range of illnesses (Fig. 12.8).

Figure 12.8 highlights the personalized and precision microbiome medicine: Customizing treatments according to individual genetics, microbiome profiles, diet, and lifestyle; next-generation probiotics and engineered strains: creating commensal and synthetic microbes with therapeutic roles; synthetic microbial consortia: standardized, multistrain communities designed to restore gut ecological balance; and Integration with Digital Health and Artificial Intelligence: utilizing wearable technology, AI, and digital health tools for adaptive therapies and predictive diagnostics.

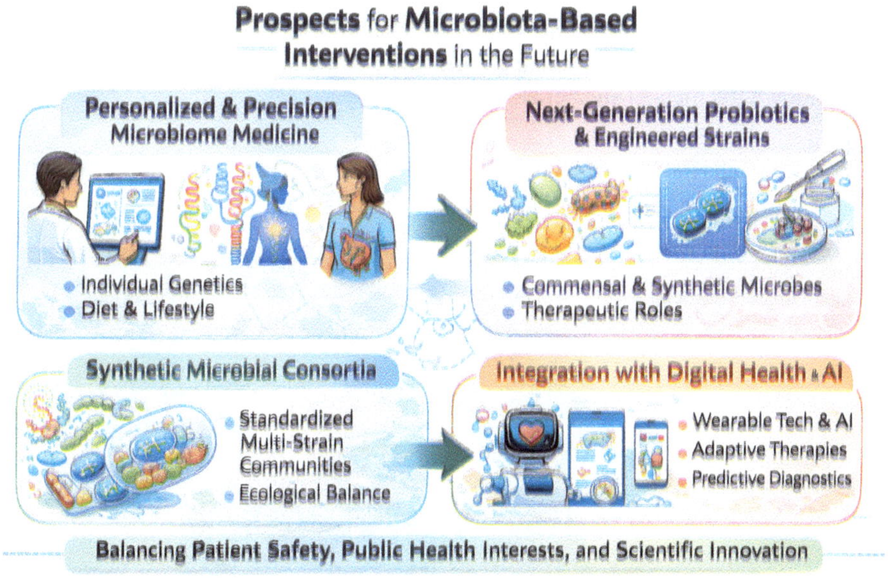

Fig. 12.8 Prospects for microbiota-based interventions in the future

12.8 Policy Recommendations for Microbiota-Based Medicine

With the rapid advancements in microbiota research from probiotics and prebiotics to engineered microbial consortia and microbiome-targeted therapies both unprecedented opportunities and challenging ethical, societal, and regulatory issues have arisen. Despite the rapid acceleration of scientific discoveries, policy development remains fragmented and lagging behind often varying significantly across regions. Without proactive, standardized, and ethically sound policies, the application of microbiota science to clinical and public health practice may face significant challenges (Marchesi et al. 2016; Simonsen et al. 2019). This section's comprehensive policy recommendations aim to balance innovation with sustainability, equity, and safety (Fig. 12.9).

12.8.1 Global Regulatory Harmonization

One of the most pressing requirements in microbiota-based medicine is globally aligned regulatory frameworks. Different countries currently classify probiotics, microbiome-derived medications and fecal microbiota transplantation differently. These categories range from dietary supplements to biologics or medications. These differences hinder international clinical trials, restrict cross-border collaboration and create uncertainty for the sector (Ouwehand et al. 2018). The Food and Agriculture Organization (FAO), the World Health Organization (WHO), and the International Council for Harmonization of Technical Requirements for Pharmaceuticals for Human Use (ICH) are examples of international organizations that could lead a harmonized approach that would provide standardized definitions, approval processes, and labeling requirements (FAO/WHO 2002). These harmonized standards would reduce testing duplication, expedite safe market access, and facilitate mutual recognition agreements (Suez et al. 2019).

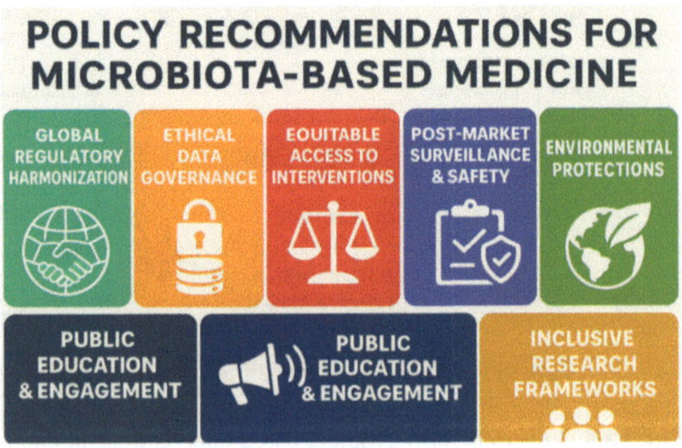

Fig. 12.9 Policy recommendation for microbiota-based medicine

12.8.2 Ethical Data Governance

Microbiome research collects and analyzes vast amounts of biological data at the individual and community levels, posing significant questions about ownership, consent and privacy. Currently, participants are not told about the possible reuse or commercialization of their microbiome data, and consent models are often unclear. Policies should require transparent, dynamic consent frameworks that allow individuals to change and manage their choices over time (Vayena and Blasimme 2021). To prevent misuse or discrimination based on microbiome profiles, particularly in industries like insurance and employment, strict data protection regulations are also necessary. In addition, benefit-sharing policies should ensure the reliability of the populations that provide microbiome data especially indigenous and underrepresented groups are fairly recognized and have the opportunity to profit from downstream applications (Simonsen et al. 2019).

12.8.3 Equitable Access to Microbiota-Based Interventions

If access is restricted to affluent populations, the potential of microbiota-based treatments will not be fulfilled. Low- and middle-income nations will be disadvantaged by the high cost and technological complexity of many new products such as personalized microbiome-based therapies or next-generation probiotics (Zmora et al. 2018). Patent pooling, open-source microbial strain libraries, tiered pricing and the incorporation of microbiota treatments into public health insurance programs are all policies that should be promoted (Khanna and Pardi 2017). In areas where infectious diseases and malnutrition continue to be public health concerns, international cooperation may also guarantee the fair distribution of crucial microbiota-based products such as safe probiotic formulations and microbial bio-fortification methods (WHO 2020).

12.8.4 Post-market Surveillance and Safety Monitoring

Unlike conventional pharmaceuticals, microbiota-based products are dynamic and can effectively alter after being administered to the human host or released into the environment. This necessitates trustworthy, long-term monitoring systems to track safety and efficacy following initial clinical trials (Eisenstein 2020). Policy frameworks should mandate the creation of microbial strain registries, real-time adverse event reporting systems, and longitudinal databases for postmarket surveillance (Cammarota et al. 2017). Such systems would allow for the early detection of unintended side effects such as pathogenic overgrowth, horizontal gene transfer or altered dynamics of microbial ecosystems. Regulatory agencies may employ adaptive licensing models granting conditional approvals contingent on continuous data collection and monitoring (Simonsen et al. 2019).

12.8.5 Environmental Safeguards

The growing use of engineered bacteria and synthetic microbial consortia in industrial, agricultural, and therapeutic contexts carries ecological risks. Genetically modified microorganisms may alter soil fertility, disrupt environmental microbiomes or elevate antibiotic resistance if they are accidentally released (de Lorenzo et al. 2021). Policymakers must enforce stringent containment measures, biosafety protocols, and ecological risk assessments prior to broad implementation (Cartagena Protocol on Biosafety 2000). International environmental treaties may regulate the flow of genetically modified microorganisms across national boundaries and their discharge into the environment. Environmental monitoring programs involving networks of scientists and citizen scientists should be incorporated into long-term policy planning (van der Lelie et al. 2017).

12.8.6 Public Education and Engagement

The success of microbiota-based medicine depends on societal acceptance which can effectively be undermined by misleading information, cultural skepticism or exaggerated marketing claims. Policymakers should support public education programs that provide easily understood, evidence-based information about microbiota research and its health implications (Sonnenburg and Sonnenburg 2019). Schools, educational institutions, and community health initiatives can effectively all place a lot of emphasis on improving microbiome literacy. In addition, active community engagement is necessary to address cultural sensitivities and ethical concerns particularly when handling situations like genetically modified microbes or fecal microbiota transplantation (Kelly et al. 2021). In addition to preventing misleading advertising by unregulated companies' partnerships between the public and private sectors may facilitate to guarantee honest communication (Eisenstein 2020).

12.8.7 Inclusive and Interdisciplinary Research Frameworks

By guaranteeing that a range of demographics such as women, children, the elderly, and people from low-resource environments are represented in clinical trials, policy should encourage inclusivity in microbiota research. As a result, microbiome-based interventions would be more equitable and broadly applicable (Khanna and Pardi 2017). Simultaneous institutionalization of interdisciplinary collaboration among microbiologists, physicians, bioinformaticians, ethicists, legal scholars, and patient advocacy groups is necessary (Marchesi et al. 2016). Given that microbiota-based medicine lies at the nexus of biology, technology, ethics and law, funding organizations and regulatory bodies ought to encourage these partnerships (Simonsen et al. 2019).

Figure 12.9 illustrates Global regulatory harmonization, ethical data governance, fair access to interventions, post-market surveillance and safety monitoring, environmental protections, public education and engagement and inclusive interdisciplinary research frameworks as the priority areas. When combined, these regulations seek to strike a balance between safety, sustainability, equity, and innovation in the clinical application of microbiota-based treatments.

12.9 Conclusion

Research on microbiota has advanced quickly, opening up new avenues for environmental sustainability, agriculture, and medicine. These developments have enormous potential to address nutritional issues, enhance health outcomes, and offer fresh approaches to environmental issues. However, despite these advancements, critically important moral, legal, social, and environmental concerns must also receive equal consideration. If microbiota-based technologies are to be trusted by the public and have a long-lasting effect, issues of consent, data privacy, equity of access, commercialization, and ecological safety cannot be ignored. One substantial obstacle to responsible progress is still the lack of long-term safety frameworks and standardized international regulations. Disparities in affordability, accessibility, and cultural acceptance, however, underscore the necessity of inclusive and socially conscious methods. Without these, microbiota interventions run the risk of making already-existing disparities worse rather than better. A multidisciplinary, internationally coordinated effort that incorporates moral consideration, strict regulation, and just policies is needed to pave the way forward. Microbiota science can effectively move from a frontier of experimentation to a basis for safe, efficient, and widely available innovations by emphasizing transparency, biosafety, and equity. The promise of microbiota-based solutions ultimately rests not only in their scientific discoveries but also in our collective obligation to use them responsibly, inclusively, and sustainably for the good of human and environmental health.

References

Aguilar-Toalá JE, Garcia-Varela R, Garcia HS, Mata-Haro V, González-Córdova AF, Vallejo-Cordoba B et al (2018) Postbiotics: an evolving term within the functional foods field. Trends Food Sci Technol 75:105–114

Bakken JS, Borody TJ, Brandt LJ, Brill JV, Demarco DC, Franzos MA, Kelly C, Khoruts A, Louie T, Martinelli LP, Moore TA, Russell G, Surawicz C (2011) Treating Clostridium difficile infection with fecal microbiota transplantation. Clin Gastroenterol Hepatol 9(12):1044–1049. https://doi.org/10.1016/j.cgh.2011.08.014

Barbey C, Latour X (2024) Molecular mechanisms of bacterial communication and their biocontrol. Int J Mol Sci 25(10):5443. https://doi.org/10.3390/ijms25105443. PMID: 38791481; PMCID: PMC11121524

Baunwall SMD, Lee MM, Eriksen MK, Mullish BH, Marchesi JR, Hvas CL (2020) Faecal microbiota transplantation for recurrent Clostridioides difficile infection: an updated systematic review and meta-analysis. EClinical Med 29:100642

Bik EM (2020) The hoops, hurdles, and hype of human microbiome research. PLoS Biol 18(2): e3000662

Bresalier RS, Chapkin RS (2020) Human microbiome in health and disease: the good, the bad, and the ugly. Dig Dis Sci 65:671–673. https://doi.org/10.1007/s10620-020-06059-y

Cammarota G, Ianiro G, Tilg H, Rajilić-Stojanović M, Kump P, Satokari R et al (2017) European consensus on faecal microbiota transplantation in clinical practice. Gut 66(4):569–580

Carlson PE Jr (2020) Regulatory considerations for fecal microbiota transplantation products. Cell Host Microbe 27(2):173–175. https://doi.org/10.1016/j.chom.2020.01.018

Cartagena Protocol on Biosafety (2000) Montreal: secretariat of the convention on biological diversity

Caulfield T, Condit C (2012) Science and the sources of hype. Public Health Genomics 15(3–4): 209–217. https://doi.org/10.1159/000336533. Epub 2012 Apr 4

Chowdhury S, Castro S, Coker C et al. (2019) Programmable bacteria induce durable tumor regression and systemic antitumor immunity. Nat Med 25:1057–1063. https://doi.org/10.1038/s41591-019-0498-z

Convention on Biological Diversity (CBD) (2011) Nagoya Protocol on access to genetic resources and the fair and equitable sharing of benefits arising from their utilization. United Nations

De Lorenzo V (2010) Environmental biosafety in the age of synthetic biology: Do we really need a radical new approach? BioEssays 32(11):926–931

de Lorenzo V, Krasnogor N, Schmidt M (2021) For the sake of the bioeconomy: define what a synthetic biology chassis is! New Biotechnol 60:44–51

de Vos WM, Tilg H, Van Hul M, Cani PD (2022) Gut microbiome and health: mechanistic insights. Gut 71(5):1020–1032. https://doi.org/10.1136/gutjnl-2021-326789. Epub 2022 Feb 1. PMID: 35105664; PMCID: PMC8995832

Eisenstein M (2020) The hunt for a healthy microbiome. Nature 577(7792):S6–S8

FAO/WHO (2002) Guidelines for the evaluation of probiotics in food. WHO, London

Heeney C, Hawkins N, de Vries J, Boddington P, Kaye J (2011) Assessing the privacy risks of data sharing in genomics. Public Health Genomics 14(1):17–25. https://doi.org/10.1159/000294150. Epub 2010 Mar 29. PMID: 20339285; PMCID: PMC2872768

Jansson JK, Hofmockel KS (2020) Soil microbiomes and climate change. Nat Rev Microbiol 18(1): 35–46

Jensen K, Murray F (2005) Intellectual property. Enhanced: intellectual property landscape of the human genome. Science 310(5746):239–240. https://doi.org/10.1126/science.1120014

Kamau E, Fedder B, Winter G (2010) The Nagoya Protocol on access to genetic resources and benefit sharing: what is new and what are the implications for provider and user countries and the scientific community? Law Dev J (LEAD) 6(3):248–263

Kapczynski A, Chaifetz S, Benkler Y, Katz Z (2005) Addressing global health inequities: an open licensing approach for university innovations. Berkley Technol Law J 20

Kelly CR, Fischer M, Allegretti JR, LaPlante K, Stewart DB, Limketkai B et al (2021) ACG clinical guidelines: prevention, diagnosis, and treatment of Clostridioides difficile infections. Am J Gastroenterol 116(6):1124–1147

Khanna S, Pardi DS (2017) Microbiota replacement therapies: innovation in gastrointestinal care. Clin Pharmacol Ther 101(1):45–53

Langdon A, Crook N, Dantas G (2016) The effects of antibiotics on the microbiome throughout development and alternative approaches for therapeutic modulation. Genome Med 8(1):39. https://doi.org/10.1186/s13073-016-0294-z. PMID: 27074706; PMCID: PMC4831151

Lloyd-Price J, Abu-Ali G, Huttenhower C (2016) The healthy human microbiome. Genome Med 8: 51. https://doi.org/10.1186/s13073-016-0307-y

Lloyd-Price J, Arze C, Ananthakrishnan AN, Schirmer M, Avila-Pacheco J, Poon TW, Andrews E, Ajami NJ, Bonham KS, Brislawn CJ, Casero D, Courtney H, Gonzalez A, Graeber TG, Hall AB, Lake K, Landers CJ, Mallick H, Plichta DR, Prasad M, Rahnavard G, Sauk J, Shungin D, Vázquez-Baeza Y, White RA 3rd, IBDMDB Investigators, Braun J, Denson LA, Jansson JK, Knight R, Kugathasan S, DPB MG, Petrosino JF, Stappenbeck TS, Winter HS, Clish CB,

Franzosa EA, Vlamakis H, Xavier RJ, Huttenhower C (2019) Multi-omics of the gut microbial ecosystem in inflammatory bowel diseases. Nature 569(7758):655–662. https://doi.org/10.1038/s41586-019-1237-9

Ma Y, Chen H, Lei R et al (2017) Biobanking for human microbiome research: promise, risks, and ethics. ABR 9:311–324. https://doi.org/10.1007/s41649-017-0033-9

Marchesi JR, Ravel J, Clemente JC et al (2016) The gut microbiota and host health: a new clinical frontier. Nat Rev Gastroenterol Hepatol 13(10):573–584

Mimee M, Citorik RJ, Lu TK (2016) Microbiome therapeutics—advances and challenges. Adv Drug Deliv Rev 105(Pt A):44–54. https://doi.org/10.1016/j.addr.2016.04.032. Epub 2016 May 5. PMID: 27158095; PMCID: PMC5093770

O'Doherty KC, Hawkins AK, Burgess MM (2012) Involving citizens in the ethics of biobank research: informing institutional policy through structured public deliberation. Soc Sci Med 75(9):1604–1611. https://doi.org/10.1016/j.socscimed.2012.06.026. Epub 2012 Jul 28

Olesen SW, Alm EJ (2016) Dysbiosis is not an answer. Nat Microbiol 1:16228

Ouwehand AC, Salminen S, Arvola T et al (2018) Classification and safety of probiotics: current perspectives. Gut Microbes 9(5):392–403

Quinn GP, Koskan A, Sehovic I, Pal T, Meade C, Gwede CK (2014) Contrasting the ethical perspectives of biospecimen research among individuals with familial risk for hereditary cancer and biomedical researchers: implications for researcher training. Genet Test Mol Biomarkers 18(7):467–473. https://doi.org/10.1089/gtmb.2013.0461. Epub 2014 Apr 30. PMID: 24786355; PMCID: PMC4094006

Ribeiro B, Shapira P (2019) Anticipating governance challenges in synthetic biology: insights from biosynthetic menthol. Technol Forecast Soc Change 139:311–320. https://doi.org/10.1016/j.techfore.2018.11.020. PMID: 30774160; PMCID: PMC6360377

Rodriguez J, Cordaillat-Simmons M, Pot B et al (2025) The regulatory framework for microbiome-based therapies: insights into European regulatory developments. Npj Biofilms Microbiomes 11:53. https://doi.org/10.1038/s41522-025-00683-0

Schultz M, Burton JP (2017) Escherichia coli Nissle 1917. In: Floch MH, Ringel Y, Allan Walker W (eds) The microbiota in gastrointestinal pathophysiology: implications for human health, prebiotics, probiotics, and dysbiosis. Academic Press, pp. 59–69. https://doi.org/10.1016/B978-0-12-804024-9.00005-7

Sherkow JS (2015) Law, history and lessons in the CRISPR patent conflict. Nat Biotechnol 33(3): 256–257. https://doi.org/10.1038/nbt.3160

Simonsen K, Baker JL, Kettner H et al (2019) Regulatory and ethical challenges in microbiome-based therapies. Trends Microbiol 27(1):2–11

Sinha R, Abnet CC, White O et al (2015) The microbiome quality control project: baseline study design and future directions. Genome Biol 16:276. https://doi.org/10.1186/s13059-015-0841-8

So AD, Sampat BN, Rai AK, Cook-Deegan R, Reichman JH, Weissman R, Kapczynski A (2008) Is Bayh-Dole good for developing countries? Lessons from the US experience. PLoS Biol 6(10): e262. https://doi.org/10.1371/journal.pbio.0060262. PMID: 18959484; PMCID: PMC2573936

Sonnenburg E, Sonnenburg J (2019) The good gut: taking control of your weight, your mood, and your long-term health. Penguin, New York

Subramanian S, Huq S, Yatsunenko T, Haque R, Mahfuz M, Alam MA, Benezra A, DeStefano J, Meier MF, Muegge BD, Barratt MJ, VanArendonk LG, Zhang Q, Province MA, Petri WA Jr, Ahmed T, Gordon JI (2014) Persistent gut microbiota immaturity in malnourished Bangladeshi children. Nature 510(7505):417–421. https://doi.org/10.1038/nature13421. Epub 2014 Jun 4. PMID: 24896187; PMCID: PMC4189846

Suez J, Zmora N, Segal E, Elinav E (2019) The pros, cons, and many unknowns of probiotics. Nat Med 25(5):716–729. https://doi.org/10.1038/s41591-019-0439-x

Timmis K, de Vos WM, Ramos JL, Vlaeminck SE, Prieto A, Danchin A, Verstraete W, de Lorenzo V, Lee SY, Brüssow H, Timmis JK, Singh BK (2017) The contribution of microbial biotechnology to sustainable development goals. Microb Biotechnol 10(5):984–987. https://doi.org/10.1111/1751-7915.12818. Epub 2017 Aug 25. PMID: 28840974; PMCID: PMC5609250 van der Lelie D, Taghavi S, Monchy S et al (2017) Microbial ecology and biosafety of genetically modified microorganisms for bioremediation. Curr Opin Biotechnol 45:197–202

van Nood E, Vrieze A, Nieuwdorp M, Fuentes S, Zoetendal EG, de Vos WM, Visser CE, Kuijper EJ, Bartelsman JFWM, Tijssen JGP, Speelman P, Dijkgraaf MGW, Keller JJ (2013) Duodenal infusion of donor feces for recurrent Clostridium difficile. N Engl J Med 368(5):407–415. https://doi.org/10.1056/NEJMoa1205037

Vayena E, Blasimme A (2021) Health research with big data: time for systemic oversight. J Law Med Ethics 49(2):253–267

Weil AA, Hohmann EL (2015) Fecal microbiota transplant: benefits and risks. Open Forum Infect Dis 2(1):ofv005. https://doi.org/10.1093/ofid/ofv005. PMID: 26034756; PMCID: PMC4438886

WHO (2020) Ending the neglect to attain the Sustainable Development Goals: a road map for neglected tropical diseases 2021–2030. WHO, Geneva

Yadav M, Chauhan NS (2021) Microbiome therapeutics: exploring the present scenario and challenges. Gastroenterol Rep (Oxf) 10:goab046. https://doi.org/10.1093/gastro/goab046. PMID: 35382166; PMCID: PMC8972995

Zmora N, Zeevi D, Korem T et al (2018) Taking it personally: personalized utilization of the human microbiome in health and disease. Cell Host Microbe 23(2):344–352

MIX
Papier aus verantwortungsvollen Quellen
Paper from responsible sources
FSC® C105338

If you have any concerns about our products,
you can contact us on
ProductSafety@springernature.com

In case Publisher is established outside the EU,
the EU authorized representative is:
Springer Nature Customer Service Center GmbH
Europaplatz 3, 69115 Heidelberg, Germany

Printed by Libri Plureos GmbH
in Hamburg, Germany